# Vibrational Spectroscopy at Electrified Interfaces

## WILEY SERIES ON ELECTROCATALYSIS AND ELECTROCHEMISTRY

**Andrzej Wieckowski**, Series Editor

*Synthetic Diamond Films: Preparation, Electrochemistry, Characterization and Applications*, Edited by Enric Brillas and Carlos Alberto Martínez-Huitle

*Fuel Cell Catalysis: A Surface Science Approach*, Edited by Marc T. M. Koper

*Electrochemistry of Functional Supramolecular Systems*, Margherita Venturi, Paola Ceroni, And Alberto Credi

*Catalysis in Electrochemistry: From Fundamentals to Strategies for Fuel Cell Development*, Elizabeth Santos and Wolfgang Schmickler

*Fuel Cell Science: Theory, Fundamentals, and Biocatalysis*, Andrzej Wieckowski and Jens Norskov

# Vibrational Spectroscopy at Electrified Interfaces

Edited by

**Andrzej Wieckowski**
**Carol Korzeniewski**
**Björn Braunschweig**

**Wiley Series on Electrocatalysis and Electrochemistry**

WILEY

Published by John Wiley & Sons, Inc., Hoboken, New Jersey
Published simultaneously in Canada



For general information on our other products and services or for technical support, please contact our Customer Care Department within the United States at (800) 762-2974, outside the United States at (317) 572-3993 or fax (317) 572-4002.

Wiley also publishes its books in a variety of electronic formats. Some content that appears in print may not be available in electronic formats. For more information about Wiley products, visit our web site at www.wiley.com.

***Library of Congress Cataloging-in-Publication Data is available.***
9781118157176

10 9 8 7 6 5 4 3 2 1

# Contents

# Preface to the Wiley Series on Electrocatalysis and Electrochemistry

This series covers recent advances in electrocatalysis and electrochemistry and depicts prospects for their contribution into the present and future of the industrial world. It aims to illustrate the transition of electrochemical sciences from its beginnings as a solid chapter of physical chemistry (covering mainly electron transfer reactions, concepts of electrode potentials, and structure of electrical double layer) to the field in which electrochemical reactivity is shown as a unique chapter of heterogeneous catalysis, is supported by high-level theory, connects to other areas of science, and includes focus on electrode surface structure, reaction environment, and interfacial spectroscopy.

The scope of this series ranges from electrocatalysis (practice, theory, relevance to fuel cell science and technology) to electrochemical charge transfer reactions, biocatalysis, and photoelectrochemistry. While individual volumes may appear quite diverse, the series promises updated and overall synergistic reports providing insights to help further our understanding of the properties of electrified solid–liquid systems. Readers of the series will also find strong reference to theoretical approaches for predicting electrocatalytic reactivity by such high-level theories as density functional theory. Beyond the theoretical perspective, further vehicles for growth are such significant topics such as energy storage, syntheses of catalytic materials via rational design, nanometer-scale technologies, prospects in electrosynthesis, new instrumentation, and surface modifications. In this context, the reader will notice that new methods being developed for one field may be readily adapted for application in another.

Electrochemistry and electrocatalysis have both benefited from numerous monographs and review articles due to their depth, complexity, and relevance to the practical world. The Wiley Series on Electrocatalysis and Electrochemistry is dedicated to present the current activity by focusing each volume on a specific topic that is timely and promising in terms of its potential toward useful science and technology. The chapters in these volumes will also demonstrate the connection of electrochemistry to other disciplines beyond chemistry and chemical engineering, such as physics, quantum mechanics, surface science, and biology. The integral goal is to offer a broad-based analysis of the total development of the fields. The progress of the series will provide a global definition of what electrocatalysis and electrochemistry are now, and will contain projections about how these fields will further evolve in time. The purpose is twofold, to provide a modern reference for graduate instruction and for active researchers in the two disciplines, as well as to document that electrocatalysis and electrochemistry are dynamic fields that are expanding rapidly, and are likewise rapidly changing in their scientific profiles and potential.

Creation of each volume required the editors involvement, vision, enthusiasm, and time. The Series Editor thanks each Volume Editor who graciously accepted his invitation. Special thanks go to Ms. Anita Lekhwani, the Series Acquisitions Editor, who extended the invitation to edit this series to me and has been a wonderful help in its assembling process.

ANDRZEJ WIECKOWSKI
Series Editor

# Foreword

Despite extensive efforts, the electrochemical interface, where central processes in electrochemical reactions occur, had long been a black box until researchers started to shed light on it in the 1960s. At the beginning, the light used was mostly that in the visible region, but information obtainable by visible light was very limited, and the use of vibrational spectroscopy, which can provide detailed information on molecules, was strongly desired. In 1974, an innovation was made in spectroelectrochemistry by the application of Raman spectroscopy. It is well known that the Raman study of pyridine adsorbed on silver electrodes led to the discovery of surface-enhanced Raman scattering (SERS). The application of infrared reflection absorption spectroscopy (IRAS) to electrochemical interfaces in 1980 also was a great achievement. Surface-enhanced infrared absorption (SEIRA), an effect similar to SERS, was discovered in the same year, and the first observation of sum frequency generation (SFG) from monolayers on solid surfaces was made in 1986, although their applications to electrochemical interfaces were somewhat delayed. During the last four decades, these surface vibrational spectroscopy techniques have been advanced greatly owing to the improvement in instrumentation and the development of experimental techniques, which are still further developing year by year.

Vibrational spectroscopy is a powerful tool to identify molecules and to study their structures and reactions, as it is often mentioned that *vibrational spectra are letters from molecules*. It is also the case at surfaces and interfaces. It provides us information on adsorbed structures and orientations of molecules. Spectra are sensitive to changes in the environment at interfaces, from which we can obtain deeper insight into chemistry and physics at the interfaces. In situ, time-resolved monitoring of reactions taking place at surfaces and interfaces is also possible. Owing to these advantages, they have gained wide application, from fundamental electrochemistry to many other related fields of science and technology, including surface science, heterogeneous electrocatalysis, energy conversion, biochemistry, nanotechnology, and sensors. However, each technique has strong and weak points. Difficulty in interpretation of the obtained spectra is another problem. For appropriate use of the techniques and for correct interpretation of spectra, sufficient fundamental understanding of the techniques is required.

This book, which features all the aforementioned four surface vibrational spectroscopy techniques and their applications to recent research topics, will provide fundamental information for nonspecialists and an up-to-date account of recent advances in this field for specialists.

Masatoshi Osawa
Catalysis Research Center, Hokkaido University, Sapporo, Japan

# Preface

Electrified interfaces play an important role in many phenomena. Electric fields that develop at junctions between different phases can align molecules and ions into configurations that greatly influence the physical and chemical nature of the interface. The molecular structure of charged interfaces impacts many practical processes, including energy conversion in batteries, solar cells, and fuel cells, corrosion at solid surfaces, chemical reactions over oxide particles in the earth and atmosphere, biochemical transformations, signal transduction in chemical sensors, and heterogeneous catalytic reactions, to name a few.

Surface vibrational spectroscopy techniques probe the structure and composition of interfaces at a molecular level. Their versatility and typically nondestructive nature often enable in situ measurements of operating devices and monitoring of interface-controlled processes under reactive conditions. This book highlights modern applications of Raman, infrared, and nonlinear optical spectroscopy in the study of charged interfaces.

Early chapters in the book provide a glimpse into the breadth of systems that can be investigated through the use of nonlinear optical techniques. Properties of interfacial water, ions, and biomolecules at charged dielectric, metal oxide, and electronically conductive metal catalyst surfaces, as probed by nonlinear optical techniques, are discussed in Part I. In addition to examples of practical experimental interest, the chapters guide readers to the latest in measurement and instrumental techniques. Part II includes coverage of Raman spectroscopy from the standpoint of sensitive approaches for detection of biomolecules at solid–liquid interfaces and the use of photon depolarization strategies to elucidate molecular orientation at surfaces. Part III reports on wide-ranging systems from small fuel molecules at well-defined surfaces to macromolecular complexes as building blocks of functional interfaces in devices that have applications in chemical sensing and electric power generation. These interfaces are amenable for infrared spectroscopy due to versatile sampling methods, that is, specular and diffuse reflectance, polarization–modulation, and total internal reflection modes.

ANDRZEJ WIECKOWSKI
CAROL KORZENIEWSKI
BJÖRN BRAUNSCHWEIG

# Contributors

**Christopher M. Berg**
School of Chemical Sciences
University of Illinois
Urbana, Illinois

**Björn Braunschweig**
Institute of Particle Technology (LFG)
Erlangen, Germany

**Christa L. Brosseau**
Department of Chemistry
Saint Mary's University, Halifax
Nova Scotia, Canada

**Angel Cuesta**
Instituto de Química Física
"Rocasolano"
CSIC
Madrid, Spain

**Dana D. Dlott**
School of Chemical Sciences
University of Illinois
Urbana, Illinois

**Jonathan Doan**
Department of Chemistry and Chemical Biology
Northeastern University
Boston, Massachusetts

**Kathrin Engelhardt**
Institute of Particle Technology (LFG)
Erlangen, Germany

**Michael D. Fayer**
Department of Chemistry
Stanford University
Stanford, California

**Juan M. Feliu**
Instituto de Electroquímica
Universidad de Alicante
Alicante, Spain

**Emily E. Fenn**
Department of Chemistry
Stanford University
Stanford, California

**Ian Kendrick**
Department of Chemistry and Chemical Biology
Northeastern University
Boston, Massachusetts

**Carol Korzeniewski**
Department of Chemistry
Texas Tech University
Lubbock, Texas

**Agata Królikowska**
Department of Chemistry
Warsaw University
Warsaw, Poland

**Robert B. Kutz**
Champaign, Illinois

**Annia H. Kycia**
Department of Chemistry
University of Guelph,
Guelph, Ontario, Canada

**Jacek Lipkowski**
Department of Chemistry
University of Guelph
Guelph, Ontario, Canada

**Prabuddha Mukherjee**
Urbana, Illinois

**Kei Murakoshi**
Department of Chemistry
Faculty of Science
Hokkaido University
Sapporo, Japan

**Hideki Nabika**
Department of Material and Biological
Chemistry, Faculty of Science
Yamagata University
Yamagata, Japan

**Fumika Nagasawa**
Department of Chemistry
Faculty of Science
Hokkaido University
Sapporo, Japan

**José Manuel Orts**
Instituto de Electroquímica
Universidad de Alicante
Alicante, Spain

**Wolfgang Peukert**
Institute of Particle Technology (LFG)
Erlangen, Germany

**Duc Thanh Pham**
Institute of Physical Chemistry
University of Bonn
Bonn, Germany

**Antonio Rodes**
Instituto de Electroquímica
Universidad de Alicante
Alicante, Spain

**Melanie Röefzaad**
Institute of Physical and Theoretical
Chemistry
University of Bonn
Bonn, Germany

**Armin Rumpel**
SwissOptic AG
Heerbrugg, Switzerland

**Andrea P. Sandoval**
Departamento de Quimica, Facultad de
Ciencias
Universidad Nacional de Colombia
Bogotá, Colombia

**Y. Ron Shen**
Physics Department
University of California, Berkeley
Berkeley, California

**Eugene S. Smotkin**
Department of Chemistry and Chemical
Biology
Northeastern University
Boston, Massachusetts

**ZhangFei Su**
Department of Chemistry
University of Guelph
Guelph, Ontario, Canada

**Mai Takase**
Catalysis Research Center
Hokkaido University
Sapporo, Japan

**Klaus Wandelt**
Institute of Physical and Theortetical Chemistry
University of Bonn
Bonn, Germany, and
Institute of Experimental Physics
University of Wroclaw
Wroclaw, Poland

**Glenn A. Waychunas**
Geochemistry Department
Earth Sciences Division
Lawrence Berkeley National Laboratory
Berkeley, California

**Andrzej Wieckowski**
Department of Chemistry
University of Illinois at Urbana Champaign
Urbana, Illinois

Part One

# Nonlinear Vibrational Spectroscopy

# Chapter 1

# Water Hydrogen Bonding Dynamics at Charged Interfaces Observed with Ultrafast Nonlinear Vibrational Spectroscopy

**Emily E. Fenn and Michael D. Fayer**
Department of Chemistry, Stanford University, Stanford, California

## 1.1 INTRODUCTION

The question of how charged species affect water structure and dynamics is relevant to many applications in chemistry, biology, geology, and industry. Biological systems are often crowded aqueous environments filled with proteins, membranes, vesicles, and other structures that often rely on the presence of ions for stability and proper functioning [1–6]. The ion–water interface is critical for ion exchange resins [7, 8], heterogeneous catalysis [9–11], electrochemistry [12], as well as processes involving mineral dissolution [13, 14] and ion adsorption [15, 16]. Because the behavior of water in the presence of ions impacts a wide range of technical and scientific fields, a great deal of literature over the years has been dedicated to studying the aqueous solvation of ions and the properties of water at charged interfaces. Studies that have examined ion–water interfaces have employed *x*-ray and neutron diffraction [17–19], Raman spectroscopy [20], ultrafast infrared spectroscopy [21–26], Fourier transform

*Vibrational Spectroscopy at Electrified Interfaces*, First Edition.
Edited by Andrzej Wieckowski, Carol Korzeniewski, and Björn Braunschweig.

infrared (FTIR) spectroscopy [20, 27], and other spectroscopic techniques [15, 28, 29]. Theoretical models [30, 31], molecular dynamics (MD) simulations [32–35], and Monte Carlo (MC) calculations [20] have also been employed. While simulations can provide some insight into the underlying dynamics, most experimental techniques only provide steady-state data. Here we utilize ultrafast infrared spectroscopy to examine the hydrogen bonding dynamics of water at several types of charged and uncharged interfaces.

Ultrafast infrared spectroscopy has been shown to be a powerful technique for elucidating dynamics in water–ion systems [21–26], other hydrogen bonding systems [36–43], protein environments [44–52], and systems that undergo chemical exchange [25, 53–57]. Here, we apply ultrafast infrared pump–probe and two-dimensional infrared (2D IR) vibrational echo spectroscopic techniques to examine the dynamics of water when it is confined in nanoscopic environments and interacting with interfaces. The question is whether the nature of confinement or the chemical composition of the interface most significantly influences the dynamics. To explore this question, the dynamics of water at charged and neutral interfaces in reverse micelles are compared. In addition, water in ionic solutions is investigated. Some water molecules are hydrogen bonded to ions, while others are hydrogen bonded to water molecules. These are in equilibrium, with water molecules bound to ions switching and becoming bound to water molecules, and vice versa. Using 2D IR chemical exchange spectroscopy, we determine the exchange time required for a water hydroxyl initially hydrogen bonded to an anion to switch to being hydrogen bonded to another water molecule.

Reverse micelles consist of a water pool surrounded by a layer of surfactant molecules and are often used as model systems for confined environments. The surfactant molecules are terminated by a hydrophilic head group that can be either charged or neutral. These hydrophilic head groups face in toward the water pool while the alkyl (hydrophobic) tails of the surfactant are suspended in a nonpolar organic phase. A schematic of a reverse micelle utilizing the surfactant Aerosol-OT, or AOT [sodium bis(2-ethylhexyl) sulfosuccinate], is shown in Figure 1.1. The AOT surfactant (Fig. 1.2) forms spherical monodispersed reverse micelles that have been well characterized. The size of the AOT reverse micelles can be easily controlled by varying the amounts of starting materials according to the $w_0$ parameter: $w_0$ = $[H_2O]$/[surfactant] [58–60]. AOT can yield sizes of $w_0$ = 0 (essentially dry reverse micelles) all the way up to $w_0$ = 60, which has a water pool diameter of 28 nm and contains ~350,000 water molecules [61]. Isooctane is a common solvent used as the nonpolar phase of AOT reverse micelle systems, but other solvents such as carbon tetrachloride, cyclohexane, and benzene can also be used with minimal changes in water pool size for a given $w_0$ [62]. A recent study has shown that the identity of the nonpolar phase has no effect on the water pool dynamics [63].

As shown in Figure 1.2, AOT has a sulfonate head group with a sodium counterion. The head group region of the reverse micelle therefore creates a charged interface that surrounds the water pool. The sodium ions will generally reside in a region close to the interface. Figure 1.1 illustrates the regions of a reverse micelle. When the total water pool diameter, *d*, is sufficiently large (≥4.6 nm) the reverse

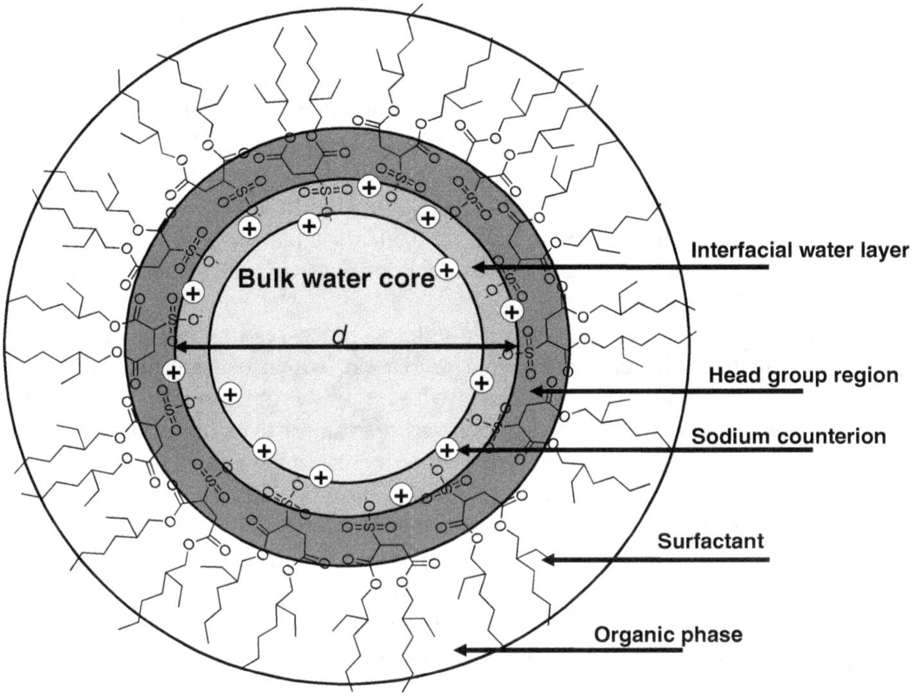


**Figure 1.1** Illustration of the reverse micelle interior. The bulk water core is surrounded by a layer of interfacial water. The total water pool diameter is denoted by *d*. The hydrophilic AOT head groups face in toward the water pool while the alkyl tails are suspended in the organic phase. The sodium counterions are dispersed in the water pool, but they generally reside close to the head group interface.

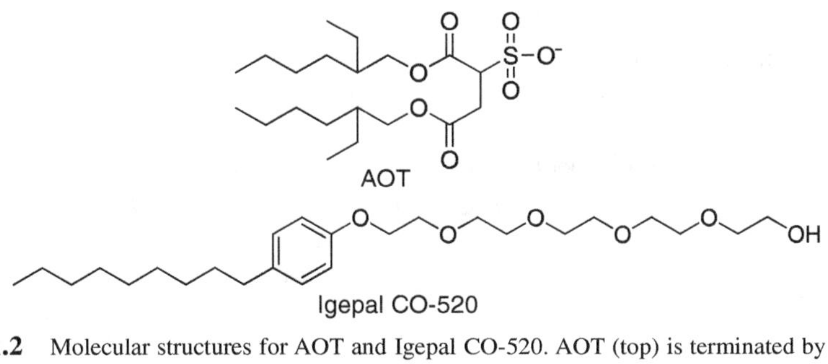


**Figure 1.2** Molecular structures for AOT and Igepal CO-520. AOT (top) is terminated by a charged sulfonate head group with a sodium counterion while Igepal (bottom) has a neutral hydroxyl head group.

micelle can support a core of water with bulklike properties. Below we will discuss how far perturbations from the charged sulfonate interfacial region extend into the water pool and what happens to the water dynamics as the size of the water pool changes in size. The chemical identity of the surfactant layer can be changed by using a neutral surfactant molecule called Igepal CO-520 (Fig. 1.2). Igepal is terminated with neutral hydroxyl head groups, so the interfacial water molecules will be exposed to a very different chemical surface compared to the AOT reverse micelle system. To what extent changes in surfactant identity, particularly charged versus neutral head group regions, and reverse micelle size affect water dynamics will be described.

Water dynamics are investigated through the processes of orientational relaxation, spectral diffusion, and vibrational relaxation, which can be measured with ultrafast infrared vibrational spectroscopy. These observables report on how the hydrogen bond network of water evolves and rearranges over time. The hydroxyl stretch of water is monitored during the experiments and is used as a reporter for hydrogen bond dynamics. During vibrational relaxation, vibrational energy dissipates by transferring into a combination of low-frequency modes, such as torsions and bath modes [64, 65]. Energy must be conserved during this process. Certain pathways that facilitate vibrational relaxation in one system may or not be present in a different system. Thus, vibrational relaxation is extremely sensitive to local environments. Orientational relaxation measures how quickly water molecules reorient by monitoring the direction of the transition dipole of the hydroxyl stretch. Molecular reorientation is involved in water hydrogen bond exchange, which leads to global hydrogen bond network reorganization [66, 67]. Bulk water consists of an extended network of hydrogen bonds that are continually rearranging and exchanging with one another. According to the theory of Laage and Hynes, water molecules exchange hydrogen bonds via a jump reorientation mechanism that involves concerted motions of water molecules in the first and second solvation shells [66, 67]. The mechanism proceeds when a molecule in the second solvation shell of another water molecule moves in toward the first solvation shell. In order to swap hydrogen bonds with the approaching water from the second solvation shell, a water molecule must pass through a five-coordinate transition state and then undergo a large-amplitude rotational motion (or "jump"). The jump allows it to switch one of its hydrogen bonds to the approaching water molecule. These large-amplitude jumps change the orientation of the transition dipole. Solutes and interfaces (such as the surfactant shell of the reverse micelles) can disrupt the jump reorientation mechanism, thus slowing down the process of reorientation [23, 68–72]. Both vibrational and orientational relaxation can be measured with ultrafast infrared pump–probe spectroscopy.

Ultrafast 2D IR vibrational echo spectroscopy is used to measure spectral diffusion of the water hydroxyl stretch. The linear infrared absorption spectrum of the hydroxyl stretch is very broad due to a large distribution in the lengths and strengths of hydrogen bonds. At the beginning of the 2D IR experiment, a hydroxyl will vibrate at a certain frequency, but due to dynamic structural evolution of the system, that frequency will change over time. This process of frequency evolution is known

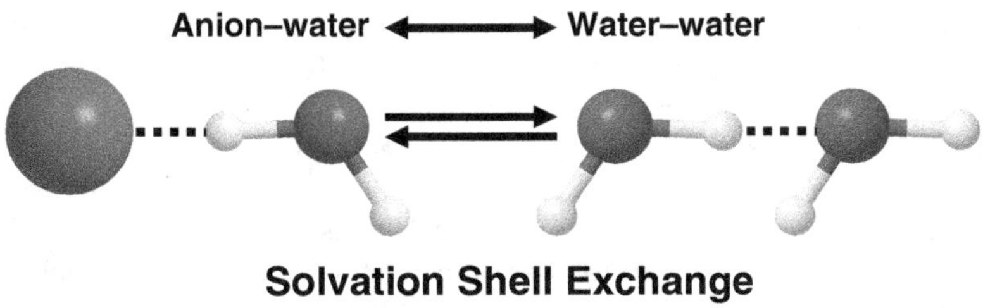


**Figure 1.3** Representation of solvation shell exchange. The time it takes for a water hydroxyl initially hydrogen bonded to an anion (left side) to switch to being hydrogen bonded to another water hydroxyl (right side) can be measured with 2D IR vibrational echo spectroscopy.

as spectral diffusion and reports on how quickly water molecules sample different structural environments.

In addition, 2D IR vibrational echo chemical exchange spectroscopy is used to examine how quickly a water hydroxyl bound to an anion will switch to being hydrogen bonded to a neighboring water hydroxyl. This process is illustrated schematically in Figure 1.3. A model system for studying water–ion exchange is a solution of sodium tetrafluoroborate ($NaBF_4$) in water because the linear IR absorption spectrum of the solution yields two resolved peaks corresponding to waters interacting with other waters and waters interacting with the tetrafluoroborate anions. It is found that the ion–water hydrogen bond switching time is ~7 ps [25]. This switching time has implications when treating the orientational relaxation dynamics of water molecules inside reverse micelles made of charged and neutral surfactants [71]. Together, these ultrafast infrared experiments involving water in reverse micelles and water–ion chemical exchange construct a dynamic picture of the behavior of water molecules at charged interfaces and interacting with ions.

## 1.2 EXPERIMENTAL METHODS

### 1.2.1 Sample Preparation

Carbon tetrachloride ($CCl_4$), cyclohexane, isooctane, $H_2O$, $D_2O$, AOT, Igepal CO-520, and $NaBF_4$ were used as received. Stock solutions of 0.5 M AOT were prepared in $CCl_4$, cyclohexane, and isooctane. A variety of solvents are necessary due to certain experimental considerations that will be discussed below. A 0.3-M stock solution of Igepal CO-520 was prepared in cyclohexane. The residual water contents of the stock solutions were measured via Karl Fischer titration. The reverse micelle samples were prepared by mass by adding appropriate amounts of a solution of 5% HOD (water with one hydrogen exchanged with deuterium) in $H_2O$ to measured quantities of the AOT or Igepal stock solutions to obtain the desired $w_0$. In the ultrafast experiments, the OD stretch of 5% HOD in $H_2O$ is probed because it not only provides an isolated stretching mode to interrogate but also prevents vibrational excitation transfer processes from artificially causing decay of the orientational correlation function and observables related to spectral diffusion [73, 74]. MD simulations demonstrate that a dilute amount of HOD does not perturb the structure and properties of $H_2O$ and that the OD stretch reports on the dynamics of water [75].

For the pump–probe experiments involving large reverse micelles presented here, the 0.5-M stock solution of AOT in isooctane was used to make samples of $w_0 = 10$, 16.5, 25, 37, and 46 (diameters of 4.0, 5.8, 9, 17, and 20 nm, respectively). It has been found that the combination of AOT and isooctane causes distortions in vibrational echo experiments, so for vibrational echo experiments on small reverse micelles the 0.5-M AOT/$CCl_4$ stock solution was used to make $w_0 = 2$, 4, and 7.5 (diameters of 1.7, 2.3, and 3.3 nm, respectively) [63]. $CCl_4$ cannot support reverse micelles larger than $w_0$ ~10 [61], so the 0.5-M AOT/cyclohexane stock solution was used to make $w_0 = 12$ and 16.5 (diameters of 4.6 and 5.8 nm, respectively) for vibrational echo experiments on larger reverse micelles. Neither $CCl_4$ nor cyclohexane cause distortions in the vibrational echo experiments. Cyclohexane can only reliably make reverse micelles of $w_0 < 20$ [76, 77], so the vibrational echo studies are limited to the lower bound of large sizes. It will be shown that both $w_0 = 12$ and 16.5 have well-defined bulk water cores and interfacial water regions, so analysis of these sizes should provide insight into the behaviors of water molecules in the even larger sizes.

To compare the effects of the chemical composition of the interface, Igepal reverse micelles with $w_0 = 12$ and 20 were also made. Like AOT, Igepal also makes monodispersed spherical reverse micelles [78]. The AOT and Igepal surfactants have different aggregation numbers, thus yielding different sizes for the same $w_0$ values. The $w_0 = 12$ Igepal reverse micelles have the same 5.8-nm diameter as AOT $w_0 = 16.5$ while the $w_0 = 20$ reverse micelles have the same 9-nm diameter as AOT $w_0 = 25$. AOT lamellar structures (sheets of AOT surfactants with water between them) were prepared by adding water to dry AOT to produce samples with various water-to-surfactant ratios, $\lambda$. These lamellar samples allow us to examine the effects of confining geometry on the water dynamics (spherical confinement versus confinement within layers).

The experimental samples are contained between two calcium fluoride windows that are separated by a Teflon spacer. The thickness of the Teflon spacer is chosen such that the optical density of the OD stretch region is ~0.1 for the vibrational echo experiments and ~0.5–0.7 for the pump–probe experiments.

For the 2D IR chemical exchange experiments, a 5.5-M solution of $NaBF_4$ in water was used. Again, the water component consisted of 5% HOD in $H_2O$ for the same reasons outlined above. The 5.5-M concentration corresponds to a system with seven water molecules per $NaBF_4$ molecule ($n = 7$). For linear IR absorption measurements, additional samples of 1.7 M, 3.1 M, and 4.3 M of $NaBF_4$ in water were made, corresponding to $n = 30$, 15, and 10, respectively.

### 1.2.2 2D IR Vibrational Echo Spectroscopy

The laser system used to generate the infrared light that excites the OD hydroxyl stretch consists of a Ti–sapphire oscillator that seeds a regenerative amplifier. The output of the regenerative amplifier pumps an optical parametric amplifier that generates near-infrared wavelengths that are difference frequency mixed in a $AgGaS_2$ crystal. The resulting mid-IR pulses are centered at ~4 μm (2500 $cm^{-1}$) but can be tuned to the peak of the absorption spectrum for a given sample (e.g., 2565 $cm^{-1}$ for $w_0 = 2$ AOT reverse micelle). The generated mid-IR beam enters a 2D IR vibrational spectrometer that can be readily converted into a pump–probe setup.

The pump–probe and 2D IR vibrational echo techniques presented here are noncollinear four-wave mixing experiments [79, 80]. In these experiments, three field–matter interactions between the sample and the incident ultrafast laser pulses create a third-order macroscopic polarization that emits a signal electric field. Depending upon the type of experiment, the signal electric field carries different types of information. As discussed in the introduction, the signal from a pump–probe experiment allows one to extract vibrational lifetimes and orientational relaxation parameters for the OD stretch of HOD in $H_2O$. The vibrational lifetime and orientational observables are sensitive to local structural and chemical environments. 2D IR vibrational echo spectroscopy is a sophisticated technique that observes how vibrational chromophores in a system evolve in frequency over time via chemical exchange, spectral diffusion, coherence transfer, or other processes. 2D IR spectroscopy can monitor these frequency changes by manipulating the quantum pathways by which the system evolves.

The experimental pulse sequence for the 2D IR vibrational echo experiments is shown in Figure 1.4*a*. The pump–probe experiment (Fig. 1.4*b*) is similar in several ways to the 2D IR experiment, but it differs in the number of input beams and the manner by which the signal is detected. In the 2D IR experiment, three time-ordered

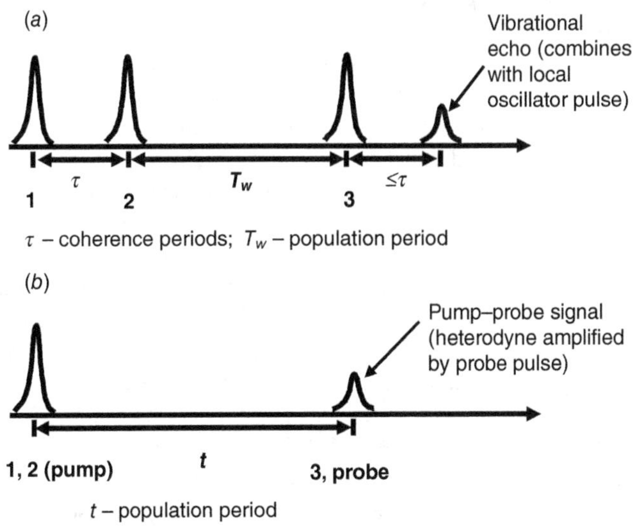


**Figure 1.4** Pulse sequences implemented for (*a*) 2D IR vibrational echo spectroscopy and (*b*) pump–probe spectroscopy. Both techniques are noncollinear four-wave mixing experiments.

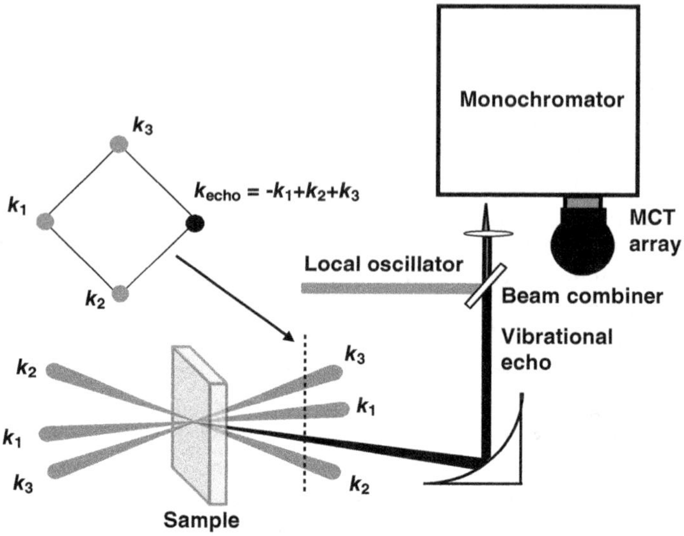


**Figure 1.5** Experimental setup for the 2D IR vibrational echo experiment. Three excitation beams in a BOXCARS geometry cross in the sample to produce the vibrational echo signal in the phase-matched direction denoted by the wave vector $k_{sig}$. The vibrational echo signal combines with a local oscillator beam for heterodyned detection. The heterodyned signal is detected by a 32-pixel MCT array detector.

beams interact with the sample (Fig. 1.5). The first pulse induces a coherence state between the ground (0) and first excited (1) vibrational levels. The vibrations are initially in phase, but inhomogeneous broadening of the absorption line and structural fluctuations cause the phase relationships to decay. After a period of time, τ, a second pulse impinges on the sample and creates a population state in either the 0 or 1 vibrational levels. A time period $T_w$ elapses (the waiting time) before the third pulse reaches the sample and induces a second coherence state, partially restoring the phase relationships between the vibrational chromophores. This rephasing process causes the vibrational echo signal electric field to emit at a time $t \leq \tau$. The vibrational echo signal propagates in the phase-matched direction according to $k_{sig} = -k_1 + k_2 + k_3$, as denoted by the BOXCARS geometry of the input beams shown in Figure 1.5. The vibrational echo signal is temporally and spatially overlapped with a local oscillator (LO) pulse for heterodyned detection. The LO is another IR pulse identical to the excitation pulses but lower in amplitude and fixed in time. The combined vibrational echo and LO beam is frequency dispersed by a monochromator, and the heterodyned signal is detected on a 32-pixel mercury–cadmium–telluride (MCT) array detector. If we denote the vibrational echo signal as $S$ and LO signal as $L$, then the quantity $|S + L|^2 = S^2 + 2LS + L^2$ is measured on the detector. Assume $S^2$ is negligible and can be ignored and $L^2$ is a constant signal and can be subtracted. The $2LS$ cross term represents the heterodyned signal of interest.

Ultimately, the 2D IR experiment obtains frequency correlation plots (2D spectra) for the second coherence period (known as the detection time period) and

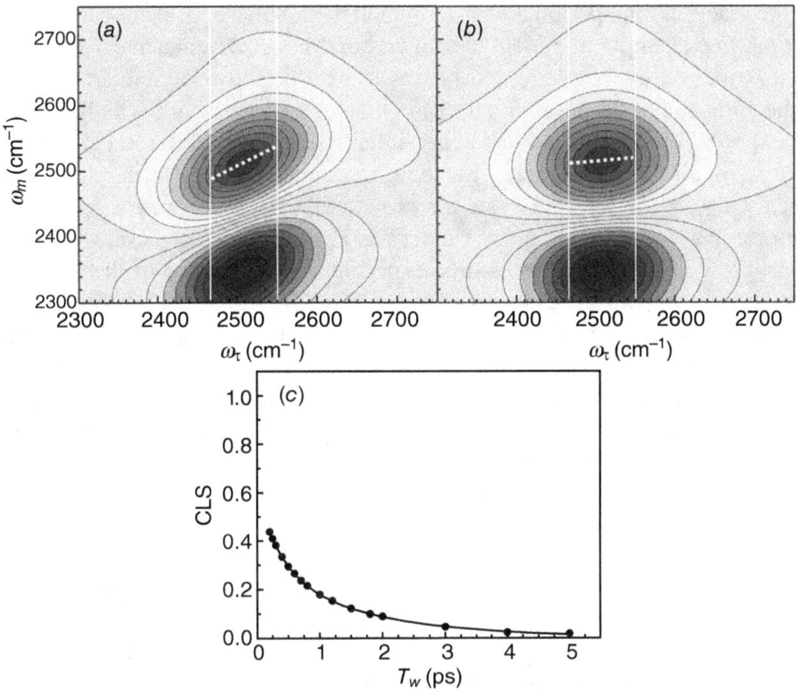


**Figure 1.6** CLS procedure: (*a*) bulk water 2D spectrum at $T_w$ = 0.2 ps, showing elongation along the diagonal and (*b*) bulk water spectrum at $T_w$ = 2 ps. Spectral diffusion has mostly completed, so the spectra become more circular. The white solid lines show the direction of cuts during the CLS procedure. The white dots indicate the peak positions through the slices (centerline data). The slope is found through each set of peak positions at each $T_w$ to obtain a plot of slope versus $T_w$, as shown in panel (*c*). The CLS curve is equal to the normalized FFCF, which is shown as the black line through the CLS data.

the first coherence period (known as the evolution time period). Each frequency axis of the correlation plot arises from Fourier transformation of each coherence period. The Fourier transform along the second coherence period is performed experimentally by the monochromator, yielding the vertical "$\omega_m$" axis while the Fourier transform along the first coherence period is obtained numerically during data processing, yielding the horizontal "$\omega_\tau$" axis. These axes are clearly marked in Figures 1.6*a*, *b*, which show correlation spectra for bulk water (5% HOD in $H_2O$). During the experiment, $\tau$ is scanned for a series of fixed $T_w$ values. As $\tau$ is scanned, the echo signal field moves in time relative to the fixed LO. The echo field goes in and out of phase with the LO field producing an interferogram. Mixing the vibrational echo signal with the LO allows interferograms to be recorded, thus providing the necessary phase information for Fourier transformation.

Qualitatively, a 2D IR vibrational echo experiment works in the following manner. The first laser pulse "labels" the initial structures of the species by establishing their initial frequencies, $\omega_\tau$. The second pulse ends the first time period $\tau$ and starts the reaction time period $T_w$ during which the labeled species undergo structural evolution. For example, the local hydrogen bond network rearranges. This ends the population period of length $T_w$ and begins a third period of length $\leq\tau$, which ends with the emission of the vibrational echo pulse of frequency $\omega_m$, which is the signal in the experiment. The vibrational echo signal reads out information about the final structures of all labeled species by their frequencies, $\omega_m$. During the period $T_w$ between pulses 2 and 3, the system's structural evolution occurs. The structural evolution and associated frequency changes as $T_w$ is increased cause new off-diagonal peaks to grow in a chemical exchange experiment or the shape of the 2D spectrum to change in a spectral diffusion experiment. The growth of the off-diagonal peaks or the change in the 2D band shapes in the 2D IR spectra with increasing $T_w$ provides the dynamical information.

The 2D IR experiments require precise timing between the excitation pulses as well as phase stability during the coherence periods. Computer-controlled precision delay lines (Aerotech ANT-50L) manipulate the delay between the excitation pulses, and a three-pulse cross-correlation measurement is used to check the timing and correct for drifts between the three excitation pulses [63]. The vibrational echo experiments are also sensitive to linear chirp in the mid-IR pulse. Linear chirp is corrected by proper insertion of materials (calcium fluoride and germanium) with opposite signs of the group velocity dispersion (GVD) [82, 83].

The 2D IR experiments presented here are used to determine the frequency–frequency correlation function (FFCF) of water molecules (OD stretch), which is a measure of spectral diffusion. The FFCF can be determined via the centerline slope (CLS) method [84, 85]. Figure 1.6 illustrates this process. In the CLS technique, the 2D IR spectrum is sliced parallel to the vertical $\omega_m$ axis over a range of frequencies surrounding the center of the 2D IR spectrum. In Figures 1.6*a*,*b*, the solid white lines show the direction of slicing and the bounds over which the slices are taken. For water systems (as shown in the figure), the range is typically $\pm$30–40 cm$^{-1}$ around the measured center frequency of each 2D IR spectrum. Each slice intercepts a frequency on the horizontal $\omega_\tau$ axis. The slices are fit to Gaussian line shape functions to determine the peak position of each slice. Only the peak positions are required. The peak positions (white dots) are plotted versus their corresponding $\omega_\tau$ frequencies, and the slope of the resulting line is calculated. This process is repeated for all $T_w$ values so that a plot of slope versus $T_w$ is obtained (Fig. 1.6*c*). Generally, the 2D plots show elongated spectra at early $T_w$, as shown by Figure 1.6*a*. By 2 ps (Fig. 1.6*b*), the spectra show a more circular shape because most of the water hydroxyl environments in the $H_2O$ inhomogeneous line shape have been sampled by the OD vibrational chromophore, indicating that spectral diffusion is nearly complete.

It has been demonstrated theoretically that the CLS curve of slope versus $T_w$ is equal to the $T_w$-dependent portion of the normalized FFCF [84, 85]. The FFCF adopts a functional form that contains both homogeneous (motionally narrowed) and inhomogeneous components,

$$C_1(t) = \langle \delta\omega_{10}(t)\delta\omega_{10}(0) \rangle = \frac{\delta(t)}{T_2^*} + \sum_i \Delta_i^2 e^{-t/\tau_i} \tag{1.1}$$

where $\langle \delta\omega_{10}(t)\delta\omega_{10}(0)\rangle$ is the correlation function for the fluctuating 0–1 transition frequency, and $\delta\omega_{10}(t) = \langle\omega_{10}\rangle - \omega_{10}(t)$. The summation term in Eq. (1.1) contains the processes that sample the inhomogeneous distribution of frequencies through structural evolution of the system. The $\Delta_i$ terms are frequency fluctuation amplitudes, and the $\tau_i$ are their associated time constants. The time constants describe different time scales that contribute to spectral diffusion. The magnitude of each $\Delta_i$ term gives the contribution to the absorption line shape from processes occurring on each time scale.

The homogeneous dephasing time $T_2$ is given by

$$\frac{1}{T_2} = \frac{1}{T_2^*} + \frac{1}{2T_1} + \frac{1}{3\tau_{\mathrm{or}}} \tag{1.2}$$

where $T_2^*$ is the pure dephasing time, $T_1$ is the vibrational lifetime, and $\tau_{\mathrm{or}}$ is the orientational relaxation time constant. The first term of Eq. (1.1) is known as the pure dephasing component. In the condensed phase, homogeneous broadening can arise from fast solvent motions [86]. The dynamics that give rise to pure dephasing occur on a very fast time scale such that the product $\Delta\tau < 1$, meaning that this portion of the total absorption line shape comes from a motionally narrowed (Lorentzian) component. The homogeneous component is often expressed as the homogeneous line width, $\Gamma = 1/(\pi T_2)$. In water, $T_2^*$ is on the order of a couple hundred femtoseconds while $T_1$ and $\tau_{\mathrm{or}}$ are a few picoseconds or longer. Therefore, the homogeneous line width is virtually completely determined by pure dephasing.

Figure 1.6*c* shows that there is a large difference between 1 and the initial CLS data points extrapolated to $T_w = 0$. This large drop is caused by fast homogeneous processes, and the difference between the $T_w = 0$ CLS data and 1 combined with the absorption spectrum permits the homogeneous line width to be determined. Slower processes sample the inhomogeneous distribution of frequencies and cause the $T_w$-dependent decay of the CLS.

The first step in extracting the FFCF involves fitting the CLS to a multiexponential decay to yield a set of amplitudes and decay constants [85]. Due to the short time approximation [85, 87, 88], the amplitude associated with the fastest of the time constants can be pushed into the homogeneous contribution. As a result, the CLS data alone cannot accurately determine the full FFCF. In order to obtain the correct homogeneous component and amplitude of the fast inhomogeneous component, it is necessary to employ the absorption line shape. The linear absorption spectrum is the Fourier transform of the linear response function, $R_1(t)$,

$$R_1(t) = |\mu_{10}|^2 e^{-i\langle\omega_{10}\rangle t} e^{-ig_1(t)} \tag{1.3}$$

where $\mu_{10}$ is the transition dipole moment of the 0–1 transition, $\langle\omega_{10}\rangle$ is the average 0–1 transition frequency, and $g_1(t)$ is the line shape function,

$$g_1(t) = \int_0^t d\tau_2 \int_0^{\tau_2} d\tau_1 \langle \delta\omega_{10}(t)\delta\omega_{10}(0)\rangle \tag{1.4}$$

Equation (1.4) contains the FFCF, as given by Eq. (1.1), showing the link between the absorption spectrum and the underlying dynamic processes. The amplitude of

**Table 1.1** Bulk Water FFCF Parameters

| $\Gamma$ (cm$^{-1}$) | $\Delta_1$ (cm$^{-1}$) | $\tau_1$ (ps) | $\Delta_2$ (cm$^{-1}$) | $\tau_2$ (ps) |
|---|---|---|---|---|
| 76 ± 14 | 41 ± 8 | 0.4 ± 0.1 | 34 ± 11 | 1.7 ± 0.5 |

the fast inhomogeneous decay and the homogeneous component are the only adjustable parameters as the absorption spectrum is fit simultaneously with the CLS data. The time constants and remaining amplitudes of the CLS multiexponential fit are accurate and are fixed during the procedure. This fitting routine has been shown to accurately determine the full FFCF including both homogeneous and inhomogeneous components [84, 85]. For bulk water, it is found that there is a fast ~400-fs decay that is attributed to fluctuations of the lengths of the hydrogen bonds [89, 90], followed by a 1.7-ps decay corresponding to global hydrogen bond network randomization [23]. In addition, there is a relatively large homogeneous contribution. The full FFCF for bulk water is listed in Table 1.1 and is shown going through the CLS points in Figure 1.6*c*.

### 1.2.3 Polarization-Selective Pump–Probe Spectroscopy

The pump–probe experiment illustrated in Figure 1.4*b* also involves three field–matter interactions. In this variation, only two beams are used (instead of the three excitation beams plus the LO in the vibrational echo experiment). A strong pump pulse and weak probe pulse are crossed in the sample. The first two field–matter interactions occur during the pump pulse so that $\tau = 0$, and the system immediately adopts a population state in either the 0 or 1 vibrational levels. The probe pulse enters at a later time to interrogate the changes in the sample due to the pump. The pump–probe signal emits in the same phase-matched direction as the probe and self-heterodynes with the probe.

The vibrational lifetime and orientational relaxation dynamics are obtained via polarization-selective pump–probe experiments. The pump is polarized at 45° relative to the probe (which is kept at horizontal polarization). Molecules whose transition dipoles are oriented at 45° will have a higher probability of being excited, while transition dipoles oriented perpendicular to the pump will have zero probability of being excited. The probe undergoes changes in absorption due to these transitions. Because the 0–1 vibrational transition is excited, there will be increased transmission of the probe as well as stimulated emission at the 0–1 transition frequency. The induced absorption of the 1–2 transition will cause a decreased transmission of the probe at the 1–2 frequency. As the molecules reorient, fewer transition dipoles will be aligned with the pump. After the sample, the probe is resolved at parallel and perpendicular polarizations relative to the pump (±45°). The resolved probe is frequency dispersed by the monochromator and detected on the 32-pixel MCT detector. The time between the pump and probe pulses is scanned to measure the time-dependent changes in absorption of the parallel and perpendicular components. The parallel and perpendicular signals are given by

$$I_{\parallel}(t) = P(t)(1+0.8C_2(t)) \tag{1.5}$$

$$I_{\perp}(t) = P(t)(1-0.4C_2(t)) \tag{1.6}$$

where $P(t)$ is population relaxation and $C_2(t)$ is the second Legendre polynomial orientational correlation function for a dipole transition [91]. Pure population relaxation (no contributions from orientation) may be obtained from

$$P(t) = \tfrac{1}{3}(I_{\parallel}(t)+2I_{\perp}(t)) \tag{1.7}$$

$P(t)$ generally adopts the form of a single or biexponential decay, depending on how many distinct OD environments a system contains:

$$P(t) = e^{-t/T_1} \tag{1.8}$$

for one component or

$$P(t) = A_1 e^{-t/T_1^1} + (1-A_1)e^{-t/T_1^2} \tag{1.9}$$

for two components, where $A_1$ and $1 - A_1$ are the fractional populations of components 1 and 2, respectively, and the $T_1^i$ terms are the vibrational lifetimes of the ODs for the $i$th component. The orientational correlation function, $C_2(t)$, can be obtained from calculating the anisotropy according to

$$r(t) = \frac{I_{\parallel}(t)-I_{\perp}(t)}{I_{\parallel}(t)+2I_{\perp}(t)} = 0.4C_2(t) \tag{1.10}$$

It is important to note that Eq. (1.10) involves dividing by the pure population relaxation [$P(t)$ of Eq. (1.7)] to obtain pure orientational relaxation dynamics. Equation (1.10) only holds for a single-component system in which the vibrational chromophores have a single vibrational lifetime and orientational correlation function. In this case, the orientational correlation function follows single exponential dynamics,

$$C_2(t) = e^{-t/\tau_{\mathrm{or}}} \tag{1.11}$$

where $\tau_{\mathrm{or}}$ is a time constant describing orientational relaxation. Sometimes $C_2(t)$ can also contain a contribution from a "wobbling-in-a-cone" mechanism [92, 93], which will be discussed further in Section 1.6.

If the system contains two distinct species, each with its own vibrational lifetime and orientational relaxation time constants, then $P(t)$ does not divide out and the anisotropy adopts a more complicated form [61, 68–72]:

$$r(t) = \frac{A_1 e^{-t/T_1^1} e^{-t/\tau_{\mathrm{or}}^1} + (1-A_1)e^{-t/T_1^2} e^{-t/\tau_{\mathrm{or}}^2}}{A_1 e^{-t/T_1^1} + (1-A_1)e^{-t/T_1^2}} \tag{1.12}$$

where the $T_1^i$ and $\tau_{\mathrm{or}}^i$ terms are the vibrational lifetimes and orientational relaxation time constants for the $i$th component, respectively. The fractional populations of each component are given as $A_1$ and $1 - A_1$.

## 1.3 LINEAR INFRARED ABSORPTION SPECTRA OF WATER NEAR CHARGED INTERFACES

### 1.3.1 Water in Salt Solutions

The hydroxyl stretch of water, in this case the OD stretch of HOD in $H_2O$, is extremely sensitive to local environments. Figure 1.7 shows the IR absorption band of the OD stretch (5% HOD in $H_2O$). The band is wide [~170 $cm^{-1}$ full width at half-maximum (FWHM)] because there is a large distribution in the lengths and strengths of hydrogen bonds. Hydrogen bonds influence the attractive part of the hydroxyl vibrational potential, opening up the potential and lowering the vibrational frequency compared to the gas phase. A strong hydrogen bond will lower the vibrational transition frequency more than a weak hydrogen bond. Relative to the line center of the OD stretch of HOD in $H_2O$, strong hydrogen bonds produce a red shift and weak hydrogen bonds produce a blue shift [21, 23, 61, 68, 72]. Raman experiments and MC calculations have suggested that the magnitude of a red or blue shift in the hydroxyl stretch frequency is actually correlated to the strength and directionality of the electric field of the hydrogen bond acceptor [20]. Figure 1.8 shows linear IR absorption spectra of the OD hydroxyl stretch in water–salt solutions at a constant ratio of 12 water molecules to 1 salt ion pair. The OD hydroxyls will interact with the anions of the salts. The salts are KF, KCl, NaCl, KBr, NaBr, KI, and NaI. As the size of the anion increases, the spectra show a greater blue shift. This trend has been observed by other groups [21, 94]. Based on the Raman and MC studies mentioned above, the blue shift arises because a more diffuse and less directed electric field is projected along the OD hydroxyl as the size of the anion increases. KF, in contrast, causes the spectrum to red shift compared to the OD stretch of HOD in bulk water. $F^-$ is a very small anion, increasing the electric field felt by the hydroxyl. The Raman and MC study also concluded that the anion only caused local changes to the hydrogen bond structure of water and that only the waters in the first solvation

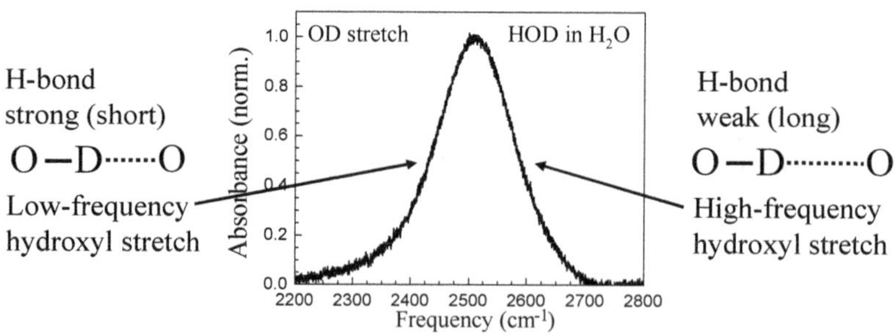


**Figure 1.7** Linear IR absorption spectrum of the OD stretch of 5% HOD in $H_2O$. The absorption band is very wide (~170 $cm^{-1}$) because there is a wide range in the distribution of lengths and strengths of hydrogen bonds in the water network. The red side of the line shape is dominated by strong hydrogen bonds that lengthen the OD bond. In contrast, the blue side of the line is dominated by weak hydrogen bonds that cause a shorter OD bond length.

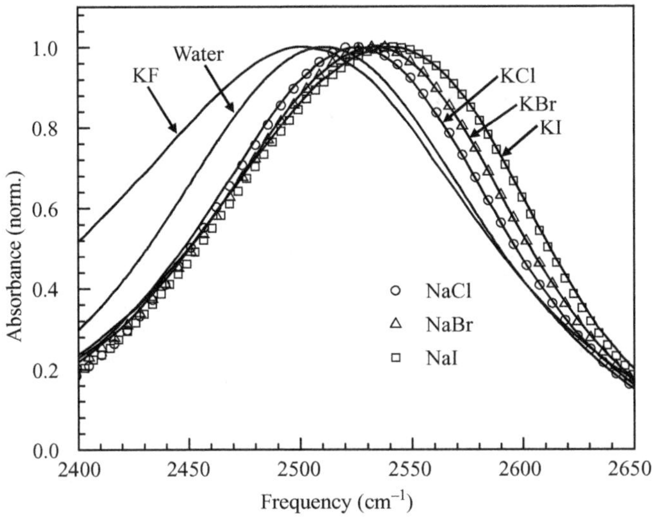


**Figure 1.8** Linear IR absorption spectra of bulk water (5% HOD in $H_2O$) and water–salt solutions at ratios of 12 : 1 water to salt. The salts are KF, KCl, KBr, KI, NaCl, NaBr, and NaI. As the anion increases in size, the spectra shift more to the blue. Large anions exert a weaker electric field along the OD hydroxyl to cause the blue shift. There appears to be no difference when the cation is changed but the anion kept the same.

shell of the anion felt significant perturbation [20]. The amount of blue or red shift is correlated to electric field strength of the anion and also may reflect the strength of the hydrogen bond. These conclusions refute the idea that salts are either kosmotropes ("structure makers") or chaotropes ("structure breakers") [95] since any changes in hydrogen bonding structure are very localized. It can also be seen from Figure 1.8 that changing the identity of the cation ($K^+$ to $Na^+$) does not affect the spectra when the anions are kept the same. The lack of change with cation identity indicates that it is the direct very local interaction of the hydroxyls and anions that determines the hydroxyl stretch frequency.

The spectra in Figure 1.8 only show one broad absorption band even though there are arguably two types of water molecules in each system: water hydrogen bonded to other water hydroxyls and water interacting with ions. The water–$NaBF_4$ system, in contrast, shows different behavior. Linear IR absorption spectra of the 1.7-M, 3.1-M, 4.3-M, and 5.5-M solutions are shown in Figure 1.9 [25]. Bulk water shows a single absorption band at ~2509 cm$^{-1}$, but at 1.7 M $NaBF_4$ concentration and higher concentrations, a second, narrower band grows in at ~2650 cm$^{-1}$. Because this second band increases in size as the $NaBF_4$ concentration increases, it is concluded that this second band is due to water hydroxyls interacting with the tetrafluoroborate anions. The main band at ~2509 cm$^{-1}$ in the spectra is attributed to water–water interactions. As discussed below, in this system, 2D IR chemical exchange spectroscopy can be used to directly measure the exchange rate between the water–ion and water–water species because cross peaks will be resolved.

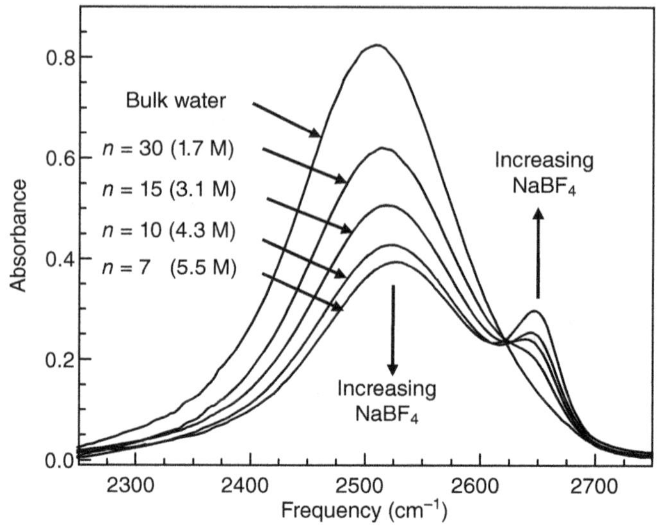


**Figure 1.9** Linear IR absorption spectra for water–$NaBF_4$ solutions at varying concentrations. The main lobe around 2509 $cm^{-1}$ corresponds to water–water hydrogen bonding environments while the smaller peak around 2650 $cm^{-1}$ corresponds to water molecules interacting with the tetrafluoroborate anions. As the $NaBF_4$ concentration increases, the main lobe decreases in size, and the smaller lobe increases.

### 1.3.2 Water in AOT Reverse Micelles

Figure 1.10 displays linear IR absorption spectra for water (5% HOD in $H_2O$) inside the full range of reverse micelle $w_0$'s examined, from $w_0 = 2$ to $w_0 = 46$. For clarity, a few $w_0$'s are omitted because they closely overlap with neighboring spectra. As the reverse micelle size decreases, the spectra systematically shift to the blue. While bulk water peaks around 2509 $cm^{-1}$, the $w_0 = 2$ spectrum peaks around 2565 $cm^{-1}$. This blue shift trend occurs because as reverse micelle size decreases, proportionally more water molecules interact with the sulfonate head groups [61, 68, 70]. The sulfonate group is a large anion, so it exerts a weak electric field along the OD hydroxyl stretch.

Figure 1.1 illustrates that distinct water environments exist in the reverse micelle. If large enough, the reverse micelle can support a core of bulklike water. The bulk water core is then surrounded by a layer of interfacial water molecules associated with the head group interface. It has been shown that the spectra in Figure 1.10 can be reproduced by a linear combination of the spectrum of bulk water and the spectrum of $w_0 = 2$ [61, 68, 70]. In the $w_0 = 2$ sample, essentially all of the water molecules interact with the head groups, and there is no bulklike core. The $w_0 = 2$ spectrum is taken to be the spectrum of interfacial water. This model of decomposing the reverse micelle spectra into bulk and interfacial contributions has been referred to as the core–shell model and is described by

$$I_{tot}(\omega) = a_1 I_1(\omega) + (1 - a_1) I_2(\omega) = S_1(\omega) + S_2(\omega) \quad (1.13)$$

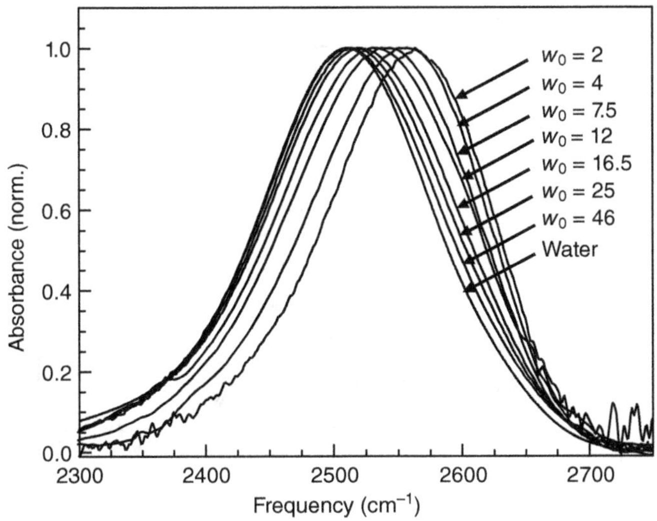


**Figure 1.10** Linear IR absorption spectra for water inside AOT reverse micelles from $w_0 = 2$ through $w_0 = 46$ along with bulk water for comparison. The spectra steadily shift to the blue as the reverse micelle size decreases.

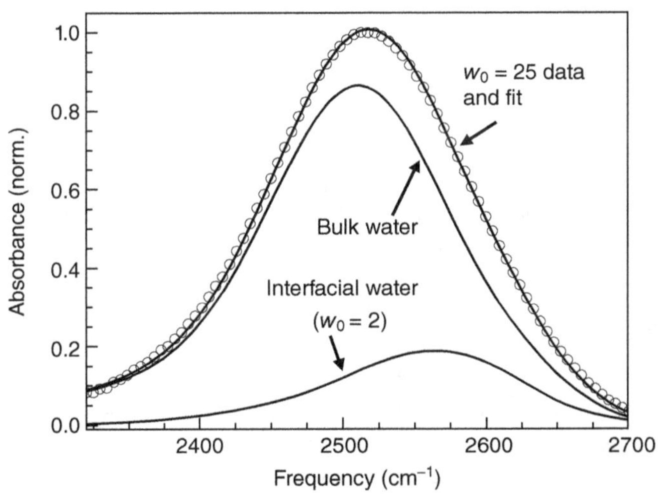


**Figure 1.11** Two-component decomposition of the water spectrum of AOT $w_0 = 25$, a large reverse micelle. The $w_0 = 25$ spectrum can be reproduced by a linear combination of the bulk water spectrum and the $w_0 = 2$ spectrum in which nearly all the water molecules interact with the head group interface.

where $I_1$ and $I_2$ are the component spectra for bulk water and $w_0 = 2$, respectively, and $a_1$ is the fractional population. When fitting the reverse micelle spectra, $a_1$ is the only adjustable parameter. Figure 1.11 displays the core–shell decomposition for AOT $w_0 = 25$. The circles are the data and the solid line through the circles is the weighted sum of the bulk water spectrum and the $w_0 = 2$ spectrum. The agreement

**Table 1.2** Fractional Population of Interfacial Water in Large AOT Reverse Micelles

| $w_0$ | Diameter (nm) | IR Fraction | Geometric Fraction |
|---|---|---|---|
| 25 | 9 | 0.17 | 0.18 |
| 37 | 17 | 0.10 | 0.10 |
| 46 | 20 | 0.08 | 0.08 |

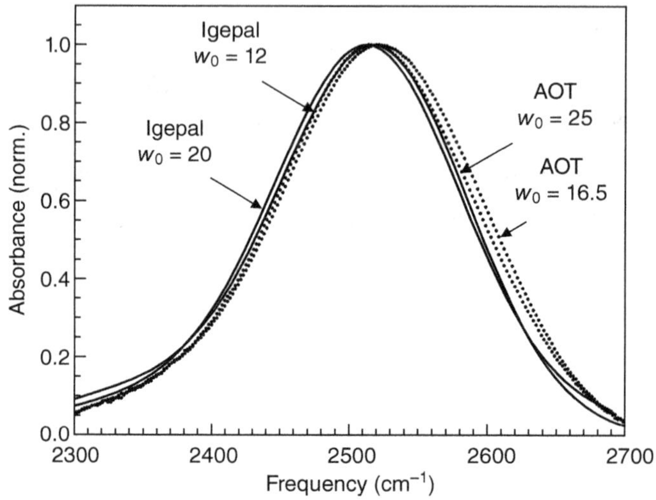


**Figure 1.12** Linear IR absorption spectra for water inside Igepal reverse micelles compared to AOT reverse micelles of the same size. Igepal $w_0 = 20$ has the same size as AOT $w_0 = 25$ (9 nm diameter), and Igepal $w_0 = 12$ has the same size as AOT $w_0 = 16.5$ (5.8 nm). Spectra for systems of the same diameter are not equivalent to each other, indicating that the hydrogen bonding interactions in the Igepal and AOT systems are different.

between the original spectrum and the linear combination is excellent. Table 1.2 [70] lists the $1 - a_1$ parameter (interfacial fraction) for AOT reverse micelles of $w_0 = 25$, 37, and 46. The numbers indicate that the fraction of interfacial water decreases as water pool diameter increases. Fractions are listed for both the IR spectral decomposition analysis as well as geometric calculations. If the interfacial region is modeled as a shell of approximately one water molecule thick (about 2.8 Å), then the interfacial fractions are nearly identical to those obtained by spectral analysis. These results strongly indicate that in large reverse micelles, the interface gives rise to a very locally perturbed region of interfacial water that does not extend far beyond the first solvation shell of the head groups [70].

### 1.3.3 Water in Igepal Reverse Micelles

The linear IR absorption spectra of water inside of Igepal reverse micelles, as shown in Figure 1.12, also show a systematic blue shift as the water pool diameter decreases [71].

Figure 1.12 compares spectra of water in Igepal reverse micelles to water in AOT reverse micelles of the same size. Igepal $w_0 = 20$ and AOT $w_0 = 25$ each have a water pool diameter of 9 nm while Igepal $w_0 = 12$ and AOT $w_0 = 16.5$ each have a water pool diameter of 5.8 nm. The Igepal spectrum of a given $w_0$ is not as blue shifted as its AOT counterpart of the same size even though both systems contain identical amounts of water. It is important to note that identical $w_0$'s of Igepal and AOT do not yield the same water pool diameters because of different aggregation numbers of the two surfactants; that is why different $w_0$'s are used to produce reverse micelles of the same size. Igepal reverse micelles, like the AOT systems, also contain bulk and interfacial water regions. In the case of Igepal, the waters interact with hydroxyl head groups, but in AOT they interact with charged sulfonate head groups. The chemical identity of the interface clearly alters the linear absorption behavior of the water molecules.

The Igepal spectra are not decomposed into core and shell regions like the AOT systems because Igepal reverse micelles are not well characterized below $w_0$ ~10 [78], so there is no low water content spectrum (akin to AOT $w_0 = 2$) to approximate the pure interfacial water spectrum. Furthermore, the Igepal reverse micelle system contains a third environment in addition to the core and interfacial water regions. The hydroxyl head groups will exchange deuterium atoms with the 5% HOD in the water pool. As a result, the interfacial region is composed of two types of hydroxyls: waters at the interface and OD head groups. The population of OD head groups is small, and their contribution to the experimental observables has been treated in detail so that the interfacial water dynamics can still be extracted (see Section 1.4.2) [71]. Figure 1.12 shows that there are significant spectral changes upon changing the chemical identity of the interface, but it will be shown below that changing the charged sulfonate head groups at the interface to neutral hydroxyl groups does not significantly affect water's hydrogen bonding dynamics.

## 1.4 POPULATION RELAXATION AND ORIENTATIONAL RELAXATION

### 1.4.1 At the Interface of Large and Intermediate AOT Reverse Micelles

The spectral analysis in Section 1.3.2 shows that large reverse micelles can be divided into two distinct regions: bulklike core water and interfacial water. This core–shell model (Fig. 1.11) has critical implications in the treatment of large reverse micelles, as it implies that the dynamics of the core and interface regions can be separated as well, with the core exhibiting the dynamics of pure water. In pure water, the vibrational lifetime of the OD stretch is 1.8 ps [using Eq. (1.8)] [68, 70], and the orientational relaxation time is 2.6 ps [using Eq. (1.11)] [96]. Consequently, the interfacial dynamics are the only unknowns in measurements on the water pools of large reverse micelles. Based on spectral analysis, waters that interact with the interface experience a blue shift in their linear IR absorption spectra. As shown by

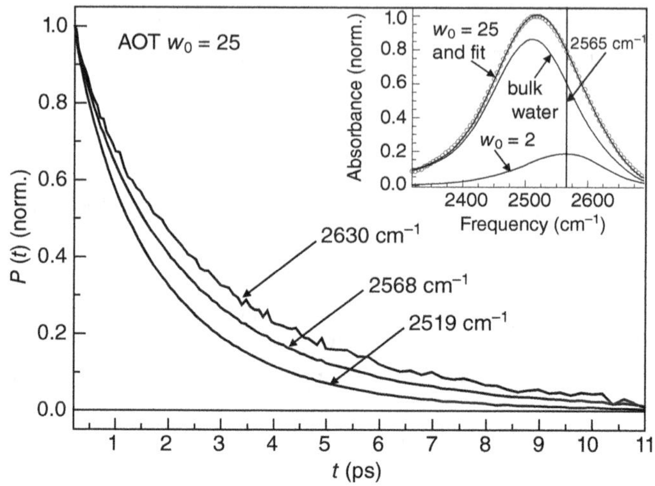


**Figure 1.13** Population relaxation data for AOT $w_0 = 25$ at three detection frequencies. The inset shows that interfacial water molecules have more population at frequencies above 2565 cm$^{-1}$. The three curves have a slight frequency dependence to them because of changing fractions of core and interfacial water.

the inset in Figure 1.13, the nearly purely interfacial $w_0 = 2$ OD–water spectrum peaks at 2565 cm$^{-1}$. The experimental pump-probe signal is frequency dispersed by a monochromator, so the dynamics at different frequencies across the absorption line shape can be monitored. For example, if the dynamics of $w_0 = 25$ were measured at 2519 cm$^{-1}$, then water molecules sampling the center of the absorption line shape would be mostly observed. Most of the spectral contribution at that frequency is due to bulklike water molecules in the core, as shown by Figure 1.11 and the inset in Figure 1.13. At higher frequencies, such as 2565 cm$^{-1}$, the water molecules at the surfactant interface make a larger contribution to the signal than they do toward the center of the absorption line. To accurately extract the dynamics at the interface from the dynamics of the core (which are known), it is useful to have the greatest amount of interfacial water contributing to the signal as possible. Because of this requirement, it is best to analyze the pump–probe signals detected at frequencies greater than or equal to 2565 cm$^{-1}$ where it is likely that a significant portion of the signal comes from interfacial water.

Figure 1.13 displays the population relaxation decay curves [Eq. (1.7)] for $w_0 = 25$ at three different frequencies: 2519 cm$^{-1}$ (near the absorption center of $w_0 = 25$), 2568 cm$^{-1}$ (near the peak absorption of $w_0 = 2$ interfacial water), and 2630 cm$^{-1}$ (far to the blue) [70]. The three curves are not the same, indicating that there is a frequency dependence to the population decay. When there are two separate water ensembles, the population decay is the weighted sum of the population decay of the two components, as given by Eq. (1.9). When the curves in Figure 1.13 are fit to Eq. (1.9), the vibrational lifetime of the first component (bulk water) is kept constant at its literature value such that $T_1^{\mathrm{w}} = 1.8$ ps at all frequencies. The super-

**Table 1.3** Interfacial Vibrational Lifetime, $T_1^{int}$, and Orientation Relaxation Times, $\tau_{or}^{int}$, for Large AOT Reverse Micelles

| $w_0$ | 12 | 16.5 | 25 | 37 | 46 |
|---|---|---|---|---|---|
| $T_1^{int}$ (ps) | $4.7 \pm 0.2$ | $4.4 \pm 0.3$ | $4.3 \pm 0.5$ | $4.6 \pm 0.5$ | $3.9 \pm 0.5$ |
| $\tau_{or}^{int}$ (ps) | $23 \pm 2$ | $17 \pm 3$ | $19 \pm 3$ | $18 \pm 3$ | $18 \pm 3$ |

script $w$ stands for the lifetime of the OD stretch of dilute HOD in bulk water. Data at all three frequencies are fit simultaneously to Eq. (1.9), and $A_1$ (the fractional population) and $T_1^{int}$ (the interfacial water lifetime) are the only adjustable parameters. It is found that this two component model for the population decay accurately describes the population relaxation dynamics for all AOT reverse micelles of $w_0 = 12$ and larger. For all reverse micelles of $w_0 \geq 12$, the vibrational lifetime of interfacial water molecules is found to be $4.3 \pm 0.5$ ps. This value of ~4.3 ps remains constant regardless of $w_0$ and regardless of frequency, indicating that the dynamics of water at the interface are quite distinct from those of the bulk water core in large reverse micelles [68, 70]. Table 1.3 lists the obtained $T_1^{int}$ values for the large reverse micelles, showing the invariability. The only parameter that changes with wavelength (leading to the three different curves in Figure 1.13) is $A_1$. As frequency increases, the amount of interfacial water increases, so $1 - A_1$ becomes larger. As $w_0$ decreases, there is more interfacial water overall, so the $1 - A_1$ values collectively increase.

Because there are two components to the population relaxation, the orientational relaxation must also be fit to a two-component model, as described by Eq. (1.12). There are quite a few variables in this equation, but population relaxation analysis allows many of the parameters in Eq. (1.12) to be fixed. $A_1$, $T_1^{w}$, $T_1^{int}$, and $\tau_{or}^{w}$ (the known bulk water reorientation time of 2.6 ps) all can be fixed, leaving the orientational relaxation time of the interface, $\tau_{or}^{int}$, as the only adjustable parameter. Figure 1.14 shows the anisotropy decays [Eq. (1.10)], for several detection wavelengths for the $w_0 = 25$ reverse micelle [70]. The plateaus in the data sets, which become apparent around 5 ps, are characteristic of two-component systems [70, 72]. The sensitivity of the shape of the anisotropy curves to the interfacial orientational relaxation time is shown in Figure 1.15. All of the parameters are held constant in these model calculations except for $\tau_{or}^{int}$. The dotted line in Figure 1.15 indicates the end of the experimentally accessible data acquisition time window. At long time all of the curves decay to zero, but the strength of the signal is limited by the vibrational lifetimes of the OD chromophores, and therefore the signal dies out before the experimental anisotropy data decays to zero.

The experimental anisotropy data at several wavelengths, like those shown in Figure 1.14, for each $w_0$ were fit simultaneously to Eq. (1.12) to determine the value for $\tau_{or}^{int}$. As shown in Table 1.3 $\tau_{or}^{int} = 18 \pm 3$ ps is obtained for $w_0$ values of 12–46. This value of 18 ps remains constant regardless of the $w_0$ or the frequency observed for these large reverse micelles. The only parameter that changes, as dictated by the population relaxation analysis, is the fractional population, $A_1$.

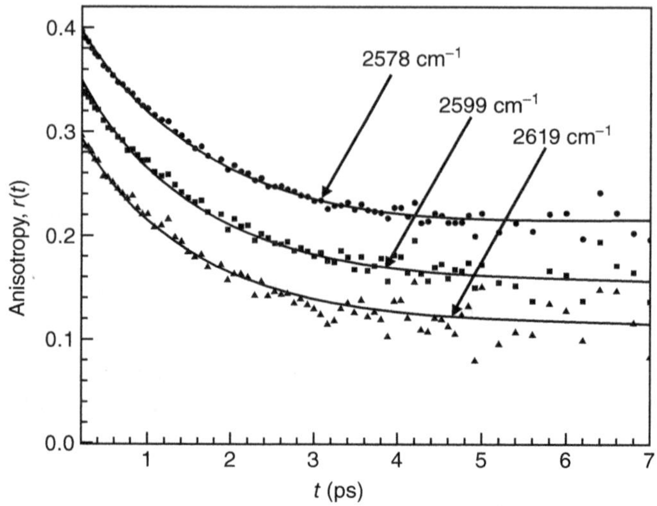


**Figure 1.14** Anisotropy curves for AOT $w_0 = 25$ at three detection frequencies. The data show plateaus around 5 ps that are characteristic of two-component systems. The curves are offset by ~0.05 for clarity. The solid lines are fits to the data using the two-component anisotropy model.

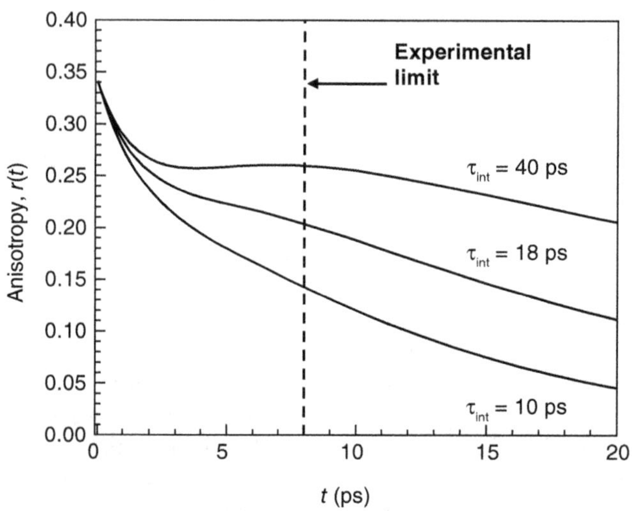


**Figure 1.15** Calculated anisotropy decays of a two-component system. The different curves are produced by using different values for the interfacial orientational relaxation time constant while keeping all other parameters in the two-component anisotropy equation constant. At long time, all of the curves decay to zero. The dashed line shows the maximum time for experimental data collection. When the anisotropy data are analyzed to this point, it appears as if there is a plateau that would normally correspond to a component that undergoes infinitely slow rotation.

**Table 1.4** Core and Interfacial Vibrational Lifetimes and Orientation Relaxation Times for Intermediate AOT Reverse Micelles

| $w_0$ | 7.5 | 10 |
|---|---|---|
| $T_1^{w}$ (ps) | 2.1 ± 0.2 | 2.2 ± 0.3 |
| $T_1^{int}$ (ps) | 5.5 ± 0.2 | 5.3 ± 0.3 |
| $\tau_{or}^{w}$ (ps) | 4.4 ± 0.2 | 4.0 ± 0.3 |
| $\tau_{or}^{int}$ (ps) | 30 ± 5 | 26 ± 3 |

Intermediate-sized reverse micelles, such as $w_0$ = 10 and $w_0$ = 7.5 also show two-component vibrational lifetime behavior that separates into core and interfacial contributions, but the core does not follow bulk water dynamics. The dynamics in both regions increase in value from their counterparts in the large reverse micelles. The core has a vibrational lifetime of ~2 ps, and the interfacial waters have a vibrational lifetime of ~5.3 ps (Table 1.4) [63, 68]. Because the core does not follow bulk dynamics, the orientational lifetimes of both the core waters and interfacial waters, $\tau_{or}^{w}$ and $\tau_{or}^{int}$ in Eq. (1.12), respectively, are unknown. Simultaneous fitting at several frequencies for each $w_0$ yield a core orientational lifetime of ~4 ps and an interfacial orientational lifetime of ~30 ps. These results indicate that the effects of confinement become more pronounced in smaller sizes. Again, the important observation is that the population and orientational lifetimes for the core and interface of intermediate reverse micelles do not change with $w_0$ or frequency and that only the fractional populations change.

The $w_0$ = 10 AOT reverse micelle has a water pool with a diameter of 4 nm. The fact that the core water no longer has bulk water properties sets a distance scale for the influence of the interface on water molecules away from the interface. As discussed above, the interface has a very substantial influence on water in the first water layer adjacent to the interface. Water molecules beyond the first layer are also perturbed. In the spherical reverse micelles, the influence of the interface is propagating in all directions. The fact that the core of the $w_0$ = 10 water pool is affected indicates that as the reverse micelles become smaller and the radius of curvature of the interface becomes smaller, the influence of the interface propagates out ~2 nm.

### 1.4.2 Neutral versus Ionic Interfaces

The AOT head group interface contains both negatively charged sulfonate head groups and positively charged sodium counterions. It is reasonable to consider how much the presence of these charged moieties affects the interfacial water dynamics. To determine the role of interfacial charges, orientational dynamics were measured for water inside of Igepal reverse micelles. Igepal (Fig. 1.2) is terminated with neutral hydroxyl head groups. Water dynamics in Igepal $w_0$ = 12 and $w_0$ = 20 were

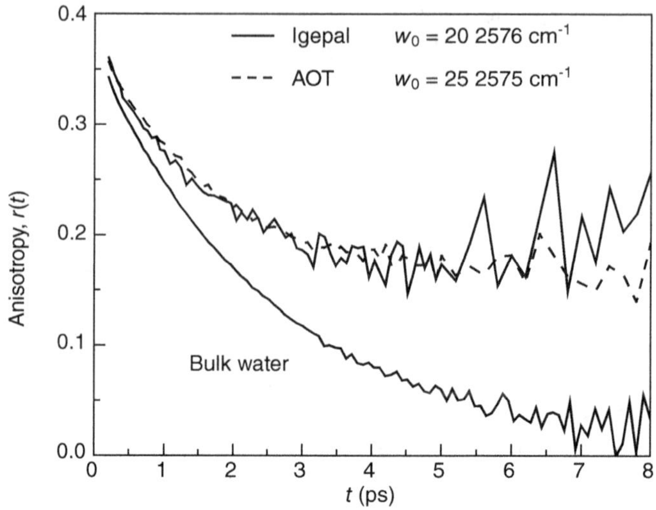


**Figure 1.16** Anisotropy data for Igepal $w_0$ = 20 and AOT $w_0$ = 25 compared to bulk water. The data for the reverse micelles are nearly identical, showing that the chemical identity and charged nature of the interface have only minor effects on water dynamics. The presence of the interface is the dominant factor.

examined. These systems have the same water pool diameters as AOT $w_0$ = 16.5 (5.8 nm) and $w_0$ = 25 (9 nm), respectively. The linear IR absorption spectra in Figure 1.12 shows that the nature of the hydrogen bonding interactions are different in the two reverse micelles, as seen by the small, but clear, red shift of the Igepal spectra from their AOT counterparts of the same size.

Anisotropy data at a detection frequency of ~2575 cm$^{-1}$ for Igepal $w_0$ = 20 and AOT $w_0$ = 25 are shown in Figure 1.16. The two curves are nearly identical, although the Igepal turns slightly up at long time, and there is also some difference at early time. Because Igepal contains hydroxyl groups that exchange with the deuterium of the HOD molecules, the Igepal data must be fit with a modified three-component model for the anisotropy and population relaxation [71]. When the data are fit in this way, the water reorientation times for Igepal $w_0$ = 12 and $w_0$ = 20 at the interface are the same within experimental error: 15 ± 3 ps and 13 ± 4 ps, respectively [71]. The reorientation time of the hydroxyls are assumed to be infinitely long (coupled to the rotation of the whole micelles), and the core adopts the bulk water reorientation time (2.6 ps). Note that the reorientation time for waters at the AOT interface are 18 ± 3 ps. The error bars of the Igepal and AOT results overlap to some extent. Therefore, the interfacial orientational relaxation in Igepal is either the same or at most slightly faster than in AOT. The orientational relaxation is directly related to the hydrogen bond dynamics. The important point of these results is that going from a charged interface to a neutral one at most makes a modest difference in the dynamics. The presence of the interface has the largest impact on the dynamics, while the chemical nature of the interface, including its charged moieties, has less of an effect.

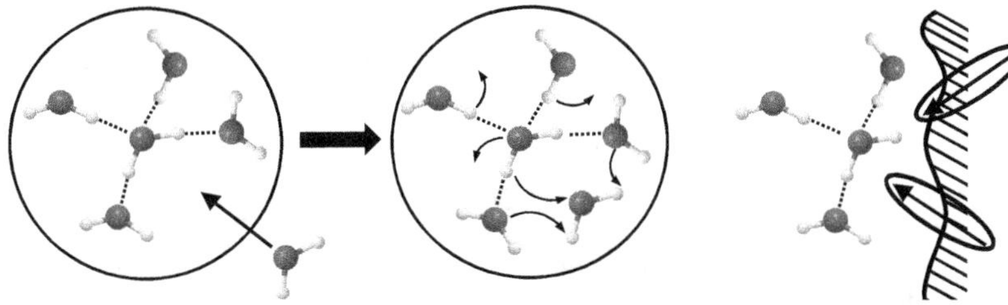


**Figure 1.17** Illustration of hydrogen bond network reorientation and reorganization through jump reorientation. The mechanism involves concerted motions of the first and second solvation shells of water molecules. The presence of an interface physically eliminates pathways by which jump reorientation can occur, regardless of its chemical composition.

Figure 1.17 illustrates schematically how water undergoes orientational relaxation in bulk water and the role that an interface plays in slowing down reorientation. Figure 1.17 is a very qualitative illustration in two dimensions of what are really three-dimensional structures and processes. In bulk water, orientational relaxation involves concerted hydrogen bond rearrangement that results in jump reorientation [66, 67]. In the left portion of the figure, a central water molecule makes four hydrogen bonds to other water molecules in its first solvation shell. A fifth water moves in from the second solvation shell. Because the water hydrogen bonding network is continually in motion, it may become favorable for the water molecule to form a hydrogen bond with the fifth water that has come into its solvation shell. After going through a five-coordinate transition state involving a bifurcated hydrogen bond with two water molecules, the central water can make a large angular jump of ~60° to switch hydrogen bonds with that fifth water molecule [66, 67]. The surrounding water molecules move and reorient concertedly to accommodate these motions such that nonhydrogen-bonded hydroxyls do not exist for any significant length of time. The right-hand portion of the figure suggests the role that an interface or large molecule, such as a protein, plays. The interface blocks many of the pathways for water to move into the first solvation shell, which greatly reduces the number of pathways that can give rise to jump reorientation. The interface eliminates an entire half-space of water molecules. In addition, the rough surface topography of the interface also inhibits water molecules from moving into the first solvation shell along a range of paths that are approximately parallel to the interface. Thus the presence of the interface or other physical hindrance, such a large molecule, reduces the rate of jump reorientation independent of the chemical nature of the blocking species.

### 1.4.3 The Role of Interfacial Geometry

In the previous sections it was shown that the presence of an interface is enough to significantly impact water orientational relaxation and hydrogen bond dynamics. The chemical composition of the interface for charged versus neutral but polar hydrogen bonding interfaces has a relatively minor impact on the dynamics. In this section, water dynamics in two different confinement geometries are compared, that is, water in AOT reverse micelles and AOT lamellar structures (2D slabs of water). Water dynamics are compared for samples of the same hydration level ($w_0$ and $\lambda$) and also samples that have the same nanoscopic dimension. The nanoscopic dimension is the surface-to-surface distance in the two types of structures: the diameter for the reverse micelles and distance between the interfacial planes for lamellar structures. The $w_0$ parameter for lamellar structures is denoted as $\lambda$. As an example, $\lambda = 46$ and $w_0 = 46$ have the same amount of water per AOT molecule, but the $w_0 = 10$ reverse micelle has the same nanoscopic dimension as $\lambda = 46$. Both distances are 4.0 nm. In all of the lamellar samples, the separation between the planar interfaces is sufficiently large that the center of the 2D water slab has bulklike water dynamics. This is true even though the separation of the planes, for example, 4 nm for $\lambda = 46$, can be in the range where for the same diameter reverse micelles, the core would not be bulk water. The difference arises because the planar slabs of water are infinite in two directions and only confined in one direction. In contrast the spherical water nanopools of reverse micelles are confined in all directions.

Figure 1.18 [69] shows the anisotropy curves for four sets of reverse micelle/lamellar systems. In all cases, the samples with the same hydration level are almost

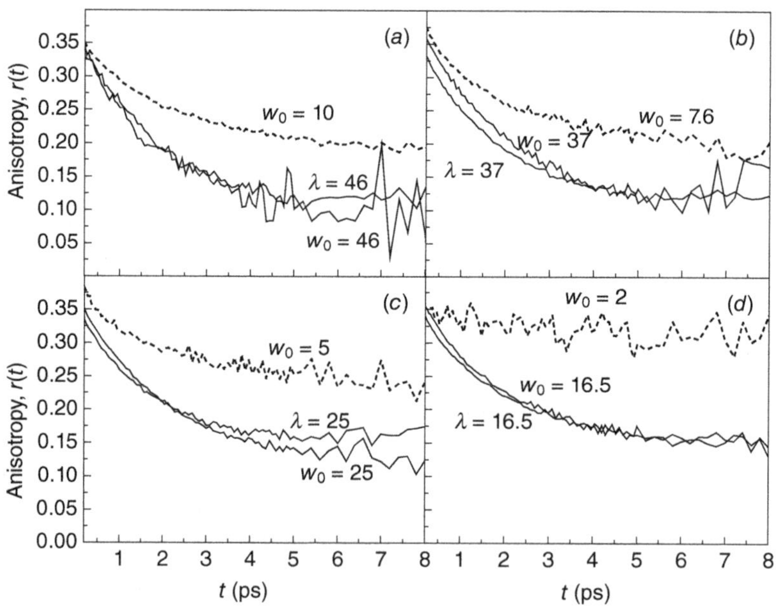


**Figure 1.18** Anisotropy data comparing water dynamics for reverse micelles ($w_0$) and lamellar structures ($\lambda$) of the same hydration level and surface-to-surface distance. Samples with the same hydration level (e.g., $w_0 = 46$ and $\lambda = 46$) have more similar dynamics than for samples with the same surface-to-surface distance (e.g., $w_0 = 10$ and $\lambda = 46$).

the same, while the samples with the same nanoscopic dimension (surface-to-surface distance) are very different. Analysis of the data shows that the vibrational lifetimes and reorientation times of water at the lamellar interfaces are the same as those in the large reverse micelles; 4.3 ps for vibrational lifetimes and 18 ps for orientational relaxation times [69]. The anisotropy decay curves for the lamellar structures are not identical to their reverse micelle counterparts of the same hydration level. This occurs because the flat geometry of the lamellar structures results in greater surface area per AOT head group. In reverse micelles, the surface area varies from <30 Å$^2$ in $w_0 = 2$ to a limiting value of 54 Å$^2$ in $w_0 = 40$ and higher [60]. In the lamellar structures, the surface area per AOT molecule is 65 Å$^2$ [97]. Therefore, there is a higher population of interfacial water in the lamellar structures for a given hydration level.

Overall, these results indicate that an interface only gives rise to a substantial perturbation of the water dynamics over a short range, approximately one solvation shell of waters at the interface. The hydration level of the reverse micelles and the lamellar structures mainly determines the dynamics. The surface-to-surface distances of the systems have little to no effect. For instance, in Figure 1.18*d*, the $\lambda = 16.5$ lamellar structure has a surface-to-surface distance of 1.7 nm (the same distance as the diameter of $w_0 = 2$), but the dynamics are almost the same as $w_0 = 16.5$ reverse micelles with a diameter of 5.8 nm. In addition, the values for interfacial water vibrational lifetime and reorientation in the lamellar structures were the same within experimental error as those for interfacial water in AOT reverse micelles (4.3 and 18 ps, respectively), indicating that the dynamics are mainly determined by the presence of an interface and how many waters interact with it.

## 1.5 ION–WATER HYDROGEN BOND SWITCHING

Two-dimensional IR chemical exchange experiments were conducted on concentrated solutions of sodium tetrafluoroborate ($NaBF_4$) [25]. Figure 1.19 shows a schematic illustration of spectra for an ideal 2D IR chemical exchange experiment. There are two species, A and B, with absorption frequencies, $\omega_A$ and $\omega_B$. At short time (top panel), prior to any chemical exchange, there are two peaks on the diagonal that arise from the 0–1 vibrational transition (black, positive going), and there are two corresponding peaks below them that occur from vibrational echo emission at the 1–2 transition frequency (gray, negative going). The 1–2 peaks are shifted to lower frequency along the $\omega_m$ axis by the amount of the vibrational anharmonicity. At long time (bottom panel), chemical exchange has occurred. Some A's have turned into B's, and the same number of B's have turned into A's. Because the system is in equilibrium, the number of A's turning into B's equals the number of B's turning into A's per unit time. In other words, the rate of A's going to B's equals the rate of B's going to A's. Chemical exchange is manifested by the growth of the off-diagonal peaks. Measurement of the time-dependent increase in the off-diagonal peaks, accounting for the vibrational lifetime and orientational relaxation that cause all peaks to decrease in amplitude, gives the time dependence of the chemical exchange

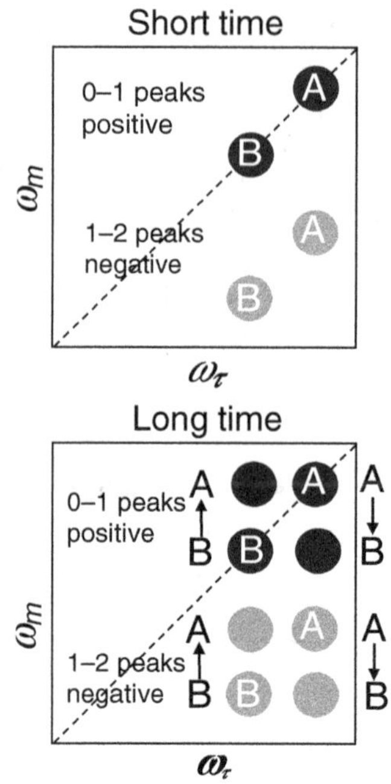


**Figure 1.19** Ideal chemical exchange system. The black diagonal peaks correspond to the 0–1 vibrational transition while the gray peaks correspond to the 1–2 vibrational emission and appear off-diagonal due to the vibrational anharmonicity. At short time, only diagonal peaks (and their 1–2 off-diagonal peaks) appear. At long time, chemical exchange causes cross peaks to grow in.

[55]. Vibrational excitation of the OD stretch does not change the equilibrium or the rates of chemical exchange processes [55].

In the schematic shown in Figure 1.19, the anharmonicity is large enough that the positive 0–1 peaks do not overlap with the negative 1–2 peaks. However, for the $NaBF_4$–water chemical exchange experiments, the anharmonicity of the OD hydroxyl bound to an anion (ha, hydroxyl/anion) is much smaller than the anharmonicity of the hydroxyl bound to a water oxygen (hw, hydroxyl/water). Therefore, positive- and negative-going peaks overlap as the off-diagonal chemical exchange peaks grow in. Figure 1.20*a* shows the 2D IR spectrum of water in a 5.5-M $NaBF_4$ solution at short time ($T_w = 0.2$ ps). At this short time, no chemical exchange has occurred. The peaks on the diagonal are the 0–1 positive-going ha and hw peaks. The negative-going 1–2 ha peak is below the diagonal ha peak. The negative-going 1–2 hw peak is not shown. In Figure 1.20*b* the spectrum is shown at 4 ps. Chemical exchange has produced large changes in the spectrum. The off-diagonal positive exchange peak to the left of the ha peak and above the hw peak can be clearly seen. The corresponding positive exchange peak that would be below the ha peak and to the right of the hw peak is on top of the negative-going 1–2 ha peak. In addition, one of the negative-going 1–2 exchange peaks appears right in the middle of the positive-going diagonal 0–1 hw peaks. It is manifested by essentially negating the center of the diagonal 0–1 hw peak. In spite of the overlapping peaks, the chemical exchange data can be readily analyzed because the frequencies at which all of the peaks will appear are precisely known [25].

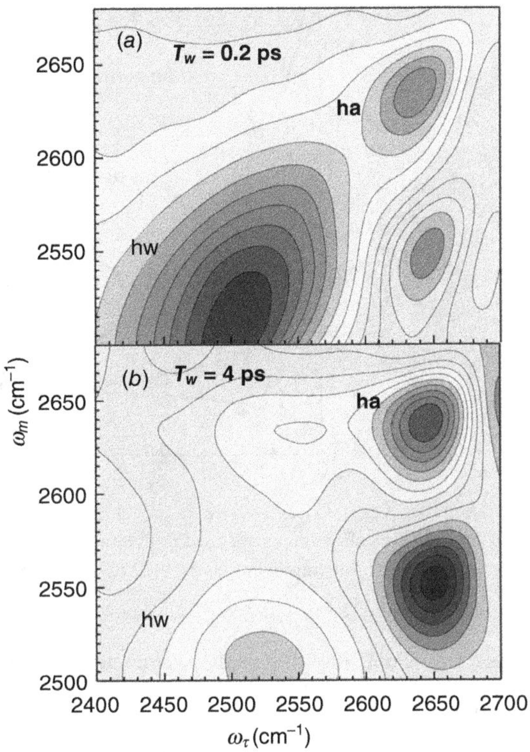


**Figure 1.20** Two-dimensional correlation plots for the 5.5-M $NaBF_4$ solution showing chemical exchange. At early $T_w$ only the peaks due to hydroxyl–water interactions (hw) and hydroxyl–anion interactions (ha) are observed. At later $T_w$ chemical exchange has occurred as seen by the growth of off-diagonal cross peaks.

To extract the exchange time in the $NaBF_4$ system, the peak volumes of the diagonal and off-diagonal peaks are measured. The peak volumes reflect the various populations. Figure 1.21 shows the results of the peak-volume analysis of the 2D IR spectra taken over a range of $T_w$'s. The diagonal peaks in Figure 1.21 decay because of chemical exchange, population relaxation, and orientational relaxation. The off-diagonal peaks grow in because of chemical exchange but decay because of the other processes. The solid curves through the data in Figure 1.21 are the result of a single adjustable parameter fit that yields the chemical exchange time. Because the system is in equilibrium, the dynamics can be characterized by a single time constant (the inverse of the rate constant for water bound to an anion going to water bound to another water). The analysis yields the time for a water molecule hydrogen bonded to a $BF_4^-$ anion to switch to being bonded to an oxygen of another water molecule. The exchange time is determined to be 7 ps [25]. This should be compared to the hydrogen bond exchange time in pure water of ~2 ps obtained from 2D IR measurements of spectral diffusion [38, 39]. The important point is that ions slow the rearrangement of hydrogen bonds, but only by a factor of 3 or 4. The relatively mild affect of ions on hydrogen bond switching may account for the relatively small difference between the interfacial dynamics of water in contact with charged versus neutral interfaces, as seen by the comparison of water dynamics in AOT and Igepal reverse micelles in Section 1.4.2.

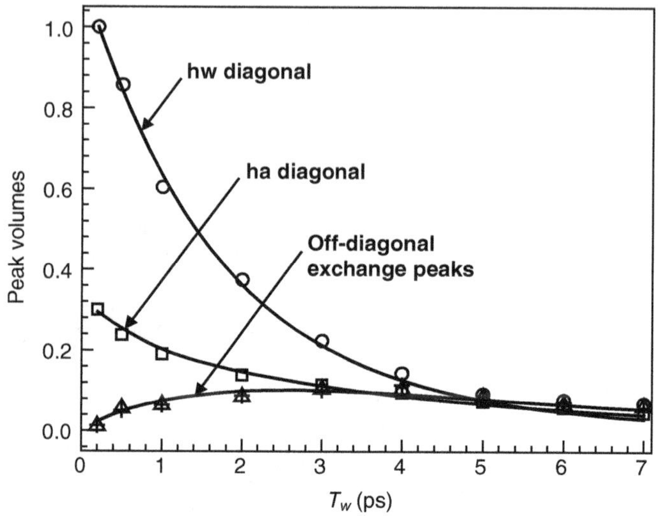


**Figure 1.21** Peak-volume data used to obtain the 7-ps exchange time for water hydrogen bonded to an anion to switch to being hydrogen bonded to a water hydroxyl.

**Table 1.5** Population Relaxation Parameters for Small AOT Reverse Micelles[a]

| Sample | Parameter | 2590 cm$^{-1}$ | 2610 cm$^{-1}$ | 2620 cm$^{-1}$ | 2640 cm$^{-1}$ |
|---|---|---|---|---|---|
| AOT $w_0 = 2$ | $A_1$ | 0.13 | 0.08 | 0 | 0 |
| | $T_1^1$ (ps) | 2.0 | 2.0 | – | – |
| | $T_1^2$ (ps) | 7.3 | 7.5 | 7.4 | 8.1 |
| AOT $w_0 = 4$ | $A_1$ | 0.22 | 0.15 | 0.13 | 0.10 |
| | $T_1^1$ (ps) | 2.0 | 2.0 | 2.0 | 2.0 |
| | $T_1^2$ (ps) | 6.4 | 6.6 | 6.9 | 7.4 |

[a]Error bars are ±0.2 ps for time constants, ±0.04 for amplitudes; $A_2 = 1 - A_1$.

## 1.6 SMALL REVERSE MICELLES

The small reverse micelles of $w_0 = 2$ and $w_0 = 4$ have diameters of 1.7 and 2.3 nm, respectively, and cannot support a distinct core region. Surprisingly, these reverse micelles present a situation where, even though the water molecules exist in a mainly interfacial environment without a core region, population relaxation still has two components, as shown by Table 1.5 [63]. For both $w_0 = 2$ and $w_0 = 4$, there is one lifetime of 2 ps and a slower lifetime of ~7 ps for $w_0 = 2$ and ~6.5 ps for $w_0 = 4$. The faster (2-ps) lifetime corresponds to OD hydroxyls that are hydrogen bonded to the oxygens of water molecules while the second, slower lifetime is attributed to ODs bound to the sulfonate head groups. Unlike the larger reverse micelles, the value of the second, longer lifetime exhibits a frequency dependence. As the frequency of observation is increased, the population of the 2-ps lifetime decreases, and the population and the value of the slower lifetime increase. At the highest frequencies for AOT $w_0 = 2$, the population decay becomes a single slow exponential because essentially all of the signal comes from ODs that are bound to head groups. The two-component behavior arises because the vibrational lifetime is extremely sensitive to local environments. The water molecule hydroxyls in the small reverse micelles are mostly hydrogen bonded to the interface, but they can also form hydrogen bonds to the oxygens of other water molecules. These water–hydroxyl interactions give rise to the faster lifetime while water–interface interactions give rise to the slower lifetime. As the frequency is tuned to the blue, the nature of the OD–head group interactions change, which is reflected in a lifetime that steadily increases as the frequency increases.

The systems discussed above, AOT and Igepal reverse micelles, AOT lamellae, and water–salt solutions, all have sufficient water in the systems so that there is bulklike water in addition to water interacting with an interface or ion. Thus, water dynamics are of two types, bulk water and interfacial water. In a reverse micelle, such as AOT, and in other systems, the situation changes as the size of the water nanopool becomes small. In AOT reverse micelles it was found that there are three regimes of water nanopools–large, intermediate, and small [68]. The large reverse micelles have cores (a substantial amount of water well separated from the interface) with bulk water characteristics.

As the size of the water nanopool becomes smaller, a larger fraction of the water molecules interacts with the interface. The influence of the interface propagates out over some distance. In the intermediate regime, the nanopool is small enough that the influence of the interface has not completely died away at the center of the nanopool. This type of system has a core of water molecules, but the core does not have bulklike properties. In connection with Figure 1.14, it was found that $w_0$ values of 46, 37, 25, 16.5, and 12 all had interfacial water orientational relaxation times of 18 ps and core relaxation times with the bulk water value of 2.6 ps. Intermediate size reverse micelles, $w_0 = 10$ and $w_0 = 7.5$, have both interfacial and core orientational relaxation times and lifetimes that are somewhat longer than those observed for the large reverse micelles.

In the $w_0 = 2$ reverse micelle (see spectrum in Fig. 1.11), there are only two water molecules per AOT molecule. Hauser et. al [98], using differential scanning calorimetry, nuclear magnetic resonance (NMR), and electron spin resonance (ESR) experiments suggest that between four and six water molecules are required to hydrate the sulfonate head group of AOT. Although there may not be a bulk water-like core, HOD molecules in small reverse micelles have their ODs experiencing a variety of environments as demonstrated by a wavelength dependence to the vibrational lifetime [63].

While small variations in the local environment of a water molecule may affect the vibrational lifetime, the long-time orientational dynamics are related to the complete structural rearrangement of the hydrogen bonding network due to hydrogen bond switching events. Anisotropy decays for HOD in $w_0 = 2$ reverse micelles are shown in Figure 1.22 with the anisotropy decay of bulk water for comparison. In the $w_0 = 2$ reverse micelles, the sample has very little water, which results in

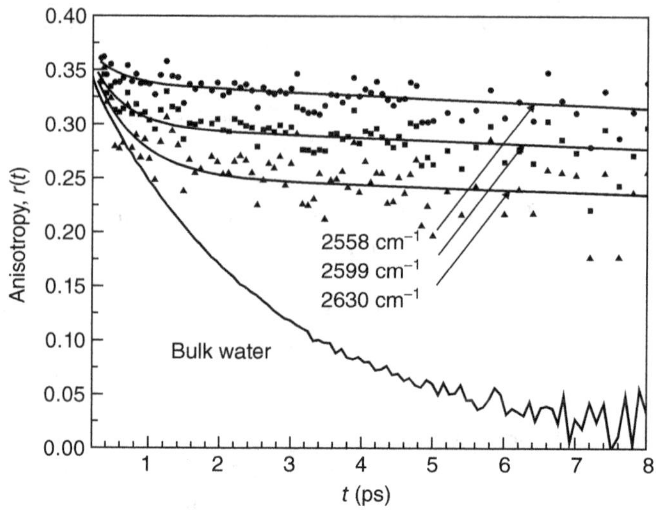


**Figure 1.22** Anisotropy curves for AOT $w_0 = 2$ at three detection frequencies. There is very little anisotropy decay on this time scale, as compared to the anisotropy decay for bulk water. Anisotropy decay for the small reverse micelles is modeled by a wobbling-in-a-cone mechanism instead of the two-component anisotropy.

decreased signal-to-noise ratio. Nonetheless, the data are adequate to draw conclusions. After a very fast initial decay, the anisotropy decay in the reverse micelle has a very slow long-time decay. The long-time decays at a variety of wavelengths are identical within experimental error [68]. This behavior is consistent with the extremely low water content, only ~50 water molecules per reverse micelle for $w_0 = 2$. Essentially, hydrogen bond rearrangement is dramatically slowed because of the energetic penalty required to disrupt the structure of sulfonate groups, water molecules, and sodium ions at the interface.

The long-time orientational dynamics are independent of frequency because hydrogen bond switching is a concerted process that depends on new hydrogen bond acceptors moving into the first solvation shell of the reorienting water (see Figure 1.17), not the strength of a particular hydrogen bond [66, 67]. In a $w_0 = 2$ reverse micelle the hydrogen bonding network is completely coupled because so few water molecules are present. Therefore, the long-time orientational dynamics are well described by a single ensemble, not a core subensemble and an interface subensemble. The long-time anisotropy decays for all frequencies in the $w_0 = 2$ reverse micelle can be fit with the same time constant of 110 ± 40 ps. The large error bars indicate the uncertainty in determining such a long-time constant from a limited time range of data. It is interesting to note that the reorientation of water in the $w_0 = 2$ reverse micelle is still an order of magnitude faster than the tumbling time of the reverse micelle itself [99], indicating that water is still able to reorient albeit slowly.

The fast anisotropy decay component seen in Figure 1.22 is not due to a fraction of the water molecules that are completely reorienting on a time scale faster than bulk water. Rather it is caused by local orientational motions of water molecules within a stable hydrogen bonding configuration. This type of restricted, incomplete orientational relaxation is referred to as wobbling in a cone [68, 93, 100]. It is associated with the wobbling of the orientation of the water molecules over a limited range of angles within an very slowly evolving hydrogen bonding network [68].

The concerted nature of water hydrogen bond rearrangement is responsible for the transition from two-component dynamics (large- and medium-sized water nanopools) to collective dynamics (very small water nanopools). As discussed above, Laage and Hynes showed that water reorientation occurs through a transition state that involves a bifurcated hydrogen bond that the rotating water molecule forms between the old and new hydrogen bond acceptors (see Fig. 1.17) [66, 67]. The new acceptor is originally in the second hydration shell of the rotating water molecule, but through concerted motions it moves into the first hydration shell, creating an overcoordinated environment for the rotating water molecule and reducing the energetic barrier for reorientation. This model shows that water reorientation is not a localized process; the reorientation of a single water molecule requires concerted rearrangement of both its first and second hydration shells involving ~16 water molecules [101]. If one or more of these 16 water molecules is itself in an environment that is not the same as bulk water, causing it to have dynamics distinct from bulk, it will affect the dynamics of its neighbors. When there are many unperturbed water molecules, such as in large reverse micelles, the effect of the perturbation dies off relatively quickly. However, in small reverse micelles, so many of the water molecules are perturbed that even water molecules not directly interacting with the interface have very slow dynamics.

## 1.7 SPECTRAL DIFFUSION MEASUREMENTS OF INTERFACIAL WATER DYNAMICS

### 1.7.1 Large and Intermediate AOT Reverse Micelles

The linear IR absorption and pump–probe data for larger reverse micelles all suggest that when the reverse micelles are large enough, two distinct environmental regions yield separable dynamics. Additional information can be obtained from measurements of spectral diffusion using 2D IR vibrational echo experiments. As representative 2D IR data for large reverse micelles, 2D IR spectra for AOT $w_0 = 16.5$ at three $T_w$'s are shown in the left column of Figure 1.23 [102]. In the spectra, the peak along the diagonal is positive going and corresponds to the 0–1 vibrational transition while the off-diagonal peak is negative going and arises from vibrational echo emission at the 1–2 transition frequency. The 1–2 peak appears off the diagonal and shifted along the $\omega_m$ axis due to the vibrational anharmonicity. At early $T_w$ values, the spectra are elongated along the diagonal, indicating that an oscillator excited at a certain frequency has a high probability of retaining that frequency after a short amount of time. The spectra become more symmetric as $T_w$ lengthens and the hydroxyl stretches of the water molecules evolve in frequency due to structural evolution of the system.

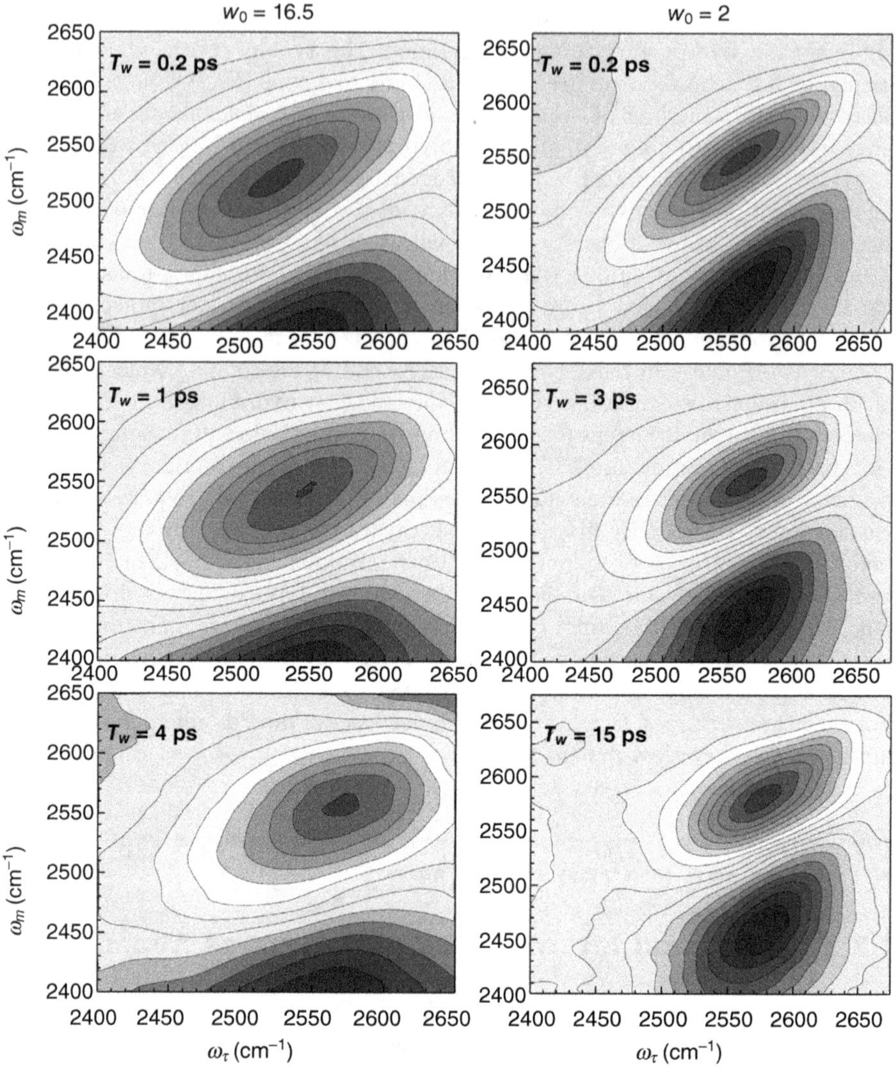


**Figure 1.23** Two-dimensional IR vibrational echo correlation plots for AOT $w_0 = 16.5$ (left column) and AOT $w_0 = 2$ (right column) at a range of $T_w$'s. Both samples show significant elongation along the diagonal at early time, and while $w_0 = 16.5$ shows faster spectral diffusion than $w_0 = 2$, neither sample completes spectral diffusion on the time scale of the experiment.

Figure 1.23 shows that the spectra maintain some elongation at long $T_w$. Therefore, spectral diffusion, which is caused by structural evolution, is not complete on the time scale of the experiment.

The large reverse micelles ($w_0$ = 12 and $w_0$ = 16.5) and intermediate reverse micelles ($w_0 = 7.5$) contain two types of water molecules: core water and interfacial water. In a 2D IR experiment, each of these ensembles generates a 2D IR spectrum such that the measured 2D spectrum is a combination of the individual spectra. When the peak separation between components is relatively small, the individual spectra are not resolved.

As discussed in the introduction, the FFCF is obtained from 2D IR spectra via CLS analysis [84, 85]. Recently, a modified CLS technique was developed to account for two-component systems [81]. A 2D IR spectrum resulting from a two-component system can be decomposed into its constituent spectra, and each of these spectra has a characteristic CLS decay. In large reverse micelles, the components are bulk water, for which the CLS is already known from previous experiments [38, 39], and interfacial water. It is therefore possible to obtain the CLS decay (and by extension, the FFCF) of the interfacial water. It was found that for the two-component 2D IR spectrum (the combined spectrum) the peak positions ($\omega^*_{mC}$) of slices parallel to the $\omega_m$ axis are weighted combinations of the peak positions ($\omega^*_{m1}$) of the bulk water spectrum and the peak positions ($\omega^*_{m2}$) of the second interfacial component:

$$\omega^*_{mC}(\omega_m, \omega_\tau, T_w) = f_1(\omega_\tau, T_w)\omega^*_{m1}(\omega_m, \omega_\tau, T_w) + [1 - f_1(\omega_\tau, T_w)]\omega^*_{m2}(\omega_m, \omega_\tau, T_w) \tag{1.14}$$

The set of peak positions for the slices are also known as centerline data. In Eq. (1.14), $f_1$ is the fractional population of bulk water, which is both $T_w$ and frequency-dependent. This fraction term can be obtained from the FTIR absorption spectra of the two components and the measured vibrational lifetimes,

$$f_1(\omega_\tau, T_w) = \frac{S_1(\omega_\tau)e^{-T_w/T_1^{\mathrm{w}}}}{S_1(\omega_\tau)e^{-T_w/T_1^{\mathrm{w}}} + S_2(\omega_\tau)e^{-T_w/T_1^{\mathrm{int}}}} \tag{1.15}$$

where $S_1$ and $S_2$ are defined by Eq. (1.13). Eq. (1.14) can be rearranged to obtain $\omega^*_{m2}$ for the interface:

$$\omega^*_{m2}(\omega_m, \omega_\tau, T_w) = \frac{\omega^*_{m2C}(\omega_m, \omega_\tau, T_w) - f_1(\omega_\tau, T_w)\omega^*_{m1}(\omega_m, \omega_\tau, T_w)}{1 - f_1(\omega_\tau, T_w)} \tag{1.16}$$

For Eq. (1.16) to be employed, the $f_1$ fraction terms and vibrational lifetimes must be known. The peak position of the interfacial water absorption spectrum must also be known or reasonably approximated. When CLS analysis is applied to the calculated interfacial water centerline data, the CLS will be calculated around the peak position of the interfacial water band, which is 2565 cm$^{-1}$ for $w_0 = 2$.

Centerline data for the interface, obtained by applying Eq. (1.16), were obtained for AOT $w_0 = 16.5$, 12, and 7.5 at the measured $T_w$'s [102]. Here, we treat $w_0 = 7.5$ as a large reverse micelle, but it will be discussed later whether that approximation is valid. The interfacial CLS curves for these three micelles are displayed in Figure 1.24. The interfacial CLS curves are similar for $w_0 = 12$ and $w_0 = 16.5$ within error. The lack of a size dependence for the large reverse micelles is consistent with a lack of size dependence for the interfacial orientational relaxation and the vibrational lifetime. The $w_0 = 7.5$ system is considerably smaller in size, and its interfacial CLS decays slower than the larger reverse micelles. It should be noted that the bulk water centerline data were used in the $w_0 = 7.5$ calculation, even though the experiments suggest that the core is not exactly bulk-water-like. The core of the $w_0 = 7.5$ reverse

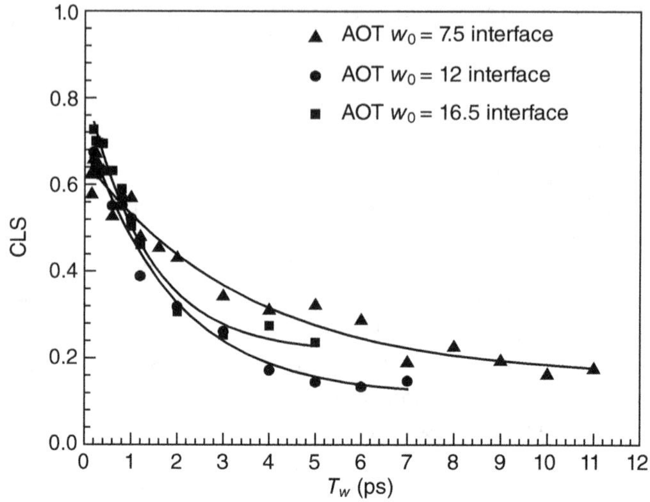


**Figure 1.24** Interfacial CLS curves for large and intermediate reverse micelles. All curves decay as a single exponential to an offset. AOT $w_0$ = 16.5 and $w_0$ = 12 show very similar interfacial CLS curves, while AOT $w_0$ = 7.5 dynamics occur slightly slower. The curves through the data are the interfacial FFCFs for each sample.

**Table 1.6** Interfacial FFCF Parameters for Large and Intermediate AOT Reverse Micelles[a]

| $w_0$ | $\Gamma$ (cm$^{-1}$) | $\Delta_1$ (cm$^{-1}$) | $\tau_1$ (ps) | $\Delta_S$ (cm$^{-1}$) |
|---|---|---|---|---|
| 7.5 | 57 | 44 | 3.5 | 24 |
| 12 | 35 | 55 | 1.9 | 21 |
| 16.5 | 28 | 50 | 1.3 | 29 |

[a]Error bars are ±5 for $\Gamma$ and $\Delta$'s; ±0.5 for $\tau_1$.

micelle cannot be measured directly, so bulk water is the best approximation for use in the calculation.

The interfacial CLS curves fit well to a single exponential decay with a ~1.6-ps time constant plus an offset. The interfacial FFCFs for $w_0$ values of 16.5, 12, and 7.5 are listed in Table 1.6 and were calculated by simultaneously fitting the CLS curve and the $w_0$ = 2 spectrum for each sample. The $w_0$ = 12 and $w_0$ = 16.5 systems yield very similar FFCF parameters. Like the rotational and population relaxation dynamics, spectral diffusion at the interface in large reverse micelles is independent of size. This result implies that local water–interface interactions and dynamics do not change with size for sufficiently large reverse micelles.

The ~1.6-ps interfacial time constant is very similar to the 1.7-ps time constant in the FFCF for bulk water that corresponds to global hydrogen bond rearrangement. Water molecules at the interface are hydrogen bonded to the sulfonate head groups but also to water molecules in the core region. The frequencies of the interfacial waters will be influenced by the motions and dynamics of the core. It is possible that the 1.6-ps decay arises from hydrogen bond rearrangement of noninterfacial waters that are making and breaking hydrogen bonds with the interfacial waters. If this proposition is correct, then water molecules immediately beyond the interfacial water layer must behave very much like bulk water. This conclusion supports the idea that an interface gives rise to a very localized strong perturbation of the water dynamics. The slower FFCF time constant of 3.5 ps for $w_0 = 7.5$ is in accord with the slowing down of core water in general in this sample, as seen by the observed retarded orientational dynamics in intermediate sizes.

The offset in the FFCFs ($\Delta_S$ terms in Table 1.6) corresponds to slow processes that cannot be measured within the experimental time window, which is limited by the vibrational lifetimes of the samples. Rotation at the interface has ~18 ps decay time, so it is likely that the offset includes effects from slow rotations. In addition, for spectral diffusion to be complete, the vibrational chromophores must sample all frequencies of the inhomogeneous line shape. This requirement means that interfacial water molecules need to exchange with core waters and diffuse to the core and vice versa. Exchange and diffusion will occur very slowly and contribute to the offset.

### 1.7.2 In Small AOT Reverse Micelles

The FFCFs for $w_0$ values of 2, 4, and 7.5 were determined by single-component CLS analysis [63]. Representative 2D IR spectra for $w_0 = 2$ are shown in the right column of Figure 1.23. Note the substantial difference between the 2D IR spectra of $w_0 = 2$ and $w_0 = 16.5$. The $w_0 = 2$ spectra are much more elongated along the diagonal at all times, and the evolution in shape is much slower than the $w_0 = 16.5$ spectra. The CLS curves for the three small AOT reverse micelles are shown in Figure 1.25. Orientational relaxation measurements for the $w_0 = 2$ and $w_0 = 4$ reverse micelles show that the structural dynamics behave essentially as a single ensemble. In the previous sections, $w_0 = 7.5$ was treated approximately using the two-component core–interface decomposition approach for the 2D IR spectral diffusion dynamics. Here $w_0 = 7.5$ is treated in the same manner as $w_0 = 2$ and $w_0 = 4$ for comparison. The experimental $T_w$ range is limited by the vibrational lifetimes of each sample. Thus, data cannot be taken at the same $T_w$ values for each sample.

The general trend of the CLS data is that spectral diffusion slows down as the water pool size decreases. The FFCF parameters determined for reverse micelles $w_0$ values of 2, 4, and 7.5 are listed in Table 1.7. The FFCF for a small reverse micelle takes the form of a biexponential decay with an offset and a motionally narrowed (homogeneous) component. All three FFCFs have a fast time constant of ~1 ps and a slower time constant of ~6–10 ps. Bulk water has a fast time constant of ~400 fs, which is attributed to fast local hydrogen bond fluctuations [89, 90]. Given the similarity in time scales and that the small reverse micelles represent very constrained environments, it is reasonable to propose that the ~1-ps time scale is also due to local hydrogen bond fluctuations. The static offset in the FFCF (not present in bulk

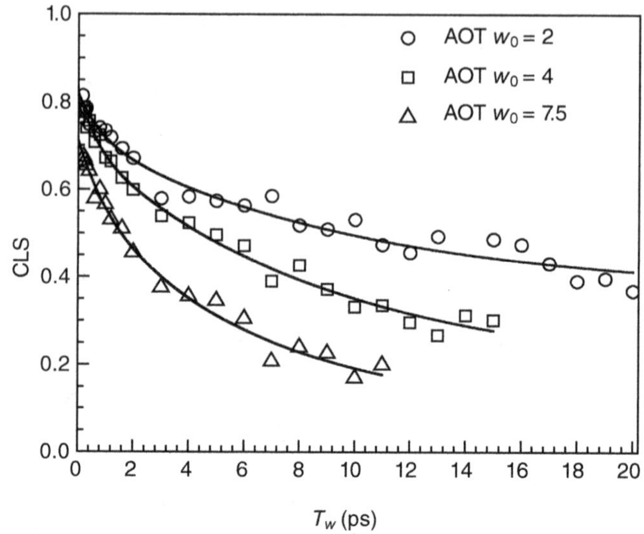


**Figure 1.25** CLS curves for intermediate and small reverse micelles. These data have not been processed with the two-component CLS model. The lines through the data are the FFCFs for each sample. AOT $w_0 = 2$ shows the slowest decay.

**Table 1.7** FFCF Parameters for Small and Intermediate AOT Reverse Micelles

| $w_0$ | $\Gamma$ (cm$^{-1}$) | $\Delta_1$ (cm$^{-1}$) | $\tau_1$ (ps) | $\Delta_2$ (cm$^{-1}$) | $\tau_2$ (ps) | $\Delta_S$ (cm$^{-1}$) |
|---|---|---|---|---|---|---|
| 2 | 33 ± 5 | 19 ± 2 | 0.9 ± 0.1 | 36 ± 1 | 10 ± 1 | 37 ± 1 |
| 4 | 33 ± 5 | 21 ± 2 | 0.7 ± 0.4 | 44 ± 5 | 8.9 ± 3 | 26 ± 1 |
| 7.5 | 57 ± 2 | 25 ± 2 | 1.1 ± 0.2 | 47 ± 3 | 6.4 ± 1 | 20 ± 2 |

water) arises from the extremely slow motions of the water molecules in the small reverse micelles. For spectral diffusion to be complete in any system (including both large and small reverse micelles), all water molecules must experience all environments. Rotations, diffusion through the reverse micelles, and exchange of populations between an OD hydroxyl bound to the interface and one bound to another water oxygen will occur very slowly in the highly confined environments of small reverse micelles.

It is possible that the intermediate 6 to 10-ps decay time scale is caused by hydrogen bond rearrangement of water molecules bound to the water directly at the interface with ones that have at least one or possibly both hydroxyls not bound to the interface. For the large reverse micelles, this mechanism was proposed because of the similarity of times for the interfacial and bulk water spectral diffusion. For the small reverse micelles the dynamics become increasingly slow as the diameter decreases. It is also possible that topography roughness of the surfactant head group interface contributes to inhomogeneous broadening, and the motions of the head groups contribute to spectral diffusion of the water molecules. MD simulations have shown the necessity of including topography roughness in order to capture accurate line shapes [103].

In addition to the time constants, the FFCF also contains frequency fluctuation amplitudes ($\Delta$ terms) and a motionally narrowed component. Each time scale, including the static offset, produces a significant amount of inhomogeneous broadening. The motionally narrowed component ($\Gamma$) is roughly the same for $w_0 = 2$ and $w_0 = 4$, but is significantly greater for $w_0 = 7.5$. It needs to be noted that $w_0 = 7.5$ is not strictly a small reverse micelle and that neither FFCF model (large or small reverse micelle) is entirely correct. The $w_0 = 7.5$ system is quite different from $w_0 = 2$ and $w_0 = 4$ in that it has a distinct core. However, that core is not bulklike, so the bulk approximation in the previous section no doubt produces some error.

## 1.8 CONCLUDING REMARKS

Ultrafast IR spectroscopy is used here to probe the population relaxation, orientational relaxation, and spectral diffusion dynamics of water molecules near charged and neutral reverse micelle surfaces. Using polarization-selective pump–probe and 2D IR techniques, water dynamics are monitored on the picosecond time scale on which they are occurring. Furthermore, these experimental techniques use the actual water molecules (5% HOD in $H_2O$) in the system as probes and therefore measure the water dynamics directly, as opposed to using other probe molecules such as fluorescent dyes. Water, both as a bulk liquid and confined in reverse micelles, has a broad absorption band, indicating a large range of heterogeneity in the hydrogen bonding environments. Water molecules near the reverse micelle interface absorb more on the blue (high-frequency) side of the line shape, a property that greatly facilitates isolating interfacial water dynamics from bulk dynamics. One of the significant results discussed here is that IR spectral analysis, population relaxation, orientational dynamics (anisotropy), and spectral diffusion all follow a two-component model in large reverse micelles such that interfacial water has very distinct and separable dynamics from the bulk water core.

Population and reorientation relaxation results show that in AOT reverse micelles (charged sulfonate head groups) that are equal to or larger than $w_0 = 12$, the dynamics of the interface remain constant with water pool size and wavelength; only the fractional populations of the core and interface change with wavelength. The decay constants for vibrational relaxation and reorientation at the interface are 4.3 and 18 ps, respectively. The equivalent parameters in bulk water are 1.8 and 2.6 ps, respectively. The interface significantly retards water dynamics, but it does not immobilize the water molecules. Spectral diffusion at the interface has a decay time constant of ~1.6 ps that goes to a long-time scale offset. The offset arises from dynamical processes at the interface that are too slow to measure within the experimental time window. The 1.6-ps time constant is the same within experimental error to the slowest component of bulk water spectral diffusion (1.7 ps), suggesting that water molecules at the interface are influenced by water molecules immediately beyond the interfacial layer in the core and that the effects of the interface are highly localized. The results suggest that water molecules just beyond the interfacial layer exhibit dynamics very similar to those of bulk water.

The results show that the presence of an interface substantially slows the hydrogen bond dynamics that are necessary for water to undergo structural evolution. An important issue is how important is the charge and chemical nature of the interface compared to just the presence of an interface of any type. In comparing the interfacial orientational relaxation at the AOT charged interface versus the Igepal neutral interface, it was found that there is at most a modest decrease in reorientation time for interfacial water molecules in Igepal reverse micelles. These results indicate that the presence of the interface is more important than its charge or chemical composition. As shown by 2D IR chemical exchange experiments on water–$NaBF_4$ solutions, the exchange time for a water hydroxyl bound to an anion to switch hydrogen bonds to another water hydroxyl is ~7 ps, which is only about a factor of 3 or 4 slower than the time for water–water hydrogen bond exchange in bulk water. This exchange behavior supports the observed small differences in interfacial water dynamics in AOT and Igepal reverse micelles.

Another question is whether the geometry of the nanoconfined system plays a role in water dynamics. Comparisons of interfacial water dynamics in large AOT reverse micelles and AOT lamellar structures demonstrates that the overall population and orientational relaxation times (4.3 and 18 ps, respectively) are the same for the two geometries. The dynamics are more similar for reverse micelles and lamellar structures of the same hydration level rather than for reverse micelles and lamellar structures of the same nanoscopic dimension (reverse micelle diameter or lamellar interplane separation). This observation further supports the conclusion that the interface only causes a strong perturbation of water dynamics in the immediate proximity of the interface.

As the reverse micelles become smaller, the water molecules start to feel the effects of nanoconfinement rather than only being influenced in the immediate presence of an interface. Intermediate sizes, such as AOT $w_0 = 10$ and $w_0 = 7.5$ show two-component behavior, but each region has slower dynamics than their counterparts in larger sizes. At very small sizes, such as $w_0 = 4$ and $w_0 = 2$, nearly all the water molecules interact with the interface and reorientation dynamics slow dramatically. The dynamics throughout the very small water pools are strongly influenced by the interface and behave more or less as a single ensemble with regard to hydrogen bond structural evolution.

The interplay between surfactant head group chemical identity, hydration level, water pool diameter, and geometry of the interface on interfacial water dynamics is complex in nature, but ultrafast IR spectroscopy has been able to provide detailed insights into the effects of these factors. Overall, slow water dynamics at the interface emerge as a very local effect that does not extend far beyond the first hydration layer for large systems in which the interfacial water is in contact with bulk water. However, when the distance scale of nanoconfinement becomes small, less than a couple of nanometers, the role of the interface changes. For very small nanoscopic pool of water molecules, the interface controls the dynamics of the entire ensemble of water molecules.

## ACKNOWLEDGMENTS

We would like to thank several individuals who participated in many of the experiments presented in this work. We would like to acknowledge Dr. David Moilanen, Daryl Wong, and Chiara Giammanco for their contributions. E.E.F. thanks the Stanford Graduate Fellowship Program for a research fellowship.

## REFERENCES

1. Bagchi, B. Water dynamics in the hydration layer around proteins and micelles, *Chem. Rev.* **2005**, *105*, 3197.
2. Lum, K.; Chandler, D.; Weeks, J. D. Hydrophobicity at small and large length scales, *J. Phys. Chem. B* **1999**, *103*, 4570.
3. Wood, T. D.; Chorush, R. A.; Wampler, F. M.; Little, D. P.; O'Connor, P. B.; McLafferty, F. W. Gas-phase folding and unfolding of cytochrome-c cations, *Proc. Natl. Acad. Sci. U.S.A.* **1995**, *92*, 2451.
4. Levy, Y.; Onuchic, J. N. Water mediation in protein folding and molecular recognition, *Annu. Rev. Biophys. Biomol. Struct.* **2006**, *35*, 389.
5. Rose, G. D.; Wolfenden, R. Hydrogen bonding, hydrophobicity, packing, and protein folding, *Annu. Rev. Biophys. Biomol. Struct.* **1993**, *22*, 381.
6. Rhee, Y. M.; Sorin, E. J.; Jayachandran, G.; Lindahl, E.; Pande, V. S. Simulations of the role of water in the protein-folding mechanism, *Proc. Natl. Acad. Sci. U.S.A.* **2004**, *101*, 6456.
7. Helferrich, F. *Ion Exchange*, McGraw-Hill: New York, **1962**.
8. Zagorodni, A. A. *Ion Exchange Materials: Properties and Applications*, Elsevier: Oxford, **2007**.
9. Narayan, S.; Muldoon, J.; Finn, M. G.; Fokin, V. V.; Kolb, H. C.; Sharpless, K. B. "On water": Unique reactivity of organic compounds in aqueous suspension, *Angew. Chem. Int. Ed.* **2005**, *44*, 3275.
10. Jung, Y.; Marcus, R. A. On the theory of organic catalysis on water, *J. Am. Chem. Soc.* **2007**, *129*, 5492.
11. Jung, Y.; Marcus, R. A. Protruding interfacial OH groups and "on-water" heterogeneous catalysis, *J. Phys.: Condens. Matter* **2010**, *22*, 284117.
12. Hubbard, A. T. Electrochemistry at well-characterized surfaces, *Chem. Rev.* **1988**, *88*, 633.
13. Brown, G. E.; Henrich, V. E.; Casey, W. H.; Clark, D. L.; Eggleston, C.; Felmy, A.; Goodman, D. W.; Gratzel, M.; Maciel, G.; McCarthy, M. I.; Nealson, K. H.; Sverjensky, D. A.; Toney, M. F.; Zachara, J. M. Metal oxide surfaces and their interactions with aqueous solutions and microbial organisms, *Chem. Rev.* **1999**, *99*, 77.
14. Morse, J. W.; Arvidson, R. S.; Luttge, A. Calcium carbonate formation and dissolution, *Chem. Rev.* **2007**, *107*, 342.
15. Jungwirth, P.; Tobias, D. J. Specific ion effects at the air/water interface, *Chem. Rev.* **2006**, *106*, 1259.
16. Benjamin, I. Chemical reactions and solvation at liquid interfaces: A microscopic perspective, *Chem. Rev.* **1996**, *96*, 1449.
17. Newton, M. D.; Friedman, H. L. A proposed neutron-diffraction experiment to measure hydrogen isotope fractionation in solution, *J. Chem. Phys.* **1985**, *83*, 5210.
18. Mancinelli, R.; Botti, A.; Bruni, F.; Ricci, M. A.; Soper, A. K. Hydration of sodium, potassium, and chloride ions in solution and the concept of structure maker/breaker, *J. Phys. Chem. B* **2007**, *111*, 13570.

19. COLLINS, K. D.; NEILSON, G. W.; ENDERBY, J. E. Ions in water: Characterizing the forces that control chemical processes and biological structure, *Biophys. Chem.* **2007**, *128*, 95.
20. SMITH, J. D.; SAYKALLY, R. J.; GEISSLER, P. L. The effect of dissolved halide anions on hydrogen bonding in liquid water, *J. Am. Chem. Soc.* **2007**, *129*, 13847.
21. BAKKER, H. J.; KROPMAN, M. F.; OMATA, Y.; WOUTERSEN, S. Hydrogen-bond dynamics of water in ionic solutions, *Phys. Scr.* **2004**, *69*, C14.
22. KROPMAN, M. F.; BAKKER, H. J. Dynamics of water molecules in aqueous solvation shells, *Science* **2001**, *291*, 2118.
23. PARK, S.; FAYER, M. D. Hydrogen bond dynamics in aqueous NaBr solutions, *Proc. Natl. Acad. Sci. U.S.A.* **2007**, *104*, 16731.
24. PARK, S.; MOILANEN, D. E.; FAYER, M. D. Water dynamics–the effects of ions and nanoconfinement, *J. Phys. Chem. B* **2008**, *112*, 5279.
25. MOILANEN, D. E.; WONG, D.; ROSENFELD, D. E.; FENN, E. E.; FAYER, M. D. Ion-water hydrogen-bond switching observed with 2D IR vibrational echo chemical exchange spectroscopy, *Proc. Natl. Acad. Sci. U.S.A.* **2009**, *106*, 375.
26. MINBIAO, J.; ODELIUS, M.; GAFFNEY, K. J. Large angular jump mechanism observed for hydrogen bond exchange in aqueous perchlorate solution, *Science* **2010**, *328*.
27. BRINK, G.; FALK, M. Infrared spectrum of HDO in aqeous solutions of perchlorates and tetrafluoroborates, *Can. J. Chem.* **1970**, *48*, 3019.
28. HOPKINS, A. J.; SCHROEDLE, S.; RICHMOND, G. L. Specific ion effects of salt solutions at the CaF2/water interface, *Langmuir* **2010**, *26*, 10784.
29. SMITH, J. D.; CAPPA, C. D.; MESSER, B. M.; DRISDELL, W. S.; COHEN, R. C.; Saykally, R. J. Probing the local structure of liquid water by x-ray absorption spectroscopy, *J. Phys. Chem. B* **2006**, *110*, 20038.
30. KERISIT, S.; ROSSO, K. M. Transition path sampling of water exchange rates and mechanisms around aqueous ions, *J. Chem. Phys.* **2009**, *131*.
31. DZOMBAK, D. A.; HUDSON, R. J. M. Ion exchange: The contributions of diffuse layer sorption and surface complexation, in *Advances in Chemistry Series; Aquatic Chemistry: Interfacial and Interspecies Processes*, C. P. HUANG, C. R. O'MELIA, and J. J. MORGAN, eds., Washington, DC, **1995**, Vol. *244*, p. 59.
32. LIN, Y. S.; AUER, B. M.; SKINNER, J. L. Water structure, dynamics, and vibrational spectroscopy in sodium bromide solutions, *J. Chem. Phys.* **2009**, *131*, 144511.
33. ARGYRIS, D.; COLE, D. R.; STRIOLO, A. Ion-specific effects under confinement: the role of interfacial water, *ACS Nano* **2010**, *4*, 2035.
34. JOUNG, I. S.; CHEATHAM, T. E., III. Molecular dynamics simulations of the dynamic and energetic properties of alkali and halide ions using water-model-specific ion parameters, *J. Phys. Chem. B* **2009**, *113*.
35. LYUBARTSEV, A. P.; LAAKSONEN, A. Concentration effects in aqueous NaCl solutions. A molecular dynamics simulation, *J. Phys. Chem.* **1996**, *100*, 16410.
36. FECKO, C. J.; EAVES, J. D.; LOPARO, J. J.; TOKMAKOFF, A.; GEISSLER, P. L. Ultrafast hydrogen-bond dynamics in the infrared spectroscopy of water, *Science* **2003**, *301*, 1698.
37. FECKO, C. J.; LOPARO, J. J.; ROBERTS, S. T.; Tokmakoff, A. Local hydrogen bonding dynamics and collective reorganization in water: Ultrafast infrared spectroscopy of $HOD/D_2O$, *J. Chem. Phys.* **2005**, *122*, 054506.
38. ASBURY, J. B.; STEINEL, T.; KWAK, K.; CORCELLI, S. A.; LAWRENCE, C. P.; SKINNER, J. L.; FAYER, M. D. Dynamics of water probed with vibrational echo correlation spectroscopy, *J. Chem. Phys.* **2004**, *121*, 12431.
39. ASBURY, J. B.; STEINEL, T.; STROMBERG, C.; CORCELLI, S. A.; LAWRENCE, C. P.; SKINNER, J. L.; FAYER, M. D. Water dynamics: Vibrational echo correlation spectroscopy and comparison to molecular dynamics simulations, *J. Phys.Chem. A* **2004**, *108*, 1107.
40. NICODEMUS, R. A.; RAMASESHA, K.; ROBERTS, S. T.; TOKMAKOFF, A. Hydrogen bond rearrangements in water probed with temperature-dependent 2D IR, *J. Phys. Chem. Lett.* **2010**, *1*, 1068.
41. COWAN, M. L.; BRUNER, B. D.; HUSE, N.; DWYER, J. R.; CHUGH, B.; NIBBERING, E. T. J.; ELSAESSER, T.; MILLER, R. J. D. Ultrafast memory loss and energy redistribution in the hydrogen bond network of liquid $H_2O$, *Nature* **2005**, *434*, 199.

42. Loparo, J. J.; Roberts, S. T.; Tokmakoff, A. Multidimensional infrared spectroscopy of water. I. Vibrational dynamics in two-dimensional IR line shapes, *J. Chem. Phys.* **2006**, *125*, 194521.
43. Loparo, J. J.; Roberts, S. T.; Tokmakoff, A. Multidimensional infrared spectroscopy of water. II. Hydrogen bond switching dynamics, *J. Chem. Phys.* **2006**, *125*, 194522.
44. Ganim, Z.; Jones, K. C.; Tokmakoff, A. Insulin dimer dissociation and unfolding revealed by amide I two-dimensional infrared spectroscopy, *Phys. Chem. Chem. Phys.* **2010**, *12*, 3579.
45. Finkelstein, I. J.; Zheng, J.; Ishikawa, H.; Kim, S.; Kwak, K.; Fayer, M. D. Probing dynamics of complex molecular systems with ultrafast 2D IR vibrational echo spectroscopy, *Phys. Chem. Chem. Phys.* **2007**, *9*, 1533.
46. DeFlores, L. P.; Ganim, Z.; Ackley, S. F.; Chung, H. S.; Tokmakoff, A. The anharmonic vibrational potential and relaxation pathways of the amide I and II modes of N-methylacetamide, *J. Phys. Chem. B.* **2006**, *110*, 18973.
47. Mukherjee, P.; Kass, I.; Arkin, I. T.; Zanni, M. T. Picosecond dynamics of a membrane protein revealed by 2D IR, *Proc. Natl. Acad. Sci. U.S.A.* **2006**, *103*, 3528.
48. Ishikawa, H.; Kwak, K.; Chung, J. K.; Kim, S.; Fayer, M. D. Direct observation of fast protein conformational switching, *Proc. Natl. Acad. Sci. U.S.A.* **2008**, *105*, 8619.
49. Tucker, M. J.; Gai, X. S.; Fenlon, E. E.; Brewer, S. H.; Hochstrasser, R. M. 2D IR photon echo of azido-probes for biomolecular dynamics, *Phys. Chem. Chem. Phys.* **2011**, *13*, 2237.
50. Chung, J. K.; Thielges, M. C.; Bowman, S. J.; Bren, K. L.; Fayer, M. D. Temperature dependent equilibrium native to unfolded protein dynamics and properties observed with IR absorption and 2D IR vibrational echo experiments, *J. Am. Chem. Soc.* **2011**, *133*, 6681.
51. Chung, J. K.; Thielges, M. C.; Fayer, M. D. Dynamics of the folded and unfolded villin headpiece (HP35) measured with ultrafast 2D IR vibrational echo spectroscopy, *Proc. Natl. Acad. Sci. U.S.A.* **2011**, *108*, 3578.
52. Thielges, M. C.; Chung, J. K.; Fayer, M. D. Protein dynamics in cytochrome P450 molecular recognition and substrate specificity using 2D IR vibrational echo spectroscopy, *J. Am. Chem. Soc.* **2011**, *133*, 3995.
53. Rosenfeld, D. E.; Kwak, K.; Gengeliczki, Z.; Fayer, M. D. Hydrogen bond migration between molecular sites observed with ultrafast 2D IR chemical exchange spectroscopy, *J. Phys. Chem. B* **2010**, *114*, 2383.
54. Kim, Y. S.; Hochstrasser, R. M. Chemical exchange 2D IR of hydrogen-bond making and breaking, *Proc. Natl. Acad. Sci. U.S.A.* **2005**, *102*, 11185.
55. Zheng, J.; Kwak, K.; Asbury, J. B.; Chen, X.; Piletic, I. R.; Fayer, M. D. Ultrafast dynamics of solute-solvent complexation observed at thermal equilibrium in real time, *Science* **2005**, *309*, 1338.
56. Zheng, J.; Kwak, K.; Xie, J.; Fayer, M. D. Ultrafast carbon-carbon single bond rotational isomerization in room temperature solution, *Science* **2006**, *313*, 1951.
57. Kwak, K.; Zheng, J.; Cang, H.; Fayer, M. D. Ultrafast 2D IR vibrational echo chemical exchange experiments and theory, *J. Phys. Chem. B.* **2006**, *110*, 19998.
58. Zulauf, M.; Eicke, H. F. Inverted micelles and microemulsions in the ternary-system $H_2O$-aerosol-OT-isooctane as studied by photon correlation spectroscopy, *J. Phys. Chem.* **1979**, *83*, 480.
59. Kinugasa, T.; Kondo, A.; Nishimura, S.; Miyauchi, Y.; Nishii, Y.; Watanabe, K.; Takeuchi, H. Estimation for size of reverse micelles formed by AOT and SDEHP based on viscosity measurement, *Coll. Surf. A—Physicochem. Eng. Aspects* **2002**, *204*, 193.
60. Eicke, H.-F.; Rehak, J. On the formation of water/oil-microemulsions, *Helv. Chim. Acta* **1976**, *59*, 2883.
61. Piletic, I. R.; Moilanen, D. E.; Spry, D. B.; Levinger, N. E.; Fayer, M. D. Testing the core/shell model of nanoconfined water in reverse micelles using linear and nonlinear IR spectroscopy, *J. Phys. Chem. A* **2006**, *110*, 4985.
62. Izquierdo, C.; Moya, M. L.; Usero, J. L.; Casado, J. Estimation of the aggregation number and core radius of microemulsions, *Monatsh. Chem.* **1992**, *123*, 383
63. Fenn, E. E.; Wong, D. B.; Fayer, M. D. Water dynamics in small reverse micelles in two solvents: Two-dimensional infrared vibrational echoes with two-dimensional background subtraction, *J. Chem. Phys.* **2011**, *134*, 054512.

64. Kenkre, V. M.; Tokmakoff, A.; Fayer, M. D. Theory of vibrational relaxation of polyatomic molecules in liquids, *J. Chem. Phys.* **1994**, *101*, 10618.
65. Egorov, S. A.; Berne, B. J. Vibrational energy relaxation in the condensed phases: quantum vs. classical bath for multiphonon processes, *J. Chem. Phys.* **1997**, *107*, 6050.
66. Laage, D.; Hynes, J. T. On the molecular mechanism of water reorientation, *J. Phys. Chem. B* **2008**, *112*, 14230.
67. Laage, D.; Hynes, J. T. A molecular jump mechanism of water reorientation, *Science* **2006**, *311*, 832.
68. Moilanen, D. E.; Fenn, E. E.; Wong, D.; Fayer, M. D. Water dynamics in large and small reverse micelles: from two ensembles to collective behavior, *J. Chem. Phys.* **2009**, *131*, 014704.
69. Moilanen, D. E.; Fenn, E. E.; Wong, D.; Fayer, M. D. Water dynamics in AOT lamellar structures and reverse micelles: Geometry and length scales vs. surface interactions, *J. Am. Chem. Soc.* **2009**, *131*, 8318.
70. Moilanen, D. E.; Fenn, E. E.; Wong, D.; Fayer, M. D. Water dynamics at the interface in AOT reverse micelles, *J. Phys. Chem. B* **2009**, *113*, 8560.
71. Fenn, E. E.; Wong, D. B.; Fayer, M. D. Water dynamics at neutral and ionic interfaces, *Proc. Natl. Acad. Sci. U.S.A.* **2009**, *106*, 15243.
72. Fenn, E. E.; Moilanen, D. E.; Levinger, N. E.; Fayer, M. D. Water dynamics and interactions in water-polyether binary mixtures, *J. Am. Chem. Soc.* **2009**, *131*, 5530.
73. Gaffney, K. J.; Piletic, I. R.; Fayer, M. D. Orientational relaxation and vibrational excitation transfer in methanol–carbon tetrachloride solutions, *J. Chem. Phys.* **2003**, *118*, 2270.
74. Woutersen, S.; Bakker, H. J. Resonant intermolecular transfer of vibrational energy in liquid water, *Nature (London)* **1999**, *402*, 507.
75. Corcelli, S.; Lawrence, C. P.; Skinner, J. L. Combined electronic structure/molecular dynamics approach for ultrafast infrared spectroscopy of dilute HOD in liquid $H_2O$ and $D_2O$, *J. Chem. Phys.* **2004**, *120*, 8107.
76. Maitra, A. Determination of size parameters of water aerosol OT oil reverse micelles from their nuclear magnetic-resonance data, *J. Phys. Chem.* **1984**, *88*, 5122.
77. Kon-No, K.; Kitahara, A. Solubility behavior of water in nonaqueous solutions of oil-soluble surfactants—effect of molecular structure of surfactants and solvents, *J. Colloid Interface Sci.* **1971**, *37*, 469.
78. Lipgens, S.; Schubel, D.; Schlicht, L.; Spilgies, J.-H.; Ilgenfritz, G.; Eastoe, J.; Heenan, R. K. Percolation in nonionic water-in-oil-microemulsion systems: A small angle neutron scattering study, *Langmuir* **1998**, *14*, 1041.
79. Mukamel, S. *Principles of Nonlinear Optical Spectroscopy*, Oxford University Press: New York, **1995**.
80. Hamm, P.; Zanni, M. T. *Concepts and Methods of 2D Infrared Spectroscopy*, Cambridge University Press: Cambridge, **2011**.
81. Fenn, E. E.; Fayer, M. D. Extracting 2D IR frequency-frequency correlation functions from two component systems, *J. Chem. Phys.* **2011**, *135*, 074502.
82. Gruetzmacher, J. A.; Scherer, N. F. Few-cycle mid-infrared pulse generation, characterization, and coherent propagation in optically dense media, *Rev. Sci. Instrum.* **2002**, *73*, 2227.
83. Demirdoven, N.; Khalil, M.; Golonzka, O.; Tokmakoff, A. Dispersion compensation with optical materials for compression of intense sub-100-fs mid-infrared pulses, *Opt. Lett.* **2002**, *27*, 433.
84. Kwak, K.; Rosenfeld, D. E.; Fayer, M. D. Taking apart the two-dimensional infrared vibrational echo spectra: more information and elimination of distortions, *J. Chem. Phys.* **2008**, *128*, 204505.
85. Kwak, K.; Park, S.; Finkelstein, I. J.; Fayer, M. D. Frequency-frequency correlation functions and apodization in 2D-IR vibrational echo spectroscopy: A new approach, *J. Chem. Phys.* **2007**, *127*, 124503.
86. Saven, J. G.; Skinner, J. L. A molecular theory of the line shape—inhomogeneous and homogeneous electronic spectra of dilute chromophores in nonpolar fluids, *J. Chem. Phys.* **1993**, *99*, 4391.
87. Roberts, S. T.; Loparo, J. J.; Tokmakoff, A. Characterization of spectral diffusion from two-dimensional line shapes, *J. Chem. Phys.* **2006**, *125*, 084502.

88. Cho, M. H.; Yu, J. Y.; Joo, T. H.; Nagasawa, Y.; Passino, S. A.; Fleming, G. R. The integrated photon echo and solvation dynamics, *J. Phys. Chem.* **1996**, *100*, 11944.
89. Lawrence, C. P.; Skinner, J. L. Vibrational spectroscopy of HOD in liquid $D_2O$. III. Spectral diffusion, and hydrogen-bonding and rotational dynamics, *J. Chem. Phys.* **2003**, *118*, 264.
90. Corcelli, S. A.; Lawrence, C. P.; Asbury, J. B.; Steinel, T.; Fayer, M. D.; Skinner, J. L. Spectral diffusion in a fluctuating charge model of water, *J. Chem. Phys.* **2004**, *121*, 8897.
91. Tokmakoff, A. Orientational correlation functions and polarization selectivity for nonlinear spectroscopy of isotropic media. 1. third order, *J. Chem. Phys.* **1996**, *105*, 1.
92. Lipari, G.; Szabo, A. Effect of librational motion on fluorescence depolarization and nuclear magnetic-resonance relaxation in macromolecules and membranes, *Biophys. J.* **1980**, *30*, 489.
93. Wang, C. C.; Pecora, R. Time-correlation functions for restricted rotational diffusion, *J. Chem. Phys.* **1980**, *72*, 5333.
94. Kropman, M. F.; Bakker, H. J. Femtosecond mid-infrared spectroscopy of aqueous solvation shells, *J. Chem. Phys.* **2001**, *115*, 8942.
95. Collins, K. D. Charge density-dependent strength of hydration and biological structure, *Biophys. J.* **1997**, *72*, 65.
96. Rezus, Y. L. A.; Bakker, H. J. Orientational dynamics of isotopically diluted $H_2O$ and $D_2O$, *J. Chem. Phys.* **2006**, *125*, 144512.
97. Fontell, K. The structure of the lamellar liquid crystalline phase in aerosol OT-water system, *J. Colloid Interface Sci.* **1973**, *44*, 318.
98. Hauser, H.; Haering, G.; Pande, A.; Luisi, P. L. Interaction of water with sodium Bis(2-ethyl-1-hexyl) sulfosuccinate in reversed micelles, *J. Phys. Chem.* **1989**, *93*, 7869.
99. Spry, D. B.; Goun, A.; Glusac, K.; Moilanen, D. E.; Fayer, M. D. Proton Transport and the water environment in nafion fuel cell membranes and AOT reverse micelles, *J. Am. Chem. Soc* **2007**, *129*, 8122.
100. Lipari, G.; Szabo, A. Model-free approach to the interpretation of nuclear magnetic-resonance relaxation in macromolecules 1. Theory and range of validity, *J. Am. Chem. Soc.* **1982**, *104*, 4546.
101. Yan, Z.; Buldyrev, S. V.; Kumar, P.; Giovambattista, N.; Debenedetti, P. G.; Stanley, H. E. Structure of the first- and second-neighbor shells of simulated water: Quantitative relation to translational and orientational order, *Phys. Rev. E: Stat., Nonlinear, Soft Matter Phys.* **2007**, *76*, 051201.
102. Fenn, E.E.; Wong, D.B.; Glammanco, C.H.; Fayer, M.D. Dynamics of water at the interface in reverse micelles: measurements of spectral diffusion with 2D IR vibrational echoes, *J. Phys. Chem. B* **2011**, *115*, 11658.
103. Pieniazek, P. A.; Lin, Y.-S.; Chowdhary, J.; Ladanyi, B. M.; Skinner, J. L. Vibrational spectroscopy and dynamics of water confined inside reverse micelles, *J. Phys. Chem. B* **2009**, *113*, 15017.

# Chapter 2

# SFG Studies of Oxide–Water Interfaces: Protonation States, Water Polar Orientations, and Comparison with Structure Results from X-Ray Scattering

**Y. Ron Shen[1] and Glenn A. Waychunas[2]**

[1]Physics Department, University of California, Berkeley, Berkeley, California
[2]Geochemistry Department, Earth Sciences Division, Lawrence Berkeley National Laboratory, Berkeley, California

## 2.1 INTRODUCTION: IMPORTANCE OF OXIDE–WATER INTERFACES

Aqueous–metal oxide interfaces are important substrates that initiate and affect chemical reactions in environmental systems, including control of many aspects of natural water chemistry and purity, and also have a host of significant technological applications including heterogeneous catalysis and photocatalysis. Further, while the substrate is not the main component of a reactive system, it is still often crucial to understand how its surface structure and chemistry affect contacting solutions and influence reactivity. Aqueous–oxide interfaces may effect control of reactions by virtue of (1) mass transfer effects, which limit the approach of reactants in solution but can greatly enhance sorbed reactant interactions at surfaces, leading to interface-controlled processes depending on surface site densities, (2) surface structure orientation effects, which can select molecules from solution with specific orientation for

*Vibrational Spectroscopy at Electrified Interfaces*, First Edition.
Edited by Andrzej Wieckowski, Carol Korzeniewski, and Björn Braunschweig.

enhanced reaction, such as a condensation reaction [1], or sort reactive species from among different candidate molecules, and (3) the acid–base and electrostatic effects of molecular interactions produced within the electrical double layer (EDL). In many aqueous environmental reactions, rate-controlling steps directly involve the substrate [2, 3], so that knowledge of interfacial chemistry is necessary to determine reaction mechanisms and predict kinetics. In this review we address what is known about the molecular structure at several aqueous–metal oxide interfaces, including near surface water, and indicate how improvements in this knowledge from new interface-sensitive methodology may aid our understanding of how these systems affect reaction progress.

## 2.2 GENERAL FEATURES OF OXIDE SURFACE STRUCTURES AND INTERFACIAL STRUCTURES OF OXIDE–WATER INTERFACES

Aqueous–oxide interfaces have several important chemically active aspects that determine their involvement in chemical reactions. The most important of these is the occurrence of surface functional groups having specific reactivity, for example, for acid–base reactions. A typical oxide functional group consists of a surface oxygen, bonded to one or more cations, that can interact with solution water molecules. An example is shown in Figure 2.1, which shows functional groups associated with two types of oxygens at the α-FeOOH (100) surface [4]. The protonation

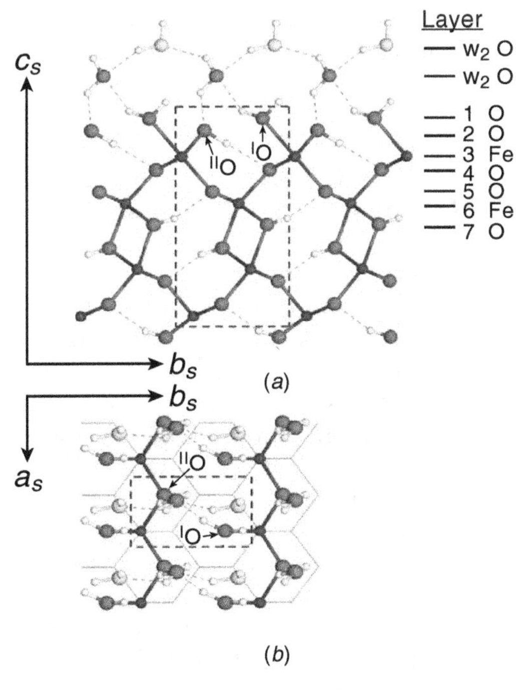


**Figure 2.1** α-FeOOH (100) interfacial structure at room temperatures in equilibrium with water from CTR analysis. (*a*) Side view of bound water bilayer. Water positions from CTR refinement and protons estimated from bond valence calculations using standard treatments. The bulk structure unit cell is outlined. (*b*) Top view of termination showing water positions. $a_s$, $b_s$, and $c_s$ refer to the surface cell used in the CTR refinement. From Ghose et al. (2010).

reactions of such surface oxygens determine the point of zero charge (pzc) pH for the surface. The position and orientations of the group can determine how water molecules may hydrogen bond to the solid surface, and evidence suggests that more than one water layer may have organization dictated in part by the distribution of these groups. In Figure 2.1, two layers of bound water are shown along with hypothesized hydrogen bonds (dashed lines). Other functional groups can be associated with electron transfer reactions and hence control redox properties. Functional groups or combinations of such groups are sometimes described as reactive sites for sorption or precipitation reactions.

Other important molecular aspects of interfacial regions include variations in atomic positions from the corresponding bulk structure (relaxation), changes in site occupancies compared to the bulk, and larger scale features such as terraces and steps, dislocations, and other defects. Steps can expose functional groups not appearing on terraces and hence may have significant effect on reactivity. Due to this complexity, determination of the complete termination of the oxide surface in equilibrium with an aqueous solution may be an important goal, though not a simple matter in practice.

Direct measurement of functional groups and molecular species at oxide–water interfaces including near-surface water has been limited due the inherent insensitivity of surface probes available within solid/bulk aqueous systems, or to the inapplicability of ultrahigh vacuum (UHV) approaches that require low interface water density (submonolayer), which does not behave like a bulk solution [5, 6]. In this chapter we consider recent approaches via X-ray scattering and sum frequency vibrational spectroscopy (SFVS) methods that specifically probe the molecular structure of the oxide–water interface with bulk solution present, allowing pH and chemical variations to be studied in detail.

## 2.3 MACROSCOPIC VERSUS MICROSCOPIC PICTURES OF OXIDE–WATER INTERFACES: ELECTRICAL DOUBLE LAYER

The charging properties of aqueous–solid interfaces depend on the interaction of surface functional groups and other surface charge sources with overlying water molecules, although models that rely on the bulk dielectric properties of the solid substrate also exist [7]. The solid surface charge can derive from permanent structural substitutions, for example, $Al^{3+}$ substituting for $Si^{4+}$, which would create a local negative charge, from surface functional groups that can enter into acid–base equilibria with exchange of $H^+$ and $OH^-$, and from adsorption of counterions from the contacting solution that respond to surface charges. The counterions may be held close to the surface or be part of a more diffuse charge region extending away from the interface. The general relationship between all of these charge sources has been developed as the EDL model. As the separation of charges produces a potential difference across the interface, the distribution of charges were originally modeled by Poisson–Boltzmann solutions for planar double layers at a flat infinite interface

[8, 9]. This original theory utilized only a diffuse swarm of counterions and did not agree well with observations made, for example, with mercury electrodes. It was superseded by an approach that allowed for the finite sizes of ions and the minimum charge separation required by ion–ion distances known as the Stern–Grahame model [10].

In this EDL model some counterions were sorbed directly to the solid surface, forming a compact layer, and this was in addition to a diffuse counterion layer whose potential was assumed to decrease linearly from the interface. This model works well at low ionic strengths and is still used today with various modifications, for treating sorption–complexation and other interface reactions. Although these models use critical length parameters normal to the interface to define planes of sorbed counterions, they are not molecular models as water is assumed to be a continuum, and the dielectric constant is assumed to be constant from the bulk up to the sorption planes. In actuality water near the interface must have a lower dielectric constant than bulk, which may affect hydrolysis reactions at the interface or alter functional group acid–base reactions from direct correspondence with analogous complexes in solution. An outstanding question is: How well do measured structural aspects of the oxide–water interface agree with the assumptions made in EDL models? This question is beginning to be approached with X-ray scattering and simulation methods [11, 12], and might be addressable with SFG approaches in the future.

## 2.4 CURRENT UNDERSTANDING OF OXIDE–WATER INTERFACES EXCLUDING INFORMATION FROM X-RAY SCATTERING AND SFVS

Early studies using surface force probes to examine interactions between atomically smooth mica surfaces demonstrated that the strength of molecular interactions varied with separation and was consistent with the spacing of water molecular layers [13]. Mica interlayer surfaces are terminated by oxygen atoms, and so this is a type of oxide surface. Subsequently, neutron scattering studies of water in clay mineral interlayers similarly indicated well-defined water structure [14], and atomic force microscopy (AFM) studies also detected hydration layering on calcite surfaces [15]. These studies suggested that water may adopt an interfacial structure different from bulk water, though such structure could also be due to confinement effects. Soon afterward, molecular dynamics (MD) simulations appeared that showed preferential water ordering on hydrophilic surfaces, and in one study [16], detailed local structures for water interaction with a protonated silicon oxide surface were observed. Since that study, numerous MD simulations have indicated well-developed ordering of water at solid surfaces [17], and, recently, MD simulations have shown notable divergence from bulk water density out to four or five molecular layers near nanoparticles of $\alpha$-$Fe_2O_3$ [18]. The physical consequences of such order include highly directional diffusion coefficients, for example, much faster along surfaces than between surface layers of water, and possible effects on particle aggregation, that is, from interacting organized water. These dynamic effects have been observed on hydrophilic surfaces in nuclear maguetic resonance (NMR) experiments [19].

Although classical MD simulations yield insight into interfacial water arrangements, they have limitations in simulating interfacial regions accurately. Interaction potentials are fixed and usually include only bulk-type water interactions. It is unclear if hydrogen bonding for surface-bound waters is accurately modeled by bulk water–water bonding interactions. Second, in MD simulations the surface chemistry (e.g., protonation of functional groups, atomic relaxations) must be assumed and fixed in the calculation. Hence, the simulation may be considering physically unrealistic interface chemistry, and the modeling of conditions such as variable pH is not possible.

Some of these limitations can be improved upon by use of ab initio simulations, such as Car-Parrinello Molecular Dynamics (CPMD), where potentials are recalculated during the simulation. However, the best approach is probably full quantum mechanical methods, such as density functional theory (DFT) calculations, which can allow minimization of the energy of the full interfacial region, including bound waters, surface atomic displacements, changes in site occupancies, and changes in protonation. Unfortunately, DFT calculations of complete interface structure are computationally demanding and represent current frontier methodology in simulations [20, 21]. Hence contemporary approaches often utilize selective "bootstrapping" of MD via small cluster DFT calculations, examples of which we will mention when discussing experimental X-ray scattering and sum frequency generation (SFG) results.

Overall, such simulations suggest two layers of bound waters at an oxide interface, the first directly connected to the surface, and the second connected via longer hydrogen bonds or via bonding to the first water layer (or possibly using both interactions). Different (*hkl*) planes of the same material may have different bonding arrangements and, hence, terminations and water arrangements. Water beyond the first two layers is generally assumed to be close to bulk water in structure.

## 2.5 X-RAY STUDIES OF OXIDE–WATER INTERFACES

More detailed interface structure information, and in particular structure information with bulk solution present, has become available as interfacial X-ray scattering studies were developed at synchrotron sources. In particular, X-ray reflectivity (XRR), which probes one-dimensional atomic structure in the $z$ direction above an interface, has revealed well-defined water structure in several types of oxide single crystals [22, 23], as well as details of the solid termination. Crystal truncation rod (CTR) scattering measurements, which include X-ray reflectivity, show the promise of determining both $z$ direction and lateral water structure character at oxide interfaces [4, 24, 25].

X-ray scattering from a real crystal with a well-defined surface termination and finite size differs in certain quantitative ways from the usually assumed diffraction subject: the infinite perfect crystal. As shown by Robinson [26], the intensity of diffraction from a small perfect cubic crystal with dimensions $N_1$, $N_2$, and $N_3$ primitive unit cells, with corresponding lattice parameters $a_1$, $a_2$, and $a_3$, and reciprocal lattice vectors $Q_1$, $Q_2$, and $Q_3$, can be written as

$$|F(Q_1, Q_2, Q_3)|^2 = \left|\sum_{j_1=1}^{N_1}\sum_{j_2=1}^{N_2}\sum_{j_3=1}^{N_3} e^{i(Q_1a_1j_1+Q_2a_2j_2+Q_3a_3j_3)}\right|^2$$
$$= \frac{\sin^2\left(\frac{1}{2}N_1Q_1a_1\right)}{\sin^2\left(\frac{1}{2}Q_1a_1\right)}\frac{\sin^2\left(\frac{1}{2}N_2Q_2a_2\right)}{\sin^2\left(\frac{1}{2}Q_2a_2\right)}\frac{\sin^2\left(\frac{1}{2}N_3Q_3a_3\right)}{\sin^2\left(\frac{1}{2}Q_3a_3\right)} \quad (2.1)$$

which is the square of the structure factor $F$. It is seen that if all the $N_i$ are large, the structure factor has sharp peaks at what are called Bragg positions, corresponding to values $Q_1a_1 = 2\pi h$, $Q_2a_2 = 2\pi k$, and $Q_3a_3 = 2\pi l$, with $h$, $k$, $l$ integers and with intensity proportional to $(N_1N_2N_3)$ [2].

If two of the conditions for Bragg diffraction are enforced but one relaxed, then the intensity is of the form

$$\left|F\left(\frac{2\pi h}{a_1}, \frac{2\pi k}{a_2}, Q_3\right)\right|^2 = N_1^2N_2^2\frac{\sin^2\left(\frac{1}{2}N_3Q_3a_3\right)}{\sin^2\left(\frac{1}{2}Q_3a_3\right)} \quad (2.2)$$

which in the limit gives

$$N_1^2N_2^2\frac{1}{2\sin^2\left(\frac{1}{2}Q_3a_3\right)} \quad \text{as} \quad N_3 \to \infty \quad \text{for} \quad Q_3a_3 \neq 2\pi l \quad (2.3)$$

This result indicates that for a finite crystal the diffracted intensity is smeared out into streaks along the $Q_3$ direction that connect the Bragg spots (Fig. 2.2*a*). The scattered intensity is proportional to $N^4$ for the smeared or "diffuse" intensity far from the Bragg peaks, which have intensity proportional to $N^6$. For a physically realistic crystal that is thick relative to the absorption length of X rays, which we might call "semi-infinite," the diffuse intensity is due only to the face of the crystal whereupon the X-ray beam is impingent. The smeared intensity is referred to as a "crystal truncation rod," or CTR, due to its form in reciprocal space, with orientation always normal to the termination surface. If we add an estimate of the X-ray penetration depth into the face of the crystal as $10^3$ unit cells, on the order of millimeters, and an X-ray coherence length, meters, of $10^2$ unit cells, this gives Bragg peak intensity proportional to $N^3m^3$ and CTR intensity proportional to $N^2m^2$, a ratio of $10^5$. This means that an experiment that seeks to quantify the diffuse intensity CTR requires a high dynamic range larger than $10^5$, and this is now readily possible using synchrotron radiation. The intensity proportional to $N^2m^2$ can be shown to be equal to the scattering from one layer of the crystal surface, by using a more quantitative kinematical theory treatment [27–29]. An example of the sensitivity of this result to changes in a single surface layer is shown in Figure 2.2*b*. Actual CTR data are shown in Figure 2.2*c*, which is the (10*L*) crystal truncation data rod collected for the surface shown in Figure 2.1.[4]

In practice, surface scattering data is analyzed by calculation of the structure factors from the bulk crystal substrate and surface region separately with slightly

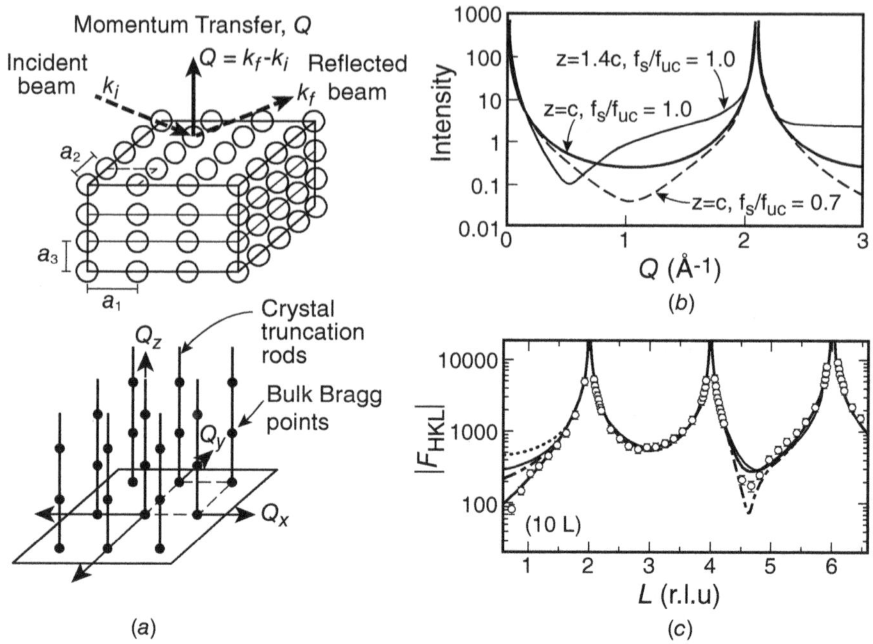


**Figure 2.2** (*a*) (*Top*) Momentum transfer in surface diffraction experiment along specular crystal truncation rod (normal to crystal surface). (*Bottom*) Reciprocal space picture of crystal truncation rods. The rod emerging from the origin is the specular rod, corresponding to X-ray reflectivity measurements, and having only $Q_z$ information. Other "nonspecular" rods emerging away from the origin have $Q_x$ and $Q_y$ components. (*b*) Kinematic diffraction model for a specular CTR from a body-centered cubic lattice showing (from top) 40% increase in surface layer to next layer distance (thin line), ideal surface termination (bold), 30% loss in surface layer occupancy (dashed line). The feature at $Q = 2$ is a Bragg reflection [29]. (*c*) Actual CTR data for (10*L*) rod from the surface shown in Figure 2.1.

different formalism [30]. Here the CTR intensity is included in the "bulk" of the substrate, and an additional surface unit cell, which may be several layers thick, is defined. The surface cell consists of one or more layers of substrate surface and/or sorbate structure having features different from the bulk that we wish to probe. Structural refinement of the atomic positions and other parameters in this cell are done by least squares fitting of the measured intensity distributions.

In this treatment the total structure factor is given by the coherent sum of surface cell and bulk (including CTR) contributions. Written in this form one sees that the actual surface diffraction that is measured is an interference between the diffuse scattering due to the CTR of the terminated crystal surface and the weak scattering from a small separately defined surface cell coherent with the crystal surface. Following Vlieg we have [30]

$$F_{\text{sum}} = F_{\text{surface}} + F_{\text{bulk}} \tag{2.4}$$

with

$$F_{\text{surface}} = \sum_j f_j \theta_j e^{-B_j Q^2/(16\pi^2)} e^{2\pi i(hx_j + ky_j + lz_j)} \tag{2.5}$$

where the summation is over the surface unit cell, and

$$F_{\text{bulk}} = \sum_{j=-\infty}^{0} F_u e^{2\pi i l j} e^{j\alpha} \tag{2.6}$$

with

$$F_u = \sum_j f_j e^{-B_j Q^2/(16\pi^2)} e^{2\pi i(hx_j + ky_j + lz_j)} \tag{2.7}$$

where the summation is over the bulk unit cell.

Here $h$, $k$, and $l$ are the Miller indices of the scattering surface, $Q$ the total momentum transfer, $B_j$ the Debye–Waller factor, and $f_j$ the atomic scattering factor, of atom $j$ at unit cell position $(xyz)_j$; $\theta_j$ is an occupancy parameter to allow for partially filled surface sites; and $\alpha$ is a parameter that is adjusted for sample X-ray penetration. The summation for the bulk is done from the top of the bulk region to –infinity using bulk unit cells. The summation for the surface uses the defined surface cells. Substitution of $F_u$ into $F_{\text{bulk}}$ yields

$$F_{\text{bulk}} = \frac{F_u}{1 - e^{-2\pi i l} e^{-\alpha}} \tag{2.8}$$

which is the parametrized form of the measured scattered intensity. These calculations and refinement are done with software such as ROD [30].

Figure 2.2*a* (bottom) shows a reciprocal space image of rod intensity and differentiates two types of rods: specular and nonspecular. The specular rod is measured in a symmetric diffraction geometry (Fig. 2.2*a*, top) by varying $Q_z$ and extends from the origin of reciprocal space through Bragg reflections of type (00*L*). These rods hold atomic positional information on the *z* direction but no lateral *x*,*y* information. Measurement along these types of rods is called "x-ray reflectivity" and is often reported normalized to the incident intensity, and hence as a fraction of the X-ray photon beam reflected. All other rods are nonspecular and require asymmetrical data collection geometry. Details of how this is done and a YouTube movie of nonspecular rod scanning are available [28, 31].

## 2.6 SFVS STUDIES OF OXIDE–WATER INTERFACES

### 2.6.1 Background

Sum frequency generation is a coherent nonlinear optical process in which two input beams at frequencies $\omega_1$ and $\omega_2$ interact in a medium and generate an output beam at $\omega_1 + \omega_2$. Sketched in Figure 2.3*a* is a case where the sum frequency (SF) wave

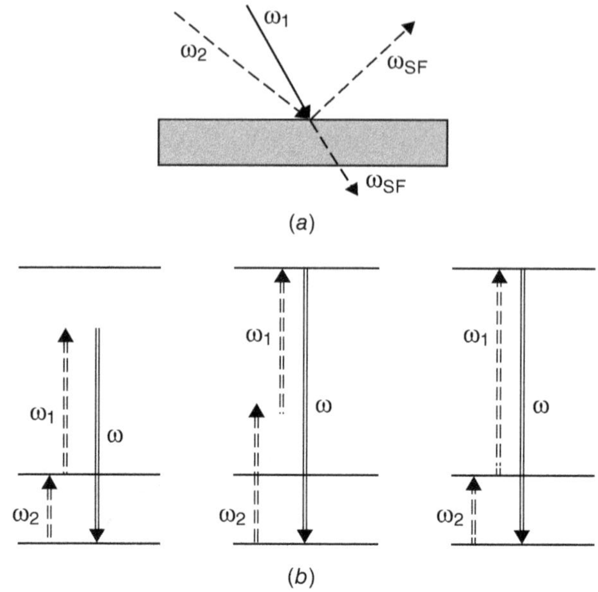


**Figure 2.3** (*a*) Geometry for SFG generation in the reflected direction. (*b*) Three possible SFG resonant cases. From left, $\omega_2$ on resonance, $\omega_1$ on resonance, $\omega_1$ and $\omega_2$ on resonances.

generated from an interfacial system appears in the reflected direction. Assume that the input fields $\vec{E}(\omega_1)$ and $\vec{E}(\omega_2)$ jointly induce a surface nonlinear polarization $\vec{P}_s^{(2)}(\omega=\omega_1+\omega_2)=\vec{\chi}_s^{(2)}:\vec{E}(\omega_1)\vec{E}(\omega_2)$ in the interfacial layer and a bulk nonlinear polarization $\vec{P}_B^{(2)}(\omega=\omega_1+\omega_2)=\vec{\chi}_B^{(2)}:\vec{E}(\omega_1)\vec{E}(\omega_2)$ in the bulk medium, with $\vec{\chi}_S^{(2)}$ and $\vec{\chi}_B^{(2)}$ denoting surface and bulk nonlinear susceptibilities, respectively. It can then be shown that the SF signal has the expression [32, 33]

$$S(\omega)\propto\left|\vec{P}_s^{(2)}(\omega)+\vec{P}_B^{(2)}(\omega)/\Delta k\right|^2=\left|\vec{\chi}_S^{(2)}(\omega)+i\vec{\chi}_B^{(2)}(\omega)/\Delta k\right|^2\left|\vec{E}(\omega_1)\right|^2\left|\vec{E}(\omega_2)\right|^2 \quad (2.9)$$

where $\Delta\vec{k}\equiv\vec{k}_1(\omega_1)+\vec{k}_2(\omega_2)-\vec{k}(\omega)$ is the wave-vector mismatch of the input and output waves. As illustrated in Figure 2.3*b*, when $\omega_1(\omega_2)$, or $\omega_1+\omega_2$, or both approach surface (bulk) resonances, $\vec{\chi}_S^{(2)}\left(\vec{\chi}_B^{(2)}\right)$ will be resonantly enhanced. SFG measurement on the system allows us to deduce $\vec{\chi}_S^{(2)}(\omega)$ and $\vec{\chi}_B^{(2)}(\omega)$, thus providing spectral information about the interface and the bulk.

For a medium without inversion symmetry, $\vec{\chi}_B^{(2)}$ is generally allowed under the electric-dipole approximation and $\left|\vec{\chi}_B^{(2)}/\vec{\chi}_S^{(2)}\right|\sim 1/d$, with $d$ being the interfacial layer thickness. Since the interface is composed of only few atomic or molecular layers and $1/|\Delta k|$ is of the order of a wavelength, we find $\left|\vec{\chi}_B^{(2)}(\omega)/\Delta k\vec{\chi}_S^{(2)}(\omega)\right|>>1$. The overwhelming contribution of the bulk contribution to SFG makes deduction of the surface $\vec{\chi}_S^{(2)}(\omega)$ extremely difficult. However, bulk and surface of a medium generally have different structural symmetry such that under the electric-dipole approximation, selective tensor elements of $\vec{\chi}_B^{(2)}$ may vanish, but those of $\vec{\chi}_S^{(2)}$ will not. One can use specific beam geometry and polarizations to access such tensor elements and will then find the bulk contribution greatly suppressed and possibly negligible in many cases. The latter is true for the oxide–water interfacial systems that have been investigated. We shall therefore neglect $\vec{\chi}_B^{(2)}$ in the discussion in the rest of the chapter.

With $\omega_2$ tunable over vibrational resonances in the infrared, SFG can yield vibrational spectra of an interface described by $\vec{\chi}_S^{(2)}$. In fact, SFG is currently the only technique that can generate vibrational spectra for buried interfaces. The surface nonlinear susceptibility $\vec{\chi}_S^{(2)}$ can yield molecular level-information about an interface as it is the averaged sum of nonlinear polarizability of molecules per unit area in the interfacial layer with the expression

$$\begin{aligned} \vec{\chi}_S^{(2)} &= N_S < \vec{\alpha}^{(2)} > \\ &= \vec{\chi}_{\mathrm{NR}}^{(2)} + \sum_q \frac{\vec{A}_q}{(\omega_{\mathrm{IR}} - \omega_q) + i\Gamma_q} \\ &= N_S \left[ < \vec{\alpha}_{\mathrm{NR}}^{(2)} > + \sum_q \frac{< \vec{a}_q >}{(\omega_{\mathrm{IR}} - \omega_q) + i\Gamma_q} \right] \end{aligned} \tag{2.10}$$

Here, the nonresonant contribution is denoted by $\vec{\chi}_{\mathrm{NR}}^{(2)} = N_S < \vec{\alpha}_{\mathrm{NR}}^{(2)} >$ with $N_s$ being the surface density of molecules, and the resonant contribution is assumed to come from discrete vibrational resonances with resonant frequencies $\omega_q$, amplitude $\vec{A}_q$, and damping constant $\Gamma_q$ for the $q$th mode. The angular bracket represents an orientation average such that

$$\vec{A}_q = N_S < \vec{a}_q > = N_S \int \vec{a}_q(\Omega) f(\Omega) d\Omega \tag{2.11}$$

where $\vec{a}_q$ is the resonant amplitude of the $q$th mode from an individual molecule, and $f(\Omega)$ is the orientation distribution function of the molecules contributing to the $q$th mode. Note that if $f(\Omega)$ has inversion symmetry, than $\vec{A}_q$ will vanish; $|\vec{A}_q|$ increases with increase of polar ordering in $f(\Omega)$. Explicitly in terms of lab coordinates $(i, j, k)$ and molecular coordinates $(\xi,\eta,\zeta)$, the tensor elements of $\vec{\chi}_S^{(2)}$ and $\vec{\alpha}^{(2)}$ are related by

$$\chi_{S,ijk}^{(2)} = N_S \sum_{\xi,\eta,\zeta} < (\hat{i} \cdot \hat{\xi})(\hat{j} \cdot \hat{\eta})(\hat{k} \cdot \hat{\zeta}) > a_{q,\xi\eta\zeta}^{(2)} \tag{2.12}$$

Similarly, we have

$$A_{q,ijk}^{(2)} = N_q \sum_{\xi,\eta,\zeta} < (\hat{i} \cdot \hat{\xi})(\hat{j} \cdot \hat{\eta})(\hat{k} \cdot \hat{\zeta}) > a_{q,\xi\eta\zeta} \tag{2.13}$$

The various tensor elements of $\chi_{S,ijk}^{(2)}$ or $A_{q,ijk}$ can be deduced from SFVS using selected input–output beam polarizations, such as SSP (denoting S-polarized SF output, S-polarized visible input, and P-polarized infrared input, respectively), SPS, PPP, and so forth. If the resonances form a continuum band, the sum over $q$ in Eq. (2.10) should be replaced by an integral [34]:

$$\vec{\chi}_S^{(2)} = \chi_{\mathrm{NR}}^{(2)} + \int \frac{\vec{A}_q \rho(\omega_q)}{\omega_{\mathrm{IR}} - \omega_q + i\Gamma_q} d\omega_q \tag{2.14}$$

with $\rho(\omega_q)$ denoting the density of resonance modes at $\omega_q$. The imaginary part of $\overleftrightarrow{\chi}_S^{(2)}$ has the expression

$$\mathrm{Im}\,\overleftrightarrow{\chi}_S^{(2)} = \frac{\overleftrightarrow{A}_q \rho(\omega_q)}{\pi}$$

which directly characterizes the continuum resonance.

One important piece of information SFVS can often provide is on the orientation of a molecular moiety at an interface [32, 33]. As seen in Eq. (2.12) or (2.13), if the number of independent, nonvanishing elements of $A_{q,\,ijk}$ is larger than that of $a_{q,\,\xi\eta\zeta}$ (or $a_{q,\,\xi\eta\zeta}$ are known by other means), then we can deduce orientation information about the moiety that contributes to the $q$th resonance mode from measurement of $A_{q,ijk}$. Take, for example, OH on a surface with $C_{1V}$ symmetry [35]. There are 10 independent, nonvanishing $A_{q,ijk}$ for this surface symmetry—$A_{q,xxz}$, $A_{q,yyz}$, $A_{q,zzz}$, $A_{q,xzx} = A_{q,\,zxx}$, $A_{q,yzy} = A_{q,zyy}$, $A_{q,xxz}$, $A_{q,yyx}$, $A_{q,zzx}$, $A_{q,xyy} = A_{q,yxy}$ and $A_{q,xzz} = A_{q,xzx}$—but only two independent nonvanishing $a_{q,\alpha\beta\gamma}$ for the OH stretch mode: $a^{(2)}_{q,\xi\xi\xi}$ and $a^{(2)}_{q,\eta\eta\xi} = a^{(2)}_{q,\zeta\zeta\xi}$. Deduction of these 10 nonvanishing $A_{q,ijk}$ from SFVS, in principle, can allow us to determine not only the two nonvanishing $a_{q,\xi\eta\zeta}$ but also a set of 8 parameters specifying the orientation distribution function for OH. If we assume a distribution function specified by only a few parameters, then these parameters are "overdetermined." Knowing orientations of hydroxyl groups on a protonated or H-passivated oxide surface can help determine the structure of the oxide surface, as we shall see.

### 2.6.2 Experimental Arrangement

The schematic of a typical experimental setup for SFVS is shown in Figure 2.4 [32]. Essential for the system is a tunable coherent infrared (IR) source. In the system of Figure 2.4, it is generated by an optical parametric generator/amplifier together with a difference frequency generation stage pumped by a picosecond mode-locked neodymium-doped yttrium aluminum garnet (Nd:YAG) laser. With proper nonlinear optical crystals for difference frequency generation, the tuning range can be extended from near IR to ~16 μm in the mid-IR. Unfortunately, tuning beyond 16 μm is difficult, thus limiting SFVS studies to vibration or phonon modes above ~650 cm$^{-1}$. As depicted in Figure 2.4, the SFG measurement is carried out by directing the tunable IR and 0.53-μm pulses to overlap at the sample surface or interface, with the SF output in reflection going into a photodetection system after proper filtering. Input and output polarizations are selected by polarizers placed appropriately in the beam path. Such a system is now commercially available. For studies of oxide–water interfaces, the input beams usually access the interfaces from the oxide side to avoid strong IR attenuation in water.

The SF signal is proportional to $|\overleftrightarrow{\chi}_S^{(2)}|^2$, but $\overleftrightarrow{\chi}_S^{(2)}$ is complex near resonances, and it is known that Im $\overleftrightarrow{\chi}_S^{(2)}$ is particularly important as it directly characterizes resonances in a way similar to the imaginary part of the optical dielectric constant for a medium. For complete determination of $\overleftrightarrow{\chi}_S^{(2)}$, we need to also measure the phase of $\overleftrightarrow{\chi}_S^{(2)}$. This can be achieved by an interference measurement [36]. As described in the

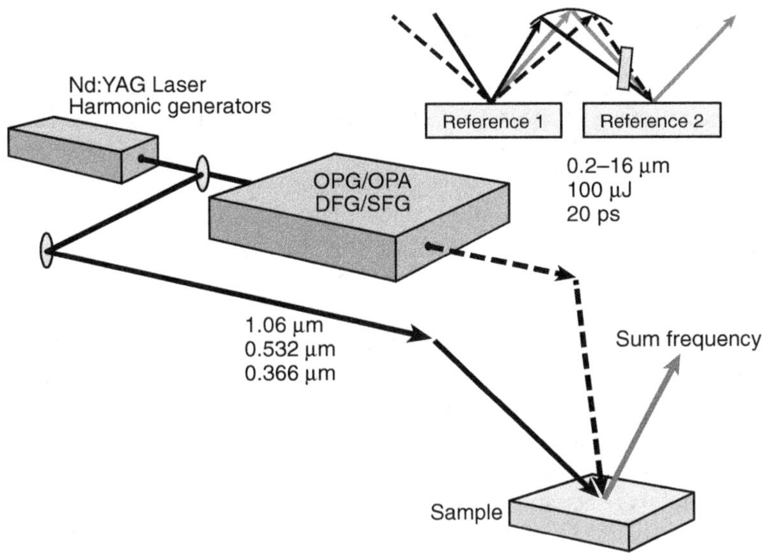


**Figure 2.4** Schematic diagram of typical SFG experimental arrangement. Inset shows geometry for a phase measurement where reference 1 is a sample and reference 2 provides a comparison phase. A variable phase plate in the SF beam produces an interference pattern at the detector from which the phase relative to the reference can be determined.

inset of Figure 2.4, the SF output from the sample is made to interfere with the SF output from a reference, and the relative phase of the two can be varied by inserting a variable-phase plate in the beams between the sample and the reference. Varying the relative phase of the two SF waves generates SF interference fringes, the measurement of which allows deduction of the relative phase of $\vec{\chi}_S^{(2)}$ with respect to that of the reference. Details on the experimental arrangement and spectral analysis of the phase-sensitive SFVS can be found in Ji and Ostroverkhov [36].

## 2.7 SFVS AND SURFACE X-RAY SCATTERING RESULTS ON DIFFERENT OXIDES

Sum frequency vibrational spectra have been applied so far to only a few oxide–air and oxide–water interfaces, mostly on fused silica, quartz, and alumina. In the following, we shall focus on SFVS studies of fused silica, alumina, and their water interfaces and only briefly mention a few others. CTR and X-ray reflectivity studies have been applied more widely, but only in the case of the $\alpha$-$Al_2O_3$ have both X-ray and SFG studies been conducted on similar samples. Some of the most detailed CTR work has been done on $\alpha$-$Fe_2O_3$ samples, whose bulk structure is identical with $\alpha$-$Al_2O_3$, and those studies are noted here for comparison.

### 2.7.1 Fused and Crystalline Quartz Interfaces

The first SFVS measurement on oxide–water interface was carried out on fused quartz–water interface in 1994 [37]. The work showed how the interfacial water structure responded to pH variation and salt in bulk water. It was a clear demonstration of the capability of SFVS to probe buried liquid–solid interfaces. Many publications have appeared on the subject since then, but most of them studied how adsorbed molecules on the silica surface can modify the surface properties and accordingly the interfacial water structures. Here, we shall focus only on unmodified quartz surfaces and interfaces.

To our knowledge, the only SFVS measurement on air–quartz interfaces is our previous work [38]. As shown in Figure 2.5, the spectrum of the fused quartz surface in air exhibits a 975 $cm^{-1}$ mode, which can be assigned to Si–OH stretch vibration because replacing H partly by octadecyltrichlorosilane (OTS) molecules greatly suppresses the mode. The α-crystalline quartz has a trigonal structure with $D_3$ point symmetry, which is noncentrosymmetric, and therefore the bulk SFG is allowed. To probe its surface with SFVS, the bulk contribution to SFG must be suppressed. Fortunately, for the (0001) surface, it has an effective sixfold rotational symmetry. Furthermore, with the beam incidence plane at an angle of $\pi/6 \pm n\pi/3$ with respect to the $\hat{a}$ axis as well as beam polarizations of SSP or PPP, the bulk contribution to SFG is forbidden, thus permitting detection of surface SFG. As shown in Figure 2.5, the SSP spectrum of α-crystalline (0001) quartz surfaces exhibits, in addition to the

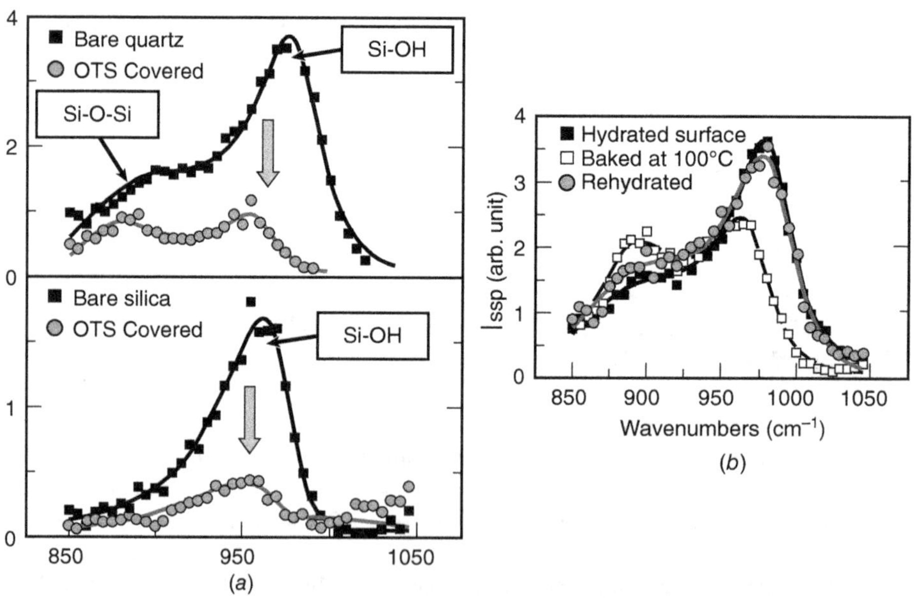


**Figure 2.5** (*a*) Comparison of quartz interface SFG with that from fused silica. (*b*) SFG of quartz showing effect of a cycle of hydration, drying, and rehydration.

975 $cm^{-1}$ mode, another mode at 870 $cm^{-1}$, which must be associated with the Si–O–Si surface structure as it is not significantly affected by replacement of H by OTS. It is seen in Figure 2.5 that annealing of the sample desorbs H and converts Si–OH to Si–O–Si as one would expect. The observed frequency of 870 $cm^{-1}$ is blue shifted from the bulk Si–O–Si symmetric phonon frequency at 795 $cm^{-1}$, suggesting that the surface has been reconstructed from the bulk-terminated structure. Although various techniques such as X-ray reflectivity [39], AFM [40–42], low-energy electron diffraction (LEED) [43], and He gas diffraction have been used to investigate various quartz surfaces [44], no clear picture on the surface structure of α-quartz (0001) has come out of such investigations.

There exist many model calculations on surface structures of quartz. The one on the unhydrated α-quartz (0001) surface that seems to predict the SFVS result the best is an ab initio calculation that finds the surface reconstructed (depicted in Figure 2.6) with the Si–O bond length and the O–Si–O bond angle the same as in the bulk, but the Si–O–Si angle at 130° different from the bulk value of 143.7° [45]. The surface Si–O–Si phonon frequency can be estimated using this reconstructed surface structure and adopting the force constants for stretch and bending of Si–O–Si in the bulk [46]. It was found to be 880 $cm^{-1}$, which is close to the observed frequency of 870 $cm^{-1}$. This lends support to the proposed reconstructed surface structure of α-quartz (0001). Upon hydration, the local surface structure will relax, but, in any case, the bonding geometry of Si–OH makes its OH lie close to the surface plane, making the stretch OH vibration difficult to observe in SFVS because of symmetry.

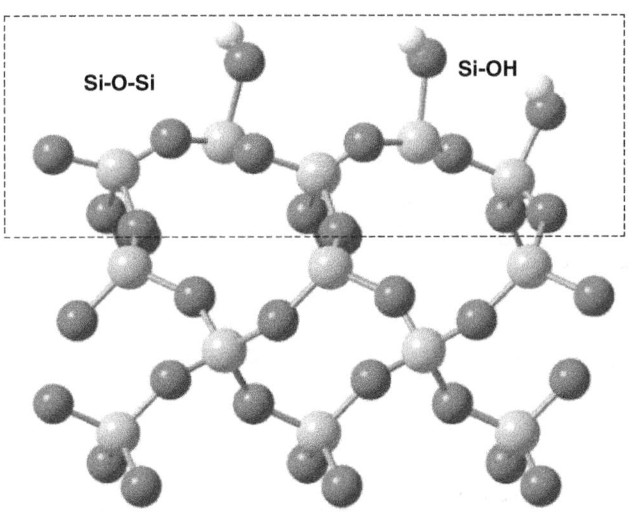


**Figure 2.6** Hypothetical structure of the unhydrated quartz (0001) surface from ab initio calculations. Darker atoms are oxygens. This structure agrees well with SFG observations and vibrational frequencies.

If immersed in water, the surface structure of quartz may undergo two changes: relaxation of the reconstructed surface structure and deprotonation of the silanol (SiOH) groups governed by the reaction SiOH ↔ $SiO^-$ + $H^+$. (Protonation reaction SiO + $H^+$ ↔ $SiOH^+$ may also occur, but only at very low pH as quartz surfaces are known to be neutral near pH~2 [47]. X-ray reflectivity measurements on $(10\bar{1}0)$ and $(10\bar{1}1)$ surfaces of quartz covered by a thin film of water detected vertical relaxation of surface Si and O atoms and an ordered monolayer of water adsorbed at the interface [48]. Unfortunately, similar measurements on the (0001) surface have not yet been reported. On the other hand, SFVS has so far only been applied to the (0001) quartz–water interface [49]. It has provided information on how the degree of deprotonation of the silanol groups that partially cover the surface varies with pH. In this respect, the surfaces of fused and crystalline quartz are at least qualitatively similar except that the interfacial water structure may appear more ordered at the crystalline interface [37, 49]. Figure 2.7 displays the $\left|\overleftrightarrow{\chi}_S^{(2)}\right|^2$ spectra of a fused quartz–water interface with different pH in the OH stretch range. The spectra originate from interfacial water; contribution from OH of the silanol groups is negligible as we mentioned earlier. It is seen that at high pH, the spectral intensity is appreciably higher. Because the surface silanol groups are nearly fully ionized at high pH, the quartz surface is strongly negatively charged, creating an EDL in the near-surface water with a field of ~$10^7$ V/cm. Such a field is strong enough to reorient water

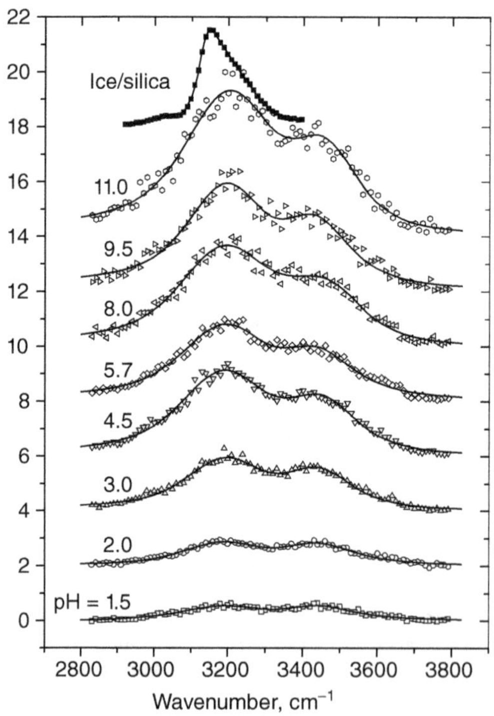


**Figure 2.7** Fused silica–water interface SFG spectra as a function of pH. The ice–silica spectrum is shown for comparison.

molecules in the EDL with their dipoles toward the surface, and the increased polar orientation of the water molecules greatly enhances the SF signal. Preferred H bonding of water molecules with H bonded to O at the surface also helps the polar orientation. At mid and low pH, less deprotonation of SiOH leads to less surface charge and surface field and corresponding weaker polar orientation of interfacial water molecules. Accordingly, the spectral intensity decreases as seen in Figure 2.7.

It has been realized that the continuum resonance spectrum of $|\overleftrightarrow{\chi}_S^{(2)}|^2$ for water is difficult to interpret because of interference between Re $\chi_S^{(2)}$, Im $\chi_S^{(2)}$, and the nonresonant $\chi_{NR}^{(2)}$. Figure 2.8*a* presents a set of Im $\chi_S^{(2)}$ spectra at different pH for the α-quartz (0001)–water interface obtained from phase-sensitive SFVS [50]. Positive Im $\chi_S^{(2)}$ at a given frequency means that the OH of water species contributing to the spectrum at that frequency has a net polar orientation of O→H toward the interface. Detailed analysis finds that each spectrum can be decomposed into two bands, a weaker icelike band at lower frequency and a stronger liquidlike band at higher

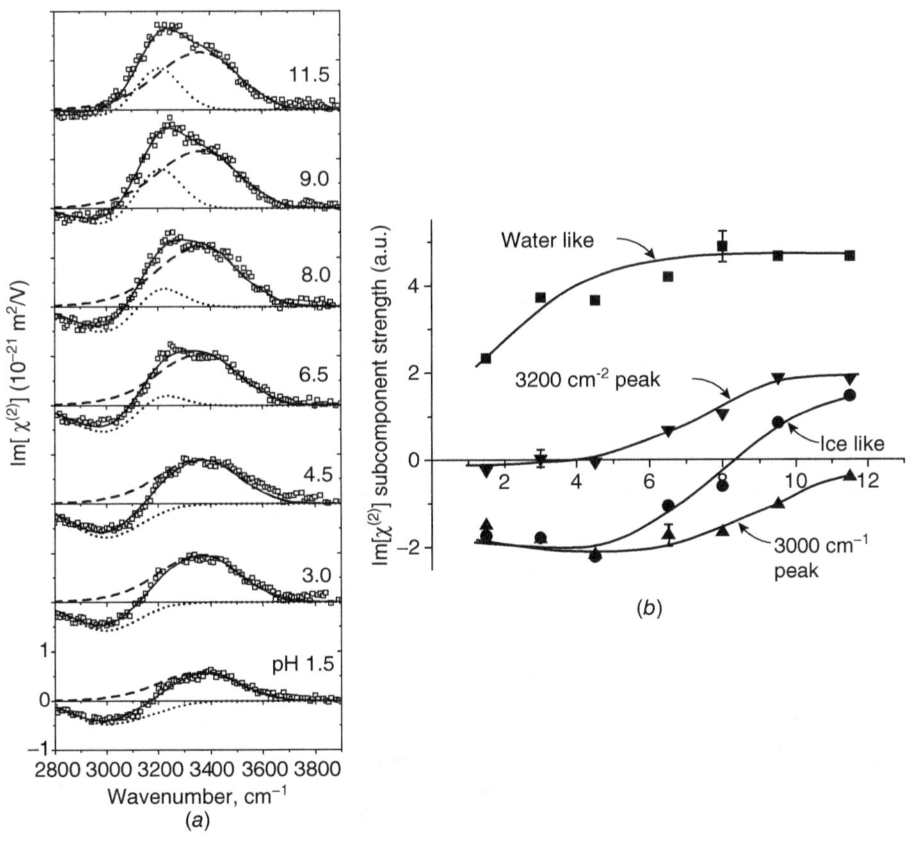


**Figure 2.8** (*a*) Decomposition of the imaginary part of the SFG signal into three contributions with different phase. Two bands at 3200 and 3000 cm$^{-1}$ make up the icelike resonance. (*b*) Trend of the intensity of these contributions with pH.

frequency; the icelike band can further be decomposed into two small subbands. Variation of the strengths of these bands with pH is displayed in Figure 2.8*b*, which shows that the icelike and liquidlike bands have very different dependences on pH. This suggests that there are probably two groups of water species associated with two distinct surface SiOH groups, one with a pK value of ~8 and the other a pK ~2. The icelike species seems to be associated with SiOH that is more difficult to deprotonate.

The observation of two pK values for the quartz surface agrees with what was observed on a fused quartz–water interface using second-harmonic generation as a probe [51]. It was found that the density of SiOH associated with the lower pK is smaller, and hence suspected that there are two types of surface silanol: isolated and H-bonded SiOH associated with lower and higher pK, respectively. Other SiOH geometries have also been proposed. More recently, it has been suggested that the lower pK originates from SiOH on sparse patches of strained surface [52]. However, the SFVS results in Figure 2.8*a* show clearly that the liquidlike band is stronger than the icelike band, indicating that the SiOH groups with lower pK must have a larger surface density. Apparently, the complexity of the surface structure of quartz makes theoretical calculations difficult, and our understanding of the quartz surface is yet to be improved.

## 2.7.2 Alumina Surfaces and Interfaces

Among the few crystalline oxides that have been probed by SFVS, alumina (or the mineral corundum, and gem-variety sapphire) is the one most extensively investigated. Crystalline alumina has trigonal $D_{3d}$ symmetry, which is centrosymmetric. Three surfaces of alumina, (0001), $(1\bar{1}02)$, and $(11\bar{2}0)$, have been most commonly studied. These have also been probed by SFVS and with X-ray scattering, with the goal to determine the exact nature of the terminated solid surface, and whether the different surface structures of the same crystal interact differently with water and have different charging properties, including the points of zero charge. In fact, the first experiment employing SFVS to determine the pzc of a solid–water interface was on alumina [53], but unfortunately the researchers did not specify which crystalline surface of alumina was used in the experiment. In the following, we describe the SFVS and X-ray studies on the three faces of alumina separately.

### C-Plane (0001) Surface of $\alpha$-$Al_2O_3$

One of the earliest CTR studies of an oxide surface with water present was that of Eng et al. [54] on $\alpha$-$Al_2O_3$ (0001), a surface having threefold symmetry. The key finding of this work was that the surface is terminated by only one kind of oxygen that is relaxed by 21% compared to the bulk structure and that the underlying Al atoms are also shifted significantly from their bulk positions, more closely resembling the positions of Al in the gibbsite structure (Fig. 2.9). Valence bond calculations indicated that the surface oxygens, part of the $Al_2O$ functional group, must be

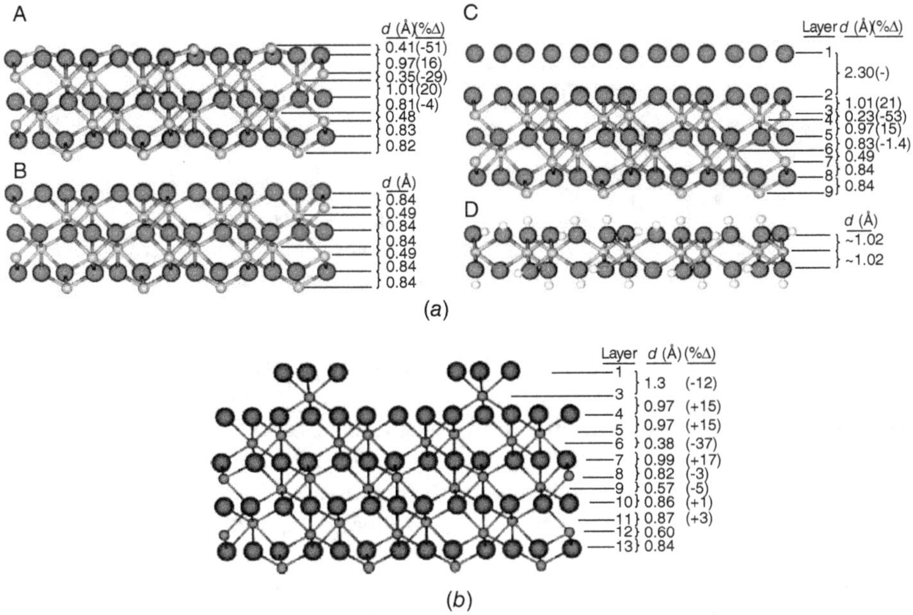


**Figure 2.9** (*a*) CTR-analysis-derived structure for the α-$Al_2O_3$ (0001) surface in equilibrium with moist air compared with other structural models. (A) Al-terminated surface model from vacuum studies. (B) Slab of analogous bulk α-$Al_2O_3$ structure. (C) CTR analysis result including overlaying water layer oxygens. (D) Slab of the bulk gibbsite structure [$Al(OH)_3$]. (*b*) Hematite (0001) surface from CTR data. Single domain model fit. Surface is terminated by both OFe and $OFe_2$ functional groups.

singly protonated. A layer of oxygens, presumed to represent a water layer, was needed for the best fit of the CTR data and had a distance from the surface oxygens of 2.3 Å. X-ray reflectivity of the same surface, with similar preparation treatment, is also reported [55]. Results mainly agree with the earlier CTR work but show surface water oxygens having a wider distribution along the $z$ direction than on other corundum surfaces, which was attributed to "weaker ordering" on this surface. However, as X-ray reflectivity has no lateral information, a wide distribution in $z$ may also be due to different types of surface binding along $x,y$ and hence not necessarily less organization. DFT calculations for this surface have also been performed and indicate surface $Al_2O$ functional groups [56].

In air, the oxygens of the $Al_2O$ group can be passivated by H to form $Al_2OH$. As a dangling OH in air, its stretch vibration is expected to have a frequency close to 3700 $cm^{-1}$, as in the air–water case. SFVS measurement of the α-$Al_2O_3$ (0001) surface indeed found a single sharp peak at 3700 $cm^{-1}$, and its polarization dependence indicates that OH is tilted at ~26° from the surface normal, close to the angle of 30° predicted by the bulk-terminated model (Fig. 2.10) [57]. Thus, the result supports the proposed relaxed bulk-terminated model for α-$Al_2O_3$ (0001).

Bonding of H to O in $Al_2OH$ is very strong, as it cannot be removed readily even when the sample is heated to 600°C. The 3700 $cm^{-1}$ OH stretch mode remains

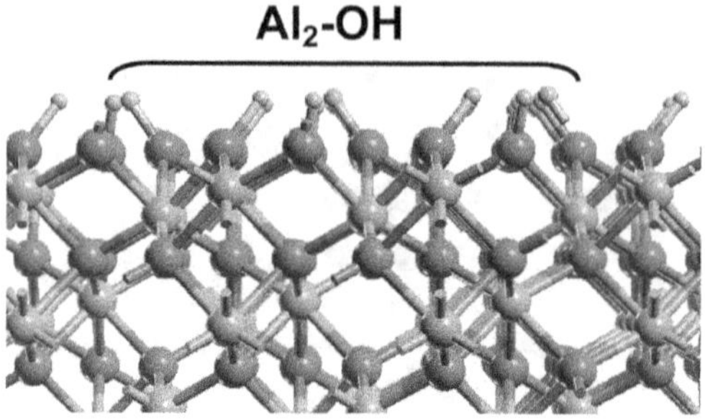


**Figure 2.10** Side view of proposed alumina (1000) interface showing $Al_2OH$ hydroxyl orientations deduced from SFG measurements.

prominent in the spectra of the α-$Al_2O_3$ (0001)–water interface (Fig. 2.11). This indicates that the dangling OH is not H bonded to a water molecule through H of $Al_2OH$. The band is actually narrower than that in the spectrum of α-$Al_2O_3$ (0001) in air, suggesting that the surface in water may have its strain relieved. The α-$Al_2O_3$ (0001) surface in contact with bulk water probably has essentially the same terminated structure with vertical relaxation as the air-equilibrated surface. It is observed that only if the pH of the bulk water solution rises above 8.2 does the dangling OH mode strength decrease appreciably. The pK value for the reaction $Al_2OH \leftrightarrow Al_2O^- + H^+$ has been estimated to be 9.7 [58]. The α-$Al_2O_3$ (0001)–water interface also exhibits two additional bands at 3200 and 3450 $cm^{-1}$ in the $|\overleftrightarrow{\chi}_S^{(2)}|^2$ spectra for different pH, presented in Figure 2.11.

They are labeled as icelike and liquidlike bands, respectively, because the former resembles OH stretch vibration of water molecules with their H symmetrically donor bonded as in ice, and the latter resembles OH stretch vibration of water molecules with their H donor bonded in a more disordered environment as in liquid water. The Im $\chi_S^{(2)}$ spectra (two of which are given in Fig. 2.12*a*) obtained from phase-sensitive SFVS show that the amplitude $A_q$ of the liquidlike band is negative at all pH, similar to that of the dangling OH of $Al_2OH$, indicating a net polar orientation of O→H pointing into the liquid. This suggests that the contributing water molecules are H bonded with one of its H to O of $Al_2OH$ and another to the next-layer water molecules near the surface. The bonding geometry is rather disordered, thus leading to the liquidlike OH stretch band. The icelike band should come from water molecules beyond the top layer since the surface structure of α-$Al_2O_3$ (0001) does not allow symmetric donor bonding of water molecules to the surface. The net polar ordering of the contributing water molecules should come mainly from induced orientation by the surface field and is expected to be weak, nearly vanishing if the surface is neutral.

Fitting of the spectra in Figure 2.11 by Eq. (2.10) with the assumption of three discrete resonances at 3700, 3450, and 3200 $cm^{-1}$ allows us to deduce their resonant amplitudes $A_q$ at various pH. For the liquidlike and icelike bands, their $A_q$ versus pH are plotted in Figure 2.12*b*. It is seen that $A_q$ of the icelike band increases with pH from negative at low pH to positive at high pH and is nearly zero at mid-pH, while $A_q$ of the liquidlike band has a similar behavior except that it is finite and negative at all pH values. The result is a reflection of the charging

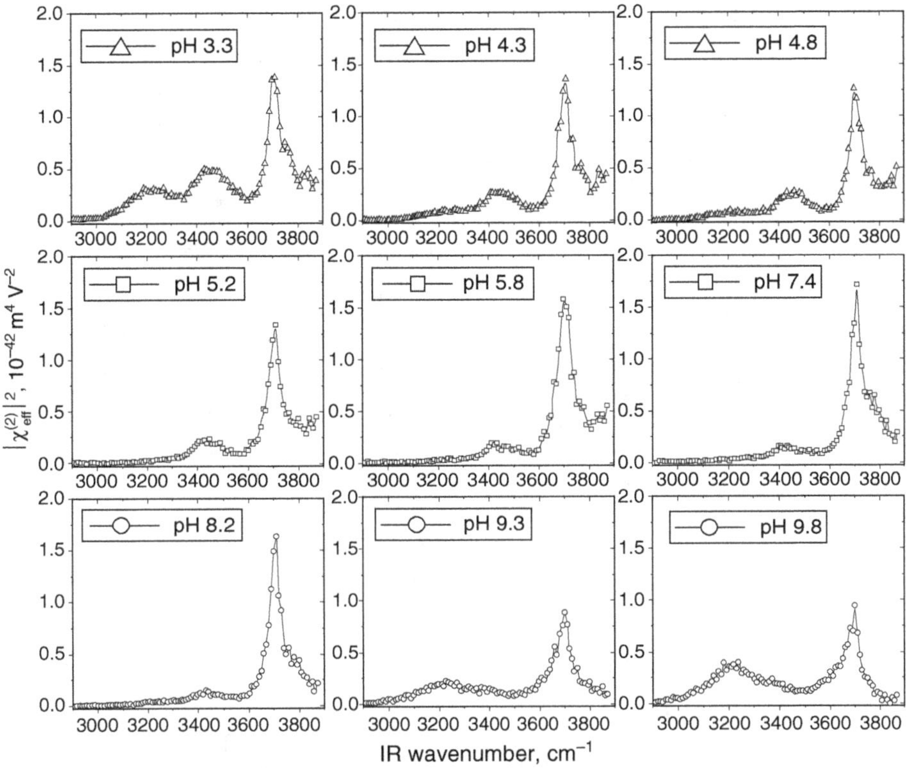


**Figure 2.11** SFG spectra for the alumina (0001) interface as a function of pH of the overlying water solution. The bands from left to right are "icelike" ca. 3200 cm$^{-1}$, "liquidlike" ca. 3450 cm$^{-1}$, and dangling O–H at 3700 cm$^{-1}$. The dangling O–H contribution is constant within error until pH exceeds 8.2.

behavior of the $\alpha$-$Al_2O_3$ (0001) surface, which can experience protonation reaction, $Al_2OH + H^+ \leftrightarrow Al_2OH_2^+$ with pK ~3, and deprotonation reaction, $Al_2OH \leftrightarrow Al_2O + H^+$ with pK = 9.7 [59]. At low pH, protonation dominates and the surface is positively charged. The surface field reorients the interfacial water molecules with O toward the interface. The protonated sites also tend to orient water molecules the same way by attracting them to H bond with O connected to H of $Al_2OH_2^+$. The result is that the amplitudes of both the icelike and the liquidlike bands become more negative with decreasing pH. The effect is reversed at high pH, where deprotonation dominates, and the surface is negatively charged. In the mid-pH range, the surface is neutral as protonation and deprotonation are balanced. The icelike band disappears because the contributing water molecules have no net polar orientation. The liquid-like band has finite negative amplitude because the water molecules H bonded to $Al_2OH$ are preferentially oriented with O toward the interface. From the plot of Figure 2.12*b*, the pzc for the $\alpha$-$Al_2O_3$ (0001)–water interface can be estimated to be at pH ~6.3. More quantitatively, at pzc, the concentration relation $[Al_2OH_2^+] = [Al_2O^-]$ holds. The

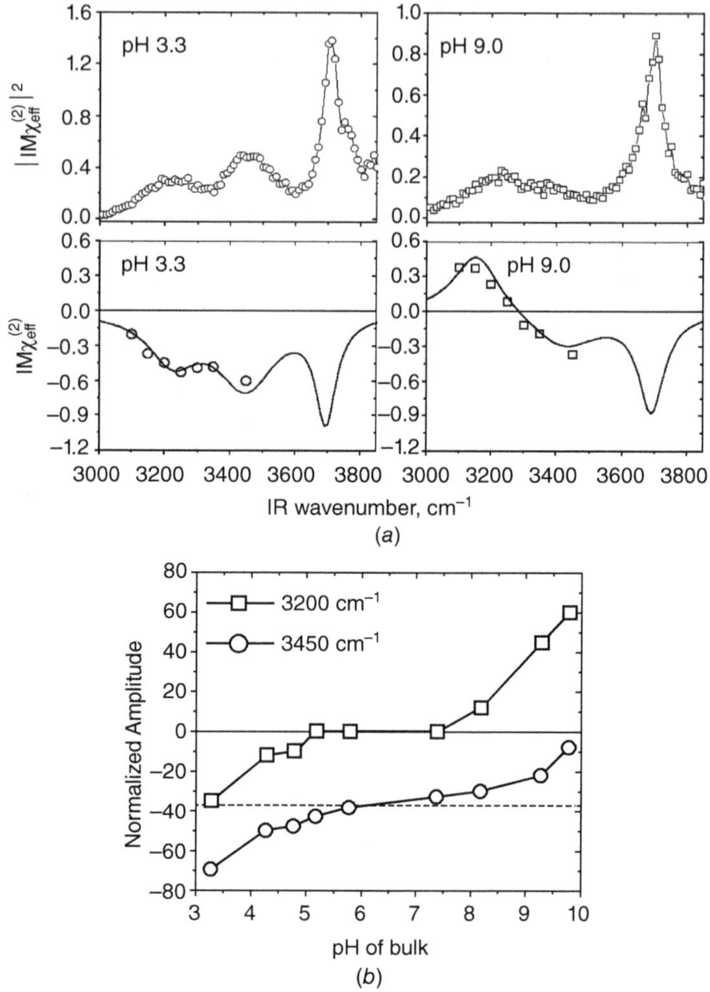


**Figure 2.12** (*a*) Imaginary nonlinear susceptibility for two pH values compared to the magnitude for the alumina (0001)–water interface. (*b*) Amplitude and sign dependence of the two water bands as a function of pH.

equilibrium constants for the protonation and deprotonation reactions are $K_P = [Al_2OH][H^+]/[Al_2OH_2^+]$ and $K_{DP} = [Al_2O^-][H^+]/[Al_2OH]$. One then finds $[H^+]^2 = (K_P)(K_{DP})$ and hence, $pH_{pzc} = (pK_P + pK_{DP})/2 = 6.35$. When the surface is neutral, salt ions dissolved in water will not accumulate at the interface to disturb the interfacial water structure. Indeed, no change was observed in the water interfacial spectrum when salt was added into water with pH around 6. Braunschweig et al. [59] studied a nanoscale roughened $\alpha$-$Al_2O_3$ (0001) surface in water using SFVS. They found a distinctly different pH dependence of the water OH bands compared to that of the smooth surface. It was believed to be due to the influence of deprotonation of hydroxyls appearing at the surface defect sites.

It is interesting to note that the net polar orientations of interfacial water molecules contributing to the liquidlike band in the cases $\alpha$-$Al_2O_3$ (0001)–water and $\alpha$-quartz (0001)–water interfaces are opposite in sign. When the surfaces are neutral, the former has preferentially O→H toward the liquid side, and the latter has O→H toward the surface. Recent calculation by Gaigeot et al. [60] shows that this should be the case because the different structures of the two surfaces lead to the different H bonding geometries of water molecules on the two surfaces. For both interfaces, however, more deprotonation from higher pH results in more O→H reoriented toward the surface as expected.

### R-Plane $(1\bar{1}02)$ Surface of $\alpha$-$Al_2O_3$

The $(1\bar{1}02)$ surface is corrugated due to its lower symmetry compared to (0001). It intersects perpendicularly with the glide plane $(\bar{1}\bar{1}20)$. CTR studies of both corundum and the Fe analog hematite show similarities with apparent nonstoichiometry of the surface, that is, layers of metal atoms are missing or low in occupancy in one model, which fits scattering data well [61, 62]. In surface science circles, the stoichiometric surface is that where the composition of the material is the same as the termination chemistry, including both top layers of cations and anions. Nonstoichiometry can result from having an excess oxygen or metal termination, as might be seen in water-equilibrated or UHV-equilibrated samples, respectively. Or the surface termination may be defective and deviate from the bulk structure in a significant way. This latter nonstoichiometry appears to occur on the $(1\bar{1}02)$ surfaces with a vacant Al layer in $Al_2O_3$ and a vacant Fe layer for $Fe_2O_3$. The differences in functional groups between these types of terminations and other termination models suggest the probable existence of unique assemblages of hydroxylation sites on the surfaces, and hence a way to verify the X-ray results using vibrational spectroscopy.

Figure 2.13 shows three possible models for the $(1\bar{1}02)$ surface, including the nonstoichiometric model with equal numbers of AlO, $Al_2O$, and $Al_3O$ functional groups (Fig. 2.13*a*), the stoichiometric model that has only $Al_3O$ groups (Fig. 2.13*b*), and the stoichiometric model with an additional oxygen layer to complete the Al coordination (Fig. 2.13*c*), having AlO and $Al_3O$ groups and suggested by X-ray reflectivity studies [63]. DFT calculations indicate that both the stoichiometric + oxygen layer termination and a relaxed nonstoichiometric termination should be stable in hydrated systems at room temperature, though they predict changes in termination at higher equilibration temperatures and the shift of Al into tetrahedral coordination at the surface [62, 64].

As these models all predict different functional groups and surface oxygen protonation states, we can see which model is more correct by using SFG spectroscopy. The stoichiometric bulk-terminated surface only allows one type of hydroxyl: $Al_3OH$. The missing-layer nonstoichiometric model allows four types: two from $AlOH_2$ and one each from $Al_2OH$ and $Al_3OH$. However, from the bond–valence

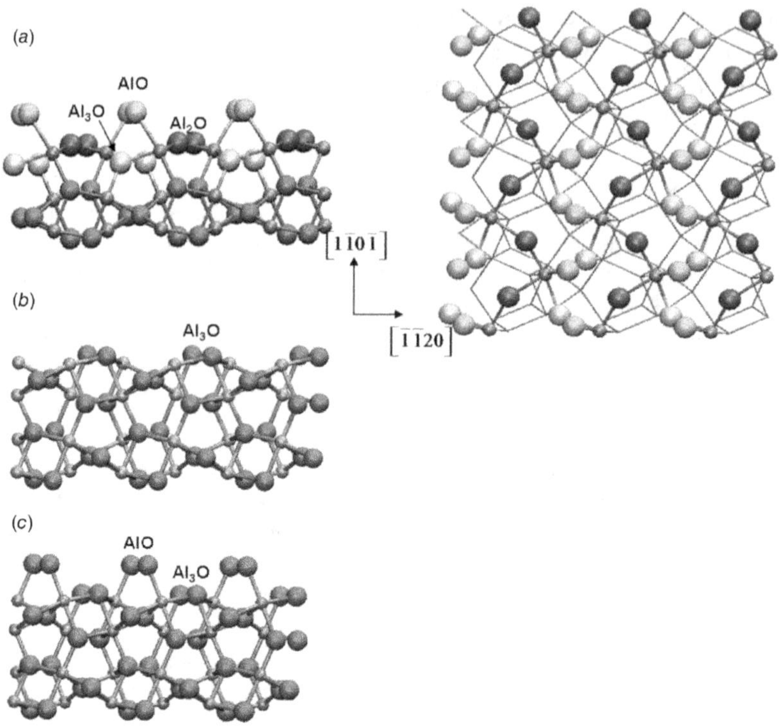


**Figure 2.13** Alumina ($1\bar{1}02$) surface termination models and characteristic functional groups. (*a*) Nonstoichiometric termination with missing Al layer and three possible functional groups. The surface is also shown on the right from above. (*b*) Stoichiometric termination model with one type of functional group. (*c*) Stoichiometric termination with an added O layer to complete the Al atom octahedral coordination. It has two types of functional groups.

argument, protonation of $Al_3O$ is not likely. The fully oxygen-terminated model can have three types of hydroxyl: two from $AlOH_2$ and one from $Al_3OH$, the latter again not being likely. Unfortunately, as the X-ray probes are not sensitive to protons, information on hydroxyls can only be deduced indirectly from observed bond length variations within known bond length systematics (e.g., the Al–O bond lengths are expected to increase in the series Al–O, Al–OH, Al–$OH_2$). On the other hand, fortunately, SFVS can detect hydroxyls and learn about their orientations by probing the OH stretch vibrations. The SFVS study of $\alpha$-$Al_2O_3$ ($1\bar{1}02$) has actually been used to prove that the missing-layer model is more reasonable, although quantitative parameters of the proposed CTR model still need to be adjusted.

Figure 2.14 shows the Im $\chi_S^{(2)}$ spectra of $\alpha$-$Al_2O_3$ ($1\bar{1}02$) taken with the incidence plane at three different azimuthal angles of $\gamma$ between the incidence plane and the direction [$1\bar{1}0\bar{1}$] in the glide plane ($1\bar{1}02$) [65]. They can be well fit by three discrete resonance modes at 3670, 3520, and 3365 cm$^{-1}$ using Eq. (2.10). This immediately indicates that the missing-layer model is most likely to be correct. From inspection

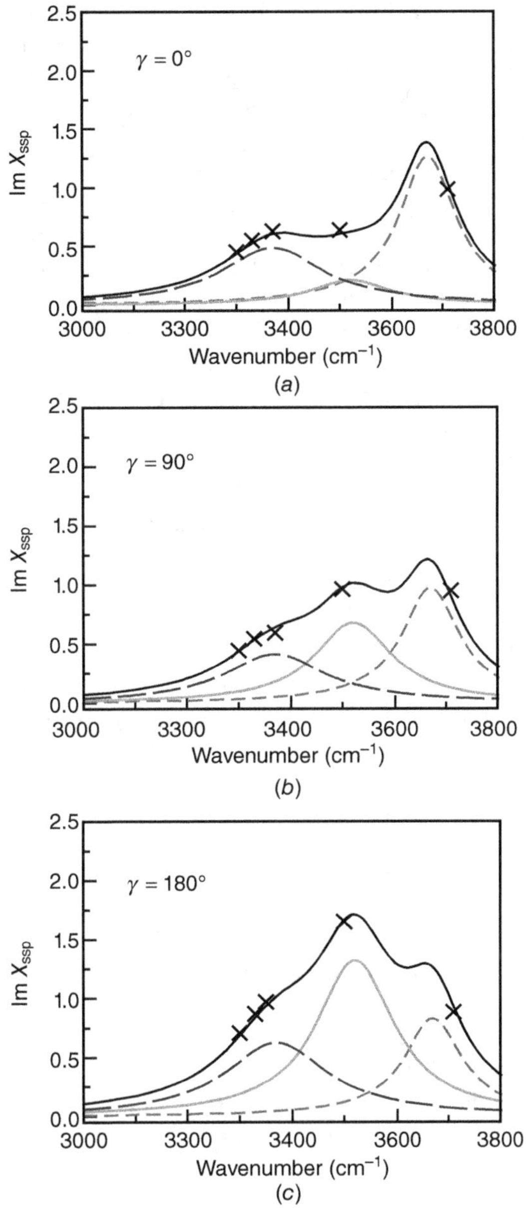


**Figure 2.14** Imaginary nonlinear amplitude functions for the hydroxyl vibrational range of the dry protonated alumina $(1\bar{1}02)$ interface taken at three different azimuthal angles (angle between incidence plane and $[1\bar{1}0\bar{1}]$ direction). Three resonances, corresponding to three types of surface hydroxyls, fit all spectra well.

of the model structure, the three modes can be identified, respectively, with the dangling OH on $AlOH_2$, OH on $AlOH_2$ H bonded to the oxygen of a neighboring $AlOH_2$, and OH of $Al_2OH$ H bonded to the oxygen of a neighboring $Al_3O$ group. However, in order for the assignment to be correct, the suggested H bonds must be strong enough to give rise to the observed red shifts of the OH stretch frequencies.

It was found that the missing-layer model constructed from the CTR study cannot create sufficiently strong H bonds for $AlOH_2$ and $Al_2OH$ with neighboring oxygens. The spacing between neighboring oxygen rows of $AlOH_2$ and layer separation between second and third oxygen layers must be adjusted to yield the expected H bonding strengths and corresponding red shifts and dictate the orientations of the surface OH groups. It is interesting to know that, as described earlier, measurement of how each mode varies in strength with the azimuthal angle $\gamma$ and the input–output polarization actually allows determination of the orientation of the corresponding OH group. The measured orientations can then be compared with the ones expected from the H bonding considerations. In the case of protonated $\alpha$-$Al_2O_3$ $(1\bar{1}02)$, it was shown that the two results agreed quite well, indicating that the adjusted missing-layer model is reasonable. This is a case demonstrating that SFVS can be complementary to X-ray probes to determine a crystalline surface structure.

The SF vibrational spectrum of the $\alpha$-$Al_2O_3$ $(1\bar{1}02)$–water interface is different and depends on pH because protonation and deprotonation of the hydroxyls at the surface can change the interfacial structure [66]. Figure 2.15 presents a representative set of Im $\chi_S^{(2)}$ spectra at different pH with $\gamma = 180°$. They can all be fit by two discrete resonance bands, one at 3230 $cm^{-1}$ and the other at 3490 $cm^{-1}$. The former can be identified as an icelike band of interfacial water molecules. The 3365 $cm^{-1}$ mode from OH of $Al_2OH$ at the surface is comparatively weak and appears negligible even when $Al_2OH$ is not deprotonated. This assignment is supported by the observation that the mode amplitude changes sign when the surface charge switches sign in Figure 2.15 (right side); interfacial water molecules can be reoriented by surface protonation or deprotonation, but hydroxyls fixed on the surface cannot. The 3230 $cm^{-1}$ mode amplitude versus pH is plotted in Figure 2.16*a*, and can be understood from the surface reactions, $Al_3O^{-0.5} + H^+ \leftrightarrow Al_3OH^{+0.5}$ and $AlOH_2^{+0.5} \leftrightarrow AlOH^{-0.5} + H^+$, estimated to have pK ~ 5.9 and between 9.9 and 11, respectively [67, 68]. At low pH, protonation of $Al_3OH$ dominates, and the surface is positively charged.

As we discussed earlier, interfacial water molecules tend to orient with O toward the protonated surface. At high pH, deprotonation of $Al_2OH$ dominates and the surface charge and water molecular orientation are reversed. The data in Figure 2.16*a* can be fit quite well by the kinetic equations for the two reactions with pK = 4.9 and 9.2, respectively, assuming that the mode amplitude (or net polar orientation of water molecules) is proportional to the surface charge density resulting from $Al_3OH^+$ and $Al_2O^-$. The solid curve in Figure 2.16*a* shows the fit. From the pK values, we can estimate the pzc for the $\alpha$-$Al_2O_3(1\bar{1}02)$–water interface to be ~7.

The 3490-$cm^{-1}$ band is believed to have contributions from both H-bonded OH of $AlOH_2$ and liquidlike interfacial water molecules. The former is red shifted by 30 $cm^{-1}$ from the same mode on the dry $\alpha$-$Al_2O_3$ $(1\bar{1}02)$ surface, and its amplitude has fixed sign corresponding to O→H pointing away from the surface. Because this OH is tilted toward the forward direction $[1\bar{1}0\bar{1}]$ on the surface, its amplitude exhibits forward ($\gamma = 0$)–backward ($\gamma = 180°$) asymmetry. The latter has amplitude that can change sign in response to sign switching of the surface field and is expected to have no forward–backward asymmetry. It is then possible to separate

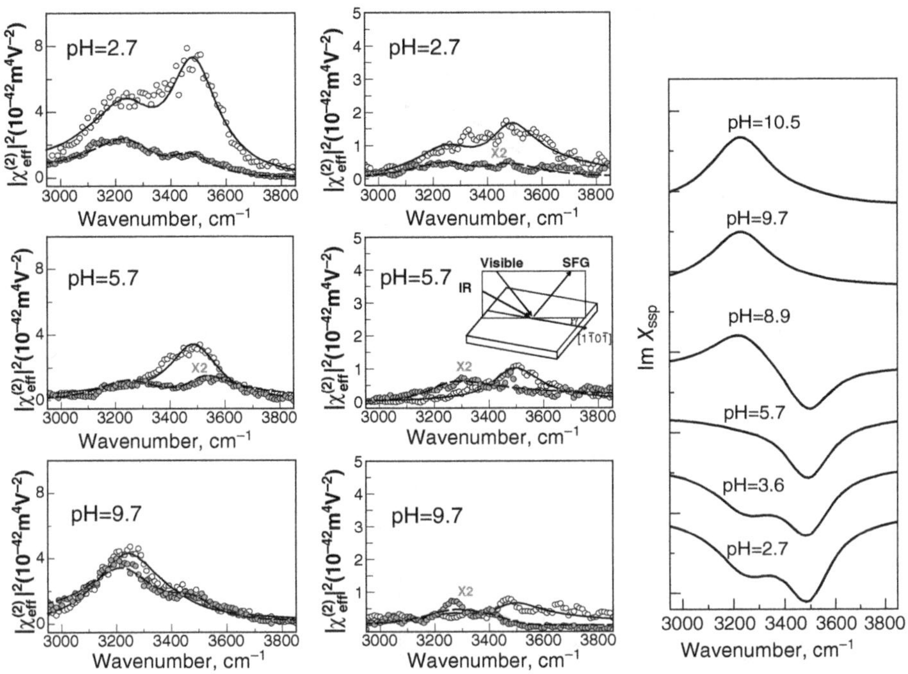


**Figure 2.15** (*Far left*) SSP polarization SFG spectra of the alumina $(1\bar{1}02)$ wet interface at azimuthal angles of 180 (open symbols) and 0 (filled symbols) degrees; (*center*) same spectra taken in SPS polarization. The spectra can be fit by two resonances at 3230 and 3490 cm$^{-1}$, identified as an icelike band, and a band composed of the two hydroxyls associated with the $AlOH_2$ surface functional group and a liquidlike water contribution. (*Right*) Imaginary spectra.

the two from the measured spectra at $\gamma = 0$ and 180°. The deduced amplitudes, $A_{OH}(\gamma = 0°) = \kappa A_{OH}(\gamma = 180°)$ for OH from $AlOH_2$ ($\kappa$ is a proportional constant) and $A_W$ from OH of water, are plotted versus pH in Figures 2.16*b*, *c*. The data for $A_{OH}(\gamma = 0°)$ and $A_{OH}(\gamma = 180°)$ can be fit by the concentration dependence of $AlOH_2$ on pH following the kinetic equation for deprotonation of $AlOH_2$: $K_{DP} = [AlOH^-][H^+] / [AlOH_2]$ with pK = 9.2 given earlier. The data for $A_W$ can be fit by the same curve that fits the data of $A_q$ versus pH for the 3230-cm$^{-1}$ band, assuming $A_W$ is proportional to surface charge density. It confirms the assignment that $A_W$ indeed originates from interfacial water.

Other hydroxyl groups attached to the $\alpha$-$Al_2O_3(1\bar{1}02)$ surface may also contribute to the spectra but could not be identified. Supposedly there are OH species associated with $Al_2OH$ that exist at all pH, with AlOH at pH $\geq 9$, with $AlOH_2$ at pH $\leq 9$ (referring to the originally dangling OH on the dry surface) and with $Al_3OH$ at pH $\leq 5$. The one associated with $Al_2OH$ is weak and negligible. The mode from originally dangling OH on AlOH and $AlOH_2$ is suppressed by H bonding to a water molecule and could become indistinguishable from other bonded OH bands. Binding of H on O in $Al_3OH$ is expected to be weak so that its stretch frequency may appear

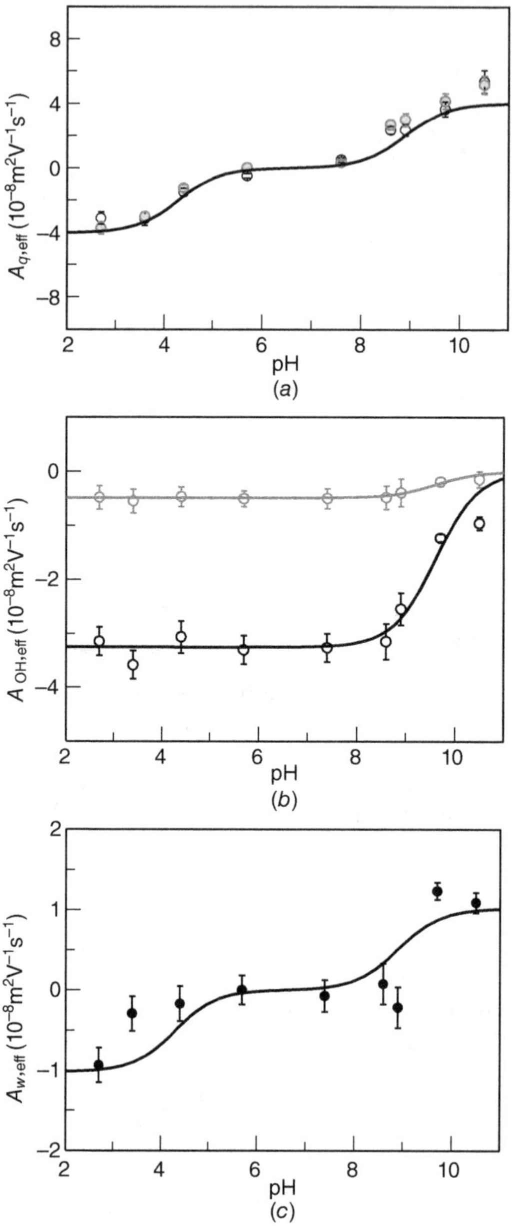


**Figure 2.16** (*a*) Amplitudes of the OH modes at 3230 cm$^{-1}$ as a function of pH. Black curve is a theoretical fit (see text). Open and filled circles are for 0° and 180° azimuthal angles, respectively. (*b*) Contributions to the 3490 cm$^{-1}$ band versus pH from OH on $AlOH_2$, and (*c*) contributions to the 3490 cm$^{-1}$ band versus pH from water. In (*b*) upper and lower data points correspond to $A_{OH}$ azimuthal angles of 0° and 180°, respectively. Solid lines are theoretical fits.

below the spectral range of investigation. The work here shows that different faces of α-$Al_2O_3$ indeed have different structures and charging properties in water, although their pzc appear fairly close, in the range between 6 and 7.

It is interesting to note that the net polar orientations of interfacial water molecules contributing to the liquidlike band are opposite at α-$Al_2O_3$ $(1\bar{1}02)$–water and α-quartz–water interfaces. The former has O→H toward the liquid side, and the latter has O→H toward the surface. Their pH dependences, however, have the same trend: starting from the neutral surface, more deprotonation leads to more net polar orientation of O→H toward the surface, although at the pzc, the net polar orientation is finite for the quartz case, but zero for the alumina case. Recent calculation shows that this is the result of different H bonding geometries of water molecules at the two surfaces because of their different surface structures [69], one having OH of water molecules donor-H bonded to surface O more vertically and the other more horizontally.

### A-Plane $(11\bar{2}0)$ Surface of $\alpha$-$Al_2O_3$

This surface is technologically important as a substrate for thin-film synthesis. X-ray reflectivity demonstrated that a high-temperature anneal produced a reconstructed surface that was stable under bulk water but that a lower-temperature anneal resulted in a stoichiometric surface with all Al atoms 6-coordinated to O [70]. The stoichiometric surface showed relaxations from bulk positions and is characterized by the presence of AlO, $Al_2O$, and $Al_3O$ surface functional groups in the ratio 1 : 1 : 1. No CTR data is available for this interface, nor are there DFT calculations available in the literature. Water appears to be well organized in two or three near-surface layers, but the lower resolution of X-ray reflectivity and lack of lateral resolution prevents extended quantitative interpretation.

A top view of the structure is sketched in Figure 2.17. When exposed to air, the surface functional groups will be protonated. SFVS has been used to identify and characterize the surface hydroxyls and confirm the proposed stoichiometric surface structure [71]. As in the cases of $\alpha$-$Al_2O_3$ (1000) and $\alpha$-$Al_2O_3$ $(1\bar{1}02)$, SFVS has also been employed to study how the hydroxyls are protonated or deprotonated in

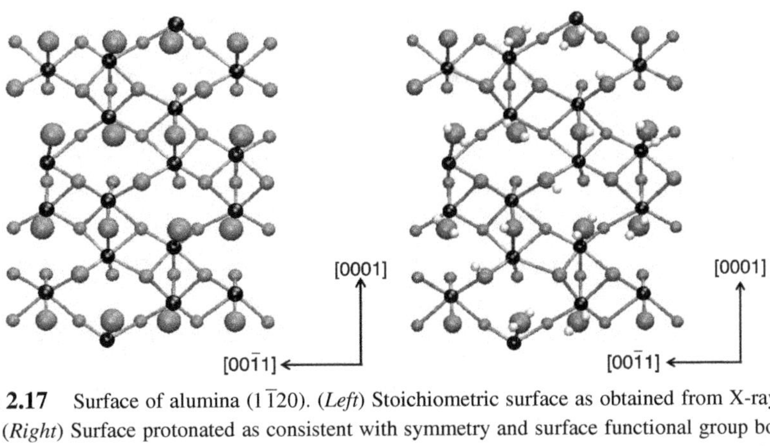


**Figure 2.17** Surface of alumina $(1\bar{1}20)$. (*Left*) Stoichiometric surface as obtained from X-ray studies. (*Right*) Surface protonated as consistent with symmetry and surface functional group bond valence. Different sizes of the dark gray oxygens represent depth into the surface with largest being uppermost. Al atoms are black. Protons are light gray.

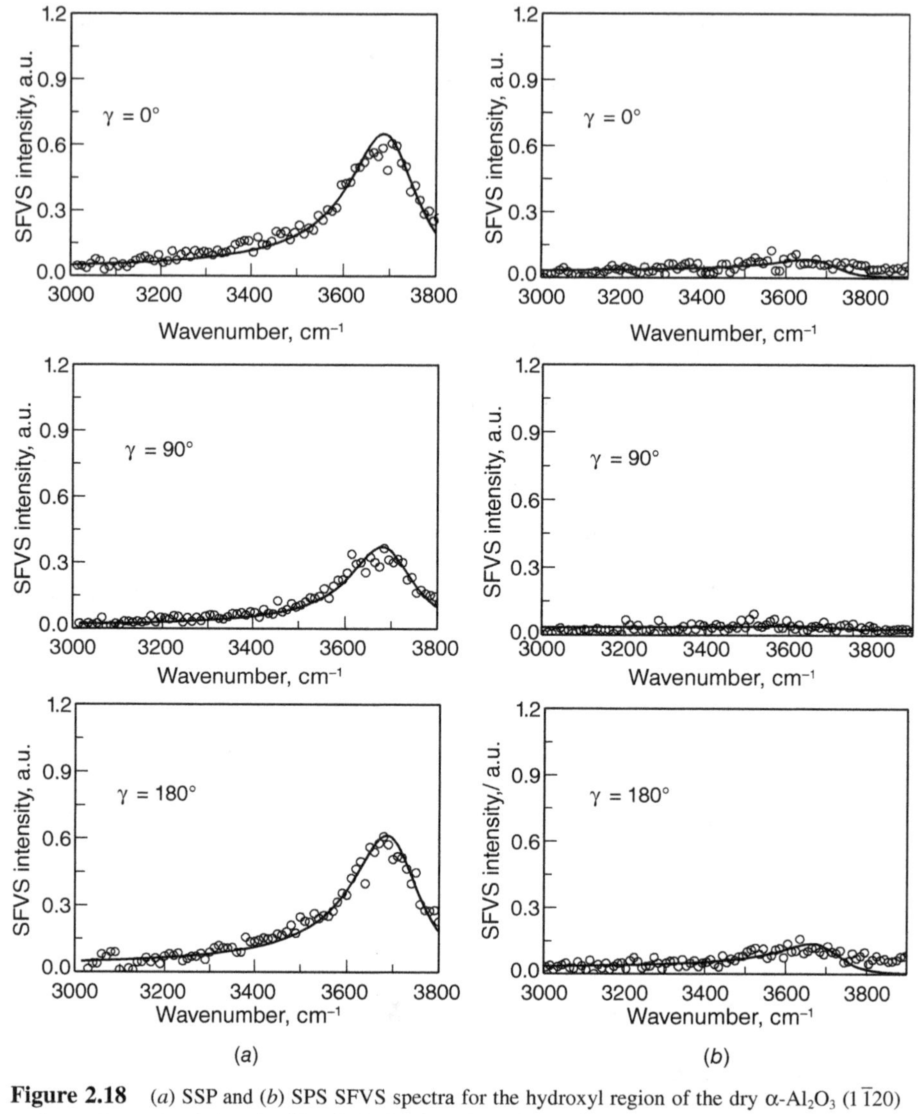


**Figure 2.18** (*a*) SSP and (*b*) SPS SFVS spectra for the hydroxyl region of the dry α-$Al_2O_3$ $(1\bar{1}20)$ surface at three azimuthal angles. The single fitted band at 3670 $cm^{-1}$ is consistent with a dangling O–H.

water of different pH and how interfacial water molecules are oriented according to the resultant surface field.

From the bond valence argument, only AlO and $Al_2O$ surface groups can be protonated into $AlOH_2$ and $Al_2OH$; protonation of $Al_3O$ is unlikely except under extremely acidic cases. This would suggest the appearance of three OH stretch modes in SFVS. The observed SSP and SPS spectra of the dry α-$Al_2O_3$ $(11\bar{2}0)$ surface, presented in Figure 2.18, however, exhibit only one band at ~3670 $cm^{-1}$ with strong polarization dependence. This band can be understood as from the dangling OHs on $AlOH_2$ and $Al_2OH$. Calculations based on the stoichiometric terminated surface structure indicate that the dangling OHs on $Al_2OH$ and on two-thirds of $AlOH_2$ point in roughly the same direction. They have nearly the same stretch frequency and should appear prominently in the SSP spectra. The OH orientation deduced from the observed polarization dependence of the band agrees well with the predicted orientations of the dangling OHs. The other hydroxyl group associated with $AlOH_2$ is H bonded to a neighboring O. The stoichiometric-terminated surface structure suggests that the group must have mirror symmetry pairs of OH inclined toward the surface. The signal of SFVS from such oriented OH is expected to be weak. Thus, the SFVS result appears to confirm the stoichiometric-terminated structure of $\alpha$-$Al_2O_3$ $(11\bar{2}0)$.

For the $\alpha$-$Al_2O_3$ $(11\bar{2}0)$–water interface, the spectrum consists of three discrete bands at 3180, 3450, and 3670 $cm^{-1}$ that change with pH. The representative Im $\chi_S^{(2)}$ spectra at three different pHs are presented in Figure 2.19. The amplitudes of the three modes versus pH deduced from fitting of spectra by Eq. (2.10) at different pH are displayed in Figure 2.20. While the amplitude of the 3670 $cm^{-1}$ mode hardly changes, those of the 3180 and 3450 $cm^{-1}$ bands vary significantly with pH. The former must originate from surface hydroxyls that are difficult to deprotonate and the latter from interfacial water molecules whose orientations are sensitive to protonation/deprotonation of the surface hydroxyls as in the cases of the other $\alpha$-$Al_2O_3$–water interfaces discussed earlier.

When immersed in water, the following reactions could occur on $\alpha$-$Al_2O_3$ $(11\bar{2}0)$ [72]:

$$\begin{aligned} &Al_2OH \leftrightarrow Al_2O^- + H^+ && \text{at} \quad pK \sim 12.9 \\ &AlOH_2 \leftrightarrow AlOH^- + H^+ && \text{at} \quad 9.9 < pK < 11 \\ &Al_3OH^+ \leftrightarrow Al_3O^- + H^+ && \text{at} \quad pK \sim 5.9 \end{aligned}$$

At very low pH, all surface oxygens are protonated, and the surface is positively charged due to protonation of $Al_3O$. Near-surface water molecules tend to be reoriented by the field with O pointed toward the surface and by preferred H bonding with O to H of the surface hydroxyls. The net polar orientation increases with decrease of pH. At high pH, the surface is negatively charged due to deprotonation of $AlOH_2$; deprotonation of $Al_2OH$ is negligible in the pH range studied. Accordingly, the net polar orientation of near-surface water molecules is reversed. At mid-pH, protonation of $Al_3O$ and deprotonation of $AlOH_2$ appear to be balanced, resulting in very weak net polar orientation of water molecules. This explains qualitatively the result shown in Figure 2.20. Quantitatively, the data in Figures 2.20*b*, *c* can be fit by assuming that the concentrations of the water species responsible for the 3450 and 3180 $cm^{-1}$ bands are proportional to the surface charge density, which can be calculated using the kinetic equations for the reactions $AlOH_2 \leftrightarrow AlOH^- + H^+$ and $Al_3OH^+ \leftrightarrow Al_3O + H^+$. The fit is good with pK = 9.2 and 4.7 for the two reactions. The steplike changes at pH ~4.7 and 9.2 reflect the pK values for the two reactions. The deviation between theory and experiment for the 3450 $cm^{-1}$ band at high pH may be due to neglect of contribution from surface hydroxyl groups. In any case, the $\alpha$-$Al_2O_3$ $(1\bar{1}20)$–water interface appears nearly neutral in the pH range

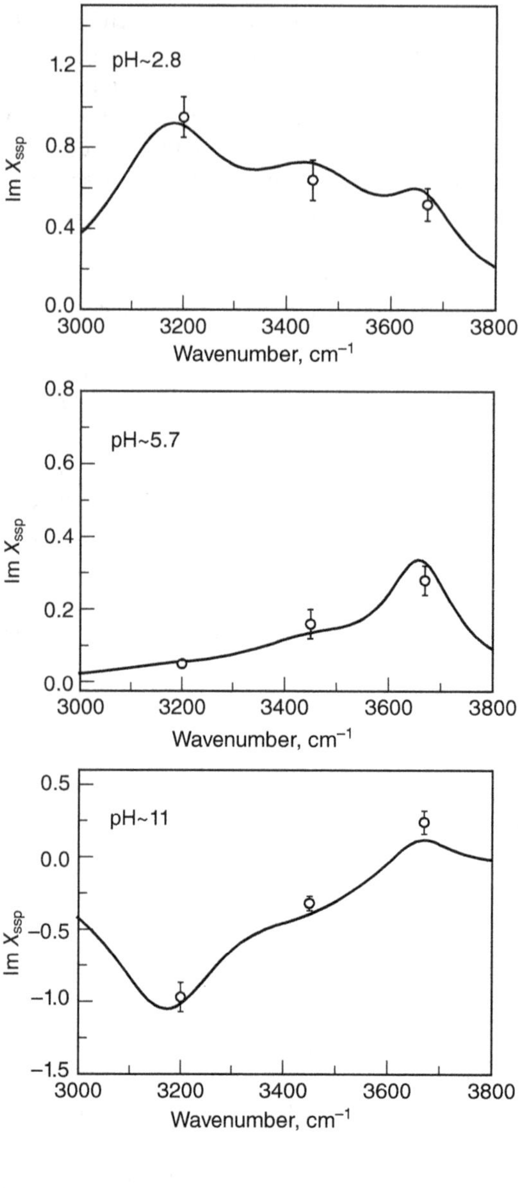


**Figure 2.19** Imaginary part of the SFVS spectrum measured at three different pH values for the $\alpha$-$Al_2O_3$ $(1\bar{1}20)$–water interface. There are amplitude and amplitude sign changes for the 3180 and 3450 cm$^{-1}$ bands, but no sign change for the 3670 cm$^{-1}$ band.

between 6 and 8, and the pzc for the interface is estimated to be ~7, again close to that of $\alpha$-$Al_2O_3$ (0001).

Overall, the above studies of alumina dry surfaces and water interfaces demonstrate that SFVS can complement X-ray scattering measurements to determine the surface structure of a crystalline solid. Further, by using the surface structures of the solid constructed from X-ray scattering measurement as a guide, SFVS can provide a clear picture on the charging properties of the solid surfaces. This is expected to be generally true for oxide surfaces having oxygen-containing functional groups

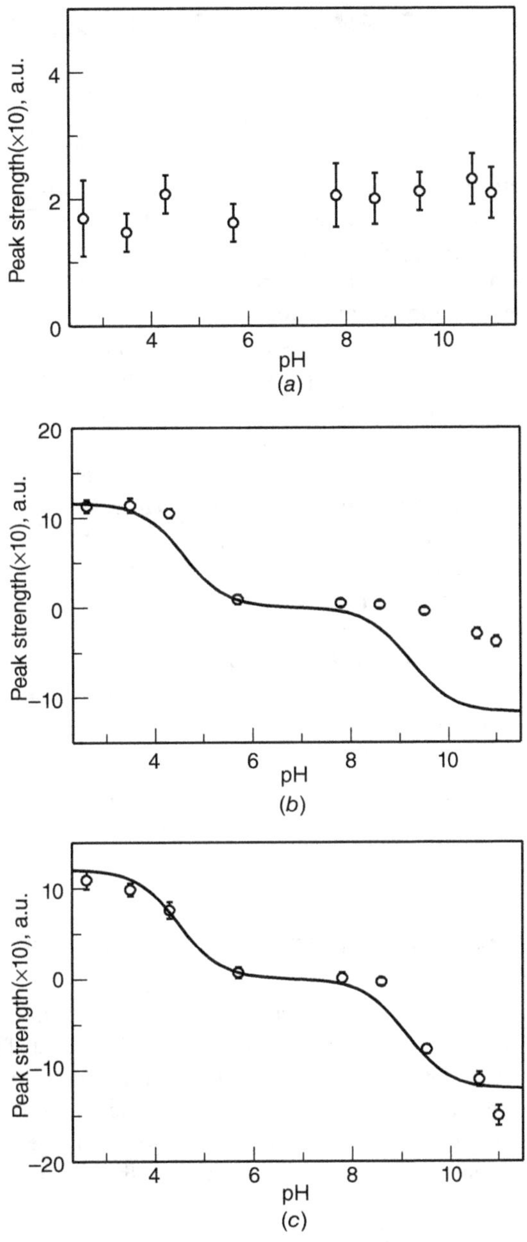


**Figure 2.20** Changes in band amplitude as a function of pH for the bands observed in Figure 2.19: (*a*) 3670 $cm^{-1}$, (*b*) 3450 $cm^{-1}$, amd (*c*) 3180 $cm^{-1}$. The solid lines in (*b*) and (*c*) represent surface charge density calculated from the known pKa values of the protonation and deprotonation reactions and the net protonation of the surface.

occupying the top surface layers. We should also remark that the charging properties of solid–liquid interfaces are often studied by other means, among which the second-harmonic generation method is closely related to SFVS [73]. However, none of the other methods are able to probe different protonation or deprotonation sites on a surface and different water species responding to the surface charging process.

## 2.8 OTHER OXIDE SURFACES AND INTERFACES

Studies of the structures of oxide surfaces and interfaces in contact with aqueous solution are extremely limited. SFVS has hardly been applied to surfaces and interfaces of oxides other than those we have discussed. Titanium dioxide is an exception, presumably because of its potentially useful photoactivities. Even so, none of the studies is on single-crystalline surfaces of $TiO_2$. Wang and co-workers [74] used SFVS to probe the interface of water and anatase $TiO_2$ nanoparticles and observed a broad band of OH stretches with multiple peaks arising from surface hydroxyls or/and interfacial water molecules. They also used methanol as probe molecules in SFVS to study surface reactivity on the $TiO_2$ nanoparticles [75]. Kataoka and co-workers [76] applied SFVS to an interface between water and a thin $TiO_2$ film on silica. They observed two bands in the OH stretch range, one icelike and the other liquidlike, presumably arising from water molecules at the interface. Both bands have a strength that varies with pH of water, strong at low and high pH, but weakest in the pH range between 4 and 6. The result suggests surface charging by protonation and deprotonation with the pzc of the interface at ~ 5.5. It will certainly be interesting to carry out the same experiment on well-defined crystalline surfaces of $TiO_2$, so that more detailed information can be deduced about the protonation–deprotonation sites on the surfaces.

Adsorption of water molecules on mica has been probed by SFVS [77]. The spectra show that the arrangement of submonolayer water molecules on mica is fairly disordered but becomes more ordered and icelike as the molecular density approaches a full monolayer. The monolayer seems to form an ordered H bonding network with no dangling OH. The result agrees well with an earlier theoretical prediction [78]. SFVS has also been used to monitor water adsorption on MgO at low temperature to form monolayer and multilayers of ice [79].

## 2.9 LIMITATIONS AND DIFFICULTIES

As we have seen, X-ray and SFVS techniques, especially the combination of the two, are very powerful for studies of oxide surfaces and interfaces. They have provided detailed structural information about silica and alumina surfaces and their interfaces with water and can be anticipated to have applications of the techniques extended to many other oxides. However, there are still difficulties and limitations in such applications. The X-ray methods, such as crystal truncation rod X-ray diffraction and X-ray reflectivity, are not sensitive to protons, and it is difficult to directly determine their positions on the surface. Analysis of the X-ray scattering data often relies on surface structural models that may not be unique, and the model calculation generally requires structural regularity (or periodicity in the lateral structure), although in one case ($\alpha$-FeOOH) quarter-cell height surface steps could be refined in the CTR analysis, and in ideal cases CTR data can be analyzed to yield terrace width between steps [4]. Thus, it is impossible to use X-ray scattering techniques to probe surfaces of amorphous solids such as fused quartz or surfaces of crystalline solids with many surface defects, such as one having a poor polish. Although X-ray scattering can be used to probe ordered water films covering a crystalline surface, controlling the pH of the water film remains difficult; and few X-ray scattering measurements including the pH dependence of solid–water interfacial structure have been reported.

Sum frequency vibrational spectroscopy also has limitations. The oxide interfaces must be accessible by all input and output light. This could be a problem for SFVS to probe surface structures of oxides in water. The spectral range covered by tunable IR sources is currently limited to less than 16 μm, restricting the surface structural study necessarily to lighter metal oxides. Spectral analysis is not simple, requiring prior knowledge of the solid surface structure as a guide. At a solid–water interface, the relaxed surface structure of the solid is often not known. Current interpretation of the interfacial spectra assumes that it is the same as that of the dry solid bulk terminated surface, but this assumption often has no proof. Understanding SFVS spectra is also not a trivial matter. Assignment of spectral features presently relies on past success on similar molecular species. Theoretical calculations on interfacial water spectra have not yet been very satisfying in identifying which water species contribute to which spectral range. There exist hardly any calculations on surface phonons of oxides in connection with the surface structure.

## 2.10 CONCLUSIONS

Despite their limitations, surface X-ray scattering and sum frequency vibrational spectroscopy are the only currently available techniques that can be employed to probe molecular-level structures of crystalline oxide surfaces and oxide–aqueous solution interfaces. The two techniques provide complementary information. X-ray scattering allows determination of the surface crystalline structures of oxides, surface structural relaxation under water, and possible ordered structures of adsorbed water layers or molecules on oxides. Its insensitivity to protons, however, prevents it from directly determining the bonding geometry of hydroxyl groups on the oxide surfaces and monitoring the protonation and deprotonation processes. Sum frequency vibrational spectroscopy, on the other hand, can generate surface phonon spectra of oxides, from which the surface structure can in principle be deduced. It can also detect different hydroxyl species at oxide surfaces through their OH stretch frequencies and their surface bonding geometry. Direct monitoring of how each hydroxyl species protonates and deprotonates and how the interfacial water molecules respond to the surface charging behavior provides unique information on the physical and chemical properties of oxide–aqueous solution interfaces at the molecular level.

Both X-ray scattering and sum frequency spectroscopy still have room to improve. CTR analysis cannot easily take advantage of the powerful X-ray structure methods employed for analysis of three-dimensional diffraction data, while interpretation of SFVS data is limited by difficulty in simulating SFVS spectra, but both issues may be improved in the future. Nevertheless, problems in surface science are often complex and generally require a multipronged approach to solve. In the case of oxide–water interfaces, experiments carried out on the same sample in a certain environment simultaneously by X-ray diffraction/reflectivity and SFVS are obviously important and desirable. It would be even more ideal if other techniques, such as ultraviolet photoemission spectroscopy, LEED, AFM, and the like can be used to characterize the same surfaces. The correlated results from different techniques certainly would provide a more coherent and correct picture on a protonated or aqueous-solution-equilibrated oxide surface or interface.

## ACKNOWLEDGMENTS

Y.R.S. acknowledges support from the U.S. Department of Energy, Office of Science, Office of Basic Energy Sciences, Materials Sciences and Engineering Division as well as the Chemical Sciences, Geosciences, and Biosciences Division, under Contract No. DE-AC03-76SF00098. G.A.W. gratefully acknowledges support from the U.S. Department of Energy, Office of Basic Energy Sciences, Division of Chemical Sciences, Geosciences and Biosciences under Contract No. DE-AC02-05CH11231.

## REFERENCES

1. Lambert, J.-F. *Orig. Life Evol. Biosph.* **2008**, *38*, 211.
2. Casey, W. H.; Rustad, J. R.; Spiccia, L. *Chemistry* **2009**, *15*, 4496.
3. Stumm, W.; Furrer, G. *Aquatic Surface Chemistry: Chemical Processes at the Particle-Water Interface*, Wiley: New York, **1987**, p. 197.
4. Ghose, S.; Waychunas, G. A.; Trainor, T. P.; Eng, P. J. *Geochim. Cosmochim. Acta* **2010**, *74*, 1943.
5. Sumito, M.; Hu, C.; Tateyama, Y. *J. Phys. Chem. C* **2010**, *114*, 18529.
6. Henderson, M. A. The interaction of water with solid surfaces: Fundamental aspects revisited, *Surf. Sci. Rep.* **2002**, *46*, 1–308.
7. Sverjensky, D. A. *Geochem. Cosmochim. Acta* **1994**, *58*, 3123.
8. Chapman, D. L. A contribution to the theory of electrocapillarity, *Phil. Mag. Ser. 6* **1913**, *25*, 475.
9. Sposito, G. *The Surface Chemistry of Soils*, Oxford University Press: Oxford, **1984**, p. 234.
10. Davis, J. A.; Kent, D. B. *Rev. Mineral. Geochem.* **1990**, *23*, 177.
11. Lee, S. S.; Fenter, P.; Park, C.; Sturchio, N. C.; Nagy, K. L. *Langmuir* **2010**, *26*, 16647.
12. Predota, M.; Zhang, Z.; Fenter, P.; Wasolowski, D. J.; Cummings, P. T. *J. Phys. Chem. B* **2004**, *108*, 12061.
13. Israelachvili, J. N.; Pashley, R. M. *Nature* **1983**, *306*, 249.
14. Skipper, N. T.; Soper, A. K.; McConnell, J. D. C.; Refson, K. *Chem. Phys. Lett.* **1990**, *166*, 141.
15. Cleveland, J. P.; Schaffer, T. E.; Hansma, P. K. *Phys. Rev. B* **1995**, *52*, R8692.
16. Lee, S. H.; Rossky, P. J. *J. Chem. Phys.* **1994**, *100*, 3334.
17. Kerisit, S.; Cooke, D. J.; Spagnoli, D.; Parker. S. C. *J. Mater. Chem.* **2005**, *15*, 1454.
18. Spagnoli, D.; Gilbert, B.; Waychunas, G. A.; Banfield, J. F. *Geochem. Cosmochim. Acta* **2009**, *73*, 4023.
19. Yoo, H.; Paranji, R.; Pollack, G. H. *J. Phys. Chem. Lett.* **2011**, *2*, 532.
20. Zhao, Y.; Truhlar, G. *Accounts Chem. Res.* **2008**, *41*, 157.
21. Kotliar, G.; Savrasov, S. Y.; Haule, K.; Oudovenko, V. S.; Parcollet, O.; Marianette, C. A. *Rev. Mod. Phys.* **2006**, *78*, 865.
22. Cheng, L.; Fenter, P.; Nagy, K. L.; Schlegel, M. L.; Sturchio, N. C. *Phys. Rev. Lett.* **2001**, *87*, 156103.

23. Zhang, Z.; Fenter, P.; Cheng, L.; Sturchio, N. C.; Bedzyk, M. J.; Machesky, M. L.; Wesolowski, D. J. *Surf. Sci.* **2004**, *554*, L95.
24. Pareek, A.; Torrelles, X.; Angermund, K.; Rius, J.; Magdans, U.; Gies, H. *Langmuir* **2008**, *24*, 2459.
25. Reedijk, M. F.; Arsic, J.; Hollander, F. F. A.; deVries, S. A.; Vlieg, E. *Phys. Rev. Lett.* **2003**, *90*, 066103.
26. Robinson, I. K. *Phys. Rev.* B **1986**, *33*, 3830.
27. Feidenhans'l, R. *Surf. Sci. Rpts.* **1989**, *10*, 105.
28. Vlieg, E.; VanDerVeen, J. F.; Gurman, S. J.; Norris, C.; Macdonald, J. E. *Surf. Sci.* **1989**, *210*, 301.
29. Fenter, P. A. *Rev. Mineral. Geochem.* **2002**, *49*, 149.
30. Vlieg, E. *J. Appl. Cryst.* **2000**, *33*, 401.
31. www.youtube.com/watch?v=ZE1Y0iBhxuU.
32. Shen, Y. R. *Surf. Sci.* **1994**, *299/300*, 551.
33. Shen, Y. R. *Proceedings of the International School of Physics "Enrico Fermi" Course CXX, Frontiers in Laser Spectroscopy*, North Holland: Amsterdam, T. W. Hansch and M. Inguscio eds., **1994**, p. 139.
34. Shen, Y. R.; Ostroverkhov, V. *Chem. Rev.* **2006**, *10*, 1140.
35. Wei, X.; Zhuang, X.; Hong, S. C.; Goto, T.; Shen, Y. R. *Phys. Rev. Lett.* **1999**, *82*, 4256.
36. Ji, N.; Ostroverkhov, V.; Chen, C. Y.; Shen, Y. R. *J. Am. Chem. Soc.* **2007**, *139*, 11056.
37. Du, Q.; Freysz, E.; Shen, Y. R. *Phys. Rev. Lett.* **1994**, *72*, 238.
38. Liu, W. T.; Shen, Y. R. *Phys. Rev. Lett.* **2008**, *101*, 016101.
39. Schlegel, M. L.; Nagy, K. L.; Fenter, P.; Sturchio, N. C. *Geochim. Cosmochim. Acta* **2002**, *66*, 3037.
40. Gratz, A. J.; Manne, A. J. S.; Hansma, P. K. *Science* **1991**, *251*, 1343.
41. Kawasaki, M.; Onuma, K.; Sunagawa, I. *J. Cryst. Growth* **2003**, *258*, 188.
42. Yanina, S. V.; Rosso, K. M.; Meakin, P. *Geochim. Cosmochim. Acta* **2006**, *70*, 1113.
43. Janossy, I. *Surf. Sci.* **1971**, *25*, 647; Bart, F.; Gautier, M. *Surf. Sci.* **1994**, *311*, L671.
44. Steurer, W.; Apfolter, A.; Koch, M.; Scarlat, T.; Sondergard, E.; Ernst, W. E.; Holst, B. *Surf. Sci.* **2007**, *601*, 4407.
45. Goumans, T. P. M.; Wander, A.; Brown, W. A.; Catlow, C. R. A. *Phys. Chem. Chem. Phys.* **2007**, *9*, 2146.
46. White, R. L.; Nair, A. *Appl. Spectrosc.* **1990**, *44*, 69.
47. Lier, R. K. *The Chemistry of Silica*, Wiley: New York, **1979**, p. 622.
48. Schlegel, M. L.; Nagy, K. L.; Fenter, P.; Sturchio, N. C. *Geochim. Cosmochim. Acta* **2002**, *66*, 3037.
49. Ostroverkhov, V.; Waychunas, G. A.; Shen, Y. R. *Chem. Phys. Lett.* **2004**, *386*, 144.
50. Ostroverkhov, V.; Waychunas, G. A.; Shen, Y. R. *Phys. Rev. Lett.* **2005**, *94*, 046102.
51. Ong, S.; Zhao, X.; Eisenthal, K. B. *Chem. Phys. Lett.* **1992**, *191*, 327.
52. Leung, K.; Nielson, I. M.; Criscenti, L. J. *J. Am. Chem. Soc.* **2009**, *131*, 18358.
53. Yaganeh, M. S.; Dougal, S. M.; Pink, H. S. *Phys. Rev. Lett.* **1999**, *83*, 1179.
54. Eng, P. J.; Trainor, T. P.; Brown, G. E.; Waychunas, G. A.; Newville, M.; Sutton, S. R.; Rivers, M. L. *Science* **2000**, *288*, 1029.
55. Catalano, J. G. *Geochem. Cosmochim. Acta* **2011**, *75*, 2062.
56. Ranea. V. A.; Carmichael, I.; Schneider, W. F. *J. Phys. Chem. C* **2009**, *113*, 2149.
57. Zhang, L.; Tian, C.; Waychunas, G. A.; Shen, Y. R. *J. Am. Chem. Soc.* **2008**, *130*, 7686.
58. Contescu, C.; Jagiello, J.; Schwarz, J. A. *Langmuir* **1993**, *9*, 1754.
59. Braunschweig, B.; Eissner, S.; Daum, W. *J. Phys. Chem. C* **2008**, *112*, 1751.
60. Gaigeot, M. P.; Sprik, M.; Salpizi, M. J. *Physics: Cond. Matt.* **2012**, *24*, 124106.
61. Trainor, T. P.; Eng, P. J.; Brown, G. E.; Robinson, I. K.; DeSantis, M. *Surf. Sci.* **2002**, *496*, 238.
62. Tanwar, K. S.; Lo, C. S.; Eng, P. J.; Catalano, J. G.; Walko, D. A.; Brown, G. E.; Waychunas, G. A.; Chaka, A. M.; Trainor, T. P. *Surf. Sci.* **2007**, *601*, 460.
63. Catalano, J. G.; Fenter, P.; Park, C. *Geochim. Cosmochim. Acta*, **2007**, *71*, 5313.
64. Mason, S. E.; Iceman, C. R.; Trainor, T. P.; Chaka, A. M. *Phys. Rev. B* **2010**, *81*, 125423.
65. Sung, J.; Zhang, L.; Tian, C. S.; Waychunas, G. A.; Shen, Y. R. *J. Am. Chem. Soc.* **2011**, *133*, 3846.

66. Sung, J.; Zhang, L.; Tian, C. S.; Shen, Y. R.; Waychunas, G. A. *J. Phys. Chem. C* **2011**, *115*, 13887.
67. Contescu, C.; Jagiello, J.; Schwarz, J. A. *Langmuir* **1993**, *9*, 1754.
68. Hiemstra, T.; Venema, P.; Van Riemsdijk, W. H. *J. Colloid Interface Sci.* **1996**, *184*, 680.
69. Gaigeot, M. P.; Sprik, M.; Sulpizi, M. *J. Phys.: Cond. Matt.* **2012**, *24*, 124106.
70. Catalano, J. G. *J. Phys. Chem. C* **2010**, *114*, 6624.
71. Sung, J.; Shen, Y. R.; Waychunas, G. A. *J. Phys.: Cond. Matt.* **2012**, *24*, 124101.
72. Bickmore, B. R.; Tadanier, C. J.; Rosso, K. M.; Monn, W. D.; Eggett, D. L. *Geochim. Coschim. Acta* **2004**, *68*, 2025 (2004).
73. Fitts, J. P.; Shang, X.; Flynn, G. W.; Heinz, T. F.; Eisenthal, K. B. *J. Phys. Chem. B* **2009**, *109*, 7981.
74. Wang, C. Y.; Groenzin, H.; Shultz, M. J. *Langmuir* **2003**, *19*, 7330.
75. Wang, C. Y.; Groenzin, H.; Shultz, M. J. J. *Phys. Chem.* B **2004**, *108*, 265.
76. Kataoka, S.; Guoro, M. C.; Albertorio, F.; Holden, M. A.; Lim, S. M.; Yang, R. D.; Cremer, P. S. *Langmuir* **2004**, *20*, 1662.
77. Miranda, P. B.; Xu, L.; Shen, Y. R.; Salmeron, M. *Phys. Rev. Lett.* **1998**, *26*, 5876.
78. Odelius, M.; Bernasconi, M.; Parrenello, M. *Phys. Rev. Lett.* **1997**, *78*, 2835.
79. Wlodarczyk, R.; Sierka, M.; Kwapien, K.; Sauer, J. *J. Phys. Chem. C* **2011**, *105*, 6764.

# Chapter 3

# Vibrational Sum Frequency Generation Spectroscopy of Interfacial Dynamics

**Christopher M. Berg and Dana D. Dlott**
School of Chemical Sciences, University of Illinois at Urbana-Champaign, Urbana, Illinois

## 3.1 INTRODUCTION

Over the past several years, vibrational sum frequency generation (SFG) spectroscopy has emerged as a powerful technique to study molecules at surfaces and interfaces [1–5]. In our laboratory, we have made advances in both SFG spectroscopy and SFG instrumentation. These developments proved especially useful for probing time-dependent processes of molecules adsorbed on metal surfaces or at buried interfaces [6, 7] such as metal–liquid or metal–solid interfaces [8–10]. In this chapter, we describe recent advances in SFG instrumentation from our laboratory and their use in two frontier research areas: the study of ultrafast dynamics of heat transfer to molecular adsorbates and the study of in situ electrochemical processes occurring at electrode–electrolyte interfaces [11–13], which are inaccessible to most surface science methods.

The technique most commonly used in our laboratory is termed broadband multiplex SFG with nonresonant (NR) suppression [6, 7], and this method and its implementation will be described here in detail. The technique allows us to obtain SFG spectra over a spectroscopic window, a few hundred wavenumbers in width, with high time resolution. The NR suppression feature is especially important when molecules are adsorbed on metal surfaces [14], which generally create the largest NR backgrounds. An NR background is not all bad. It is sometimes useful for

*Vibrational Spectroscopy at Electrified Interfaces*, First Edition.
Edited by Andrzej Wieckowski, Carol Korzeniewski, and Björn Braunschweig.

providing a phase reference for determining the absolute orientation of interfacial molecules, but frequently the NR background interferes with high signal-to-noise detection of resonant (R) molecular transitions. So it is useful to have the ability to both detect and to suppress NR signals.

As explained below, the factor that ultimately limits the time needed to acquire an SFG spectrum is the inverse line width of the molecular transition [8]. For example, the stretching transition of CO adsorbed on a Pt electrode under a liquid electrolyte has a line width $\Delta\nu$ of about 20 $cm^{-1}$ [11, 15]. Using the relation, $\Delta\nu = (\pi c T_2)^{-1}$, this line width corresponds to a dephasing time constant $T_2 = 530$ fs. The ultimate limit for obtaining an SFG spectrum of CO on Pt is the duration of the vibrational polarization, which decays with this time constant. The CO SFG spectrum could therefore, in principle, be obtained in a time window that is a few multiples of 530 fs, say 2 ps. However, in practice, due to the small number of molecules at an interface, obtaining high-quality SFG spectra requires the averaging of signals from multiple laser shots. This need for averaging allows us to time-resolve interfacial molecular processes in two ways. First, it is possible to study molecular processes with a time resolution of picoseconds or less using pump–probe techniques that reconstruct the dynamics [5]. In pump–probe measurements, a laser pulse is used to induce an interfacial process, which is then studied with a time series of SFG spectra. For instance, in this chapter we describe studies of the rearrangements of interfacial molecules caused by a temperature jump (T-jump) created by a laser "pump" pulse [8, 10, 16–19]. In this method, the T-jump rearrangement experiment is repeated many thousands of times at each pump–probe delay, each time using either a fresh sample or by allowing the sample to relax back to its initial state before the laser fires again. Such measurements may take minutes or hours in their entirety, but by temporally reconstructing the spectra, the time resolution that can be obtained is ultimately limited solely by $T_2$. Second, because the lasers we use generate about 1000 pulses per second, it is possible to obtain good spectra in seconds, sometimes even in a few hundred milliseconds. We can consequently time-resolve—in real time in this case as opposed to reconstructed time—processes such as potential-dependent changes in the concentration of molecular species or molecular intermediates at an electrode–electrolyte interface, provided the processes occur slower than, say, 0.1 s [11, 15]. The time resolution in this real-time case is not a fundamental limitation. Using lasers that operate at higher repetition rates, the real-time resolution could be proportionately increased.

In the rest of this chapter, we describe the instrumentation for broadband multiplex NR-suppressed SFG spectroscopy used in our laboratory. We have two such instruments. One is an "ultrafast system" optimized for pump–probe measurements, and the other is a "compact system" optimized for real-time measurements of electrochemistry. In order to understand how these instruments work, in particular the NR suppression feature, we present a time-domain picture of SFG spectroscopy [6, 8, 10] that is more useful in this context than the usual frequency-domain picture [20]. We then describe the application of the ultrafast SFG system to understand the dynamics of self-assembled monolayers (SAMs) that have been flash heated by laser pulses, and the application of the compact SFG system to study the development, during potential scanning, of solid–electrolyte interfaces (SEI) in lithium ion battery systems.

## 3.2 SUM FREQUENCY GENERATION SPECTROSCOPY

In SFG, an infrared (IR) laser pulse $\omega_{IR}$ is combined with a secondary pulse $\omega_{vis}$, conventionally called "visible," at the sample, and the coherent sum frequency signal is detected [20, 21]. With Ti:sapphire lasers, which are today the most common pump source for broadband multiplex SFG, the visible pulse is 800 nm, in the near-IR. The coherent SFG emission is at frequency $\omega_{SFG} = \omega_{IR} + \omega_{vis}$. When $\omega_{IR}$ is resonant with an SFG-active vibrational transition, $\omega_{SFG}$ is enhanced [20]. Thus an SFG spectrum possesses features arising from vibrational transitions, similar to IR or Raman spectroscopies. In the dipole approximation, the SFG spectrum is proportional to the Fourier transform of a time correlation function of the form [22] $<\alpha(t)\mu(0)>$, where $\alpha$ and $\mu$ are the polarizability tensor and dipole moment. Thus a molecular transition is SFG active only if it is simultaneously IR and Raman active. This criteria is satisfied only if the molecular moiety exists in an environment that is noncentrosymmetric on the length scale of the visible wavelength [23, 24].

Consider the case of metal nanospheres coated with some SAM of interest, say 4-nitrobenzenethiol (NBT) on Au, being packed in a cube having dimensions of the visible wavelength. In the limit of the particles being infinitely small, the system behaves as an isotropic liquid where there is a strong probability that the orientation of the dipole from one molecule will be exactly opposite of another. Consequently, in the dipole approximation, no SFG signal will be generated. However, as the spheres approach the dimensions of the cube or even larger, we can only probe a certain section of a single sphere. Taken to the extreme, an infinitely large sphere (in respect to the wavelengths being considered) would look like a flat metal surface. All the molecular dipoles will be oriented in a given direction and will not cancel each other, which supports SFG signal production. SFG helps solve the two biggest problems of interfacial molecular spectroscopy: sensitivity and selectivity. Sensitivity derives from the efficiency of detecting a coherent visible signal as a result of nonlinear optical generation by intense ultrashort laser pulses. Selectivity derives from the centrosymmetric selection rule. At the interface between two centrosymmetric media, for instance, oil and water [25], only interfacial molecules would be seen by SFG.

The original SFG experiments [26, 27] were performed with picosecond pulse lasers. SFG spectra were obtained by stepping $\omega_{IR}$ point by point through vibrational transitions, and a single spectrum could take minutes or even hours to acquire. The concept of a broadband multiplex detection technique was well known from its use in coherent anti-Stokes Raman spectroscopy (CARS) [28–31], and in 1996, two groups independently introduced broadband techniques to SFG [23, 32]. The method introduced by Richter and co-workers [23], which used a tabletop Ti:sapphire system rather than a free-electron laser, was an important development that has been frequently copied and improved. In the Richter method, $\omega_{IR}$ was a broadband femtosecond

IR (BBIR) pulse that simultaneously probed all transitions within its spectral bandwidth, $\omega_{vis}$ was a picosecond pulse with a narrower bandwidth ($NB_{vis}$) that determined the spectroscopic resolution, and the SFG signal was detected by a multichannel spectrograph. Besides greatly decreasing spectral acquisition times, the broadband multiplex technique made time-resolved studies much easier due to the laser arrangement promoting straightforward pump pulse synchronization [5]. The Richter SFG apparatus was tested [23] by obtaining the well-known spectrum [27, 33] of a SAM of *n*-octadecylthiol (ODT) on Au in the CH-stretching region near 2900 $cm^{-1}$. This spectrum had a broad NR background from Au with the same spectral width as the BBIR pulses, plus three narrower [~10 $cm^{-1}$ full width at half-maximum (FWHM)] dips against that background. The same spectrum taken in our lab is shown in Figure 3.1*b*. The three dips were due to the resonant CH stretch transitions of the terminal

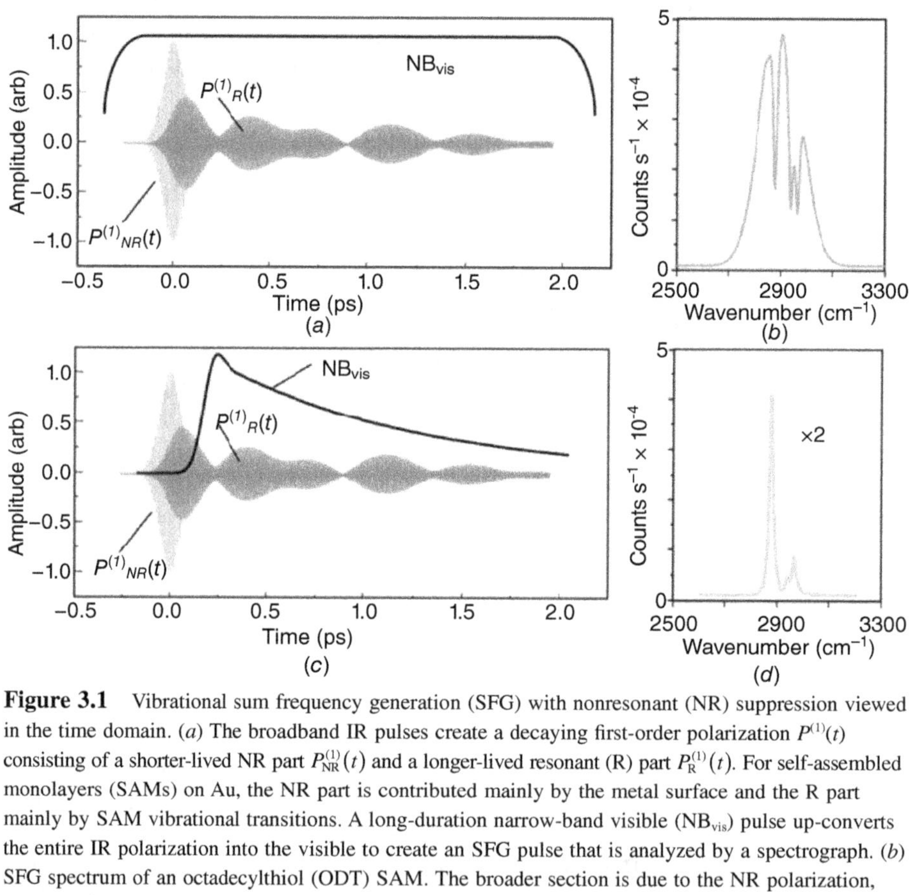


**Figure 3.1** Vibrational sum frequency generation (SFG) with nonresonant (NR) suppression viewed in the time domain. (*a*) The broadband IR pulses create a decaying first-order polarization $P^{(1)}(t)$ consisting of a shorter-lived NR part $P_{NR}^{(1)}(t)$ and a longer-lived resonant (R) part $P_{R}^{(1)}(t)$. For self-assembled monolayers (SAMs) on Au, the NR part is contributed mainly by the metal surface and the R part mainly by SAM vibrational transitions. A long-duration narrow-band visible ($NB_{vis}$) pulse up-converts the entire IR polarization into the visible to create an SFG pulse that is analyzed by a spectrograph. (*b*) SFG spectrum of an octadecylthiol (ODT) SAM. The broader section is due to the NR polarization, and the three sharper dips are a result of CH stretch transitions of the terminal methyl groups. (*c*) A time-asymmetric $NB_{vis}$ pulse can be time delayed beyond $P_{NR}(t)$, so the NR contribution is suppressed and not converted into a sum frequency signal. (*d*) ODT SAM spectrum with NR suppression showing only the three CH-stretch transitions. Reproduced with permission from Ref. [7], copyright 2010, Elsevier.

methyl ($CH_3$) groups. The remarkable properties of SFG are illustrated by the observation that almost no signals were seen from the methylene ($-CH_2-$) groups, even though methylene outnumbers methyl by 17:1 in ODT. The methylene suppression occurs because these groups, in the aggregate, form a nearly centrosymmetric crystal lattice. Another factor leading to this suppression is that the dipole moments of the methylene groups in the well-ordered SAM are almost parallel to the metal surface [34, 35]. By contrast, in reflection–absorption IR spectroscopy of the same sample, the methylene groups dominate due to their 17 times greater concentration [36]. The R transitions appear as dips against the NR background because the nonlinear polarization generated from the ODT $CH_3$ stretch transitions is opposite in phase from the nonlinear polarization produced from the Au surface, causing the R and NR signals to interfere destructively. As an aside, if the methyl groups were somehow flipped 180°, the R transitions would become peaks rather than dips. This is the basis of a useful technique in electrochemistry where the relative phases of the R and NR signals of molecules adsorbed on an electrode undergo potential-dependent reorientation [37, 38].

## 3.3 TIME-DOMAIN PICTURE AND NONRESONANT SUPPRESSION

In the frequency-domain picture originally used to describe SFG spectroscopy [20, 39], IR and visible pulses interact via the second-order nonlinear susceptibility $\chi^{(2)}$ to generate a second-order nonlinear SFG polarization $P_{SFG}^{(2)}(\omega)$ with a resonant (R) part from molecular vibrations and a nonresonant (NR) component [39, 40] from surface electronic states and distant vibrational resonances. As a result, the intensity of the outputted SFG is as follows:

$$I_{SFG}(\omega) \propto \left|P_{SFG}^{(2)}(\omega)\right|^2 = \left|P_R^{(2)}(\omega) + P_{NR}^{(2)}(\omega)\right|^2 = \left|[\chi_R^{(2)}(\omega) + \chi_{NR}^{(2)}(\omega)]E_{IR}E_{vis}\right|^2 \quad (3.1)$$

where $\chi^{(2)}$ is a third-rank tensor and $E$ denotes an electric field. Equation (3.1) shows that the R and NR polarizations, which are complex numbers, may undergo constructive or destructive interference.

The NR suppression method we developed replaces the picosecond time-symmetric visible pulse used by Richter and co-workers [23] with a similar pulse that is time asymmetric [6]. As further described, this time-asymmetric pulse is created in a particularly simple manner by transmitting a femtosecond visible pulse through a Fabry–Perot étalon. The use of an étalon to narrow the visible pulses in replacement of interference filters or diffraction gratings was a crucial improvement introduced by us. To understand how this works, consider SFG polarizations in the time domain [8, 39] as illustrated in Figure 3.1. As mentioned above, in the dipole approximation, the SFG process is determined by the time correlation function $\langle\alpha(t)\mu(0)\rangle$, so it looks like an IR interaction followed by a Raman process. A ~200-fs duration BBIR pulse arriving at $t = 0$ creates coherent first-order polarizations $P_{NR}^{(1)}(t)$ and $P_R^{(1)}(t)$, which undergo free-induction decays. These polarizations radiate coherent signals in the IR, which are not detected in SFG. The $P_{NR}^{(1)}(t)$, originating mainly from metal surface electronic transitions, has a broad spectrum and an ultrashort coherence lifetime $T_2$. The $P_{NR}^{(1)}(t)$, therefore, exists only when BBIR pulses are present. For the sake of discussion, we will model the SFG-active molecular vibrational resonances as consisting of identical Lorentzian line shapes having an FWHM $\Delta\nu$, which is an adequate description of the $CH_3$ stretch transitions of ODT. Each resonance will be coherently driven by the BBIR pulses to an extent determined by $\vec{\mu}\cdot\vec{E}$. $P_{R}^{(1)}(t)$ will have a complicated beat pattern (Fig. 3.1*a*) that decays overall as $\exp(-t/T_2)$. When $P_{NR}^{(1)}(t)$ and $P_{R}^{(1)}(t)$ interact with $\omega_{vis}$, a coherent signal pulse having a complicated temporal structure will be generated at $\omega_{SFG}$. In the dipole approximation, the interaction of the vibrational polarization with the visible pulse is the technical equivalent of coherent anti-Stokes Raman scattering [20]. Thus all IR-active vibrational transitions contribute to the free-induction decay (FID), but the SFG signal is generated only for those that are both IR and Raman active, which—in the dipole approximation—is possible only in noncentrosymmetric media. The resulting SFG intensity spectrum is the square modulus of the Fourier transform of the emitted signal pulse. The Fourier transformation is performed by the spectrograph and multichannel detector array.

Provided the spectrograph has enough resolving power to not broaden the visible pulse spectrum, the spectroscopic resolution is defined by the frequency bandwidth of the $NB_{vis}$, which is narrower in bandwidth and longer in duration than the BBIR pulses [7]. In the frequency-domain picture, the visible pulse provides a range of frequencies to interact with the first-order polarizations. In the time-domain picture [8]; $\omega_{vis}$ can interact with $P^{(1)}(t)$ only for a finite time period, resulting in a finite duration $P_{SFG}^{(2)}$ output with broadened features. Figure 3.1*a* illustrates the case where $\omega_{vis}$ has a narrow bandwidth much less than $\Delta\nu$, or equivalently, a transform-limited duration much longer than $T_2$. Consequently, the FID looks like an infinite duration continuous plane wave $\exp(-i\omega_{vis}t)$. The emitted SFG pulses will then have a time dependence of $\left[P_{NR}^{(1)}(t)+P_{R}^{(1)}(t)\right]\exp(-i\omega_{vis}t)$. The SFG spectrum will be the Fourier transform of the IR polarization upshifted to the frequency $\omega_{IR} + \omega_{vis}$. The SFG resonant transition widths in this case will be determined solely by $T_2$. However, if the visible pulse has a finite bandwidth and a finite pulse duration, as in Figure 3.1*c*, the visible intensity will decline during the later stages of the FID. This phenomenon has the effect of reducing the emitted SFG signal intensity at longer times, which is equivalent to apodization of the time response by the $\omega_{vis}$ prior to the Fourier transform [8, 41]. The finite pulse duration of $\omega_{vis}$ broadens SFG transitions. This property was put to use in an interesting way by Weeraman and co-workers [42], who utilized an "inverted" visible pulse. In contrast with the waveforms employed by us, which decay exponentially, the intensity of this pulse increased with time. Inverted waveforms interacted more efficiently with the longer time tail of the FID, producing a longer duration SFG pulse. As a result, the inverted visible pulse actually increased the resolution of the SFG measurement slightly beyond the limit imposed by the vibrational $T_2$.

As illustrated in Figure 3.1*c*, a time-delayed time-asymmetric visible pulse can deeply suppress the NR contribution via time windowing. This visible pulse can be time delayed so its rising edge is just past the NR polarization $P_{NR}^{(1)}(t)$, while maintaining most of its interaction with $P_{R}^{(1)}(t)$. Figure 3.1*b* shows the SFG spectrum of ODT on Au [6, 9], where the timing of the visible pulses was adjusted to maximize the NR background, and Figure 3.1*d* demonstrates the spectrum of ODT on Au where timing of the visible pulses was altered to suppress the NR signal. Since there is no longer destructive interference between the ODT molecular resonances and the NR background, the ODT $CH_3$ stretch transitions appear as peaks rather than dips.

## 3.4 FABRY–PEROT ÉTALON

The use of an étalon in replacement of the more usual optical bandpass filter or zero dispersion spectrograph as the visible bandwidth limiting element allows for NR suppression [6]. In contrast to those other methods, which generate time-symmetric pulses, an étalon creates a *time-asymmetric pulse*. An étalon (Fig. 3.2) consists of two parallel mirrors with partial reflectivity $\Re$ separated by an air gap *d*. The transmission spectrum is a series of peaks spaced by the free-spectral range FSR = *c*/2*d*.

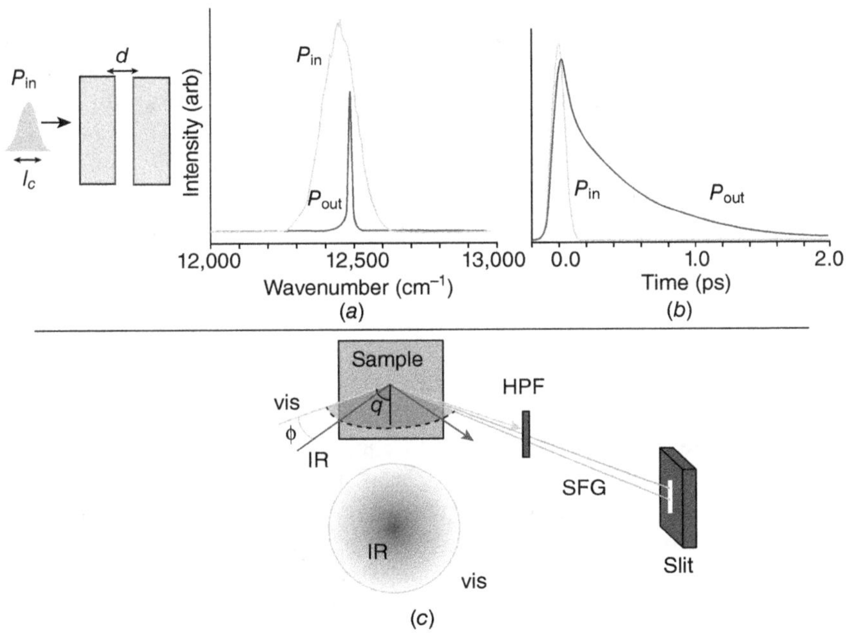


**Figure 3.2** Converting a femtosecond pulse $P_{in}$ with coherence length $\ell_c$ into a time-asymmetric picosecond pulse $P_{out}$ using an étalon. (*a*) Spectrum of femtosecond visible and étalon-filtered $NB_{vis}$ pulses. (*b*) Computed time dependence of the incident ($P_{in}$) and transmitted ($P_{out}$) pulses. (*c*) SFG beam geometry with a horizontally mounted sample and a high-pass optical filter (HPF). Due to wave vector matching, tuning the BBIR causes only a small movement parallel to the spectrograph slit. The visible beam diameter is larger than the IR beam. Reproduced with permission from Ref. [7], copyright 2010, Elsevier.

The transmitted spectrum width $\delta\nu$ = FSR/$F$, where the finesse $F = (\pi\,\Re^{1/2})/(1-\Re)$. Figure 3.2*a* shows the spectrum of the initial femtosecond visible pulses ($P_{in}$) and the time-asymmetric picosecond narrow-band pulses after the étalon ($P_{out}$) [6]. Although the highest energies would be generated by tilting the étalon to transmit at the Ti:sapphire laser amplifier spectral peak of 803 nm, we usually fix the visible pulse at 800.0 nm, as shown in the figure, to simplify wavenumber calibration. Without direct knowledge of the visible probe beam wavelength, the absolute frequencies of measured transitions cannot be gauged. Consequently, a fixed grating portable spectrometer (Ocean Optics, Dunedin, FL) was employed to accurately determine the wavelength of the visible pulses. Furthermore, the étalon FSR should be more than twice the 150-cm$^{-1}$ FWHM of the input visible pulse spectrum to avoid transmitted sidebands. Our étalon (TecOptics, Merrick, NY) has an air gap $d$ = 11.1 µm giving an FSR = 450 cm$^{-1}$. In addition, the mirror reflectivity $\Re$ = 93%, so the étalon's finesse $F$ = 43. The transmitted pulses through our optic have $\delta\nu$ = 10.4 cm$^{-1}$.

A short-duration pulse incident on an étalon generates a train of output wave packets separated by the mirror cavity round-trip time. This pulse train decays exponentially with a cavity lifetime $\tau_c = (2\pi\,\delta\nu)^{-1}$. However, our étalon is used in the limit where the gap $d$ is less than the pulse coherence length $\ell_c = t_p c$, where $t_p$ is the pulse duration. The pulse coherence length for a 130-fs pulse is ~40 µm [6], and the étalon gap is ~10 µm. In this case, the individual pulses in the transmitted train overlap in time. The outputted signal is therefore not a series of discrete wave packets but a train of overlapping pulses, which combine to produce a smooth exponentially decaying envelope (Fig. 3.2*b*). This envelope is free from high-frequency interference effects because the pulses in the train are automatically phase locked by the étalon. The transmitted pulse intensity $I_{tr}(t)$ can be approximately represented as the convolution of the input pulse with a decaying single-sided exponential [6],

$$I_{\mathrm{tr}}(t) = (1-\Re)^2\, G_g(t, t_p) * \exp(-\tau_c t) \tag{3.2}$$

where * denotes convolution. A computed intensity profile of the étalon-filtered pulses $P_{out}$, using a Gaussian input pulse $G_g(t, t_p)$ with a duration $t_p$ = 100 fs (FWHM), is shown in Figure 3.2*b* [6]. The measured frequency-domain spectrum of the time-asymmetric pulse $P_{out}$, shown in Figure 3.2*a*, is the Fourier transform of Eq. (3.2), so a resulting Voigt line shape, which is the convolution of a Gaussian and a Lorentzian, is generated. But since $\tau_c >> t_p$, the pulse spectrum is quite close to Lorentzian [6]. One possible drawback of our NR suppression technique, pointed out by Laaser and co-workers [43], results from this near-Lorentzian spectrum of the $\omega_{vis}$ pulse. Since Lorentzians have long, extended wings, SFG spectra obtained with these pulses invariably have extended Lorentzian wings. The SFG vibrational line shapes consequently appear to be Lorentzians even if they are not.

It is useful to keep in mind that, all things being equal, the intensity of the SFG signal is a *quadratically increasing function of the étalon bandpass* [7]. For this reason, the étalon should be designed to generate the broadest possible bandpass compatible with the needed spectral resolution. The spectral bandpass of an étalon

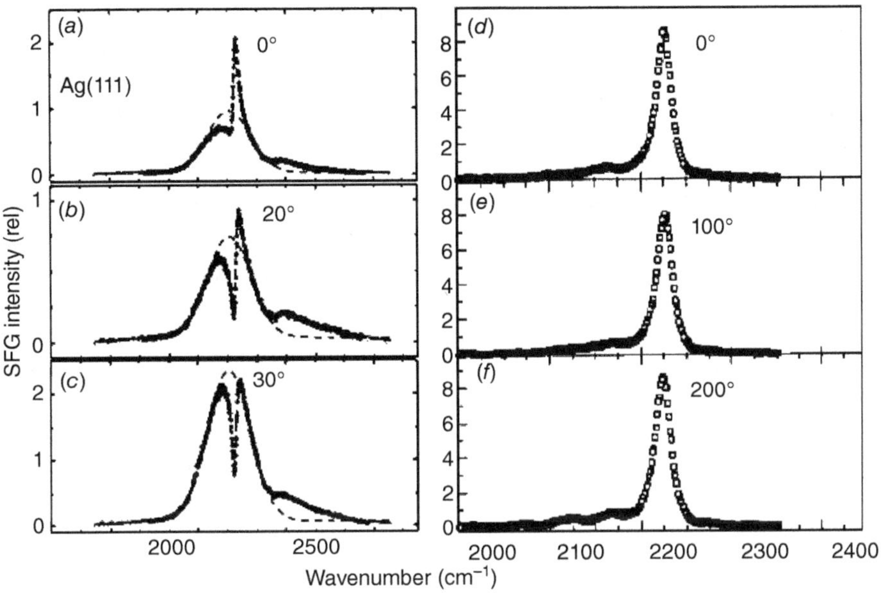


**Figure 3.3** SFG spectra of a 4-cyanobenzenethiol SAM on a single-crystal Ag(111) surface, in the CN stretch region, as a function of crystal rotation about the azimuth. The dashed curves indicate the IR pulse spectrum. (*a*)–(*c*) Without NR suppression, the NR amplitude and phase vary strongly with azimuthal rotation, making it difficult to determine the R amplitude. (*d*)–(*f*) With NR suppression, the R signal amplitude is seen to be independent of azimuthal rotation. The weaker peaks on either side of the prominent transition are caused by spectral chirping of the intense laser pulses as they pass through optical elements. Adapted with permission from Ref. [14], copyright 2009, American Chemical Society.

with a fixed gap increases as the étalon finesse decreases. Decreasing the finesse by a factor of 2 therefore doubles the étalon bandpass. This doubles the transmitted energy and also decreases the transmitted pulse duration by one-half. Thus, doubling the étalon bandpass increases the visible pulse intensity and the SFG signal intensity by a factor of 4. For example, if the materials of interest have vibrational transitions with FWHM, let us say, of 10 cm$^{-1}$, then an étalon bandpass of 8 cm$^{-1}$ will not broaden the transitions by more than 3 cm$^{-1}$. On the other hand, an étalon bandpass of 4 cm$^{-1}$ will broaden the transitions by no more than 1 cm$^{-1}$, but this 2 cm$^{-1}$ reduction in spectral width will come at a cost of a factor of 4 in the signal intensity.

An example of the benefits of NR suppression is shown in Figure 3.3, where SFG spectra were obtained from a SAM of 4-cyanobenzenethiol on a single-crystal Ag(111) surface [14] with and without NR suppression. The BBIR pulses were tuned into the 2230 cm$^{-1}$ CN stretch transition [14]. SAMs of this sort typically exist in rotational domains that are much smaller than the IR beam diameter of 200 μm or more [44]. The domains are related to each other by a rotation about the surface normal, so spectroscopic measurements sample a large number of equivalent domains [45]. One way to learn about the orientation of the SAM molecules and the SAM domain structure is to measure the spectrum as a function of Ag crystal rotation about the azimuth. If the domains were randomly oriented, then the R amplitude of SAM transitions would be invariant to azimuthal rotation. However, in prior works, other researchers observed surprising strong azimuthal angle dependences of the R amplitudes from related monolayers such as octadecanethiol [45, 46], which led them to propose novel structures for the SAMs at odds with what had been measured by other techniques [47, 48].

In Figure 3.3, the 4-cyanobenzenethiol SAM spectrum is dominated by a single resonance and an NR background. As the Ag single crystal was rotated azimuthally over a 30° range, the SFG resonant spectrum changed from a peak to a derivative shape to a dip (Figs. 3.3*a–c*) because the NR intensity and phase depend strongly on crystal orientation [14]. The large orientation dependence of the NR signal makes it difficult to study the orientation intensity dependence of resonant SAM transitions. When time windowing was used to suppress the NR background, as shown in Figures 3.3*d–f*, the resonant signal became a single peak [14]. The weaker tails at lower wavenumbers result from modulation of both the IR pulses as they interact with atmospheric $CO_2$ and the laser system's intense femtosecond pulses as they pass through various optics on the way to generating the IR via pumping an optical parametric amplifier. These effects created a minor component to the IR pulse that had a complicated temporal dependence, which hindered the complete suppression of the NR signals. However, these modulations can be eliminated by taking care to purge the spectrometer with dry, $CO_2$-free air and by slightly decreasing the IR pulse intensity. From Figures 3.3*d–f*, it is easy to see that the resonant signal intensity is actually independent of orientation [14]. The complicated SAM structures previously proposed to explain the orientation-dependent SFG intensities are not needed. The few-percent variation in the signal intensity in Figures 3.3*d–f* is attributed to variations in optical alignment as the crystal was rotated [14].

## 3.5 INSTRUMENTATION FOR SFG SPECTROSCOPY

Our SFG spectrometers are based on femtosecond chirped-pulse amplified Ti: sapphire lasers that produce tunable mid-IR pulses by pumping optical parametric amplifiers (OPAs) with difference frequency mixing (DFM) stages. Our lasers were manufactured by Quantronix Corp. (Santa Clara, CA) and the OPAs by Light Conversion, Inc. (Vilnius, Lithuania). Similar lasers and OPAs are available from several manufacturers, and although the construction and design differs among manufacturers, the output pulses are quite similar. The Ti: sapphire lasers in our two SFG systems output 800-nm pulses at a 1-kHz repetition rate. The ultrafast system [8, 9, 16, 19, 34], depicted schematically in Figure 3.4*a*, produces 4.5-mJ pulses with a duration of 90 fs. The compact system [7], depicted schematically in Figure 3.5, produces 2.0-mJ pulses with a duration of 120 fs. The extra energy of the ultrafast system is needed to generate the pump pulses. However, the SFG section of the two systems are similar: The laser beam is split into two parts. One part pumps the OPA to produce the BBIR pulses, and the other is transmitted through an étalon to create the $NB_{vis}$ pulses. After a variable delay stage, the beams are made spatially and

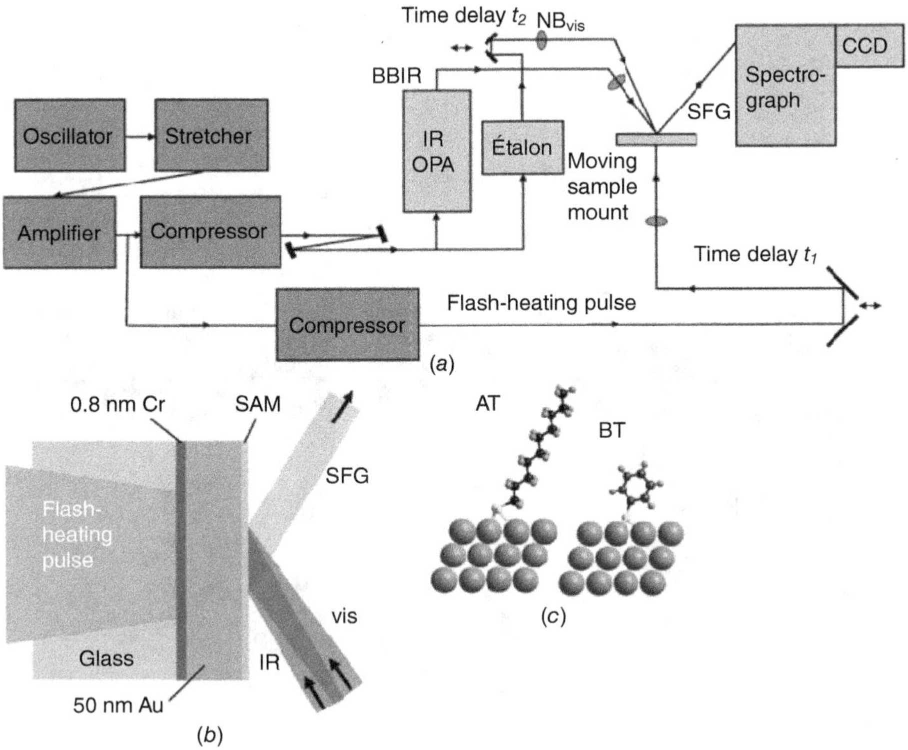


**Figure 3.4** (*a*) Block diagram of the ultrafast laser apparatus used for pump–probe measurements to determine the effects of temperature jumps (T-jumps) on SAMs. (*b*) Schematic of the T-jump substrate, flash-heating pulse, SAM layer, and SFG probe. (*c*) Structure of a single molecule of an alkanethiol (AT) or benzenethiol (BT) SAM on an Au surface. Although the figure shows only a single SAM molecule, the experiment, on each shot, probes ~$10^{11}$ molecules of a densely packed SAM. Adapted with permission from Ref. [8], copyright 2009, American Chemical Society.

approximately temporally coincident on the sample surface, and the SFG signal is directed into a spectrograph with a charge-coupled device (CCD) detector. The compact system was described in detail by Lagutchev and co-workers [7]. A photograph of the compact system is shown in Figure 3.5. Including a spacious sample chamber large enough for a spectroelectrochemical cell or a crystal goniometer, it fits on an optical bench $2 \times 0.75\ m^2$ [7].

The only parts in a broadband multiplex SFG system that need to move once an experiment is set up are the motors that tilt the mirrors and crystals in the OPA to tune the BBIR wavelength. As this wavelength is tuned, there are variations in the IR beam pointing and diameter at the sample. In order to reduce the sensitivity of the SFG signal to these variations, we make the visible beam diameter (referring to $1/e^2$ intensity points) somewhat larger than the IR (Fig. 3.2*c*) [7]. The visible lens was selected to give a focused diameter of ~375 μm, and the IR lens was chosen to produce a focused beam that, depending on wavelength in the 3- to 12-μm range,

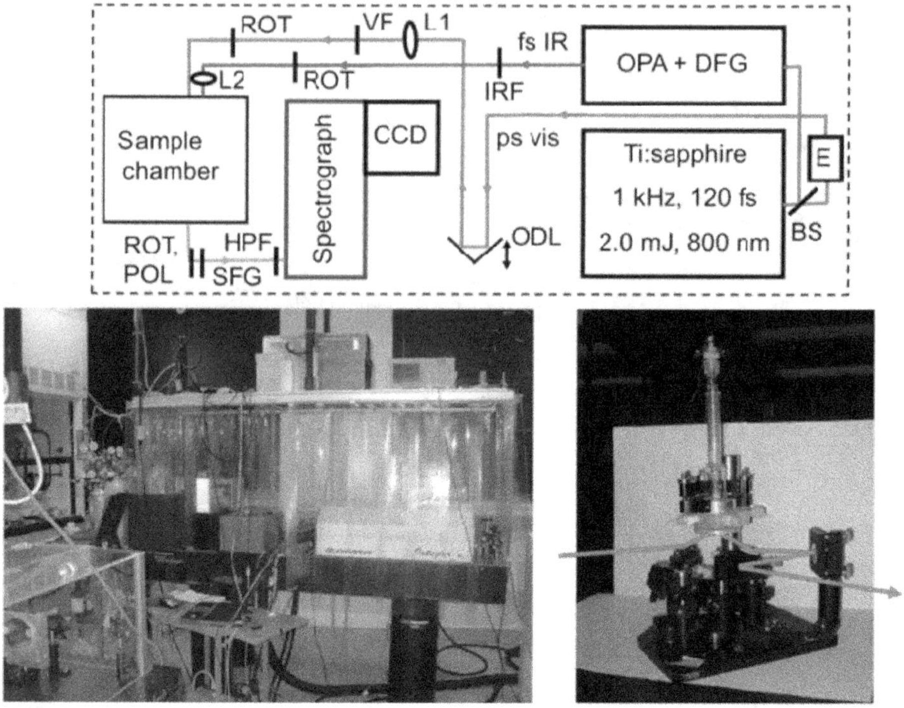


**Figure 3.5** (*Top*) Block diagram of a compact SFG spectrometer. Key: OPA: optical parametric amplifier; DFG: difference frequency generator; L1 and L2: lenses; ROT: polarization rotator; VF: filter that transmits only visible light; IRF: filter that transmits only IR light; BS: beam splitter; E: étalon; ODL: optical delay line; POL: polarizer. (*Bottom left*) Photo of the spectrometer on a $2 \times 0.75\ m^2$ optical bench. (*Bottom right*) Photo of the spectroelectrochemical cell utilized for lithium ion battery studies. The inserted line denotes the path of the SFG probe beams and the generated signal. Reproduced with permission from Ref. [7], copyright 2010, Elsevier.

varied from about 125 to 250 μm [7]. The IR focusing lens was antireflection-coated ZnSe with a 10-cm focal length. It is important to not use IR lenses of materials such as $CaF_2$ or $MgF_2$. Their low refractive indices mandate large lens curvatures, resulting in large spherical aberrations and poor focusing.

The IR and visible beams make an angle of $\theta$ ~60° with the surface normal, as depicted in Figure 3.2*c*. Relative to the wave vectors of the reflected visible and IR pulses, the SFG wave vector is given by

$$\vec{k}_{\mathrm{SFG}} = \vec{k}_{\mathrm{IR}} + \vec{k}_{\mathrm{vis}} \tag{3.3}$$

The IR wave vector is *much smaller* than that of the visible beam. *Therefore, regardless of the IR parameters, the SFG beam wave vector and divergence are always quite similar to that of the visible beam* [7].

The Richter SFG arrangement [23] employed a dichroic IR–visible beam splitter to make the two input beams collinear. We found this arrangement loses quite a bit of pulse energy, enhances the difficulty of proper focusing by utilizing a common lens for the two beams, and generates a spurious SFG signal at the beam splitter surface. It is preferable to use separate focusing optics for IR and visible beams due to their very different wavelengths and beam properties. This preference necessitates the beams to be noncollinear. However, when the IR and visible beams are noncollinear, tuning the IR causes the direction of SFG signal emission to vary, according to Eq. (3.3), even if the IR beam position does not wander at all. Since the SFG signal is typically invisible to the naked eye, these directional changes can hinder efforts to maintain alignment with the detector. For instance, tuning the IR could cause the SFG signal to walk off the narrow entrance slit of the spectrograph. In order to mitigate the detection system sensitivity to IR wavelength tuning, we rely on the fact that the spectrograph slit is narrow in width but much larger in height. We use the noncollinear noncoplanar arrangement shown in Figure 3.2*c* [7]. The focused visible beam is reflected off the sample surface, through a collimating lens, and into a lens that directs and focuses the beam onto the spectrograph slit. We then bring the IR beam at a small (5–10°) angle $\phi$ above (or below) the plane of visible incidence and reflection. The SFG signal then always lies in the plane defined by the visible and IR beams and will be quite close to the visible beam (Fig. 3.2*c*). Consequently, when the visible beam is aligned to pass through the spectrograph slit, the SFG signal will also pass. Most importantly, when the IR wavelength is varied, the SFG beam moves only slightly, and this motion is *along the height axis of the slit*. But the SFG signal always remains centered on the slit width axis and thus always enters the spectrograph. Since the CCD detector is 2.0 mm in height, it is possible to be certain that the SFG beam will always land on the active region of the detector no matter what IR wavelength is used [7]. Finally, a high-frequency pass filter (Omega Optical, Brattleboro, VT) is inserted before the slit to block 800.0 nm and transmit the SFG signals.

## 3.6 INTERFACIAL DYNAMICS VIA ULTRAFAST TEMPERATURE JUMP

The experiments described in this section used SFG to investigate the response of SAM molecules adsorbed on a metal surface, after the metal was perturbed by a large-amplitude temperature jump (T-jump). With the flash-heating technique employed here, after a transient period lasting a few picoseconds, the metal surface equilibrated at an elevated temperature that persisted for a few tens of nanoseconds [10, 18, 19]. In this way, as illustrated in Figure 3.4*c*, heat can be input to the molecules at a specific location, namely through the Au–S adbond, and the arrival of the heat at another location on the molecule, for instance, at terminal methyl groups of alkane chains, can be monitored. These experiments address our lack of knowledge about heat transfer at the molecular level [10], which is crucial to support applications such as in molecular electronics, where individual molecules may function as conductive wires, diodes, transistors, and so on [49–53].

We have developed an apparatus to perform these ultrafast flash thermal conductance measurements [10] as diagrammed in Figure 3.4*b*. On a large area ($5 \times 5$ cm$^2$)

glass substrate, we deposit a metal layer and SAM of interest (e.g., the alkane thiols or benzenethiols depicted in Figure 3.4*c*). Usually, we use 50-nm Au layers with a thin 0.8-nm Cr adhesion layer [18]. The flash-heating pulses are directed through the glass onto Cr, which is partially transmitting, so the pulses heat both Cr and Au. The flash-heating pulses, however, do not heat the SAM directly. To detect the molecular response, SFG probe pulses are directed onto the SAM adsorbed on Au. Due to the high laser intensities and large-amplitude T-jumps, the large-area (25-cm$^2$) sample is kept in continuous motion, in a Lissajous scan pattern, so the laser does not dwell very long on any individual spot.

To assist us in interpreting these experiments, we [10] and our collaborators in the Hase group [54, 55] have performed molecular dynamics simulations of flash-heated alkylthiolate SAMs as depicted in Figure 3.6. These chains at ambient temperature form well-ordered layers, but a few picoseconds after the T-jump, the chains

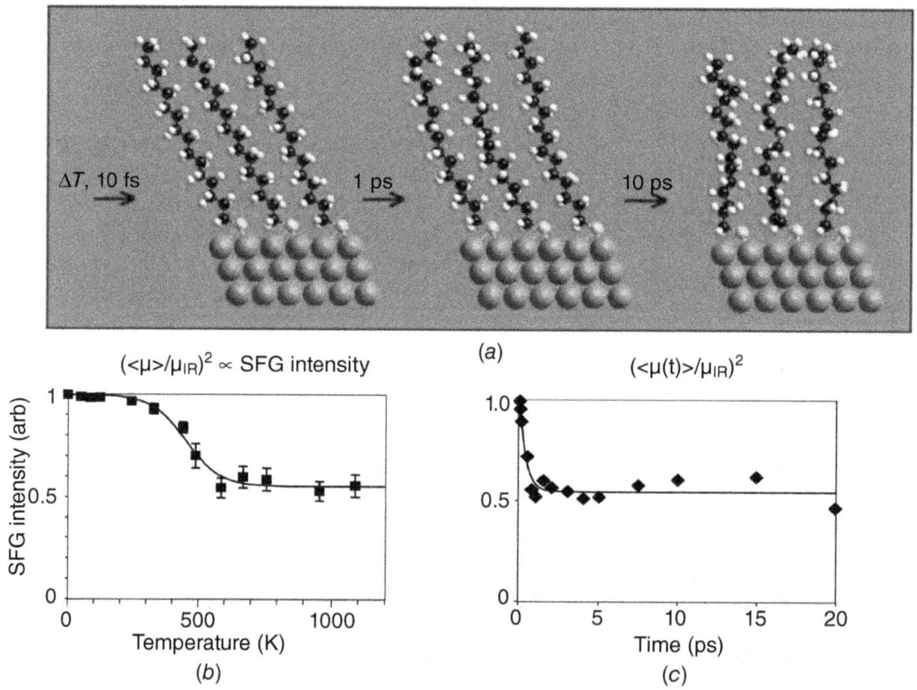


**Figure 3.6** Results from molecular simulations of alkanethiol SAMs. (*a*) Structure of an alkanethiol SAM with 16 carbon atoms. Simulations were performed on a unit cell of 27 alkanes with periodic boundary conditions. When the system was jumped to a high temperature, the methyl head groups became orientationally disordered. (*b*) The SFG intensity for the $\nu_s CH_3$ transition (symmetric methyl CH stretch) is approximately proportional to the square of the normalized ensemble-average IR dipole moment, $(\langle\mu\rangle/\mu IR)^2$, which is temperature dependent and is predicted to decrease linearly with temperature in the 300–500 K range. (*c*) With an instantaneous temperature jump, the methyl head groups become orientationally disordered in less than 2 ps. Reproduced from Ref. [10], copyright 2007, American Association for the Advancement of Science.

become disordered (Fig. 3.6*a*). Recall that with alkylthiolate SAMs, SFG probes the terminal methyl groups only, and the SFG signal is proportional to the square of the ensemble-average molecular hyperpolarizability. As the chains undergo heat-induced disorder, the SFG signal decreases because the terminal methyl groups adopt a variety of conformations, reducing the ensemble average. Our simulations [10] calculated the square of the ensemble-average dipole moment,

$$\frac{\langle \mu \rangle^2}{\mu_{IR}^2} = \left( \frac{(1/n)\sum_{i=1}^{n} \mu_i}{\mu_{IR}} \right)^2 \tag{3.4}$$

as a surrogate for hyperpolarizability, and they show (Fig. 3.6*b*) that as temperature is raised from 300 to 600 K, the SFG signal would be expected to decrease almost linearly with increasing temperature. Therefore, SFG of the alkane SAMs is a kind of molecular thermometer. Our simulations also demonstrate (Fig. 3.6*c*) that if we instantaneously T-jump all the atoms (metal surface and SAM together), the chain response is ~1 ps. Therefore, the chains can function as molecular thermometers with an intrinsic response time of ~1 ps, and an observed response slower than 1 ps should be attributed to heat flow rather than the finite response time of the thermometer [10]. Figure 3.7 shows NR-suppressed SFG spectra of ODT in the CH-stretch regime in response to a flash-heating pulse arriving at the Au surface at $t = 0$. These NR-suppressed spectra have much better signal-to-noise ratios than obtained in our original experiments [10], which were performed prior to the development of NR suppression. The T-jump causes the SFG transitions to lose intensity in a process that takes tens of picoseconds. The SFG intensity loss represents time needed for

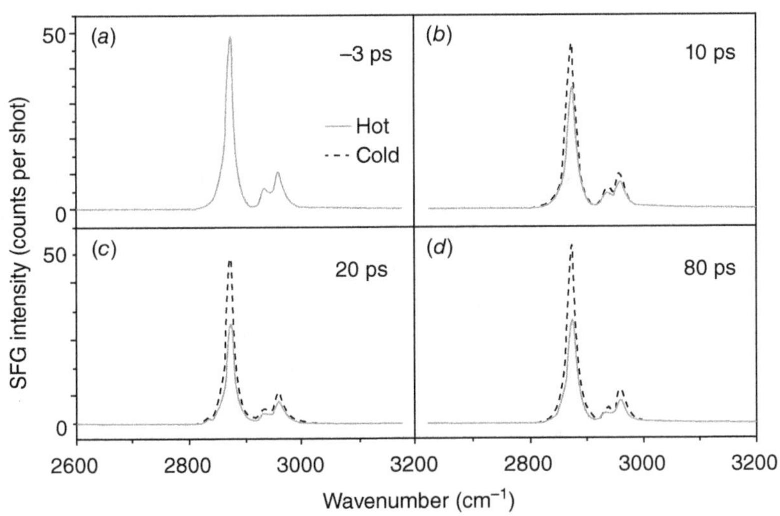


**Figure 3.7** Spectra from temperature jump measurements of an ODT SAM on Au using NR suppression. The flash-heating pulses arrive at $t = 0$. The spectra denoted "cold" were obtained without flash heating.

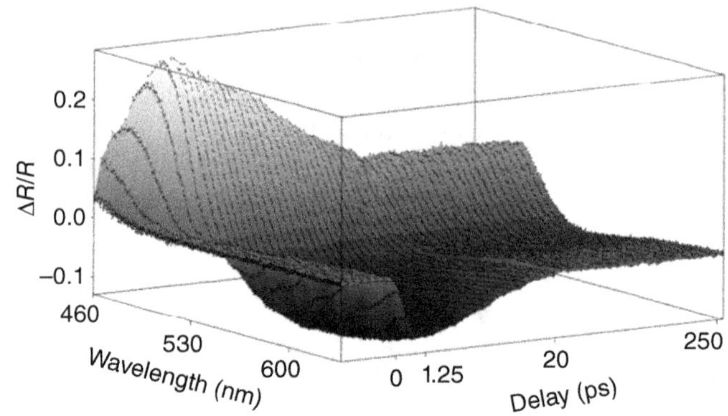


**Figure 3.8** Time-resolved reflectivity changes of a Au surface (0.8 nm Cr + 50 nm Au on a glass substrate) irradiated with 400-nm pulses at a fluence of 12.7 mJ $cm^{-2}$, which produced an equilibrium temperature jump of 200 K. Temperature measurements are usually made at 600 nm, in the Drude region, away from the Au interband transition near 510 nm.

ambient temperature alkyl chains to thermally equilibrate with the flash-heated Au surface.

In order to interpret these measurements, transient reflectance experiments were performed to determine the temperature of the Au surface [8]. We find it useful to monitor the reflectivity at 600 nm and showed that the fractional reflectivity changes at this wavelength, $\Delta R/R$, off Au surfaces decreased with temperature according to the optical thermometry equation,

$$\Delta R / R = -1.32(\pm 0.02)\times 10^{-4}\Delta T \tag{3.5}$$

in agreement with Beran [56]. This optical thermometry calibration for the electron-beam-deposited Cr/Au samples used in our laboratory was achieved by monitoring reflectivity with the substrate in a resistively heated oven in contact with a thermocouple. The reflectance change $\Delta R/R$ is not linear in $\Delta T$ at all wavelengths but that is the case at 600 nm. As shown in Figure 3.8, the flash-heating process was studied by time-resolved reflectivity measurements using femtosecond white-light continuum pulses. Figure 3.8 shows that $\Delta R/R$ has a transient part induced by the pump pulses that is complicated and wavelength dependent. The reflectance change then equilibrates at a (wavelength-dependent) constant value after a few picoseconds. The nature of the transient part is understood in terms of a two-temperature model [57, 58]. In that model, light pulses couple to the Au electrons but not the lattice. The light-driven electrons equilibrate among themselves very rapidly, so the electrons can be described by an electronic quasi-temperature $T_e$. At shorter times, Au is not in thermal equilibrium. The electrons are much hotter than the lattice. The lattice is subsequently heated by these hot electrons as a result of electron–phonon coupling. Since the electron heat capacity is much less than the lattice heat capacity, with a short flash-heating pulse that ultimately produces an equilibrium T-jump $\Delta T$, $T_e$ is initially much greater than $\Delta T$. The initial value of $T_e$ depends on the flash-heating pulse intensity and duration, but in our experiments where $\Delta T$ is on the order of 100 K, the initial value of $T_e$ is ~10,000 K [57, 58].

The reflectivity transients at shorter times are associated with the combination of temporarily very hot electrons and a much colder lattice. The reflectivity transients might also have a contribution from heat conduction from the back (irradiated) surface to the front surface, but this phenomenon is expected to be much faster than the electron cooling and lattice heating. The reflectivity changes during this transient period do not indicate actual temperatures. According to Figure 3.8, the transient period, which represents the time for electron–phonon equilibration and which depends on $\Delta T$ and the thickness of the Cr adhesion layer [18], is characterized by a time constant of a few picoseconds. After this time, the reflectivity transients can be associated with an equilibrium temperature jump.

A better view of the T-jump dynamics of the Au layer is found by taking slices, specifically at 600-nm wavelength, of data such as that in Figure 3.8. As shown in Figure 3.9, we can control $\Delta T$ (proportional to the longer time, >10 ps, reflectance changes) by varying the flash-heating pulse energies or wavelengths. We are limited by practical considerations in the size of the $\Delta T$ we can make due to the need to signal-average many laser shots to obtain good data. After a spot of the Au/Cr coating has been exposed to a few thousands of flash-heating pulses, it begins to change color, which is symptomatic of multiple-pulse optical damage of the metal thin film. Because the sample is constantly in motion, it takes millions of pulses before this damage becomes noticeable, but in order to get good data, we have to keep the pump pulse energies below the values that create even small amounts of coating damage. To make sure we have a generous margin of safety, the maximum

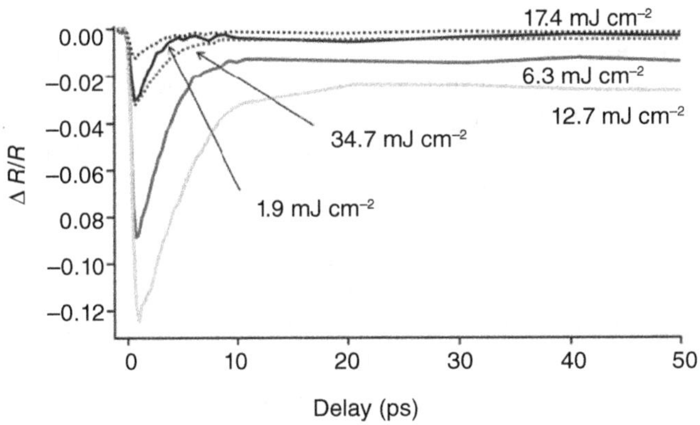


**Figure 3.9** Time-dependent reflectivity changes of flash-heated Au (0.8 nm Cr + 50 nm Au on a glass substrate) pumped on the Cr side by 800-nm (dashed curves) or 400-nm (solid curves) pulses at indicated fluences. The shorter-time transients are caused by the temporarily diminished reflectivity due to nonequilibrium hot electrons. After a few picoseconds, the Au electrons and phonons are in thermal equilibrium, and this equilibrium high temperature persists for tens of nanoseconds. The largest temperature jump was 200 K.

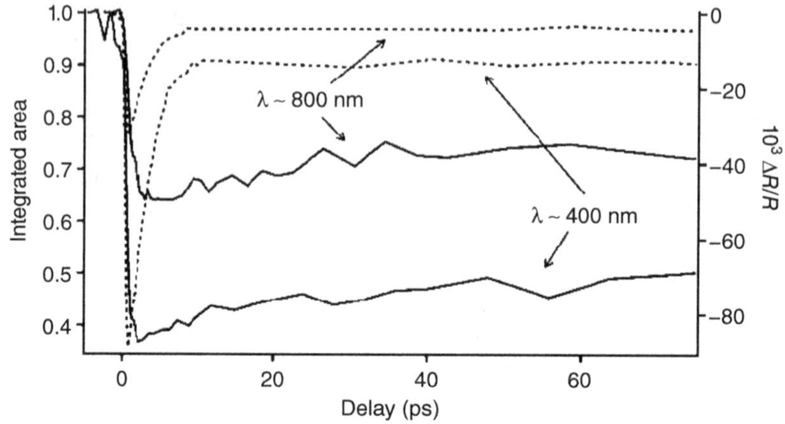


**Figure 3.10** Dynamics of a 4-nitrobenzenethiolate (NBT) monolayer induced with 800- and 400-nm flash-heating pulses and measured via SFG of the nitro symmetric stretch near 1340 $cm^{-1}$ (solid curves, left axis). With 800-nm pulses (27.9 mJ $cm^{-2}$), $\Delta T = 30$ K. With 400-nm pulses (7.45 mJ $cm^{-2}$), $\Delta T = 120$ K. Also shown are the corresponding Au surface reflectivity transients (dashed curves, right axis). During the initial period, when hot nonequilibrium electrons are present in Au, the SFG signal from the SAM overshoots the value reached when the Au surface is in thermal equilibrium at the elevated temperature.

$\Delta T$ with 800-nm pumping was kept below 35 K, and with 400-nm pumping, 200 K was achievable. In some previous works prior to our development of the transient thermoreflectance apparatus, we used only 800-nm flash-heating pulses. Our estimates of $\Delta T$ for these works were too large [10, 18], but now we can accurately measure $\Delta T$ with any flash-heating pulses.

Figure 3.10 shows the response to a T-jump of a SAM comprised of 4-nitrobenzenethiolate (NBT) molecules, where SFG was used to probe the symmetric $NO_2$ stretching transition at 1340 $cm^{-1}$. The T-jumps were 30 K when 800-nm pulses were used and 120 K when 400-nm pulses were utilized. The associated Au surface reflectivity transients are also plotted in Figure 3.10. As with the alkyl SAMs, T-jump causes the SFG signals to lose intensity, presumably due to NBT SAM thermal disordering. But something unusual is observed during the first 10 ps. The SFG signal drops to a minimum and then recovers somewhat before an equilibrium is reached at the new elevated temperature. We call this effect “overshoot” [19]. As discussed previously, overshoot is associated with the transient population of hot electrons created by the flash-heating process. When the very hot electrons are present, they can tunnel out of the metal surface onto vacant SAM molecular orbitals and then return to Au [19]. The return process is very fast because the electrons are strongly attracted by the positive image charges created on the metal surface, so the electrons spend very little time on the SAM molecules. Although most electron tunneling and return processes are elastic, having little net effect on the SAM molecules, occasionally there is an inelastic process where the electrons transfer energy to the SAM molecule vibrations. The electron–SAM energy transfer process can affect the

SFG signal in two ways. First, the transient vibrational population can thermalize and heat the SAM, causing it to become disordered. In this case, the SAM could temporarily be hotter than the Au lattice. Second, if the vibration being probed (in this case the nitro stretch) becomes temporarily excited, the SFG signal will temporarily decrease because the vibrational ground states that generate the SFG signals are depleted. It is this second effect that we believe is the cause of overshoot in NBT and other benzenethiols we have studied [19]. The overshoot emergence in NBT occurs as the hot electrons temporarily excite the nitro stretch being probed, and the overshoot decay, which occurs with a time constant of ~10 ps, results when the nitro stretch vibrational energy is redistributed into other NBT vibrations, primarily those involving the phenyl ring.

In order to get a better idea of how the electron heating mechanism works, we purchased or synthesized molecules to fabricate the alkylbenzenethiolate SAMs depicted in Figure 3.11 [19]. We call these "phenyl–linker" molecules, with the "linker" being the alkyl chain [19]. With these SAMs, SFG near 3100 $cm^{-1}$ probes a vibration of only the C and H atoms at the para position of the phenyl groups [16]. The hyperpolarizabilities of the two CH groups at the meta position are identical and opposite in sign, and the same is true for the two CH groups at the ortho position. Consequently, these four atomic groups do not contribute to the SFG signal [16]. This series of phenyl–linker molecules allows us to systematically increase the

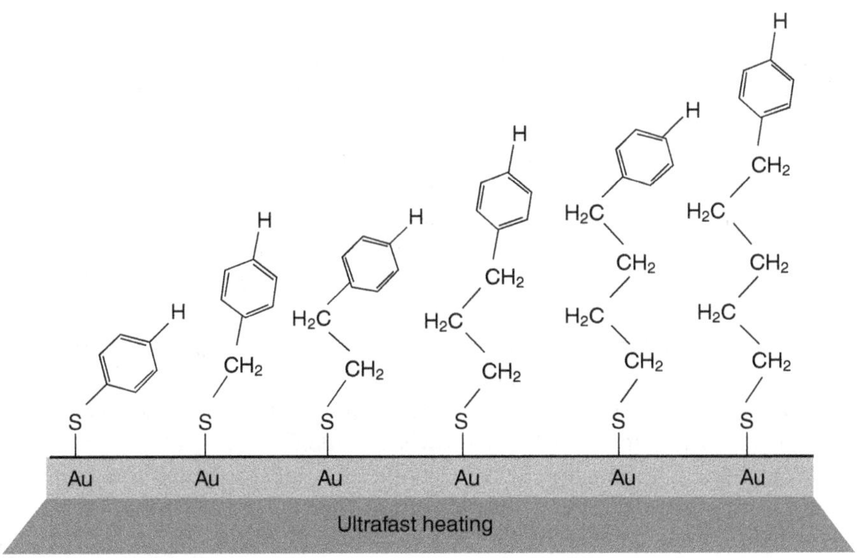


**Figure 3.11** Structures of phenyl–linker molecules employed to fabricate SAMs in order to study the distance dependence of SAM heating by nonequilibrium electrons from the Au surface. With these molecules, SFG probes the CH stretch transition originating from the CH group at the para position. Adapted with permission from Ref. [19], copyright 2009, American Chemical Society.

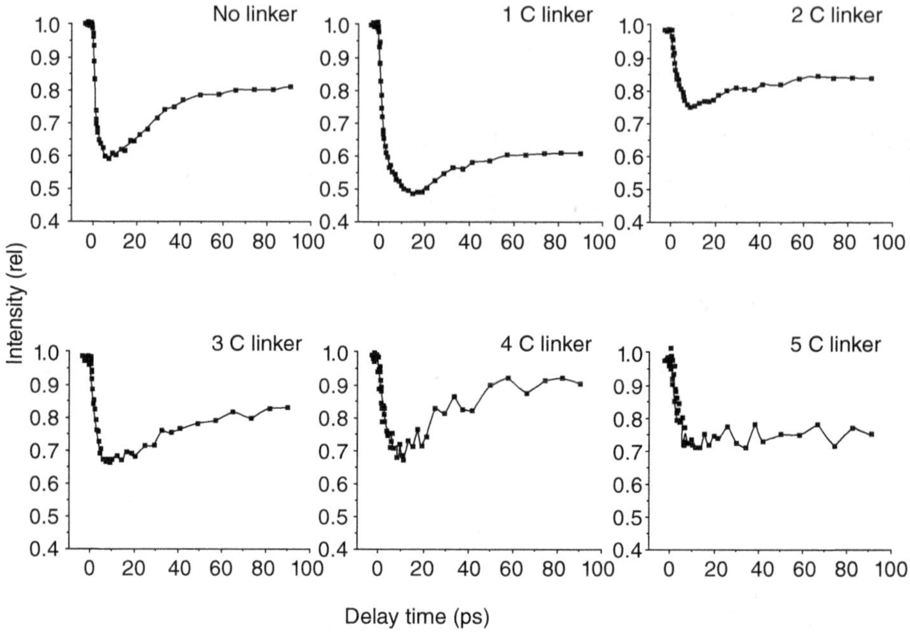


**Figure 3.12** SFG transients originating from flash heating a Au surface coated by a SAM of the phenyl–linker molecules illustrated in Figure 3.11. With no linker or a one-carbon linker (i.e., the linker is $-CH_2-$) an overshoot is prominent. When the linker has five carbon atoms, the overshoot is absent. Reproduced with permission from Ref. [19], copyright 2009, American Chemical Society.

distance of the probed CH moieties from the Au surface. The distance increases in a staggered manner depending on whether there is an even or odd number of carbon atoms in the alkane chain. The SFG pump–probe data are shown in Figure 3.12 [19]. The overshoot due to electron excitation of the CH groups is prominent when there are zero or one $CH_2$ linker groups. However, the overshoot fades in prominence with increasing stand-off distance, and when there are five carbon atoms between Au and phenyl, the overshoot is gone. These measurements indicate that the hot surface electrons cannot effectively excite molecular moieties more than four carbon atoms away from the surface.

We made detailed studies of the chain length dependence on even-carbon-atom alkyl chains, where the chain length ranged from 6 to 24 carbon atoms [10]. The T-jump was induced by 125-fs, 800-nm flash-heating pulses that produced temperature jumps of 30 K. Figures 3.13*a*,*b* show some representative data from an ODT (18-carbon) SAM. The data are in the form of a vibrational response function (VRF) [10]. Because the original SFG intensities and the T-jump-induced SFG signal reductions were different in magnitude for the various length alkane chains, we computed the VRF for each chain length to facilitate comparison. The VRF is just a normalized version of the time-dependent SFG signal intensity, which varies from zero to unity.

In Figure 3.13*a*, the small peak at $t = 0$ is not due to the SAM response to a T-jump. It is a coherent artifact due to interactions of the pump and probe pulses. It

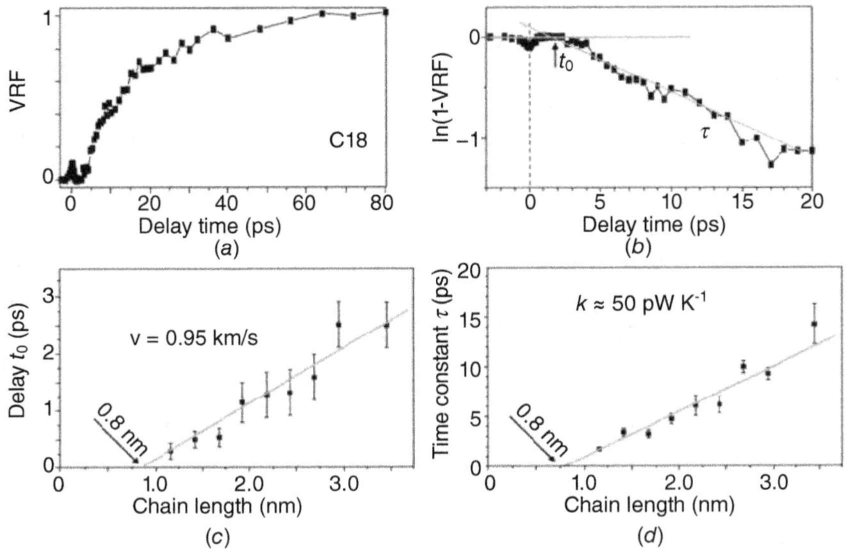


**Figure 3.13** Temperature jump dynamics probed via SFG of alkanethiol SAMs on Au. (*a*) Vibrational response function (VRF) for the symmetric CH stretch mode of an 18-carbon SAM (ODT). The small peak near zero delay is an artifact resulting from coupling between the pump and IR probe pulses that is used as a fiducial marker. (*b*) Semilog plot of 1-VRF from (*a*) showing the onset delay time $t_0$ and the exponential decay time constant $\tau$. (*c*) Dependence of $t_0$ on chain length for chains ranging from 6 to 24 carbon atoms. The slope gives a ballistic velocity for heat bursts traveling along alkane chains of 0.95 km/s. (*d*) Dependence of $\tau$ on chain length. The slope can be used to compute an interface thermal conductance of 50 pW K$^{-1}$. Reproduced with permission from Ref. [10], copyright 2007, American Association for the Advancement of Science.

is a convenient fiducial marker to denote time zero, which is when the pump pulses arrive at the Au surface [10]. By definition, the VRF starts at zero and rises to unity. Figure 3.13*b* is a semilog plot of 1-VRF for ODT, which is a function that decays in time. This plot shows that the VRF does not start to decay until a few picoseconds after time zero, and, subsequently, the decay can be approximated by an exponential function. This characteristic time dependence was observed to be true for all even carbon chain lengths [10]. So we fit the VRFs with a function having two adjustable parameters $t_0$ and $\tau$,

$$\mathrm{VRF}(t) = 0 \quad \text{for} \quad t < t_0$$

and

$$\mathrm{VRF}(t) = \left[1 - \exp\left(\frac{t - t_0}{\tau}\right)\right] \quad \text{for} \quad t \geq t_0 \tag{3.6}$$

The parameter $t_0$ represents the time delay between the flash-heating pulse arrival and the time when the terminal methyl groups of the SAMs begin to respond. The parameter $\tau$ represents the time needed for the SAM to equilibrate with the hot Au surface. Figures 3.13*c*,*d* give the measured values of $t_0$ and $\tau$ for chains whose lengths varied from 6 to 24 carbon atoms, which corresponded to chain lengths of 1.2–3.5 nm [10]. Both $t_0$ and $\tau$ were observed to increase linearly with chain length but for different reasons. In the interface thermal conductance model [59, 60], heat flows across the Au–SAM interface at a rate proportional to the temperature difference. So longer chains heat up more slowly because they have a larger heat capacity. More heat has to flow across the interface to heat a longer chain to the final equilibrium temperature. Since the chain heat capacity increases almost linearly with chain length, $\tau$ increased linearly with chain length as well [10].

The delay time $t_0$ emerges from our ability to selectively and impulsively input heat to the base of the chain (i.e., the sulfur atom and the first few carbon atoms) and to selectively and rapidly probe the terminal methyl groups at the ends of the chains [10]. Measurement of this heat travel time is not possible with conventional thermal measurements such as those made by Wang and co-workers [61]. A heat burst from the Au substrate travels along the chains, but only after the leading edge reaches the terminal methyl groups does the SFG optical thermometer begin to register an effect. Thus, $t_0$ is interpreted as the time for heat to travel from the Au surface to the ends of the alkane chains. Our experiment is an ultrafast version of a flash thermal conductance apparatus. The linear dependence of $t_0$ on chain length indicates that the leading edge of the heat burst propagates ballistically, as opposed

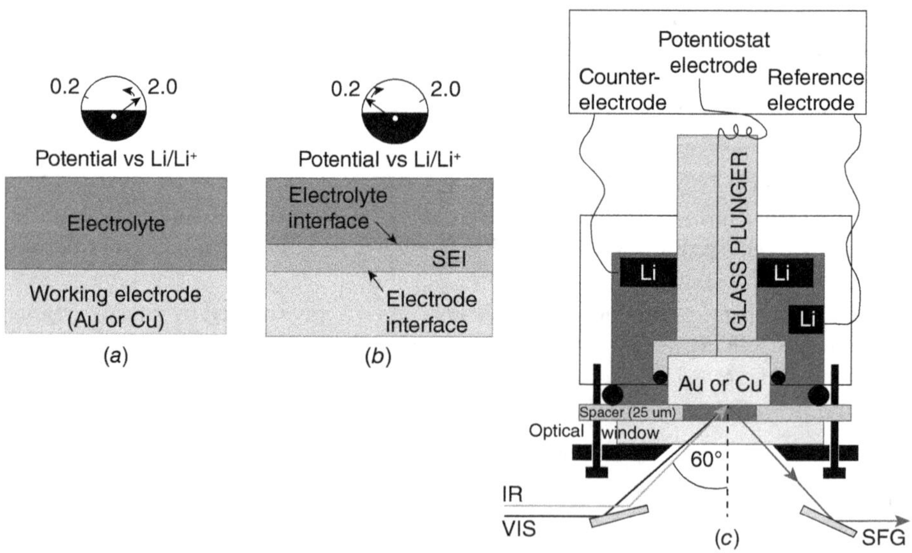


**Figure 3.14** (*a*) Electrolyte–electrode interface under potential control near the open-circuit potential. (*b*) As the potential is reduced, electrolyte reduction and lithium underpotential deposition create a solid-electrolyte interphase (SEI) layer on the electrode. When probed by SFG, two interfaces will be seen, the SEI–electrolyte interface (the "electrolyte interface") and the electrode–SEI interface (the "electrode interface"). (*c*) Schematic of the spectroelectrochemical cell employed for the SFG probing experiment. A photo of the cell is shown in Figure 3.5. Reproduced from Ref. [63], copyright 2012, The Materials Research Society.

to diffusively, along the chains, and the slope of the data in Figure 3.13*c* gives a velocity of 0.95(±0.1) nm/ps = 0.95 km/s. In contrast to diffusive heat flow through macroscopic conductive wires, our results indicate that heat is flowing ballistically at the molecular level through the alkane chains.

The data sets in Figures 3.13*c*,*d* have an additional interesting feature. The linear fits to the data both intercept the *x* axis at a chain length of 0.8 nm, which is close to the length of a chain with four carbon atoms [10]. Taken literally, this means that the time needed for heat to be transported along chains four carbon atoms or fewer is zero. Our interpretation of this result is that the Au surface transfers energy to the chains not only at the attachment point (the Au–S bond) but also to the first few carbon atoms [10]. Based on the phenyl-linker molecule experiments, one possible explanation is that hot electrons from the Au surface primarily heat the alkane chains along the first four carbon atoms [50]. A second possible explanation emerges from a quantum mechanical treatment of heat flow from hot Au to an adsorbed SAM developed by Segal and co-workers [50]. A physical picture of the surface–SAM coupling has the heavy Au atoms on the surface vibrating as a result of the T-jump. There is a poor impedance match between the heavy Au atoms and the lighter carbon atoms. The mass mismatch between Au and C means that the coupling needed to transfer vibrational energy from the surface to the base of the chains would increase if energy was transferred between Au and several carbon atoms in aggregate. This effect is due to the combined masses of the several carbon atoms being more closely matched to the Au atoms. However, such a coupling would also decrease with distance from the surface. The maximum coupling, as expressed by the atomic motions associated with the modes responsible for the most efficient energy transport, is somewhere around four carbon atoms, which is a compromise between the need to increase mass and to decrease interaction distance.

## 3.7 LITHIUM BATTERIES

We have used SFG to study the in situ development of a solid electrolyte interphase (SEI) during successive charge–discharge cycles in a model system relevant to the lithium ion battery [62, 63]. SFG has not previously been used to study SEI formation, although we should mention the study by Liu and co-workers [64]. They utilized the related technique of second-harmonic generation (SHG) to study propylene carbonate adsorption on $LiCoO_2$ without potential control and vibrational selectivity [64]. A lithium ion battery is a high stored-energy system, which has many potentially undesirable reactions that can degrade the capacity or even cause the battery to burn up or explode. The SEI plays a crucial role in preventing such reactions. The SEI is a nanoscopic layer that develops on Li battery anodes due to electrodecomposition of the electrolyte [65]. The formation of an SEI protects the remaining electrolyte from further decomposition, helps stabilize the electrodes that undergo repeated cycles of volume expansion and collapse as lithium ions are forced into and out of the electrodes, and improves the battery capacity.

Solid electrolyte interphase structure has been studied extensively over the last decade [66], but a detailed understanding of SEI formation and dynamic behavior is needed to develop a fundamental scientific understanding of Li ion battery chemistry, Li transport processes, and the roles of stabilizing additives. It has been recognized that in situ studies can be particularly valuable [66]. SEI growth on an anode is potential dependent as illustrated [62] in Figure 3.14. Prior to potential cycling (Fig. 3.14*a*), a pristine electrode–electrolyte interface is present. As the potential is lowered to 0.2 V (Fig. 3.14*b*), constituents of the electrolyte such as organic carbonates and lithium salts will be reduced to form an SEI on the anode. During SEI formation two interfaces are created, an electrode–SEI interface (the "electrode interface") and an SEI–electrolyte interface (the "electrolyte interface") [12]. The experimental arrangement to study SEI formation via SFG is depicted in Figure 3.14*c*, and a photo of the spectroelectrochemical cell is shown in Figure 3.5.

We obtained our first data on a model system based on work by the Zhang group [67] at Berkeley. The electrolyte was ethylene carbonate (EC) dissolved in tetrahydrofuran (THF) with $LiClO_4$. Battery electrolytes typically include a combination of EC and other organic carbonates that are liquids at ambient temperature. We used EC as the only organic carbonate because it is known that reduction of EC at the anode generates a primary product that has a known structure. Furthermore, the main component of the SEI layer from the EC–THF electrolyte is lithium ethylene dicarbonate (LiEDC) [68], which has been synthesized and characterized spectroscopically. However, EC is a solid and requires the use of THF solvent. THF is fortunately nominally unreactive in the 2.0–0.2 V (versus $Li/Li^+$) potential range of interest. Serendipitously, THF proved to be a useful probe of the mobility of small molecules through the SEI as explained below. The working electrode was Au, and on Au underpotential deposition of Li is known to occur and accompany organocarbonate reduction. As a control, we also studied Cu electrodes where Li underpotential deposition does not occur, but those studies will not be discussed here. The counter electrodes and reference electrodes were Li metal.

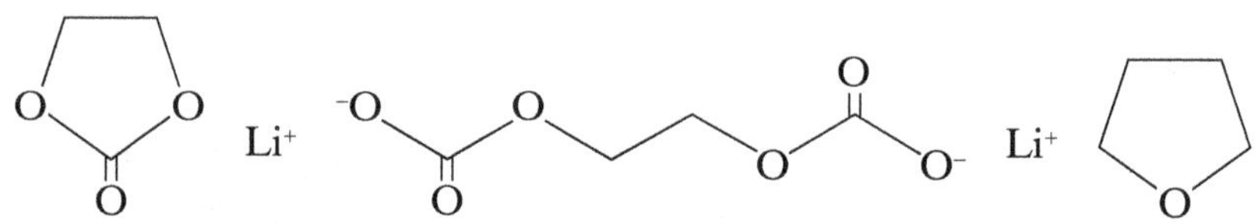

Ethylene Carbonate (EC) — Lithium Ethylene Dicarbonate (LIEDC) — Tetrahydrofuran (THF)

Figure 3.15 shows the Fourier transform infrared (FTIR) spectra of a 25-μm-thick electrolyte layer (1.0 M EC + 1.0 M $LiClO_4$ in THF) and 25 μm of pure THF for reference. Also shown is an ex situ FTIR spectrum of the SEI, which was obtained by cycling the Au electrochemical cell for six potential round trips, disassembling the cell, and scraping off the material accumulated on the electrode surface [69]. This powdery SEI material was spread on a $CaF_2$ window that could then be placed in the FTIR.

Based on the ex situ FTIR spectra of the recovered SEI material in Figure 3.15, the BBIR source was tuned to 1300 $cm^{-1}$ to acquire in situ data from the SEI–electrode interfaces. One such SFG spectrum [62] is shown in Figure 3.15, and it

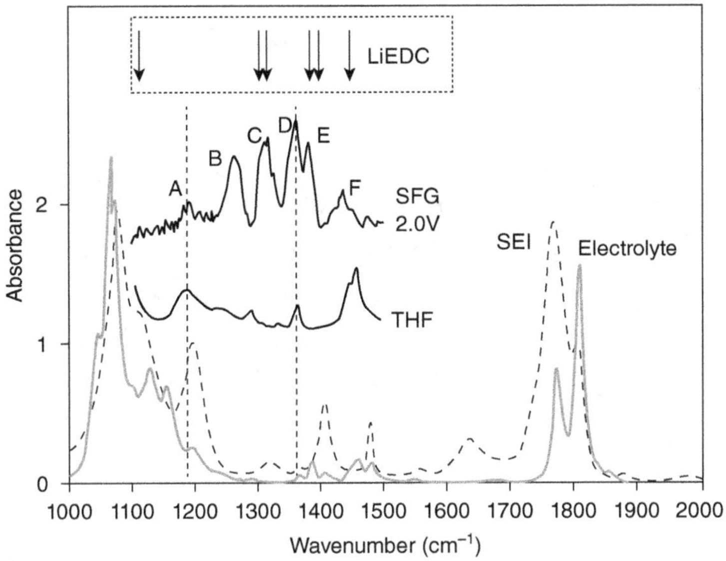


**Figure 3.15** FTIR spectra of 25-μm-thick layers of electrolyte (1 M EC + 1 M $LiClO_4$ in THF) and pure THF (offset for clarity). The spectrum marked SEI is an FTIR spectrum from material scraped from the Au electrode after six potential cycles. Also shown is an SFG spectrum from the Au electrode at 2.0 V after one potential cycle (offset for clarity). The six observed SFG transitions are labeled *A–F*. The dotted lines show transitions *A* and *D* coincident with THF transitions. The arrows are known locations of lithium ethylene dicarbonate (LiEDC) resonances. Peaks *C*, *E*, and *F* are identified as LiEDC transitions. Resonance *B* is tentatively assigned to ethylene oxide. Reproduced with permission from Ref. [62], copyright 2012, The Electrochemical Society.

has six peaks labeled *A–F*. Also shown in Figure 3.15 is a box with arrows marking previously known locations of strongly IR-active transitions of LiEDC. Based on spectral coincidences, we have assigned five of the six transitions as follows: *A* and *D* are the ring-stretching and $CH_2$-wagging modes of THF [70]. *C*, *E*, and *F* are quite close to the known locations of LiEDC resonances, which correspond to the previously assigned modes of $CH_2$ wagging, OCOO asymmetric stretching plus $CH_2$ twisting, and $CH_2$ deformation (scissoring), respectively [68]. Peak *B* is believed to be the $\nu_3$ band of ethylene oxide formed as a side product of EC reduction [71]. However, we cannot rule out the $\delta_s(CH_2)$ transition [72] of ethylene formed due to the reaction 2EC + 2e$^-$ + 2Li → LiEDC + ethylene.

Sum frequency generation acquisitions in this spectral regime were obtained over the course of six consecutive potential cycles, using a time delay of 0.7 ps between BBIR and NBvis pulses to suppress NR signals from the electrochemical cell [62]. The spectra obtained during the first cycle are shown in Figure 3.16. The dashed curve "laser" in this figure is the spectrum of the BBIR pulses. The SFG measurements are insensitive to transitions outside the laser region. It is clear from the figure that various transitions respond differently to potential cycling. At the beginning of the first cycle, the intensities of peaks *B* and *D* decline more than the

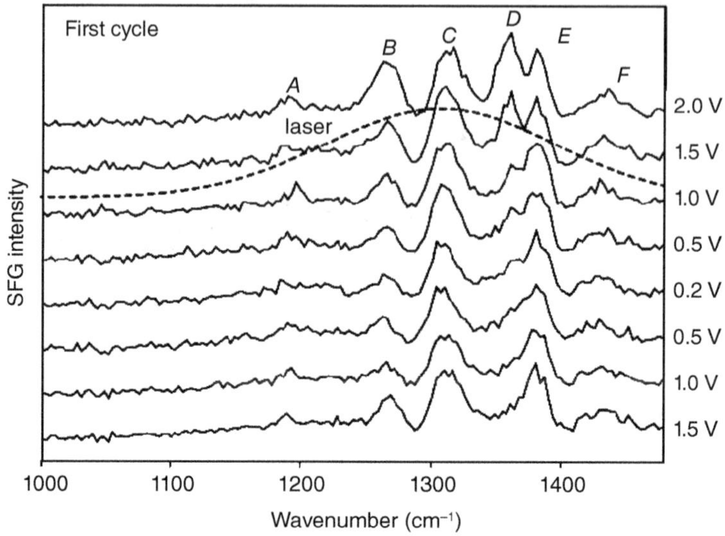


**Figure 3.16** SFG spectra of the Au electrode–SEI interface during the first potential cycle scanned from 2.0 V to 0.2 V versus Li/Li$^+$ (top to bottom). Six distinct transitions labeled *A–F* were observed. The dashed curve "laser" represents the spectrum of the broadband IR pulses used to obtain the data. Reproduced with permission from Ref. [62], copyright 2012, The Electrochemical Society.

others. Furthermore, peak *B* has the largest frequency shifts. In order to quantify these characteristics, we have plotted the potential dependent integrated intensities and peak shifts [62] of each of the six spectral transitions in Figures 3.17 and 3.18. The peak width data that also emerged from our analysis were noisy but showed little variation in width during potential cycling.

Peaks *A* and *D* (Figs. 3.17*a,d*) have intensity oscillations with maxima at 0.2 V that increase in amplitude during the first four cycles and then level out. The frequency shifts of *A* and *D* (Figs. 3.18*a,d*) are quite small and the data are noisy, but it is conceivable that they mimic the intensity oscillations, with the same phase (peak *A*) and with the opposite phase (peak *D*). Consequently, it appears that transitions *A* and *D* are associated with the same chemical species based on the similarity of the intensity variations during potential cycling.

The potential dependence of the integrated intensities for *C*, *E*, and *F* (Figs. 3.17*c,e,f*) is also quite similar. These dependences are complicated and appear to have maxima of similar intensities at both 0.2 and 2.0 V along with an overall gradual decrease over the six cycles. However, the frequency shifts are quite different, with peak *C* (Fig. 3.18*c*) showing larger oscillations peaked at 2.0 V and a gradual net blue shift, peak *E* (Fig. 3.18*e*) demonstrating almost no oscillations or overall shifting and peak *F* (Fig. 3.18*f*) showing damped oscillations peaked at 2.0 V and a net red shift.

The intensities of peak *B* during potential scanning (Fig. 3.17*b*) are on the surface similar to *C*, *E*, and *F*, but there is one clear difference. Throughout the initial

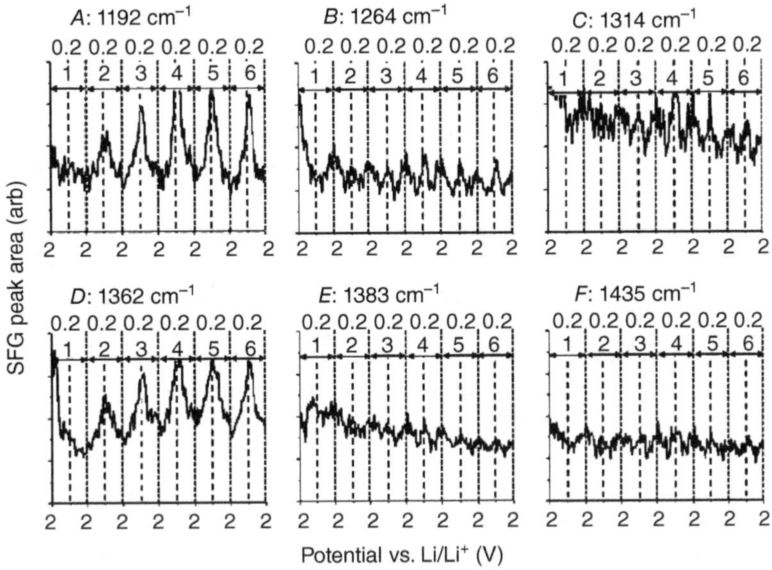


**Figure 3.17** Peak areas of the six transitions *A–F* (Fig. 3.16) in the SFG spectra of the Au electrode–SEI interface acquired during six repetitions of potential cycling at 1 mV/s from 2.0 to 0.2 V (versus Li/Li$^+$). Reproduced with permission from Ref. [62], copyright 2012, The Electrochemical Society.

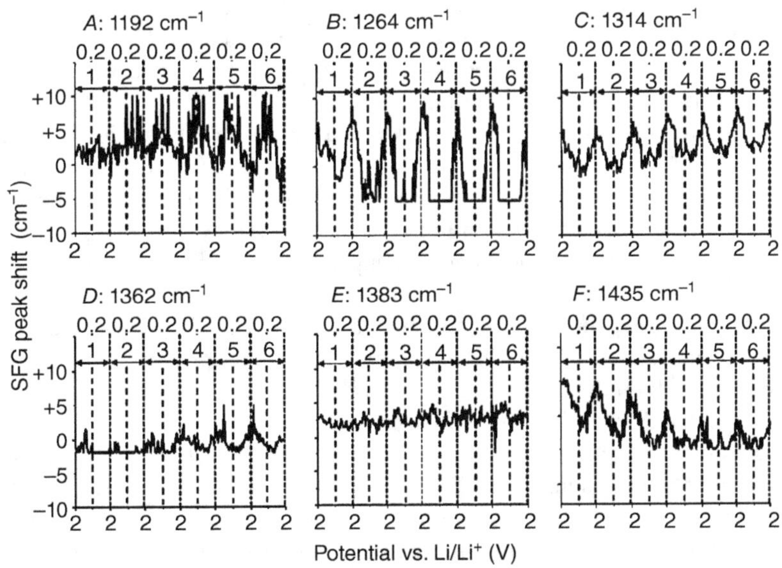


**Figure 3.18** Frequency shifts of transitions *A–F* (Fig. 3.16) in the SFG spectra of the Au electrode–SEI interface obtained during six repetitions of potential cycling at 1 mV/s from 2.0 to 0.2 V (versus Li/Li$^+$). Reproduced with permission from Ref. [62], copyright 2012, The Electrochemical Society.

potential drop from 2.0 to 0.2 V during the first cycle, peak *B* shows a dramatic intensity loss not seen with peak *C*, *E*, or *F*. Peak *B* also has the largest potential dependent frequency fluctuations (Fig. 3.18*b*).

Our assignments of transitions *A*–*F* are reinforced by the peak area data in Figure 3.17. The two peaks *A* and *D* assigned to THF have similar and highly characteristic potential dependences. They have oscillations peaking at 0.2 V that increase in intensity and then level out after about four cycles. These unique qualities confirm our assessment that peaks *A* and *D* belong to the same species. The three peaks *C*, *E*, and *F* assigned to LiEDC also have similar characteristic potential dependences. They have quite small oscillations, and the overall intensities decline slightly during the initial two to three cycles and then level out. The peak *B* assigned to ethylene or ethylene oxide also has a unique attribute that is different from the THF and LiEDC transitions. At the beginning of the first potential scan, peak *B* has a large-amplitude change that peaks *C*, *E*, and *F* do not have.

The before-mentioned intensity oscillations of THF resonances probed at the electrode interface are suggested to be associated with THF mobility in the SEI layer. We believe that at all potentials THF exists near the electrode, but being the smallest molecule in the electrolyte allows a degree of mobility in the SEI. Furthermore, THF is drawn to the electrode more strongly at 0.2 V than at 2.0 V. When the THF concentration near the electrode reaches its maximum, so does its SFG signal. In this view, the changing magnitudes of the THF intensity oscillations would be indicative of variations in THF mobility within the SEI [62]. If this were to be the case, the data in Figures 3.17*a*,*d* report that during the first cycle the THF at the electrode is relatively immobile, but in the subsequent two to three cycles, the SEI loosens up a bit and allows the THF increased limited movement within the SEI. Then the THF mobility stabilizes after four or so cycles. This is consistent with what is seen in Li ion batteries, where the capacity grows during the first few charge–discharge cycles and then levels off as the SEI stabilizes.

The LiEDC intensity profiles of peaks *C*, *E*, and *F* in Figures 3.17*c*,*e*,*f* refer to LiEDC at the electrode interface. These profiles say that the LiEDC concentration, mobility, and orientation at the electrode undergo small changes during the first cycle, but those changes become minimal in successive cycles. Keep in mind that SFG is sensitive to potential dependent changes in the structure and composition of the electrode interface. Apparently, the LiEDC density at the electrode interface decreases a bit during the first cycle or two, but except for the wandering THF, this interface is stable throughout SEI development. In contrast, peak *B* has a large intensity drop at the beginning of the first cycle, which is consistent with *B* being a by-product created when LiEDC is first formed.

Taken together, these observations suggest the following picture about SEI formation. A relatively dense and compact LiEDC layer is formed by EC reduction on Au in the presence of lithium. We see the presence of LiEDC transitions (peaks *C*, *E*, and *F*) and the putative by-product of LiEDC formation, peak *B*, quite early on in the first potential cycle. Even before much lithium is deposited, LiEDC is seen, presumably because SFG is sensitive and needs just a monolayer to produce a full signal. These LiEDC signals decrease during the first cycle as the SEI is formed.

The increase in LiEDC at the anode is not seen. Rather SFG is sensitive only to the interfacial concentration, so the first-cycle decrease in intensity is attributed to LiEDC becoming disordered at the electrode interface as the SEI develops. After the first cycle, despite the growth and shrinkage of the SEI, the LiEDC and the putative ethylene oxide or ethylene at the SEI interfaces appear to exist in a stable state, but the THF spectrum begins to evolve in time. During the first two cycles, THF peaks *A* and *D* have almost constant intensity, which then starts to oscillate with potential cycling. The oscillations grow and level out around cycle 4. We interpret the THF behavior as a measure of the THF mobility causing the THF concentration at the electrode interface to vary with potential. Initially, this mobility is poor, so little THF is drawn to the electrode interface. However, as repeated potential cycling results in an SEI whose density is reduced due to recurring transits of Li ions and THF through its bulk, the THF movement within the SEI increases. Lithium ion movement through the SEI is a critical feature of successful batteries. To the extent that ion mobility can be associated with THF mobility, our results suggest that the lithium ion mobility increases for the first few potential cycles and then levels off by cycle 6. These results are consistent with lithium battery charge–discharge behavior, since batteries subjected to charge and discharge cycles in an effort to develop a robust SEI reach their maximum capacity around six cycles.

## 3.8 SUMMARY AND OUTLOOK FOR FUTURE WORK

In this chapter, we have described recent advances in SFG instrumentation that enable us to study time-resolved dynamics of interfaces. The NR suppression technique, in particular, has greatly improved our ability to study molecules on metal or metal electrode surfaces. Flash heating has been developed to study ultrafast thermal processes at interfaces, including thermal conduction along molecular wires and thermally induced molecular rearrangements. The NR suppression technique in addition enabled us to study SEI formation in lithium ion systems with excellent signal-to-noise ratios.

Several directions are envisioned for future research. We do not see a high likelihood of making radical improvements in our present SFG technology. We believe that it is possible to incrementally improve our apparatus by using lasers that produce more average power at higher repetition rates, but the rest of the apparatus appears to be quite well optimized. However, there are some new developments that could lead to major improvements in SFG technologies. One possibility is a laser system where the "visible" pulses are tunable. The intensity of Raman scattering increases as the fourth power of the laser frequency, so all things being equal, the SFG intensity should increase as the visible wavelength is blue shifted. For instance, shifting from 800 to 450 nm should, in principle, boost the SFG signal by an order of magnitude. However, the blue-shifted light might decompose the sample or interact in difficult-to-interpret ways with electronic transitions. Furthermore, the NR background of metal surfaces can be highly dependent on visible wavelength. For instance, with 800-nm "visible" light, Au produces a large NR signal but Pt has a minimal background. It therefore might be worthwhile tuning the visible pulses to wavelengths where the NR background is minimized.

Another important recent development is heterodyne detection [73]. The SFG techniques described here are homodyne methods where a square-law optical detector records the SFG signal from the sample. In the heterodyne scheme developed by Stiopkin and co-workers [73], the SFG experiment was performed by combining the SFG pulse from the sample with a reference SFG pulse, sometimes called the "local oscillator." The reference pulse was created in a nonlinear medium with large second-order susceptibility and no resonant vibrational transitions. The resulting local oscillator is much more intense than surface or interface signal pulses. If the SFG signal field is denoted $E_{SFG}$ and the (much larger) local oscillator signal field is $E_{LO}$, then the intensity of the combined pulses at the detector becomes

$$I_{SFG} \propto |E_{SFG} + E_{LO}|^2 \approx |E_{LO}|^2 + E_{SFG}E_{LO}\cos(\theta). \tag{3.7}$$

When the phase angle $\theta$ is set to zero and the constant value from the local oscillator is subtracted away, the SFG signal field is amplified by the local oscillator. In addition, the SFG signal is linear in $E_{SFG}$, rather than the quadratic dependence associated with homodyne detection. This means the SFG intensity with heterodyne detection is linearly proportional to molecular number density, whereas the signal is proportional to the square of molecular number density with homodyne detection [73]. Thus, heterodyne methods have increased sensitivity to lower-concentration molecular species.

One of the biggest drawbacks with heterodyne detection is the need to maintain a constant phase difference between the SFG signals and the local oscillator. The phase difference is very sensitive to environmental effects such as temperature changes and vibration, since a variation in the two optical paths by a fraction of a wavelength causes the phase to shift. Tahara and co-workers [74–76] have developed a scheme that minimizes these fluctuations. They have additionally shown that molecular reorientation at an interface due to an imposed electric field can be detected by measuring how the phase of the SFG signal changes in reference to the constant phase of the local oscillator [74–76].

An exciting recent innovation is the heterodyne time-domain scheme by Laaser and co-workers [43], which uses a computer-controlled IR acoustooptic pulse shaper [77] that can actively control the phase of the BBIR pulses and can further time-shift these pulses while maintaining constant phase. The optical arrangement used in those experiments also minimizes differences in the optical paths of the signal and local oscillator, leading to improved phase stability [43]. With this scheme, Laaser and co-workers [43] have shown, by obtaining high-quality SFG spectra of SAMs, that their scheme has quite a few advantages over the usual methods, including removal of NR backgrounds by phase cycling the IR pulses, a linear dependence of signal on concentration, and vibrational line shapes with high fidelity. The phase-cycling method works because the R and NR signals are 90° out of phase. On successive shots, the IR phase is shifted by 180°, and the signals from the successive shots are subtracted. This procedure causes the R signals to build up and the NR signals to vanish [43]. In addition, with minor changes to the Laaser setup, Zanni and co-workers demonstrated the first two-dimensional SFG measurements [78], which are the SFG analog of two-dimensional (2D) IR or nuclear magnetic resonance (NMR) spectroscopy. Utilizing this technology, data were obtained from CO bound to a metal surface. This setup, where a computer actively controls the IR pulses, is extremely versatile and has the potential of improving SFG by utilizing many of the methods that have made NMR such a powerful spectroscopic tool.

The ultrafast temperature jump technique discussed in this chapter allowed us to study the flow of vibrational energy across interfaces and through molecules. One place where improvements seem possible is to redesign the T-jump substrates to promote faster lattice heating and suppress the generation of hot surface electrons. The existing Cr–Au substrates worked well with long-chain alkanes precisely because the chains were long. The transient period when hot electrons were present was a shorter duration than the heat transfer process. However, the same substrates gave the undesirable hot-electron overshoot effect on benzenethiols, where the phenyl groups were close to the surface. With these short molecules, the groups being probed by SFG were also immediately excited by hot electrons. It might be possible to improve the substrates using metals other than Cr and Au, or to develop strategies to suppress hot-electron generation using multiple metal layers. In this scheme, one of the layers would have a much greater electron–phonon coupling than Au along with facile heat transfer to the layer that provides the foundation for the SAM adsorbates.

The initial lithium ion results are encouraging. One issue will be transitioning from model systems such as the Au electrode to realistic battery systems. Batteries ordinarily use carbon electrodes that consist of carbon particles with a polymer binder, and it is a challenge to fabricate such electrodes to create optical-quality surfaces. Another challenge is to understand the roles of additives. In many lithium ion systems, relatively small quantities of additives can have a major effect on lithium ion battery performance, so the application of heterodyne SFG methods with their ability to detect smaller concentration species could lead to important advances in lithium ion battery studies.

## ACKNOWLEDGMENTS

The SFG development efforts described here were based upon work supported by the National Science Foundation under award DMR 0504038, the Air Force Office of Scientific Research under award FA9550-06-1-0235, and the U.S. Army Research Office under awards W911NF-04-1-0178, W911NF-05-1-0345, and W911NF-06-1-0171. The flash-heating studies described here were based upon work supported by the National Science Foundation under award DMR 0855259, the Air Force Office of Scientific Research under award FA9550-09-1-0163, and the U.S. Army Research Office under awards W911NF-04-1-0178, W911NF-05-1-0345, and W911NF-06-1-0171. The lithium battery efforts were based upon work supported as a part of the Center for Electrical Energy Storage–Tailored Interfaces, an Energy Frontier Research Center funded by the U.S. Department of Energy, Office of Science, Office of Basic Energy Sciences under award number DE-AC02-06CH11 (subcontract no. 9F-31921). We thank the Center for Microanalysis of Materials at the University of Illinois at Urbana-Champaign, supported by award DEFG02-91ER45439. We also gratefully acknowledge the efforts of many colleagues, former postdocs, and former students who contributed to these efforts.

## REFERENCES

1. Chen, Z.; Shen, Y. R.; Somorjai, G. A. Studies of polymer surfaces by sum frequency generation vibrational spectroscopy, *Annu. Rev. Phys. Chem.* **2002**, *53*, 437.
2. Williams, C. T.; Beattie, D. A. Probing buried interfaces with non-linear optical spectroscopy, *Surf. Sci.* **2001**, *500*, 545.
3. Richmond, G. L. Molecular bonding and interactions at aqueous surfaces as probed by vibrational sum frequency spectroscopy, *Chem. Rev.* **2002**, *102*, 2693.
4. Vidal, F.; Tadjeddine, A. Sum-frequency generation spectroscopy of interfaces, *Rep. Prog. Phys.* **2005**, *68*, 1095.
5. Arnolds, H.; Bonn, M. Ultrafast surface vibrational dynamics, *Surf. Sci. Rep.* **2010**, *65*, 45.
6. Lagutchev, A.; Hambir, S. A.; Dlott, D. D. Nonresonant background suppression in broadband vibrational sum-frequency generation spectroscopy, *J. Phys. Chem. C* **2007**, *111*, 13645.
7. Lagutchev, A.; Lozano, A.; Mukherjee, P.; Hambir, S. A.; Dlott, D. Compact broadband vibrational sum-frequency generation spectrometer with nonresonant suppression, *Spectrochim. Acta A* **2010**, *75*, 1289.
8. Carter, J. A.; Wang, Z.; Dlott, D. D. Ultrafast nonlinear coherent vibrational sum-frequency spectroscopy methods to study thermal conductance of molecules at interfaces, *Acct. Chem. Res.* **2009**, *42*, 1343.
9. Patterson, J. E.; Lagutchev, A. S.; Huang, W.; Dlott, D. D. Ultrafast dynamics of shock compression of molecular monolayers, *Phys. Rev. Lett.* **2005**, *94*, 015501.
10. Wang, Z.; Carter, J. A.; Lagutchev, A.; Koh, Y. K.; Seong, N.-H.; Cahill, D. G.; Dlott, D. D. Ultrafast flash thermal conductance of molecular chains, *Science* **2007**, *317*, 787.
11. Lu, G. Q.; Lagutchev, A.; Dlott, D. D.; Wieckowski, A. Quantitative vibrational sum-frequency generation spectroscopy of thin layer electrochemistry: CO on a Pt electrode, *Surf. Sci.* **2005**, *585*, 3.
12. Shen, Y. R. Surfaces probed by nonlinear optics, *Surf. Sci.* **1994**, *299/300*, 551.
13. Tadjeddine, A. Spectroscopic investigation of surfaces and interfaces by using infrared-visible sum and difference generation, *Surf. Rev. and Lett.* **2000**, *7*, 423.
14. Shaw, S. K.; Lagutchev, A.; Dlott, D. D.; Gewirth, A. A. Sum-frequency spectroscopy of molecular adsorbates on low-Index Ag surfaces: Effects of azimuthal rotation, *Anal. Chem.* **2009**, *81*, 1154.
15. Lagutchev, A.; Lu, G. Q.; Takeshita, T.; Dlott, D. D.; Wieckowski, A. Vibrational sum frequency generation studies of the $(2 \times 2) \rightarrow (\sqrt{19} \times \sqrt{19})$ phase transition of CO on Pt(111) electrodes, *J. Chem. Phys.* **2006**, *125*, 154705.
16. Carter, J. A.; Wang, Z.; Dlott, D. D. Spatially resolved vibrational energy transfer in molecular monolayers, *J. Phys. Chem. A* **2008**, *112*, 3523.
17. Carter, J. A.; Wang, Z.; Lagutchev, A.; Fang, Y.; Seong, N.-H.; Cahill, D. G.; Dlott, D. D. Ultrafast shock wave coherent dissociation and spectroscopy of materials, *AIP. Confer. Proc.* **2007**, *995*, 1221.
18. Wang, Z.; Cahill, D. G.; Carter, J. A.; Koh, Y. K.; Lagutchev, A.; Seong, N.-H.; Dlott, D. D. Ultrafast dynamics of heat flow across molecules, *Chem. Phys.* **2008**, *350*, 31.
19. Carter, J. A.; Wang, Z.; Fujiwara, H.; Dlott, D. D. Ultrafast excitation of molecular adsorbates on flash-heated gold surfaces, *J. Phys. Chem. A* **2009**, *113*, 12105.
20. Shen, Y. R. *The Principles of Nonlinear Optics*, Wiley: New York, **1984**.

21. EISENTHAL, K. B. Liquid interfaces probed by second-harmonic and sum-frequency spectroscopy, *Chem. Rev.* **1996**, *1343*.
22. PERRY, A.; NIEPERT, C.; SPACE, B.; MOORE, P. B. Theoretical modeling of interface specific vibrational spectroscopy: Methods and applications to aqueous interfaces, *Chem. Rev.* **2006**, *106*, 1234.
23. RICHTER, L. J.; PETRALLI-MALLOW, T. P.; STEPHENSON, J. C. Vibrationally resolved sum-frequency generation with broad-bandwidth infrared pulses, *Opt. Lett.* **1998**, *23*, 1594.
24. EISENTHAL, K. B. Second harmonic spectroscopy of aqueous nano- and microparticle interfaces, *Chem. Rev.* **2006**, *106*, 1462.
25. SCATENA, L. F.; RICHMOND, G. L. Orientation, hydrogen bonding, and penetration of water at the organic/water interface, *J. Phys. Chem.* **2001**, *105*, 11240.
26. ZHU, X. D.; SUHR, H.; SHEN, Y. R. Surface vibrational spectroscopy by infrared-visible sum frequency generation, *Phys. Rev. B* **1987**, *35*, 3047.
27. HARRIS, A. L.; CHIDSEY, C. E. D.; LEVINOS, N. J.; LOIACONO, D. N. Monolayer vibrational spectroscopy by infrared-visible sum generation at metal and semiconductor surfaces, *Chem. Phys. Lett.* **1987**, *141*, 350.
28. LEE, I.-Y. S.; WEN, X.; TOLBERT, W. A.; DLOTT, D. D. Direct measurement of polymer temperature during laser ablation using a molecular thermometer, *J. Appl. Phys.* **1992**, *72*, 2440.
29. HARE, D. E.; DLOTT, D. D. Picosecond coherent Raman study of solid-state chemical reactions during laser polymer ablation, *Appl. Phys. Lett.* **1994**, *64*, 715.
30. HARE, D. E.; FRANKEN, J.; DLOTT, D. D.; CHRONISTER, E. L.; FLORES, J. Dynamics of a polymer shock optical microgauge studied by picosecond coherent Raman spectroscopy, *Appl. Phys. Lett.* **1994**, *65*, 3051.
31. EESLEY, G. L. *Coherent Raman Spectroscopy*, Pergamon: Oxford, **1991**.
32. van der HAM, E. W. M.; VREHEN, Q. H. F.; ELIEL, E. R. Self-dispersive sum-frequency generation at interfaces, *Opt. Lett.* **1996**, *21*, 1448.
33. BAIN, C. D.; DAVIES, P. B.; ONG, T. H.; WARD, R. N. Quantitative analysis of monolayer composition by sum-frequency vibrational spectroscopy, *Langmuir* **1991**, *7*, 1563.
34. LAGUTCHEV, A. S.; PATTERSON, J. E.; HUANG, W.; DLOTT, D. D. Ultrafast dynamics of self-assembled monolayers under shock compression: Effects of molecular and substrate structure, *J. Phys. Chem. B* **2005**, *109*, 5033.
35. PATTERSON, J. E.; DLOTT, D. D. Ultrafast shock compression of self-assembled monolayers: A molecular picture, *J. Phys. Chem. B* **2005**, *109*, 5045.
36. NUZZO, R. G.; ZEGARSKI, B. R.; DUBOIS, L. H. Fundamental studies of the chemisorption of organosulfur compounds on Au(111). Implications for molecular self-assembly on gold surfaces, *J. Am. Chem. Soc.* **1987**, *109*, 733.
37. SHAW, S. K.; LAGUTCHEV, A.; DLOTT, D. D.; GEWIRTH, A. A. Electrochemically driven reorientation of three ionic states of *p*-amino benzoic acid on Ag(111), *J. Phys. Chem. C* **2009**, *113*, 2417.
38. CHOU, K. C.; KIM, J.; BALDELLI, S.; SOMORJAI, G. A. Vibrational spectroscopy of carbon monoxide, acetonitrile, and phenylalanine adsorbed on liquid|electrode interfaces by sum frequency generation, *J. Electroanal. Chem.* **2003**, *554–555*, 253.
39. ROKE, S.; KLEYN, A. W.; BONN, M. Time- vs. frequency-domain femtosecond surface sum frequency generation, *Chem. Phys. Lett.* **2003**, *370*, 227.
40. SHEN, Y. R. Surface properties probed by second-harmonic and sum-frequency generation, *Nature* **1989**, *337*, 519.
41. STIOPKIN, I. V.; JAYATHILAKE, H. D.; WEERAMAN, C.; BENDERSKII, A. V. Temporal effects on spectroscopic line shapes, resolution, and sensitivity of the broad-band sum frequency generation, *J. Chem. Phys.* **2010**, *132*.
42. WEERAMAN, C.; MITCHELL, S. A.; LAUSTEN, R.; JOHNSTON, L. J.; STOLOW, A. Vibrational sum frequency generation spectroscopy using inverted visible pulses, *Opt. Express* **2010**, *18*, 11483.
43. LAASER, J. E.; XIONG, W.; ZANNI, M. T. Time-domain SFG spectroscopy using mid-IR pulse shaping: Practical and intrinsic advantages, *J. Phys. Chem. B* **2011**, *115*, 9920.
44. SIEPMANN, J. I.; MCDONALD, I. R. Domain formation and system-size dependence in simulations of self-assembled monolayers, *Langmuir* **1993**, *9*, 2351.

45. Yeganeh, M. S.; Dougal, S. M.; Polizzotti, R. S.; Rabinowitz, P. Interfacial atomic structure of a self-assembled alkyl thiol monolayer/Au(111): A sum-frequency generation study, *Phys. Rev. Lett.* **1995**, *74*, 1811.
46. Follonier, S.; Miller, W. J. W.; Abbott, N. L.; Knoesen, A. Characterization of the molecular orientation of self-assembled monolayers of alkanethiols on obliquely deposited gold films by using infrared-visible sum-frequency spectroscopy, *Langmuir* **2003**, *19*, 10501.
47. DuBois, L. H.; Nuzzo, R. G. The synthesis, structure and properties of model organic surfaces, *Annu. Rev. Phys. Chem.* **1992**, *43*, 437.
48. Schreiber, F. Structure and growth of self-assembling monolayers, *Prog. Surf. Sci.* **2000**, *65*, 151.
49. Segal, D.; Nitzan, A. Heating in current carrying molecular junctions, *J. Chem. Phys.* **2002**, *117*, 3915.
50. Segal, D.; Nitzan, A.; Hänggi, P. Thermal conductance through molecular wires, *J. Chem. Phys.* **2003**, *119*, 6840.
51. Segal, D.; Nitzan, A. Heat rectification in molecular junctions, *J. Chem. Phys.* **2005**, *122*, 104704.
52. Segal, D.; Nitzan, A. Molecular heat pump, *Phys. Rev. E* **2006**, *73*, 026109.
53. Galperin, M.; Ratner, M. A.; Nitzan, A. Molecular transport junctions: Vibrational effects, *J. Phys.: Condens. Matter* **2007**, *19*, 103201.
54. Zhang, Y.; Barnes, G. L.; Yan, T. Y.; Hase, W. L. Model non-equilibrium molecular dynamics simulations of heat transfer from a hot gold surface to an alkylthiolate self-assembled monolayer, *Phys. Chem. Chem. Phys.* **2010**, *12*, 4435.
55. Manikandan, P.; Carter, J. A.; Dlott, D. D.; Hase, W. L. Effect of carbon chain length on the dynamics of heat transfer at a gold/hydrocarbon interface: Comparison of simulation with experiment, *J. Phys. Chem. C* **2011**, *115*, 9622.
56. Beran, A. The reflectance behaviour of gold at temperatures up to 500°C, *Miner. Petrol.* **1985**, *34*, 211.
57. Brorson, S. D.; Fujimoto, J. G.; Ippen, E. P. Femtosecond electronic heat-transport dynamics in thin gold films, *Phys. Rev. Lett.* **1987**, *59*, 1962.
58. Hohlfeld, J.; Wellershoff, S.-S.; Güdde, J.; Conrad, U.; Jähnke, V.; Matthias, E. Electron and lattice dynamics following optical excitation of metals, *Chem. Phys.* **2000**, *251*, 237.
59. Cahill, D. G. Analysis of heat flow in layered structures for time-domain thermoreflectance, *Rev. Sci. Instrum.* **2004**, *75*, 5119.
60. Gundrum, B. G.; Cahill, D. G.; Averback, R. S. Thermal conductance of metal-metal interfaces, *Phys. Rev. B* **2005**, *72*, 245426.
61. Wang, R. Y.; Segalman, R. A.; Majumdar, A. Room temperature thermal conductance of alkanedithiol self-assembled monolayers, *Appl. Phys. Lett.* **2006**, *89*, 173113.
62. Mukherjee, P.; Lagutchev, A.; Dlott, D. D. In situ probing of solid-electrolyte interfaces with nonlinear coherent vibrational spectroscopy, *J. Electrochem. Soc.* **2012**, *159*, A244.
63. Mukherjee, P.; Lagutchev, A.; Dlott, D. D. Solid electrolyte interfaces and interphases in lithium batteries: In situ studies using nonlinear optical probes, *Mater. Res. Soc. Symp. Proc.* **2012**, *1388*.
64. Liu, H.; Tong, Y.; Kuwata, N.; Osawa, M.; Kawamura, J.; Ye, S. Adsorption of propylene carbonate on the LiCoO2 surface investigated by nonlinear vibrational spectroscopy, *J. Phys. Chem. C* **2009**, *113*, 20531.
65. Balbuena, P. B.; Wang, Y. *Lithium-Ion Batteries. Solid-Electrolyte Interphase*; Imperial College Press: London, **2004**.
66. Amalraj, S. F.; Aurbach, D. The use of in situ techniques in R&D of Li and Mg rechargeable batteries, *J. Solid State Electrochem.* **2011**, *15*, 877.
67. Zhang, X. R.; Kostecki, R.; Richardson, T. J.; Pugh, J. K.; Ross, P. N. Electrochemical and infrared studies of the reduction of organic carbonates, *J. Electrochem. Soc.* **2001**, *148*, A1341.
68. Zhuang, G. R. V.; Xu, K.; Yang, H.; Jow, T. R.; Ross, P. N. Lithium ethylene dicarbonate identified as the primary product of chemical and electrochemical reduction of EC in 1.2 m LiPF6/EC : EMC electrolyte, *J. Phys. Chem. B* **2005**, *109*, 17567.
69. Zhuang, G. R. V.; Ross, P. N. Analysis of the chemical composition of the passive film on Li-ion battery anodes using attentuated total reflection infrared spectroscopy, *Electrochem. Sol. State Lett.* **2003**, *6*, A136.

70. EYSTER, J. M.; PROHOFSKY, E. W. The normal vibrations of tetrahydrofuran and its deuterated derivatives, *Spectrochim. Acta A: Mol. Spectr.* **1974**, *30*, 2041.
71. ZHANG, X.; PUGH, J. K.; ROSS Jr., P. N. Evidence for epoxide formation from the electrochemical reduction of ethylene carbonate, *Electrochem. and Solid-State Lett.* **2001**, *4*, A82.
72. ITOH, K. I.; KIYOHARA, T.; SHINOHARA, H.; OHE, C.; KAWAMURA, Y.; NAKAI, H. DFT calculation analysis of the infrared spectra of ethylene adsorbed on Cu(110), Pd(110), and Ag(110), *J. Phys. Chem. B* **2002**, *106*, 10714.
73. STIOPKIN, I. V.; JAYATHILAKE, H. D.; BORDENYUK, A. N.; BENDERSKII, A. V. Heterodyne-detected vibrational sum-frequency generation spectroscopy, *J. Am. Chem. Soc.* **2008**, *130*, 2271.
74. MONDAL, J. A.; NIHONYANAGI, S.; YAMAGUCHI, S.; TAHARA, T. Structure and orientation of water at charged lipid monolayer/water interfaces probed by heterodyne-detected vibrational sum frequency generation spectroscopy, *J. Am. Chem. Soc.* **2010**, *132*, 10656.
75. MONDAL, J. A.; NIHONYANAGI, S.; YAMAGUCHI, S.; TAHARA, T. Three distinct water structures at a zwitterionic lipid/water interface revealed by heterodyne-detected vibrational sum frequency generation, *J. Am. Chem. Soc.* **2012**, *134*, 7842.
76. NIHONYANAGI, S.; YAMAGUCHI, S.; TAHARA, T. Direct evidence for orientational flip-flop of water molecules at charged interfaces: A heterodyne-detected vibrational sum frequency generation study, *J. Chem. Phys.* **2009**, *130*.
77. SHIM, S. H.; STRASFELD, D. B.; FULMER, E. C.; ZANNI, M. T. Femtosecond pulse shaping directly in the mid-IR using acousto-optic modulation, *Opt. Lett.* **2006**, *31*, 838.
78. ZANNI, M. T.; HOCHSTRASSER, R. M. Two-dimensional infrared spectroscopy: A promising new method for the time resolution of structures, *Curr. Opin. Struc. Biol.* **2001**, *11*, 516.

# Chapter 4

# Spectroscopy of Electrified Interfaces with Broadband Sum Frequency Generation: From Electrocatalysis to Protein Foams

**Björn Braunschweig,[1] Prabuddha Mukherjee, Robert B. Kutz, Armin Rumpel,[2] Kathrin Engelhardt,[1] Wolfgang Peukert,[1] Dana D. Dlott,[3] and Andrzej Wieckowski[4]**

[1]Institute of Particle Technology (LFG), Erlangen, Germany
[2]SwissOptic AG, Heerbrugg, Switzerland
[3]School of Chemical Sciences, University of Illinois, Urbana, Illinois
[4]Department of Chemistry, University of Illinois at Urbana Champaign, Urbana, Illinois

## 4.1 INTRODUCTION

The molecular processes that occur at electrically charged interfaces are at the very heart of current research in the fields of fundamental electrochemistry [1], electrocatalysis [2], photocatalysis (e.g., water splitting) [3], biomolecular interfaces (e.g., membrane proteins) [4], and colloid science [5], to mention only a few. Because these processes can dominate both the microscopic and macroscopic properties of the system, increasing our understanding of them is of utmost importance.

Although electrified interfaces are of great importance in many fields of research, we know only very little about the molecular structure of these interfaces. In fact,

*Vibrational Spectroscopy at Electrified Interfaces*, First Edition.
Edited by Andrzej Wieckowski, Carol Korzeniewski, and Björn Braunschweig.

concepts for the electric double layer such as the Gouy–Chapman model are nearly a century old and are still incomplete, as they account only for some aspects of the interfacial chemistry. For example, they fail to explain chemical reactions in an external electric field, specific ion adsorption, colloid stability phenomena such as Hofmeister effects [6], and many more observations [5]. We do know that although the composition and structure of various charged interfaces can be very different, they have an important commonality—a strong electric field at the interface that drives solvent molecules, ions, and other solvated molecules into ordered structures of the interfacial region [5]. Yet a more detailed description of electrically charged interfaces including effects of the solvent polarization, inhomogeneous charge distributions, and effects of the finite size of ions has so far suffered mainly from two problems: the lack of inherently interface-specific spectroscopic techniques that provide information on a molecular level and from the lack of ab initio simulations that can be compared with the experimental results. In recent years, however, new experimental [4, 7–13] and theoretical methods [14–20] for solid–electrolyte and liquid–gas interfaces have become available. In particular, broadband sum frequency generation spectroscopy synchronized with electrochemical experiments has evolved into a powerful optical tool for studying chemical reactions at electrode–electrolyte interfaces [21]. In this chapter, we will focus on the application of vibrational sum frequency generation to study the molecular structure of electrode–electrolyte and electrified liquid–gas interfaces.

## 4.2 PRINCIPLES OF VIBRATIONAL SUM FREQUENCY GENERATION AND EXPERIMENTAL DETAILS

Vibrational sum frequency generation (SFG) is inherently interface specific because a second-order process such as SFG is not allowed in the bulk of materials with inversion symmetry. Such materials include those with cubic or diamond crystalline structure, amorphous materials, liquids, and gases. Interfaces, however, break the prevailing bulk symmetry and give rise to SFG-signals that originate from the interface only. In broadband SFG, tunable broadband infrared (BBIR) laser pulses with the frequency $\Omega$ and a bandwidth $\Delta$ of ca. 200 $cm^{-1}$ or more are mixed at the interface with narrowband (<10 $cm^{-1}$) "visible" laser pulses where they coherently generate photons whose frequencies are the sum of the visible and IR frequencies. In fact, using broadband SFG, vibrational spectra can be recorded within the entire bandwidth of the BBIR pulse, which eliminates the need for tuning the IR frequency and allows for rapid acquisitions of vibrational spectra within the limit of the BBIR bandwidth (Fig. 4.1). A schematic setup of such a broadband SFG spectrometer is presented in Figure 4.2. In the first step, femtosecond laser pulses with nanojoule pulse energy, megahertz repetition rate, and a wavelength of 800 nm are generated in a Ti: sapphire oscillator and subsequently amplified to several millijoule pulse energies in a Ti: sapphire regenerative amplifier with typically a 1-kHz repetition rate. Spectral filtering of a small fraction of the pulse energy from the amplifier with

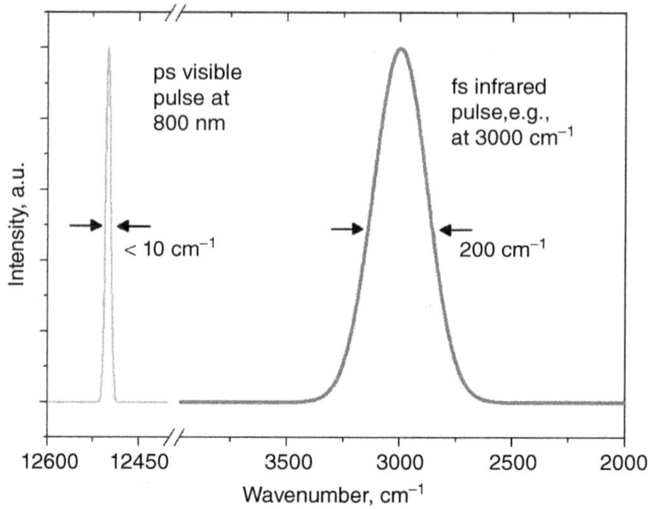


**Figure 4.1** Typical spectra of an etalon-filtered "visible" and a broadband mid-IR pulse, which are used for broadband SFG spectroscopy.

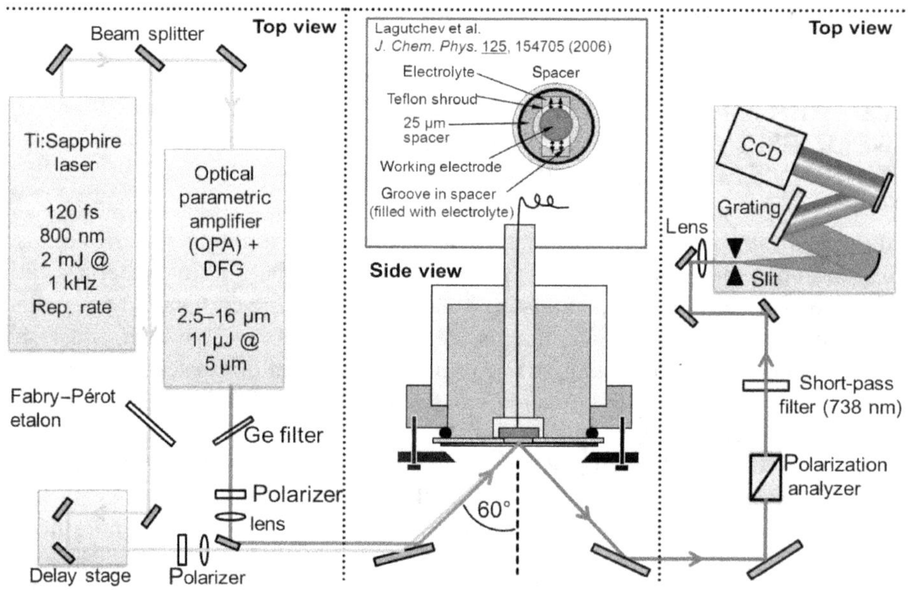


**Figure 4.2** Schematic setup of a spectrometer for broadband sum frequency generation (SFG), which is synchronized with voltammetry. Broadband IR pulses are generated in an OPA with a subsequent DFG (difference frequency generation) stage. The narrowband visible pulse is generated from a broadband femtosecond pulse by narrowing the line width with an etalon. The detection of the SFG output is performed with a spectrograph and a CCD camera. The side view shows schematically an electrochemical cell where a 25-μm electrolyte layer is formed between a $CaF_2$ window and the electrode. Reproduced with permission from Ref. [28], copyright 2012, Elsevier.

a Fabry–Pérot etalon generates a narrowband pulse in the frequency domain with a highly asymmetric intensity profile in the time domain. In fact, etalon filtering leads to a rise time of the pulse duration, which is comparable to the rise time of the initial femtosecond pulse, but with a picosecond tail (see Fig. 4.5*b*). These time-asymmetric pulse shapes are particularly helpful to suppress nonresonant responses from a sample (see below). The remainder of the pulse energy from the amplified laser system is used to pump an optical parametric amplifier (OPA).

The OPA generates tunable near-infrared pulses with wavelengths of 1.2–2.6 μm. An additional mixing stage for phase-matched difference frequency generation (DFG) of idler and signal photons from the OPA extends the tuning range to the mid-infrared (mid-IR). The tuning range that is commonly used in an experiment for vibrational SFG is from 2 to 12 μm wavelength (4000–833 $cm^{-1}$). In many SFG spectrometers the difference frequency of idler and signal pulses is generated in a collinear geometry and signal, and idler and pump photons have to be removed from the mid-IR beam with a germanium window that leads to some attenuation due to reflectance and transmittance losses of the Ge filter. Here, noncollinear DFG setups are advantageous because the noncollinear generation leads to a spatial separation of signal, idler, and the generated BBIR beam, where the unwanted photons can simply be blocked without distorting the broadband mid-IR pulses. This has particular advantages for pulse durations of 100 fs or less since pulse broadening and an additional chirp of the IR pulse, which is due to different group velocity dispersions of the individual frequencies within the broadband pulse, can be avoided.

### 4.2.1 Detection Unit in Broadband SFG Spectrometer

In a broadband SFG spectrometer the detection unit comprises a polarization analyzer, a spectrograph, and a charge-coupled device (CCD) camera. Conventional CCDs (e.g., Andor iDus, Andor Newton or Princeton Instruments PIXIS detectors) or intensified CCDs (iCCD) with a generation III image amplifier (e.g., Andor iStar 320, Princeton Instruments PI-Max4) have been used so far. Both detectors have their advantages. A conventional CCD has an exceptionally high quantum efficiency (QE) of >90%, but no gated acquisition mode is possible, which results in an accumulation of stray light in addition to the actual SFG signal. For a 1-kHz laser system the SFG signal reaches the detector only every millisecond, and in the time between laser pulses the stray light from incompletely blocked lights in the laboratory or from the experimental setup itself can accumulate on the CCD sensor. In order to decrease the background signal, extra measures have to be taken, but a noticeable background cannot completely be avoided since it is necessary to operate the spectrograph with a rather wide slit width of typically 300 μm. The strength of the background signal and therefore the quality of stray light suppression dominates in this setup the signal-to-noise ratio. The latter is not the case for an iCCD detector since the image amplifier allows setting a gate of a few nanoseconds around the SFG pulse, which efficiently suppresses background light from any source. Although the

QE of an iCCD detector has a relative low maximum of ~40%, the possible decrease in signal-to-noise ratio is partially compensated by the low background signal.

### 4.2.2 Spectroelectrochemical Cells

For electrochemical SFG spectroscopy the sample compartment is equipped with an electrochemical cell that allows for synchronized spectroscopic and electrochemical characterization (Fig. 4.2). As previously reported by Wieckowski and co-workers [8, 22] a well-defined electrolyte gap of ≥25 μm between an optical window and a thick metal electrode is needed to avoid depletion of chemical species in the gap since this geometry does not suffer from diffusion limitations from the bulk into the thin-layer electrolyte. Furthermore, effects like the distortion of the electrode potential due to the higher resistance of the thin-layer electrolyte—often referred to as IR or Ohmic drop effects—are also minimized if the electrolyte layer is equal to or exceeds a thickness of 25 μm and voltammetric sweep rates are kept ≤5 mV/s. The use of the latter geometry in a spectroelectrochemical experiment is only possible when the electrode–electrolyte interface is interrogated with broadband SFG. That is because the dramatic losses in mid-IR intensity due to the substantial optical absorption in an 25-μm-thick aqueous electrolyte layer can only be tolerated if the initial intensity of the broadband IR pulse is huge. In fact, compared to a picosecond IR pulse in a traditional scanning SFG setup or in infrared absorption spectroscopy (IRAS) the intensity of a femtosecond mid-IR pulse is orders of magnitude higher.

Although the use of this setup is greatly beneficial for surface science studies of single crystals immersed in an aqueous electrolyte, it does have its disadvantages because investigations of the interfacial water molecules are extremely challenging if not impossible. For that purpose one has to resort to thin-layer electrodes that are deposited on optically transparent substrates [23]. Thin metal films are often less ordered and have a high defect density, but they allow for the transmittance of both the mid-IR and the visible pulse through the electrode, which consequently avoids the strong absorption of the electrolyte. SFG measurements of water layers in contact with a thin-film Pt electrode have been recently reported by Noguchi et al. [24].

### 4.2.3 Theoretical Description and Interpretation of SFG Spectra

For an interface of two centrosymmetric materials the SFG intensity is proportional to the square of the interfacial second-order susceptibility [9, 12, 25, 26]:

$$I(\omega_{\mathrm{SF}}) \propto \left| \chi_{\mathrm{NR}}^{(2)} + \sum_{q} \chi_{q}^{(2)} \right|^{2} I_{0}^{\omega_{vis}} I_{0}^{\omega_{\mathrm{IR}}} \cdot \exp\left[ -4 \ln 2 \frac{(\omega_{\mathrm{IR}} - \Omega)^{2}}{\Delta^{2}} \right]$$

$$\text{with } \chi_{q}^{(2)} = \frac{A_{q} e^{i\theta_{q}}}{\omega_{q} - \omega_{\mathrm{IR}} + i\Gamma_{q}} \qquad (4.1)$$

Here, $\chi^{(2)}_{NR}$ and $\chi^{(2)}_q$ correspond to the nonresonant and resonant parts of the second-order electric susceptibility $\chi^{(2)}$, respectively. The resonant part of $\chi^{(2)}$ is related to the amplitude $A_q$, frequency $\omega_q$, bandwidth $\Gamma_q$, and phase $\theta_q$ of the $q$th vibrational mode, while $\chi^{(2)}_{NR}$ is due to nonresonant electronic excitations at the interface. The sensitivity of SFG toward interfacial molecular order combined with an inherent interfacial sensitivity arises from the coherent nature of the SFG process. Unlike linear-optical vibrational spectroscopy, the amplitude $A_q$ of the sum frequency polarization is an orientational average of the Raman polarizability and the dynamic dipole moment. $A_q = N\langle\alpha_q\mu_q\rangle$ correlates, therefore, directly with the molecular order at the interface and with the molecular density $N$. In fact, a perfectly polar-ordered adlayer results into the highest amplitude and SFG intensity while a layer of randomly distributed molecules without long-range order or a nearly centrosymmetric layer, but identical molecular density $N$ has negligible SFG intensities. An additional advantage of SFG over other vibrational techniques is that background-free measurements at any given electrode potential can be performed. Although, the nonresonant contribution $\chi^{(2)}_{NR}$ is nearly independent of the IR frequency and manifests as a featureless SFG contribution, it is highly dependent on the electronic structure of the surface. In Figure 4.3, potentiodynamic SFG spectra of the Au nonresonant contribution are shown as a function of the electrode potential and are compared with the corresponding cyclic voltammogram (CV). The CV of a polycrystalline Au electrode in dilute perchloric acid solution has a featureless double-layer region for potentials <0.7 V and is dominated by well-known surface oxidation features at potentials >0.7 V and a negative-going feature centered at 0.87 V due to the reductive removal of the previously formed surface and bulk oxides. A close comparison of the changes in the $\chi^{(2)}_{NR}$ of Au and the oxidation and reduction currents in the cyclic voltammogram of Figure 4.3 demonstrates that the strength of the interfacial $\chi^{(2)}_{NR}$ contribution is directly

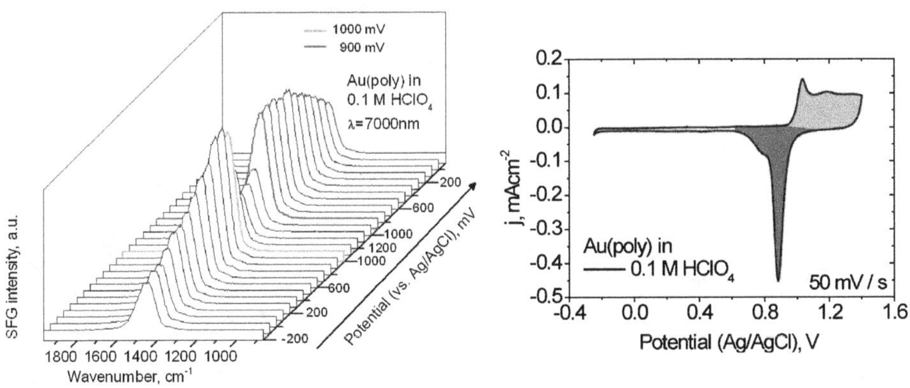


**Figure 4.3** SFG spectra of a polycrystalline Au electrode in blank 0.1 M $HClO_4$ solution and hanging meniscus cyclic voltammogram of the same electrode. Note that $\chi^{(2)}_{NR}$ of the Au electrode is the only contribution to the SFG spectrum while its shape is determined by the Gaussian spectrum of the broadband IR pulse.

related to the oxidation state of the Au surface. Obviously, substantial but reversible changes in $\chi_{NR}^{(2)}$ are observed at the onset of surface oxidation and reduction. In general, modifications such as a change in the oxidation state of a surface or the presence of strongly adsorbed molecules can alter the interfacial electronic structure and influence $\chi_{NR}^{(2)}$ considerably [27, 28].

In order to address the interfacial electronic structure directly, second-harmonic generation is often used and offers information complementary to SFG [29–32]. Thus, measurements of $\chi_{NR}^{(2)}$ can provide useful information on the interfacial electronic structure and $\chi_{NR}^{(2)}$ may be also used as a phase reference for determining the orientations of adsorbed molecular species. In some cases, however, resonant and nonresonant contributions can change simultaneously, which makes a separation of the latter extremely difficult.

In such a situation it is often helpful if one can discriminate better between nonresonant and the resonant behavior. Here, a $\chi_{NR}^{(2)}$ suppression technique that was developed by Dlott and co-workers can be an extremely powerful tool for disambiguating resonant and nonresonant signals [26, 27, 33]. The technique relies on the different time dependencies of resonant and nonresonant polarizations at metal interfaces. The decay of the nonresonant response is extremely fast and follows the femtosecond BBIR pulse in the time domain, whereas the resonant polarization as described by the free induction decay of the vibrational mode has life and dephasing times on the order of picoseconds. For that reason, the time delay between the arrival of a femtosecond BBIR pulse and a picosecond etalon-filtered visible pulse at the sample surface can be adjusted to mostly exclude the femtosecond nonresonant response component and nearly suppress the nonresonant signal. An example where the described suppression technique is very helpful can be found in Figure 4.4. In

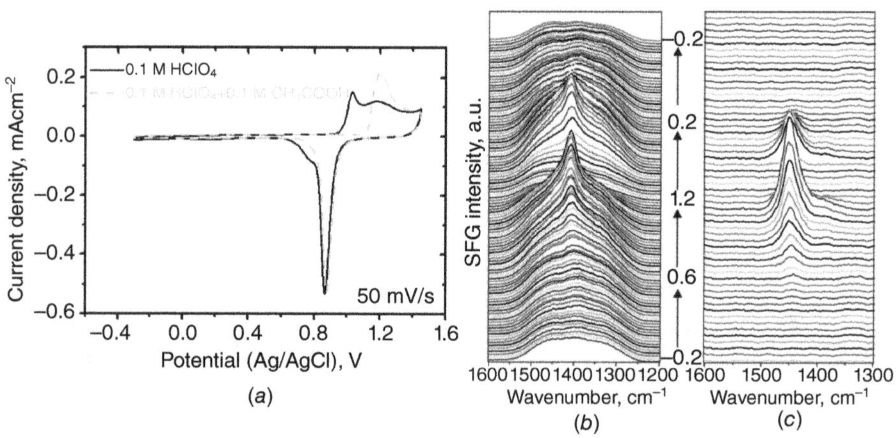


**Figure 4.4** (*a*) Cyclic voltammograms of a polycrystalline Au electrode in 0.1 M $HClO_4$ with and without 0.1 M acetic acid. (*b*) Au nonresonant and resonant contribution from adsorbed acetate to the SFG intensity of a polycrystalline Au electrode in contact with a 0.1 M $HClO_4$ and 0.02 M acetic acid electrolyte. (*c*) SFG spectra of the same system as in (*b*), but with $\chi_{NR}^{(2)}$ efficiently suppressed. Reproduced with permission from Ref. [27], copyright 2010, American Institute of Physics.

blank perchloric electrolytes the polycrystalline Au surface oxidizes at potentials >0.95 V versus Ag/AgCl, while in 0.1 M $HClO_4$ electrolyte with 0.02 M acetic acid the onset of Au surface oxidation is shifted to 1.12 V due to the competition of acetate adsorption and surface oxidation (Fig. 4.4*a*). Figure 4.4*b* presents broadband SFG spectra of polycrystalline Au in contact with 0.1 M $HClO_4$ electrolyte with 0.1 M acetic acid. As discussed above the broad Gaussian-shaped feature of the SFG spectra is due the $\chi^{(2)}_{NR}$ response of the Au–electrolyte interface. For electrode potentials >0.6 V a vibrational band centered at approximately 1401 $cm^{-1}$ is observed and constructively interferes with $\chi^{(2)}_{NR}$ contributions (Fig. 4.4*b*). This vibrational band is attributable to carboxylate ($-COO^-$) symmetric stretching vibrations of adsorbed acetate on a polycrystalline Au electrode.

Au surface oxidation at potentials >1.13 V leads to a considerable decrease of both vibrationally resonant and electronically nonresonant contributions (Fig. 4.4*b*) and clearly demonstrates that a distinction between resonant and nonresonant excitations is at least in the present case a serious issue in interpreting vibrational SFG spectra. However, suppressing the nonresonant response bypasses this issue and facilitates direct measurements of the vibrational band without the here undesired $\chi^{(2)}_{NR}$ contributions (Fig. 4.4*c*). An increased time delay between the BBIR and the visible pulse does also lead to substantial decay of the resonant intensity, and one has to make a compromise between a complete $\chi^{(2)}_{NR}$ suppression and maintaining a relatively good signal-to-noise ratio of the observed vibrational bands. Delaying the visible pulse versus. the infrared pulse also leads to some changes in the appearance of the resonant contributions in the SFG spectrum. This results from a considerable apodization of the spectrum in the time domain. These effects are shown schematically for an arbitrary SFG spectrum in Figure 4.5, where the spectrum in the frequency

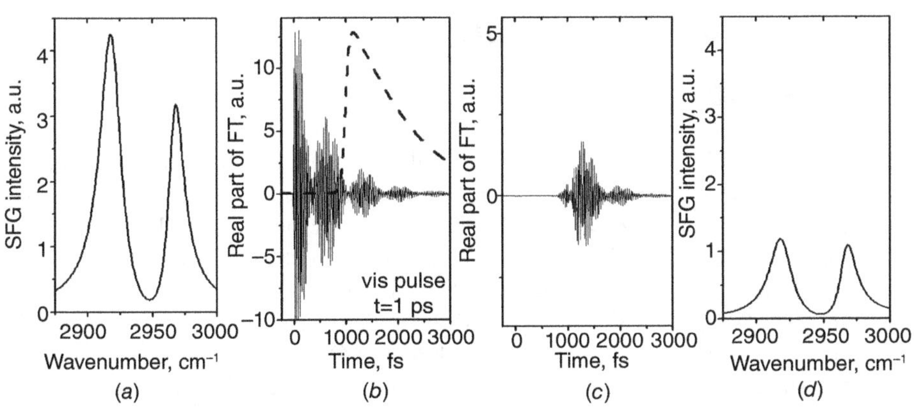


**Figure 4.5** (*a*) SFG spectrum of resonances at 2919 $cm^{-1}$ ($A_q = 2$; $\Gamma_q = 10\ cm^{-1}$) and 2965 $cm^{-1}$ ($A_q = 1.2$; $\Gamma_q = 7.4\ cm^{-1}$). (*b*) Real part of the Fourier transformed spectrum in (*a*) and a time-asymmetric visible pulse with a delay of 1 ps. (*c*) Real part of the Fourier transform of (*a*) folded with the visible pulse in (*b*). (*d*) SFG spectrum with apodization that was obtained by Fourier transformation of real and imaginary parts (not shown) of (*c*).

domain is converted into the time domain by Fourier transformation. The apodization of the time-domain response with the time profile of an etalon-filtered visible pulse removes considerable portions of spectrum because the induced polarization at times where the intensity of the visible pulse is negligible is not converted into the sum frequency polarization and thus cannot contribute to the SFG spectrum (Fig. 4.5*c*). Back transformation of the spectrum in the time domain consequently leads to a substantial distortion of the resulting SFG spectra, which can be a serious issue if several vibrational bands are present in the investigated spectral region. For example, in Figure 4.5 the intensity ratio of the vibrational bands with slightly different line widths has changed after apodization due to the different life times of these modes. Note that the line width is directly proportional to the inverse of life and dephasing times. These effects have to be taken under consideration when SFG spectra are recorded at high IR–visible delays.

Besides the $\chi^{(2)}_{NR}$ contribution, which contains useful information and helps to track changes in the interfacial electronic structure, the phase $\theta_q$ of a resonant contribution relative to the phase of $\chi^{(2)}_{NR}$ or another vibrational band can be of particular interest. The phase contains information on the orientation of molecular species at the interface and dramatically influences the line shape of vibrational bands in SFG spectra. In Section 4.5 we will give an example of how the isoelectric point of an interface can be estimated from the polar ordering and orientation of interfacial water molecules, where the latter is deduced from the phase of the observed water bands. In that example it is sufficient to determine relative phase changes between two vibrational resonances, however, in some cases a more elaborate phase-sensitive SFG setup is required, which has been invented by Shen and co-workers [34]. A recent perspective from the Allen group highlights conventional and phase-sensitive SFG for studies on the water organization at aqueous interfaces [35].

## 4.3 SPECIFIC ADSORPTION OF (BI)SULFATE ANIONS ON PT(111) SURFACES STUDIED WITH BROADBAND SFG

The specific adsorption of bisulfate and sulfate anions on Pt(111) single-crystal surfaces represents an important model system in electrochemical surface science and has been shown to change reaction pathways and efficiency of electrochemical reactions significantly [7, 13, 27, 28, 36–38]. The surface structure of Pt(111) in the potential range of sulfate and bisulfate adsorption has been studied with scanning tunneling microscopy (STM) [39, 40] and theoretical simulations [16–18]. The CV of a Pt(111) single crystal immersed in 0.1 M $H_2SO_4$ is presented in Figure 4.6*a*. The reversible CV comprises four potential regions that are indicative of a well-ordered surface with large atomically flat terraces. The pair of plateaulike current features at −0.2 to 0.07 V versus Ag/AgCl is due to hydrogen deposition on (111) surface sites and hydrogen stripping for cathodic and anodic sweeps, respectively. The adsorption of sulfate or (bi)sulfate on Pt(111) terraces leads to a broad feature centered at 0.16 V and to a subsequent sharp spike at 0.21 V. The necklike current

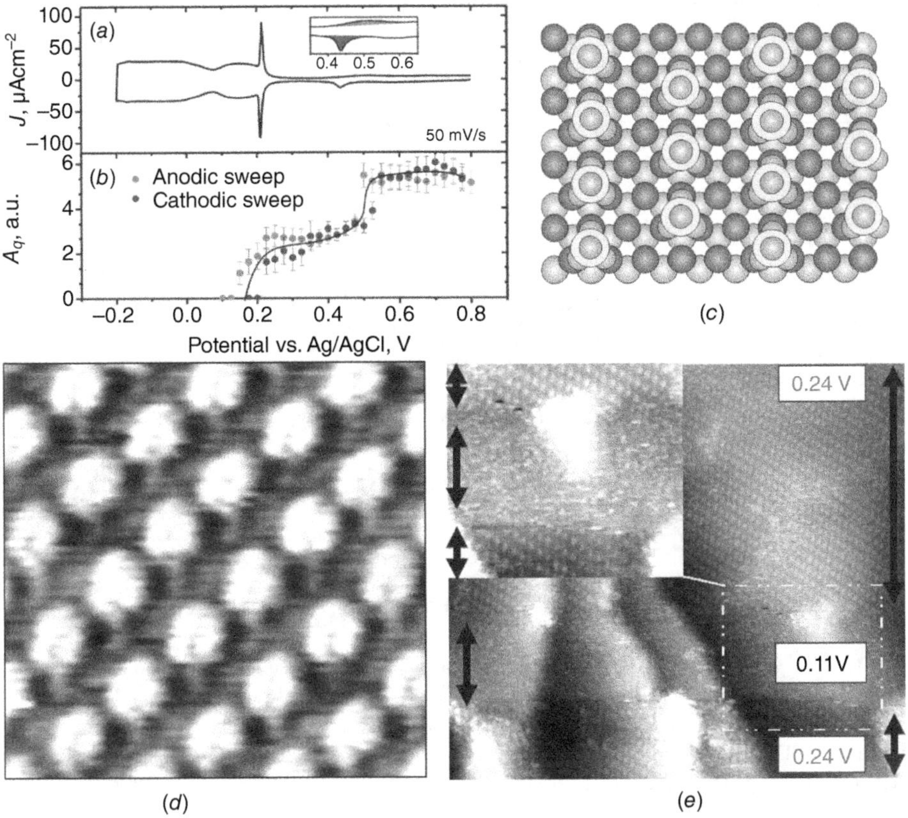


**Figure 4.6** (*a*) Hanging-meniscus cyclic voltammogram of a well-ordered Pt(111) electrode in 0.1 M $H_2SO_4$. (*b*) SFG amplitude $A_q$ of the vibrational band in (*a*) as a function of the applied potential $E$. (*c*) Structure model of the $(\sqrt{3}\times\sqrt{7})$R19.1° (bi)sulfate adlayer on Pt(111) (not shown: co-adsorbed water molecules). (*d*) 2.7 nm × 2.7 nm STM of the (bi)sulfate adlayer at an electrode potential of 0.31 V. (*e*) Potentiodynamic STM image of the Pt(111) crystal where the electrode potential was stepped between 0.24 and 0.11 V vs. Ag/AgCl. (*a*), (*b*) are reproduced with permission from Ref. [41], copyright 2010, American Chemical Society, and (*c*)–(*e*) are reproduced with permission from Ref. [40], copyright 2009, American Chemical Society.

feature for potentials higher than the sharp spike arises from double-layer charging only and is followed by a broad current wave centered at 0.5 V with a much narrower desorption feature at 0.43 V (inset of Fig. 4.6*a*). Potentiodynamic SFG spectra of Pt(111) in 0.1 M $H_2SO_4$ solution, which were synchronized with a CV from −0.2 to 0.8 V are presented in Figure 4.7*a*. At potentials $E < 0.2$ V, featureless SFG spectra are observed while for $E > 0.2$ V a vibrational band at 1245–1290 $cm^{-1}$ dominates the SFG spectra and is assigned to symmetric S–O stretching vibrations of adsorbed (bi)sulfate. Frequency $\omega_q$, bandwidth $\Gamma_q$, and amplitude $A_q$ of this vibrational band strongly depend on the applied potential (Figs. 4.6 and 4.7). Further analysis of our SFG spectra was performed by fitting the SFG spectra presented in Figure 4.7*a* with

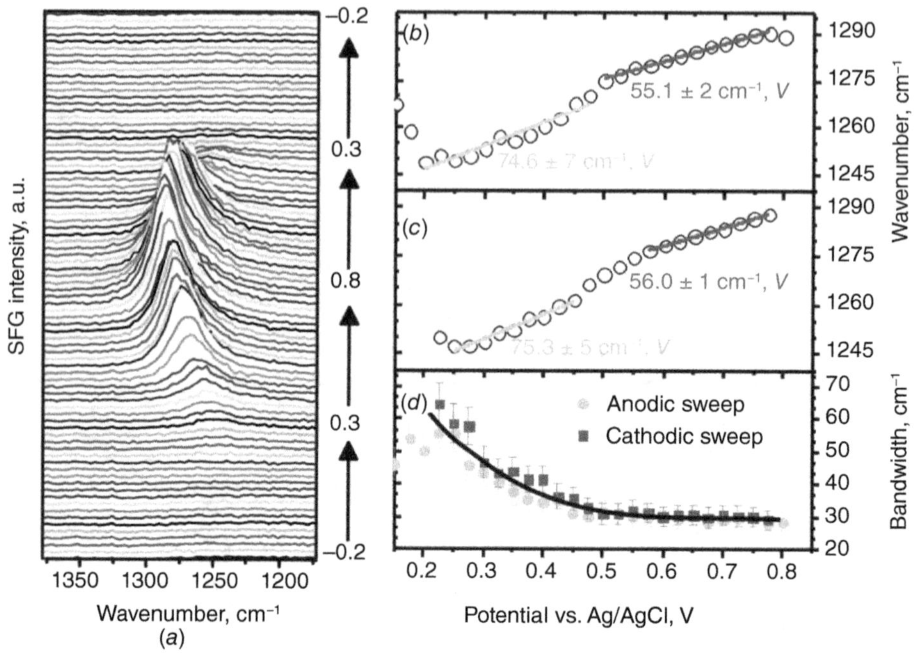


**Figure 4.7** (*a*) Consecutive series of broadband SFG spectra of (bi)sulfate adsorption on Pt(111) in 0.1 M $H_2SO_4$. Electrode potentials *E* (V) are indicated on the right-hand side. (*b*) and (*c*) Stark tuning for anodic and cathodic sweeps of the band in (*a*), respectively. (*d*) Bandwidth $\Gamma_q$ of the (bi)sulfate band in (*a*). SFG spectra were recorded with 5 s acquisition time and during a 5-mV/s potential sweep. Reproduced with permission from Ref. [41], copyright 2010, American Chemical Society.

model functions of the interfacial second-order susceptibility $\chi^{(2)}$ according to Eq. (4.1) and with $A_q$, $\omega_q$ and $\Gamma_q$ as adjustable parameters.

The potential dependence of $A_q$ is presented in Figure 4.6*b*. While $A_q$ is only weakly potential dependent in the double-layer region and for $E > 0.5$ V, two abrupt changes occur at 0.21 and 0.5 V. These changes in $A_q$ are accompanied by similarly abrupt changes in the Stark tuning rate of the frequency $\omega_q$ (Figs. 4.7*b*,*c*) and changes in bandwidth $\Gamma_q$ (Fig. 4.7*d*). In particular, the tuning of $\omega_q$ decreases from 75 cm$^{-1}$/V at potentials <0.5 V to 55 cm$^{-1}$/V at $E > 0.5$ V. In addition to the modified Stark tuning, a sudden blue shift of $\omega_q$ by 5 cm$^{-1}$ at 0.5 V is revealed by a close inspection of Figures 4.7*b*,*c*. The bandwidth $\Gamma_q$ decreases strongly between 0.2 and 0.5 V and remains at a constant value of 30 cm$^{-1}$ for $E > 0.5$ V (Fig. 4.7*d*). This bandwidth is notably higher than the experimental resolution of <10 cm$^{-1}$. All parameters in Figures 4.6*b* and 4.7*b*–*d* show within the experimental scatter a reversible behavior for anodic and cathodic sweeps. Note that the observed changes to $A_q$, $\omega_q$, and $\Gamma_q$ at 0.21 and 0.5 V occur at potentials where prominent current features are observed in the corresponding CV (Fig. 4.6*a*).

It is most revealing to relate the observed potential dependencies of the (bi) sulfate band to surface transformations of Pt(111) observed in previous STM studies

[39, 40]. The broad (bi)sulfate adsorption feature in Figure 4.6*a* indicates a (bi) sulfate coverage close to saturation when $E$ is increased to potentials close to the sharp spike (Fig. 4.6*a*). Yet, this adlayer does not give rise to measurable SFG intensities (Figs. 4.6*b* and 4.7*a*). According to recent density functional theory (DFT) simulations by Santana et al. [16], (bi)sulfate adsorbs in a bidentate geometry and, therefore, has a negligible dynamic dipole moment perpendicular to the Pt surface. As a consequence, this twofold coordinated state is SFG inactive by virtue of the dipole selection rule for metal surfaces. It is, therefore, reasonable to conclude that our SFG spectra at potentials of ~0.21 V reveal an abrupt transformation of the twofold into a threefold coordinated state of (bi)sulfate anions. Apparently, this transformation is associated with the sharp spike in CV (Fig. 4.6*a*) and accompanied with a similarly abrupt transformation from a laterally disordered (bi)sulfate adlayer into the ordered $(\sqrt{3}\times\sqrt{7})$R19.1°, which can be observed with electrochemical STM (see Fig. 4.6). Once the ordered (bi)sulfate layer is established, sweeping the electrode potential to more positive values shifts the frequency of the (bi)sulfate band considerably (Fig. 4.7*a*) and is indicative of a pronounced Stark tuning of (bi)sulfate stretching vibrations. The strong decrease of $\Gamma_q$ from 60 to 30 $cm^{-1}$ (Fig. 4.7*d*) and the slight increase of $A_q$ (Fig. 4.6*b*) between 0.2 and 0.5 V could be indicative of further ordering of (bi)sulfate structures on the Pt(111) surface and, consequently, to a decrease of inhomogeneous line width. Increasing the electrode potential could induce an increase of interfacial long-range order with a transformation of different rotational $(\sqrt{3}\times\sqrt{7})$R19.1° domains, into a nearly single-domain adlayer. The second abrupt increase in SFG amplitude at 0.5 V seems at first to suggest that the (bi)sulfate concentration at the interface increases further. Such a scenario is, however, rather unlikely since a closed-packed (bi)sulfate adlayer has already been established at lower potentials and the corresponding adsorption feature in the CV (Fig. 4.6*a*) at 0.5 V involves only a very low charge transfer, which is inconsistent with the magnitude of the change in SFG amplitude (Fig. 4.6*b*). A transformation of co-adsorbed water molecules in the $(\sqrt{3}\times\sqrt{7})$R19.1° layer into surface hydroxyl (Pt–OH) via the reaction $2H_2O + Pt \rightarrow Pt{-}OH + e^- + H_3O^+$ is much more likely. The latter scenario leads to a weaker hydrogen bonding of the (bi)sulfate adsorbate with Pt–OH species compared to the hydrogen bonding of co-adsorbed $H_2O$, which would be consistent with the observed blue shift of the (bi)sulfate band at an electrode potential of 0.5 V (Figs. 4.7*b*,*c*). Additional support comes from recent simulations [17]. Furthermore, Pt–OH and additional $H_3O^+$ cations from the Pt–OH formation can be incorporated into the double layer and screen the interfacial electric field while the (bi)sulfate layer persists on the Pt(111) surface. As a result of a decreased interfacial electric field, the electrochemical Stark tuning rates for potentials $E > 0.5$ V (Figs. 4.7*b*,*c*) is reduced. Thus, the origin of the substantial increase in amplitude $A_q \propto \langle \alpha_q\ \mu_q \rangle$ is likely related to modifications of the polarizability $\alpha_q$ and the dynamic dipole moment $\mu_q$ of (bi)sulfate vibrations due to a different local electric field and an increase of charge density in the S–O bond that is induced by a decreased hydrogen bonding. The loss in long-range order above 0.5 V as observed in previous in situ STM studies [39, 40] could be explained by the absence of stabilizing co-adsorbed water molecules at these potentials, which further corroborates

our hypothesis on the origin of the spectral characteristics of the sulfate band above 0.5 V.

## 4.4 BROADBAND SFG STUDIES OF ELECTROCHEMICAL REACTIONS: ELECTROCATALYSIS OF ETHANOL

### 4.4.1 Introduction

Ethanol has become globally recognized as an important source of energy because it is abundant, has a high energy density per molecule, and is easily produced from plant matter. The majority of the ethanol production in most industrial nations is used as a fuel additive for combustion engines and serves as a means to reduce the dependency on foreign oil. However, using combustion engines to extract energy from ethanol fuel is inefficient due to the inherent thermodynamic losses. Ethanol fuel cells are theoretically capable of bypassing these thermodynamic losses by converting chemical energy stored in the ethanol molecule directly to electrical energy. Furthermore, both ethanol and fuel cells are clean and compactable, expanding the potential of ethanol fuel to power a wider variety of devices including portable electronics. Despite these advantages, the design of efficient ethanol fuel cell catalysts has proven to be highly challenging. Polycrystalline Pt catalysts provide a good model system for studies of ethanol electrooxidation because they are widely known as highly active but inefficient catalysts for the ethanol oxidation reaction (EOR) [10, 28, 42–44]. In order to establish the necessary characteristics of more efficient catalysts, we are interested in a molecular-level understanding of the EOR that reveals all possible reaction steps, which could decrease the overall efficiency. SFG, as an inherently surface-specific technique, provides direct observations of only surface-adsorbed species without spectral interference from identical species in the bulk electrolyte and, thus, provides additional information to more established methods such as IR [11, 42, 45–48] and Raman spectroscopy [10] or differential electrochemical mass spectroscopy (DEMS) [44].

### 4.4.2 Ethanol Electrooxidation: CO Pathway

Potentiodynamic SFG spectra presented in Figures 4.9 and 4.10 were recorded during the electrooxidation of both regular and isotopically labeled ethanol, respectively. Both figures reveal intense CO vibrational bands that indicate predominant adsorption of CO as the main surface-adsorbed intermediate at electrode potentials of <0.6 V. Increasing the electrode potential leads to an oxidative removal of CO and to the predominant adsorption of acetate and, in the case of a $H_2SO_4$ electrolyte, to the co-adsorption of (bi)sulfate anions (Fig. 4.10*b*). On the basis of our spectroscopic and voltammetric observations, we propose a reaction mechanism for ethanol

Cathode: $3O_2 + 12\,H^+ + 12e^- \rightarrow 6H_2O$

Anode: $CH_3CH_2OH + 3H_2O \rightarrow 12\,H^+ + 12e^- + 2CO_2$

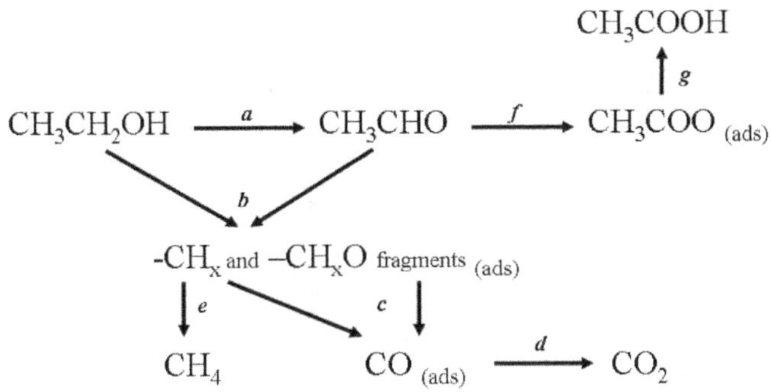


**Figure 4.8** Reaction equations for a complete oxidation of ethanol to $CO_2$ and the necessary reduction of oxygen at the cathode side. Possible intermediates and dead ends of the ethanol oxidation reactions that have been observed so far [10, 42, 44] and proposed reaction pathways for the electrooxidation of ethanol on a platinum surface. Each pathway is discussed in detail in the text. Reproduced with permission from Ref. [28], copyright 2011, Elsevier.

electrooxidation on polycrystalline Pt, which will be substantiated and discussed in the text below. The reaction pathways presented in Fig. 4.8 include oxidation of ethanol by dehydrogenation and the formation of acetaldehyde in step *a*. In step *b* adsorbed acetaldehyde and adsorbed ethanol (parallel to *a*) quickly decompose into surface-adsorbed $–CH_x$ [10] and $–CH_xO$ fragments. Note that no direct observation of a $–CH_xO$ fragment has been made so far. This is possibly due to a short lifetime of this species. It is, however, reasonable to assume that after the ethanol C–C bond is broken, a $–CH_xO$ fragment has to exist on the catalyst surface as a logical precursor to more oxidized species such as CO or $CO_2$. Both $–CH_x$ and $–CH_xO$ fragments can oxidize and form adsorbed CO (step *c*, Fig. 4.8) at potentials as low as –0.2 V, although oxidation of $–CH_x$ is likely to be more difficult.

For electrode potentials ≥0.4 V, adsorbed CO molecules are oxidized further to $CO_2$ (Fig. 4.8), which desorbs from the surface and diffuses into the bulk electrolyte. Not all single-carbon fragments of ethanol oxidize to CO and subsequently to $CO_2$, however. As shown in step *e* of Fig. 4.8, adsorbed $–CH_x$ fragments can be electroreduced at sufficiently low potentials to $CH_4$, which subsequently desorbs from the surface and diffuses in the bulk electrolyte. In the likely case that the C–C bond is not broken, acetaldehyde can oxidize to adsorbed acetate (Fig. 4.10*b*). At this point, the adsorbed acetate can only diffuse from the surface by conversion to acetic acid (step *g*). Further oxidation of acetic acid to $CO_2$ at fuel cell relevant potentials is not possible. Therefore, the formation of this product represents an undesirable "dead end" of the ethanol oxidation.

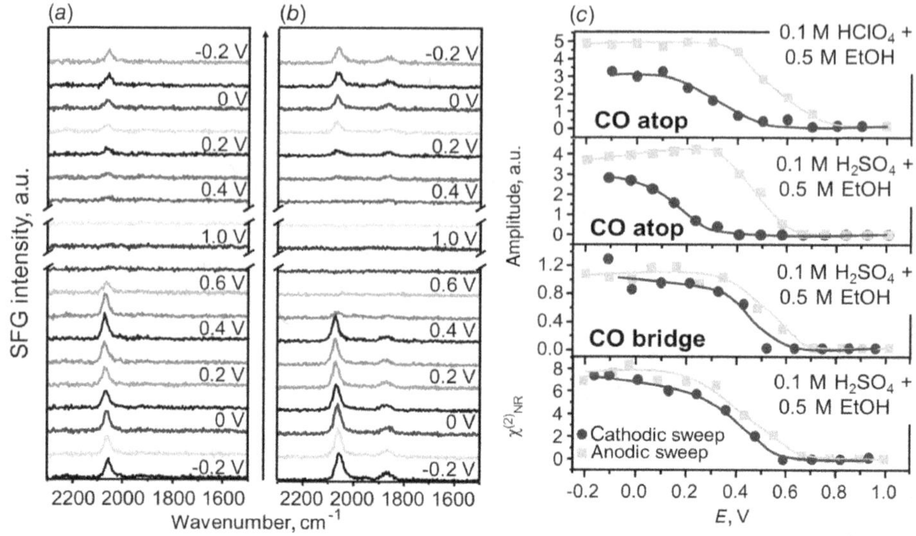


**Figure 4.9** Potentiodynamic SFG spectra of atop and bridge-bonded CO vibrational bands recorded during 0.5 M ethanol ($^{12}CH_3{}^{12}CH_2OH$) electrooxidation on a polycrystalline Pt electrode in (*a*) 0.1 M $HClO_4$ and (*b*) 0.1 M $H_2SO_4$. Electrode potentials *E* (V) vs. Ag/AgCl were as indicated in the figure. The frequency $\Omega$ of the BBIR pulse was centered at 1961 $cm^{-1}$. (*c*) Shows the potential dependence of modeled SFG amplitudes for atop and bridge-bonded CO vibrational bands at ~2080 and ~1850 $cm^{-1}$, respectively, and the potential dependence of the nonresonant contribution. Curves are a guide to the eye. Reproduced with permission from Ref. [28], copyright 2011, Elsevier.

### 4.4.3 Electrooxidation of $-CH_x$ and $-CH_xO$ Fragments from Ethanol

The formation of CO from both $-CH_x$ and $-CH_xO$ fragments (step *c*) is confirmed by our spectroscopic data on the electrooxidation of regular (Fig. 4.9) and isotopically labeled ethanol ($^{12}CH_3{}^{13}CH_2OH$): In Figure 4.10*a* CO vibrational bands of both $^{12}CO$ and $^{13}CO$ are clearly resolved and indicative of the oxidation of ethanol fragments in step *c*. The coverages of $^{13}CO$ and $^{12}CO$ are not identical, however, as we observe a greater quantity of $^{13}CO$ formed from the $-^{13}CH_xO$ fragment. Because the latter fragment is already partially oxidized, further oxidation to CO is far more facile than the oxidation of $-^{12}CH_x$ fragments, as is evidenced by the SFG amplitudes shown in Figure 4.9*c*. Here, the amplitude of the $^{13}CO$ vibrational band is considerably higher than the amplitude of the $^{12}CO$ vibrational band. Such a difference strongly implies that the surface coverage of adsorbed $^{13}CO$ molecules is much higher than the surface coverage of adsorbed $^{12}CO$ molecules and consequently indicates that the $-^{12}CH_x$ fragment oxidizes to $^{12}CO$ to a far lesser extent. At this point it should be noted that dipole–dipole coupling effects arising from interactions between the different CO species could be the cause for different $^{13}CO$ and $^{12}CO$ oscillator strengths in our SFG spectra. These coupling effects, which are well-

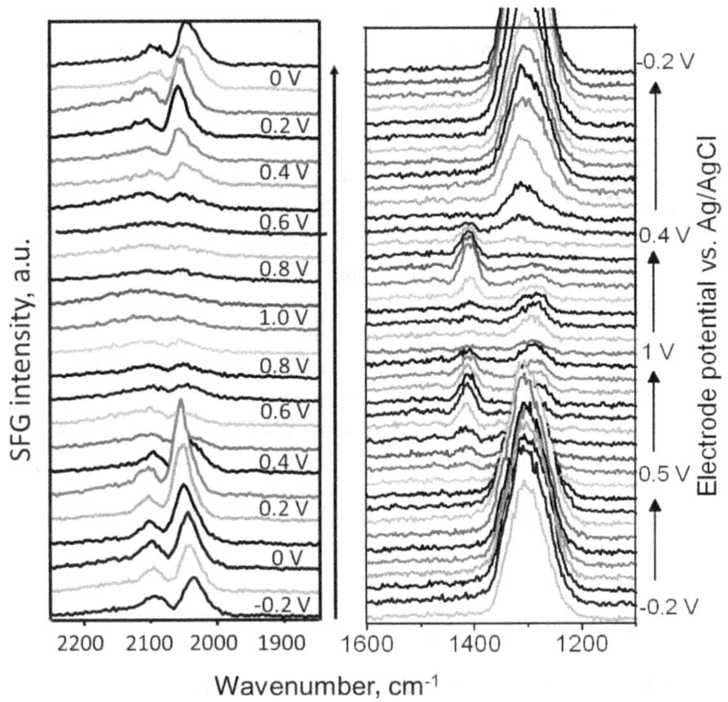


**Figure 4.10** Consecutive series of potentiodynamic SFG spectra of (*a*) $^{12}CO/^{13}CO$, (*b*) (bi)sulfate adsorbed and acetate (symmetric –$COOH^-$ stretch) vibrational bands as well as $\chi^{(2)}_{NR}$ contributions. SFG spectra in (*a*) were recorded during electrooxidation of 0.2 M isotopically labeled ethanol ($^{12}CH_3{}^{13}CH_2OH$) in 0.1 M $HClO_4$ and in (*b*) for oxidation of 0.5 M regular ethanol in 0.1 M $H_2SO_4$. Electrode potentials *E* (V) are indicated in the figure. The frequency Ω of the BBIR pulse was centered at (*a*) 2174 $cm^{-1}$ and (*b*) 1428 $cm^{-1}$. Reproduced with permission from Ref. [28], copyright 2011, Elsevier.

known phenomena common to closed-packed CO adlayers, may shift vibrational intensity from the $^{12}CO$ to the $^{13}CO$ vibrational band and consequently result in a virtually lower $^{12}CO$ intensity. If such a shift in intensity were present in Figure 4.10*a*, then the higher amplitude observed in the $^{13}CO$ band would not necessarily correspond to higher $^{13}CO$ coverages. However, this scenario is rather unlikely since a previous analysis by Severson et al. [49] of dipole-coupling effects with mixtures of $^{12}CO/^{13}CO$ on Pt(111) shows the opposite behavior. In their work, Severson et al. reported that for a well-defined mixture of >17% $^{12}CO$ with $^{13}CO$, the $^{12}CO$ vibrational band dominated the infrared spectra. It, therefore, appears clear that there is an intensity shift from the $^{13}CO$ vibrational band to the $^{12}CO$ band. Thus, if our Pt catalyst has an equal coverage of $^{13}CO$ molecules (formed from –$CH_xO$ fragments) and $^{12}CO$ molecules (formed from –$CH_x$) fragments, we would expect much higher intensities of the $^{12}CO$ vibrational band. The fact that we see the opposite, a higher intensity of $^{13}CO$ vibrational bands, suggests that the surface coverage of $^{12}CO$ molecules is less than 17%. This is further supported by a Fourier transform infrared

(FTIR) study of ethanol ($^{12}CH_3{}^{13}CH_2OH$) oxidation on Pt(332) surfaces performed by Souza-Garcia et al. [50], where the existence of an adsorbed –$CH_x$ species is inferred by comparing the notably different intensities for $^{12}CO_2$ and $^{13}CO_2$ vibrational bands (1 : 3, respectively). We, therefore, conclude that the oxidation of adsorbed –$CH_x$ species to CO is significantly more difficult than the oxidation of –$CH_xO$ species. In addition, the difficulty of oxidizing –$CH_x$ species has been demonstrated in recent DFT model calculations [51], which show a much higher potential barrier for the oxidation of –$CH_x$ compared to –$CH_xO$ species. This leads to an accumulation of adsorbed –$CH_x$ on the catalyst surface. Additional support comes from surface-enhanced Raman spectroscopy (SERS) [10] studies where the presence of surface-adsorbed –$CH_x$ species has been suggested. In fact, Abd-El-Latif et al. point out in their recent DEMS study that the remaining fraction of the –$CH_x$ species, which has not been converted to CO, may be oxidized in the oxygen region only [44]. Furthermore, at potentials lower than 0 V, adsorbed –$CH_x$ species can be electroreduced to methane, as shown in step *e* of Fig. 4.7 and has been observed by prior DEMS studies [44]. The electrooxidation of adsorbed –$CH_x$ species to CO can be expected to be less efficient and competes with the electroreduction of –$CH_x$ to $CH_4$ at lower potentials. Since –$CH_x$ species block available adsorption sites at fuel cell relevant potentials, oxidation of –$CH_x$ has to be considered as a serious issue in the performance of ethanol-based fuel cells.

### 4.4.4 Ethanol Electrooxidation: Acetate Pathway

Our spectroscopic results presented in Figures 4.9 and 4.10 show that as soon as the electrode potential is increased to 0.4 V versus Ag/AgCl there is a steady drop in the intensity of the CO vibrational bands, as is most clearly seen in the corresponding SFG amplitudes in Figure 4.9*c*. This drop in amplitude originates from the oxidative removal of CO from the Pt surface and the formation of $CO_2$. Conversely, on the cathodic potential sweep, we see the reappearance of the CO vibrational band at $E < 0.4$ V, where the Pt oxide layer is stripped and ethanol is oxidized to CO, which subsequently poisons active Pt sites. There is a slight hysteresis observed in the recovery of the CO atop oscillator strength at $E < 0.4$ V on the cathodic sweep, seen in Figure 4.9*c*. This is likely attributable to the slow diffusion of ethanol into the 25-μm gap and subsequent accumulation of CO atop Pt sites. We do not observe such a slow rise in the vibrational band amplitude at the beginning of the anodic sweep because the electrode is held at 0 V for several minutes prior to the scan, which is sufficient time to achieve saturation coverage of CO molecules on the catalyst surface. The recovery of bridge-bonded CO is more reversible compared to CO, which is consistent with a preferential binding of CO to bridge sites when the overall CO coverage is low. The $\chi^{(2)}_{NR}$ amplitude is also highly reversible. A close comparison of the amplitudes in Figure 4.9*c* reveals a nearly identical potential dependence of the nonresonant contribution $\chi^{(2)}_{NR}$ [Eq. (4.1)] compared to the potential dependence of the resonant contribution due to adsorbed CO.

In fact, Figure 4.10*b* shows substantial changes in the contribution of $\chi^{(2)}_{NR}$ as soon as the CO adlayer starts to oxidize. We recall that the $\chi^{(2)}_{NR}$ contribution is purely electronic in origin and is caused by the nonlinear response of free electrons at the metal side of the interface. The strong chemisorption of CO on most metal surfaces localizes the electrons in the adsorbate–metal bonds and decreases $\chi_{NR}^{(2)}$ contributions. Consequently, the observed changes in $\chi_{NR}^{(2)}$ can be attributed to changes in CO coverage. In addition to the potential-induced changes in CO coverage, a significant effect of $H_2SO_4$ and $HClO_4$ supporting electrolytes on CO amplitudes and CO binding behavior is observed. Figure 4.9 shows that the vibrational band assigned to CO molecules bound to bridge sites of the Pt electrode is present in a sulfuric acid electrolyte only. The existence of CO molecules on bridge sites is indicative of an overall lower surface coverage of CO in sulfuric acid electrolyte and is in agreement with overall lower current densities in sulfuric acid solutions as seen by chronoamperometry and cyclic voltammetry [28]. Further support for this conclusion comes from the observation of a (bi)sulfate band in our SFG spectra (Fig. 4.10*b*). Obviously, adsorbed (bi)sulfate anions block adsorption sites on the Pt surface and, therefore, inhibit ethanol oxidation. The latter results in a lower CO coverage and in significantly reduced reaction efficiency.

After the CO adlayer has been stripped from the Pt surface at potentials >0.4 V, Pt surface sites become available for other more weakly adsorbing molecular species such as acetate or (bi)sulfate. In fact, a vibrational band of adsorbed acetate arises, as shown in Figure 4.10*b* at these potentials. The formation of acetate is represented as step *f* in the reaction of Figure 4.8. The acetate anion, however, does not oxidize within the potential range in which fuel cells operate and thus either diffuses from the surface as acetic acid (step *g* in Fig. 4.8) or blocks Pt catalyst sites. In contrast to the competing adsorption of (bi)sulfate and acetate on Au surfaces [27], co-adsorption of (bi)sulfate on Pt surfaces leads to acetate adsorption in a broader potential window (Fig. 4.11). Here, the interaction of co-adsorbed (bi)sulfate with acetate anions may modify the potential dependence of acetate on Pt electrodes and lead to an excess in acetate coverage at potentials of <0.45 and >0.7 V. Moreover, co-adsorption of (bi)sulfate limits the possible variety of structural arrangements of the interfacial acetate layer and causes additional ordering and may consequently decrease the inhomogeneous line broadening. This is evidenced from the notably different bandwidths $\Gamma_q$ of 52 and 38 $cm^{-1}$ from acetate vibrational bands in $HClO_4$ and $H_2SO_4$ solutions, respectively.

Therefore, the stabilization of acetate on the surface by (bi)sulfate co-adsorption [27, 28] additionally discourages the use of dilute sulfuric acid as an electrolytic medium for fuel cells. Since the reaction pathway in which acetate and acetic acid are produced yields only four electrons per ethanol molecule, it is highly undesirable.

### 4.4.5 Electrooxidation of Ethanol in Alkaline Electrolytes

As discussed above, the development of direct ethanol fuel cells demands an in-depth understanding of ethanol's electrooxidation reactions. Most studies of ethanol electrooxidation on Pt-based catalysts have been conducted in acidic electrolytes, yet

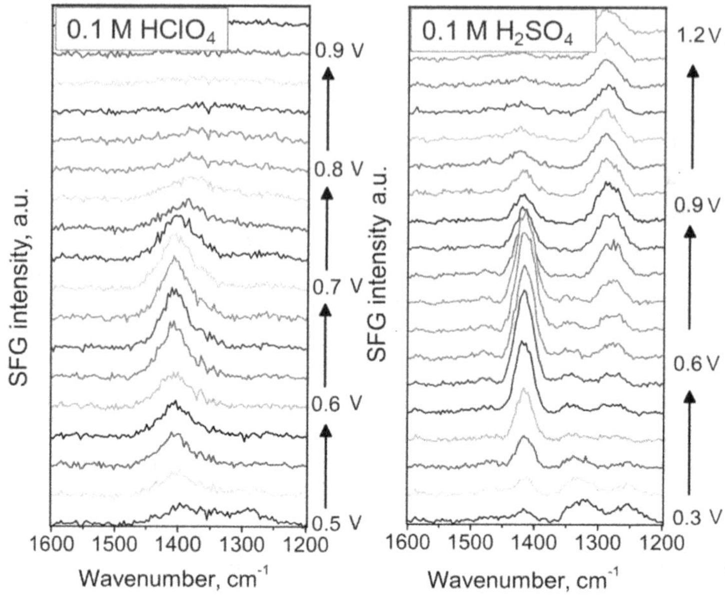


**Figure 4.11** Consecutive series of potentiodynamic BB-SFG spectra of a polycrystalline Pt electrode immersed in perchloric and sulfuric acid solution with 0.1 M acetic acid. The spectra (25 s acquisition time) were measured simultaneously with a cyclic voltammogram (1 mV/s). The electrode potential $E$ (V) vs. Ag/AgCl is indicated on the right-hand side. Reproduced with permission from Ref. [27], copyright 2010, American Institute of Physics.

alkaline electrolytes offer the promise of more complete conversion to $CO_2$ at lower overpotentials. With recent advances in the practical design of alkaline fuel cells, a thorough understanding of the ethanol electrooxidation mechanism in alkaline media is now more important than ever. In this chapter, we provide novel insights into this mechanism by examining new data on surface-adsorbed reaction intermediates at fuel cell relevant conditions.

Consecutive SFG spectra were recorded in 0.1 M NaOH with 0.5 M ethanol (Fig. 4.12). Here, a vibrational band centered at 2030 $cm^{-1}$ dominates the SFG spectra and can be assigned to stretching vibrations of CO molecules adsorbed atop Pt sites. Note that the significantly lower frequency and intensity of the CO band compared to acidic conditions (see Section 4.2) indicates a much lower surface coverage of CO, which is consistent with the lower overpotential for CO oxidation in alkaline electrolytes. For potentials >0.35 V the amplitude of the CO band decreases with increasing surface oxide concentration. However, a substantial recovery of the CO band at potentials between 0.75 and 1.25 V is observed in an ethanol-containing solution (Fig. 4.12*a*). In order to address the possible origin of the CO recovery, which appears to belie the relative ease of oxidizing CO in alkaline media, we have tracked the oxidation of the different carbon species that are immediate products of ethanol C–C bond cleavage. In particular, subsequent oxidation of the

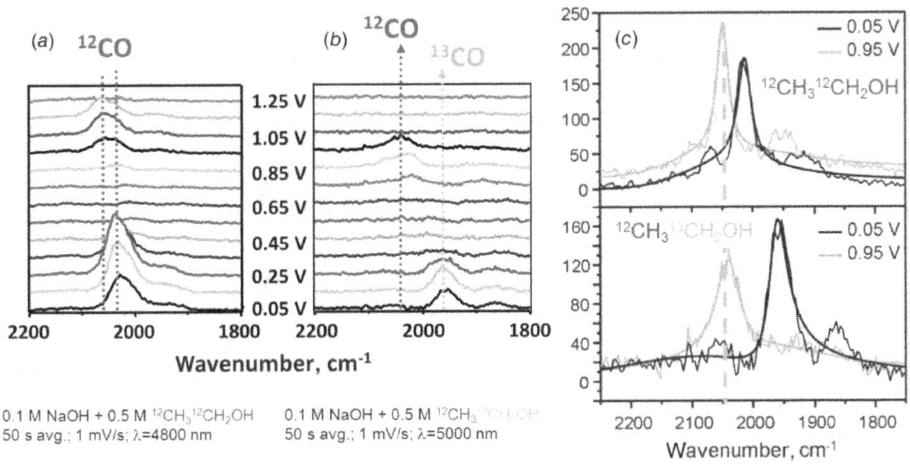


**Figure 4.12** Potentiodynamic series of SFG spectra in 0.1 M NaOH with (*a*) 0.5 M ethanol ($CH_3CH_2OH$) and (*b*) 0.5 M isotopically labeled ethanol ($^{12}CH_3{}^{13}CH_2OH$). Potentials were as indicated in the figure. (*c*) Comparison of the spectra for CO vibrational bands in (*a*) with the spectra in (*b*) at potentials of 0.05 and 0.95 V. Modeled spectra according Eq. (4.1) are bolded and superimposed over their respective raw spectra to clearly indicate the position of each CO band. Reproduced with permission from Ref. [43], copyright 2011, American Chemical Society.

resulting $-^{12}CH_x$ and $-^{13}COH_x$ fragments from isotopically labeled ethanol ($^{12}CH_3{}^{13}CH_2OH$) results in surface-adsorbed $^{12}CO$ and $^{13}CO$. These two molecular species can be identified directly with SFG spectroscopy due to their inherently different CO stretching frequencies of approximately 1995 and 2035 cm$^{-1}$ for $^{13}CO$ and $^{12}CO$, respectively. Therefore, this surprising potentiodynamic behavior of the CO intermediate was investigated in more detail in a similar experiment, but with 0.5-M isotopically labeled ethanol ($^{12}CH_3{}^{13}CH_2OH$) (Fig. 4.12*b*). Oxidation of different carbon species from isotopically labeled ethanol is expected to give rise to CO vibrational bands with different vibrational frequencies that can be used as fingerprints for the relative $^{12}C$ and $^{13}C$ coverages, while information on possible oxidation pathways is additionally gained. Figure 4.12 presents the spectral region of CO vibrations for an experiment where isotopically labeled ethanol is oxidized. Consistent with previous results with regular ethanol, oxidation of isotopically labeled ethanol leads to an identical decrease and sudden increase of the CO band at higher potentials: At 0.05 V, a band at 1995 cm$^{-1}$ attributable to $^{13}CO$ is observed, increases in intensity with increasing potential, and begins to diminish at 0.35 V (Fig. 4.12*b*). However, during the recovery of the band at ~0.75 V, a significant and abrupt blue shift of CO stretching frequencies to ~2030 cm$^{-1}$ takes place, while the SFG intensity near 1995 cm$^{-1}$ rapidly diminishes. A close comparison of the vibrational frequencies of the observed bands in Fig. 4.12*c* clearly indicates that the frequency shift between the electrode potentials of 0.05 and 0.95 V is much more pronounced for isotopically labeled ethanol. A potential-induced Stark tuning of the 1995 cm$^{-1}$ band as a possible origin for the observed dramatic blue shift is therefore highly unlikely. Consequently, the band at 0.75 V must be composed of two bands: one at ~2030 $cm^{-1}$ and a second band at 1995 $cm^{-1}$ attributable to $^{12}CO$ and $^{13}CO$ species, respectively. In Figure 4.12*b*, a sudden shift at 0.75 V in SFG intensity from the lower-frequency $^{13}CO$ band to the higher-frequency $^{12}CO$ band, further clarified in Figure 4.12*c*, unequivocally demonstrates the oxidative stripping of $^{13}CO$ from the Pt surface and the simultaneous deposition of $^{12}CO$. In Section 4.3, we have demonstrated for ethanol oxidation in acidic media that oxidation of partially oxidized $–CH_xO$ fragments occurs more rapidly at lower overpotentials compared to $–CH_x$ fragments that resist further oxidation and persist on the Pt surface as a catalyst poison. It is reasonable to assume that the absence of catalytic sites for efficient oxidation of $–CH_x$ on polycrystalline Pt, exists in alkaline electrolytes as well. This conclusion is corroborated by our SFG spectra during electrooxidation of $^{12}CH_3{}^{13}CH_2OH$, where at low potentials the formation of $^{13}CO$ is observed only while a $^{12}CO$ band due to partial oxidation of $–^{12}CH_x$ fragments is absent. A complete change of the situation takes place at high anodic potentials where $^{13}CO$ is absent and $^{12}CO$ dominates the SFG spectra. Therefore, the observed CO band at potentials of 1.05–1.15 V originates from the oxidation of $–CH_x$ in alkaline electrolytes where subsequent oxidation of CO to $CO_2$ at these potentials occurs. In conclusion, our results provide evidence that an anodic potential sweep to potentials of >0.65 V yields oxidation of $–CH_x$. Here, the supporting electrolyte may play a crucial role and act as a co-catalyst that lowers the overpotential for the latter reaction.

### 4.4.6 Ethanol Electrooxidation: Summary and Conclusions

Recent advances in vibrational broadband sum frequency generation (SFG) have allowed us to address electrochemical adsorption and electrocatalytic reactions in unprecedented detail. SFG, as an inherently surface-specific technique, provides a direct observation of only those molecular species that are adsorbed to either single crystalline or polycrystalline substrates without spectral interference from the bulk electrolyte. The use of a Teflon spacer creates a fixed 25-μm gap between an optical window and the electrode surfaces, permitting synchronous application of voltammetric sweeps at rates of <5 mV/s without detrimental Ohmic drop effects. Our data thus offer gives insight into the chemical identity of surface-adsorbed intermediates. In particular, rapid acquisitions of broadband SFG spectra synchronized with voltammetric scans have allowed us to perform real-time studies of (bi)sulfate anions and ethanol oxidation intermediates such as CO and acetic acid anions on Pt catalyst surfaces. We report new information on the mechanism of ethanol decomposition using isotopically labeled ethanol and clarify the behavior of adsorbed single-carbon intermediates that point to the presence of strongly adsorbed $–CH_x$ species. At low electrode potentials, CO is the predominantly adsorbed species. Increasing the potential results in an oxidative removal of CO and to the adsorption of acetate anions. In $H_2SO_4$ solutions specific adsorption of acetate and co-adsorbed (bi) sulfate anions is observed but is absent under basic conditions. SFG spectra during electrooxidation of isotopically labeled ethanol ($^{12}CH_3{}^{13}CH_2OH$) reveal further information on the carbon–carbon bond cleavage and the subsequent CO formation. In particular, the methyl part ($-^{12}CH_3$) of the ethanol molecule produces far less $^{12}CO$ than the preoxidized $-^{13}CH_2OH$ part, which is presumably caused by methyl electroreduction to methane and/or the persistence of $-CH_x$ on the Pt surface. Our conclusions—specifically the persistence of an adsorbed $-CH_x$ intermediate in a potential range beyond the CO oxidation overpotential—spur vital new considerations on more efficient Pt–catalyst design.

## 4.5 SUM FREQUENCY GENERATION OF PROTEINS AT ELECTRIFIED AIR–WATER INTERFACES

Foams are materials of particular importance since they have a broad range of applications such as metal foams for light-weight structures, polymer foams for thermal insulation, as well as in textured food products, just to mention a few. Although the chemical composition of these foams is largely different, they share common similarities for foam formation and stabilization. The latter is controlled by adsorption processes at the interface between the gaseous phase and the surrounding solid or liquid phase (Fig. 4.13). Therefore, it is of great importance to understand foam stabilization processes at interfaces in order to design advanced materials with tunable properties. For that purpose we have addressed the surface chemistry of the protein bovine serum albumin (BSA) at the air–water interface and the effect of pH on BSA foams, which represent a role model for foams with a certain tunability.

In aqueous electrolytes proteins are charged because functional groups on the protein surface such as carboxyl or amine groups can be protonated or deprotonated as a function of the electrolyte pH. Consequently, the adsorption of proteins at air–water interfaces leads to a strong electric field and to the formation of an electric double layer that can be very similar to the double layers known from electrode–electrolyte interfaces. To gain further insights into the molecular structure of the interface, vibrational SFG spectra were recorded in the spectral region of 1000–3700 $cm^{-1}$.

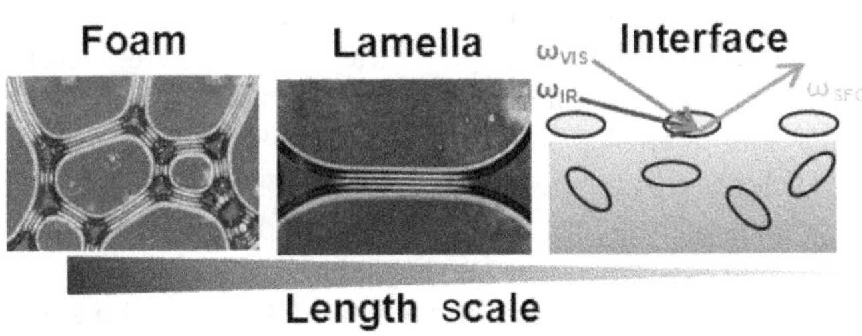


**Figure 4.13** Micrographs of foam structures and their hierarchy with ubiquitous air–water interfaces (schematically shown). Reproduced with permission from Ref. [52], copyright 2012, American Chemical Society.

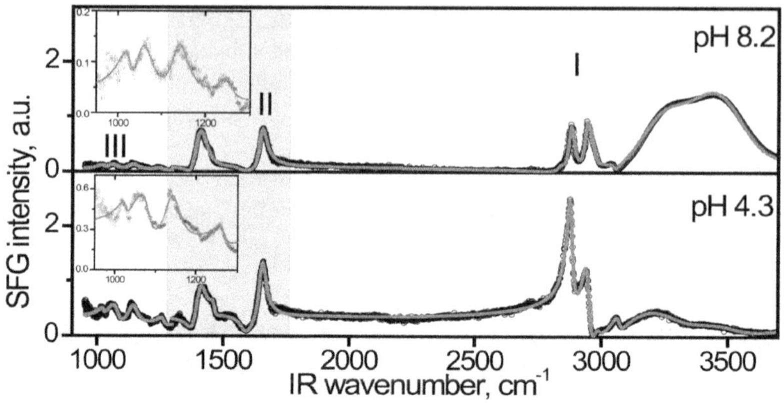


**Figure 4.14** Vibrational SFG spectra of BSA at air–water interfaces for pH 8.2 and 4.3. Solid lines are fits to the experimental data as explained in the text. Details to the spectral regions I–III can be found in the text. Reproduced with permission from Ref. [52], copyright 2012, American Chemical Society.

This approach allows one to identify vibrational fingerprints of BSA and interfacial water molecules and to select specific regions of interest where changes due to the electrolyte pH are most pronounced. Figure 4.14 shows vibrational SFG spectra of BSA adsorbed to the air–water interface for pH values of 8.2 and 4.3. A close comparison of the spectra reveals that they are dominated by vibrational bands in unique spectral regions where substantial changes as a function of the electrolyte pH can be observed:

1. Functional groups of adsorbed BSA give rise to strong vibrational bands at ~2875 and ~2936 cm$^{-1}$ that are attributable to $CH_3$ symmetric stretching vibrations and to the $CH_3$ Fermi resonance, respectively. Furthermore, $CH_3$ asymmetric, symmetric $CH_2$, and aromatic CH stretching vibrations give rise to bands at 2964, 2850, and 3060 cm$^{-1}$, respectively [53]. Strong bands between 3100 and 3800 cm$^{-1}$ originate from OH stretching vibrations of interfacial hydrogen-bonded water molecules [12, 54, 55].
2. In a second spectral region, additional vibrational bands of interfacial BSA at ~1654 and 1410 cm$^{-1}$ are observed and can be attributed to amide I [56] and to carboxylate (R-COO$^-$) symmetric stretching vibrations, respectively (Fig. 4.11). The amide I band originates from R-C=O carbonyl vibrations of molecular groups in the protein backbone where they can form hydrogen bonds to neighboring amide groups (Fig. 4.15). The strength of the hydrogen bonds influences the frequency of the amide I band greatly, and this band is, therefore, strongly dependent on the secondary structure of the protein. In previous IR studies, changes in the position of this band were often referred to denaturation, unfolding, or aggregation processes. The so-called amide II band at ~1550 cm$^{-1}$, which can be easily observed with linear IR spectroscopy [57], contributes, however, only weakly to the SFG intensity. Prereq-

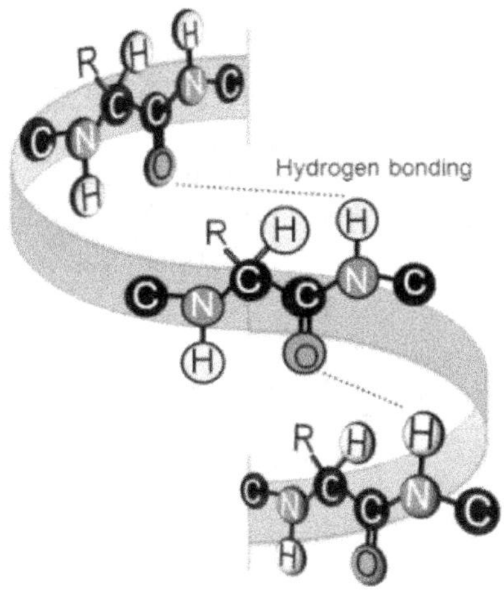


**Figure 4.15** Schematic representation of amide groups and their local orientation in a protein with an α-helix secondary structure. Reproduced with permission from Ref. [52], copyright 2012, American Chemical Society.

uisite for an SFG active mode is both Raman as well as IR activity. Since the amide II mode does lead to weak Raman resonances only, weak resonant SFG contributions from this band are also expected. Furthermore, the amide II band is a combination of C–N stretch and N–H bending vibrations and has a dynamic dipole moment that is perpendicular to the dipole moment of the amide I band. As a result, the intensity of this band is additionally weakened due to weaker excitations by p-polarized IR light.

3. In the spectral region below 1400 $cm^{-1}$ a band centered at ~1250 $cm^{-1}$ is observed and has been attributed to amide III vibrations [58]. Three weak bands at ~1140, ~1070, and ~1020 $cm^{-1}$ are also observed in the spectra, which could be related to –C–O– stretching modes, N–H deformation, and CH aromatic bending vibrations. Since substantial changes as a function of the bulk pH are observed for CH and OH vibrations (Fig. 4.14), further pH-dependent SFG studies were focused on the latter vibrations.

In order to reveal changes in the interfacial molecular structure, SFG spectra of BSA adsorbed to the air–water interface were recorded and analyzed for pH values of 3–10. Figure 4.16 shows representative SFG spectra in this pH range where OH and CH bands have strong pH dependencies. At a pH of 5.5 the SFG intensity of the $H_2O$ bands is negligible but increases substantially for lower and higher pH values than 5.5. In particular for the OH stretching bands of interfacial water molecules, we observe dramatic changes in the SFG intensity (Fig. 4.16). Further analysis of SFG bands as a function of pH was performed by fitting our measured spectra with model functions according to Eq. (4.1). We have used the amplitude, the resonance frequencies, and the nonresonant contribution as adjustable parameters in our fitting procedures. In order to determine the nonresonant contribution to the SFG intensity in Figure 4.16 the overview spectra in Figure 4.14 are extremely helpful since the nonresonant response and its influence on the dispersion of the vibrational

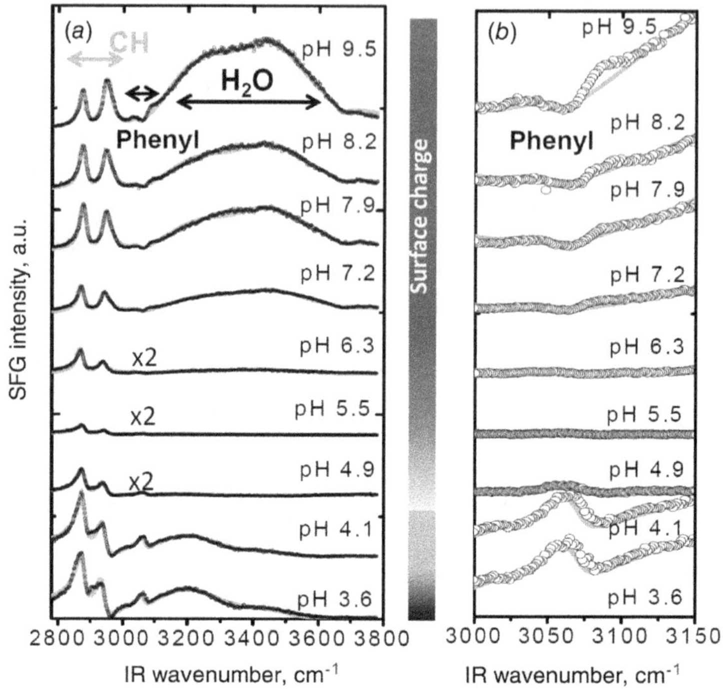


**Figure 4.16** Vibrational SFG spectra of BSA at air–water interfaces as a function of the electrolyte pH: (*a*) shows an overview of the CH and OH stretching region and (*b*) enhances the spectral region that is dominated by the resonance due to phenyl-like residues of BSA. The pH was as indicated in the figure. Solid lines are fits to the experimental data according to Eq. (4.1). The color scale represents schematically the interfacial charge density as a function of the pH. Reproduced with permission from Ref. [52], copyright 2012, American Chemical Society.

resonances can be easily deduced. In Figure 4.17*b* the results according to the described fitting procedures are presented for the absolute value of the OH oscillator strength. The pH dependence shows a local minimum around pH 5. While the amplitudes of amide I (Fig. 4.17*c*) and carboxylate bands (not shown) have a narrow minimum at pH 5 and are only slightly smaller for more alkaline or acidic pH values, the amplitude of OH vibrations is also minimal at pH 5 [52] but varies in a much broader pH range (Fig. 4.17*b*).

We will now compare the observed pH effects of adsorbed BSA adlayers from ellipsometric measurements with the discussed results from SFG spectroscopy. It is obvious that the bulk isoelectric point of BSA at pH 5.2 [52] with a zeta potential equal to zero clearly corresponds to a minimum in SFG amplitudes of the BSA and interfacial $H_2O$ related bands (Figs. 4.14 and 4.16) and to a maximum in the thickness of adsorbed BSA adlayers (Fig. 4.17*a*). In fact, the results from ellipsometry in Figure 4.17*a* indicate the presence of multilayers for pH values near the bulk isoelectric point and are in agreement with previous neutron reflection studies of BSA at air–water interfaces. In their report, Lu et al. [59] have estimated the adsor-

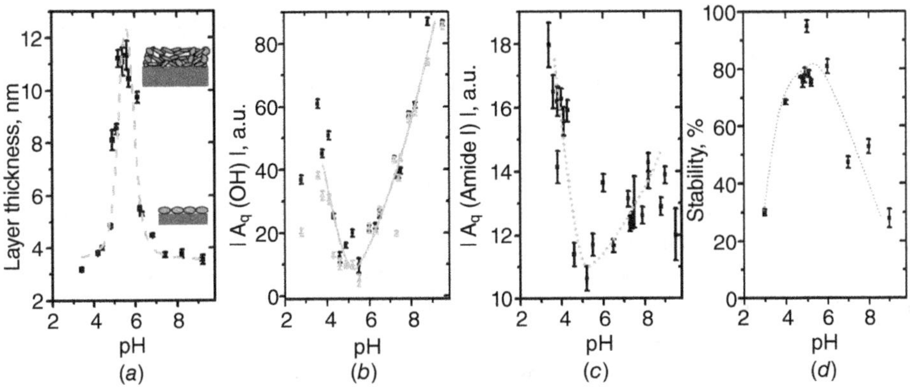


**Figure 4.17** (*a*) Thickness of adsorbed BSA layers at the air–water interface as a function of the electrolyte pH. The thickness was determined with ellipsometry under the assumption of a simple water ($n$ = 1.33)–BSA ($n$ = 1.40)–air ($n$ = 1.00) layer model. The dashed line is a guide to the eye. (*b*) and (*c*) pH dependence of the absolute value of the amplitudes from OH [3200 $cm^{-1}$ (■) 3400 $cm^{-1}$ (▲)] and amide I bands, respectively. Dotted lines are guide to the eye. (*d*) Foam stability (*b*) as a function of pH. Reproduced with permission from Ref. [52], copyright 2012, American Chemical Society.

bate thickness to approximately 7 nm and a reduced thickness of approximately 4 nm for pH values higher or lower than the bulk isoelectric point. Modeling the geometric dimensions of BSA by a simple incompressible ellipsoid, thicknesses of ~4 nm in Figure 4.17*a* point to a monolayer of adsorbed BSA proteins with their long axis oriented parallel to the interface. Threefold thicker layers at a pH near 5 are also observed (Fig. 4.17*a*) and suggest the formation of multilayers.

Although a seemingly opposite behavior of SFG amplitudes and adsorbate thickness is resolved, it can be concluded that the isoelectric points of bulk and interfacial BSA proteins are identical. The pH values at the isoelectric point of the interface lead to a highly disordered electrolyte subphase and BSA (multi)layers with a low degree of order: In Section 2.1 it has been already pointed out that the SFG amplitude $A_q = N\langle\alpha_q\mu_q\rangle$ is dependent on both the number density of interfacial molecules and their molecular order due to the orientational average of dynamic dipole moment and Raman polarizability. In the present case, the contribution of the number density to the SFG amplitudes is only minor, otherwise the amplitude should reach a maximum at a pH where the thickness of the BSA adlayer is also maximal. Since we observe the opposite (Figs. 4.17*b*,*c*), signals in SFG spectra must be dominated by the interfacial molecular order. At a pH near the bulk isoelectric point, the net charge of BSA proteins is negligible and a macroscopic electric field at the interface is absent. As the pH increases or decreases with respect to the isoelectric point of BSA, the charge density and, consequently, the electric field of the interface increase. As a result, polar ordering of interfacial water molecules and BSA is induced by the interfacial electric field. Further support for this hypothesis comes from SFG experiments at electrified oxide interfaces where electric field-induced polar ordering of interfacial water molecules was shown [12, 54, 55]. It is also interesting to compare the dispersion of the band at 3050 $cm^{-1}$ attributable to phenyl-type residues of the protein adsorbate in Figure 4.16*b*. At acidic pH values, the band appears as a positive-going feature versus the broad contributions from hydrogen-bonded water molecules centered at 3200 and 3450 $cm^{-1}$, while at alkaline pH of >7 the polarity of this band is apparently reversed. These changes can be attributed to a phase change of the water bands due to a reversal of the interfacial electric field. As a consequence of electric field-induced polar ordering of interfacial $H_2O$, the net orientation of interfacial water changes by 180° when the pH crosses the isoelectric point. We recall that the phase of a vibrational band is directly correlated with the molecular orientation and, therefore, necessarily also changes by $\pi$ when the molecular orientation is reversed. In vibrational SFG, observations of field-induced polar ordering and phase changes of the water bands are, consequently, directly related to the strength of the interfacial electric field and allow an estimate of the isoelectric point of the interface. At this point, it should be pointed out that electrokinetic zeta potential measurements and SFG spectroscopy actually probe different physical properties that, however, have their physical origin both in the charge distribution at the surface. The zeta potential refers to the electric potential at the hydrodynamic shear (or slipping) plane. The slipping plane separates ions (if present) and solvent molecules which travel with the migrating protein from those in the diffuse layer that do not travel with the charged protein in an external electric field. In contrast, SFG probes the average orientation of molecules within the interfacial electric field. Since the amide I band originates from molecular groups in the interior of the protein (see above), pH effects on polar ordering must be related to the net charge of the protein surface, which is also determined by the charge of possible bound counterions in the adjacent Stern layer. The physical origins of polar ordering as seen in SFG and in the zeta potential are, thus, identical.

For pH values near the isoelectric point, BSA adsorption is controlled by a gain in entropy and the formation of attractive noncovalent interactions such as van der Waals forces and hydrogen bonds of the BSA hydration shell. The situation is different for pH values <4.6 and >6.7 where electrostatic forces dominate and lead to a polar-ordered BSA monolayer. In order to establish multiple BSA layers at the interface, the lateral interactions between individual BSA proteins have to be attractive. In fact, the absence of a strong electric charge at the BSA surface, as it is suggested by the zeta potential of BSA in the bulk, leads to weak repulsive electrostatic interactions and consequently to an agglomerated BSA adlayer. At the isoelectric point the orientation of adsorbed BSA is, however, not completely random since the nonzero SFG amplitudes of amide I (Fig. 4.17*c*) and carboxylate vibrations indicate that to some extent a preferential orientation of BSA is maintained. This observation can be rationalized in terms of hydrophilic–hydrophobic interactions, where the hydrophobic parts of the BSA proteins tend to protrude into the gas phase. In contrast to the protein layer, the interfacial water molecules are highly disordered and lead to negligible SFG amplitudes (Figs. 4.16 and 4.17*b*).

Having established the interfacial molecular structure, we can now deduce (to some extent) macroscopic properties such as foam stability and foam capacity. Prerequisite for good foam formation is a fast diffusion of the proteins to the interface, where they can build a viscoelastic adsorbed layer around the gas bubbles that prevents destabilization of the foam lamella and coalescence of gas bubbles. As we have shown, at the interfacial isoelectric point of BSA a network of agglomerated proteins is formed and held together by weak attractive forces, while at a pH more alkaline or more acidic than the isoelectric point BSA forms monolayers with repulsive interactions. It is therefore likely that producing BSA foams at the isoelectric point will lead to foams with higher stability. This hypothesis was tested with macroscopic foams from BSA dilutions [52]. The results from foam stability measurements are shown in Figure 4.17*d*. The foam stability, which is defined by the loss in foam volume after a certain time period, shows a clear maximum around pH 5. Obviously, a network of agglomerated protein multilayers can encapsulate the air and prevent the foam from drainage much more efficiently than ordered protein monolayers with repulsive interactions.

## ACKNOWLEDGMENTS

B.B. gratefully acknowledges stimulating discussions with Professor Winfried Daum (Clausthal University of Technology) and his careful reading of this chapter.

The research on ethanol oxidation pathways was supported by the U.S. Army Research Office under award W911NF-08-10309. D.D.D. acknowledges partial support from the U.S. Air Force Office of Scientific Research under award FA9550-09-1-0163.

B.B. gratefully acknowledges support by Professor Martin Gruebele and financial support from a Feodor Lynen fellowship of the Alexander von Humboldt Foundation. Support by the German Science Foundation (DFG) for studies on protein adsorption and foam stability is gratefully acknowledged by B.B., K.E., and W.P.

## REFERENCES

1. Climent, V.; Feliu, J. M. Thirty years of platinum single crystal electrochemistry, *J. Solid State Electrochem.* **2011**, *15*, 1297–1315.
2. Antolini, E. Catalysts for direct ethanol fuel cells, *J. Power Sources* **2007**, *170*, 1–12.
3. Henderson, M. A. A surface science perspective on $TiO_2$ photocatalysis, *Surf. Sci. Rep.* **2011**, *66*, 185–297.
4. Ataka, K.; Kottke, T.; Heberle, J. Thinner, smaller, faster: IR techniques to probe the functionality of biological and biomimetic systems; *Angewandte Chem.* **2010**, *49*, 5416–5424.
5. Lyklema, J. *Fundamentals of Interface and Colloid Science*, Volume 2, *Solid-Liquid Interfaces*, Academic: San Diego, **1995**.
6. Zhang, Y.; Cremer, P. S. Interactions between macromolecules and ions: The Hofmeister series, *Curr. Opinion Chem. Biol.* **2006**, *10*, 658–663.
7. Lachenwitzer, A.; Li, N.; Lipkowski, J. Determination of the acid dissociation constant for bisulfate adsorbed at the Pt(111) electrode by subtractively normalized interfacial Fourier transform infrared spectroscopy, *J. Electroanal. Chem.* **2002**, *532*, 85–98.
8. Lu, G. Q.; Lagutchev, A.; Dlott, D. D.; Wieckowski, A. Quantitative vibrational sum-frequency generation spectroscopy of thin layer electrochemistry: CO on a Pt electrode, *Surf. Sci.* **2005**, *585*, 3–16.
9. Shen, Y. R. Basic theory of surface sum-frequency generation, *J. Phys. Chem. C* **2012**, *116*, 15505–15509.

10. Stanley, C. S.; Lai, S. E. F.; Kleyn, V. R.; Koper, M. T. M. Mechanism of the dissociation and electrooxidation of ethanol and acetaldehyde on platinum as studied by SERS, *J. Phys. Chem. C* **2008**, *112*, 19080–19087.
11. Chen, Y. X.; Miki, A.; Ye, S.; Sakai, H.; Osawa, M. Formate, an active intermediate for direct oxidation of methanol on Pt electrode, *J. Am. Chem. Soc.* **2003**, *125*, 3680–3681.
12. Richmond, G. L. Molecular bonding and interactions at aqueous surfaces as probed by vibrational sum frequency spectroscopy, *Chem. Rev.* **2002**, *102*, 2693–2724.
13. Magnussen, O. M. Ordered anion adlayers on metal electrode surfaces, *Chem. Rev.* **2002**, *102*, 679–725.
14. Wieckowski, A.; Neurock, M. Contrast and synergy between electrocatalysis and heterogeneous catalysis, *Adv. Phys. Chem.* **2011**, doi:10.1155/2011/907129.
15. Gao, W.; Keith, J. A.; Anton, J.; Jacob, T. Theoretical elucidation of the competitive electro-oxidation mechanisms of formic acid on Pt(111), *J. Am. Chem. Soc.* **2010**, *132*, 18377–18385.
16. Santana, J. A.; Cabrera, C. R.; Ishikawa, Y. A density-functional theory study of electrochemical adsorption of sulfuric acid anions on Pt(111), *Phys. Chem. Chem. Phys.* **2010**, *12*, 9526–9534.
17. Yeh, K.-Y.; Restaino, N. A.; Esopi, M. R.; Maranas, J. K.; Janik, M. J. The adsorption of bisulfate and sulfate anions over a Pt(111) electrode: A first principle study of adsorption configurations, vibrational frequencies and linear sweep voltammogram simulations, *Catalysis Today*, http://dx.doi.org/10.1016/j.cattod.2012.03.011.
18. Saravanan, C.; Koper, M. T. M.; Markovic, N. M.; Head-Gordon, M.; Ross, P. N. Modeling base voltammetry and CO electrooxidation at the Pt(111)-electrolyte interface: Monte Carlo simulations including anion adsorption, *Phys. Chem. Chem. Phys.* **2002**, *4*, 2660–2666.
19. Jungwirth, P.; Tobias, D. J. Specific ion effects at the air/water interface, *Chem. Rev.* **2006**, *106*, 1259–1281.
20. Roudgar, A.; Gross, A. Water bilayer on the Pd/Au(111) overlayer system: Coadsorption and electric field effects, *Chem. Phys. Lett.* **2005**, *409*, 157–162.
21. Guyot-Sionnest, P. The mature years of Sum-Frequency Generation are ahead, *Surf. Sci.* **2005**, *585*, 1–2.
22. Lagutchev, A.; Lu, G. Q.; Takeshita, T.; Dlott, D. D.; Wieckowski, A. Vibrational sum frequency generation studies of the $(2\times2)$-$(\sqrt{19}\times\sqrt{19})$ phase transition of CO on Pt(111) electrodes, *J. Chem. Phys.* **2006**, *125*, 1–10.
23. Braunschweig, B.; Mitin, A.; Daum, W. Pt(111) thin-layer electrodes on alpha-Al2O3(0001): Morphology and atomic structure, *Surf. Sci.* **2011**, *605*, 1082–1089.
24. Noguchi, H.; Okada, T.; Uosaki, K. Molecular structure at electrode/electrolyte solution interfaces related to electrocatalysis, *Faraday Disc.* **2008**, *140*, 125–137.
25. Shen, Y. R. Surface properties probed by second-harmonic and sum-frequency generation, *Nature* **1989**, *337*, 519–525.
26. Lagutchev, A.; Lozano, A.; Mukherjee, P.; Hambir, S. A.; Dlott, D. D. Compact broadband vibrational sum-frequency generation spectrometer with nonresonant suppression, *Spectrochim. Acta A* **2010**, *75*, 1289–1296.
27. Braunschweig, B.; Mukherjee, P.; Kutz, R. B.; Wieckowski, A.; Dlott, D. D. Sum-frequency generation of acetate adsorption on Au and Pt surfaces: Molecular structure effects, *J. Chem. Phys.* **2010**, *133*, 1–8.
28. Kutz, R. B.; Braunschweig, B.; Mukherjee, P.; Behrens, R. L.; Dlott, D. D.; Wieckowski, A. Reaction pathways of ethanol electrooxidation on polycrystalline platinum catalysts in acidic electrolytes, *J. Catalysis* **2011**, *278*, 181–188. Braunschweig, B.; Hibbitts, D.; Neurock, M.; Wieckowski, A. Electrocatalysis: A direct alcohol fuel cell and surface science perspective, *Catalysis Today*, http://dx.doi.org/10.1016/j.cattod.2012.08.013.
29. Zhu, X. D.; Daum, W.; Xiao, X.-D; Chin, R.; Shen, Y. R. Coverage dependence of surface optical second-harmonic generation from CO/Ni(110): Investigation with a nonlinear-interference technique, *Phys. Rev. B* **1991**, *43*, 11571–11580.
30. Tom, H. W. K.; Mate, C. M.; Zhu, X. D.; Crowell, J. E.; Heinz, T. F.; Somorjai, G. A.; Shen, Y. R. Surface studies by optical second-harmonic generation: The adsorption of $O_2$, CO, and sodium on the Rh(111) surface, *Phys. Rev. Lett.* **1984**, *52*, 348–351.

31. Pozniak, B.; Mo, Y.; Stefan, I. C.; Mantey, K.; Hartmann, M.; Scherson, D. A. In situ time-resolved second-harmonic generation from Pt(111) microfacetted single-crystal platinum microspheres, *J. Phys. Chem. B* **2001**, *105*, 7874–7877.
32. Daum, W. Optical studies of Si/$SiO_2$ interfaces by second-harmonic generation spectroscopy of silicon interband transitions, *Appl. Phys. A* **2007**, *87*, 451–460.
33. Lagutchev, A.; Hambir, S. A.; Dlott, D. D. Nonresonant background suppression in broadband vibrational sum-frequency generation spectroscopy, *J. Phys. Chem. C* **2007**, *111*, 13645–13647.
34. Ji, N.; Ostroverkhov, V.; Chen, C.-Y; Shen, Y.-R. Phase-sensitive sum-frequency vibrational spectroscopy and its application to studies of interfacial alkyl chains, *J. Am. Chem. Soc.* **2007**, *129*, 10056–10057.
35. Verreault, D.; Hua, W.; Allen, H. C. From conventional to phase-sensitive vibrational sum frequency generation spectroscopy: probing water organization at aqueous interfaces, *J. Phys. Chem. Lett.* **2012**, *3*, 3012–3028.
36. Garcia-Araez, N.; Climent, V.; Rodriguez, P.; Feliu, J. M. Elucidation of the chemical nature of adsorbed species for Pt(111) in $H_2SO_4$ solutions by thermodynamic analysis, *Langmuir* **2010**, *26*, 12408–12417.
37. Ayato, Y.; Kunimatsu, K.; Osawa, M.; Okadaa, T. Study of Pt electrode/nafion ionomer interface in $HClO_4$ by in situ surface-enhanced FTIR spectroscopy, *J. Electrochem. Soc.* **2006**, *153*, A203–A209.
38. Climent, V.; Feliu, J. M. Thirty years of platinum single crystal electrochemistry, *J. Solid State Electrochem.* **2011**, *15*, 1297–1315.
39. Funtikov, A. M.; Stimming, U.; Vogel, R. Anion adsorption from sulfuric acid solutions on Pt(111) single crystal electrodes, *J. Electroanal. Chem.* **1997**, *428*, 147–153.
40. Braunschweig, B.; Daum, W. Superstructures and order-disorder transition of sulfate adlayers on Pt(111) in sulfuric acid solution, *Langmuir* **2009**, *25*, 11112–11120.
41. Braunschweig, B.; Mukherjee, P.; Dlott, D. D.; Wieckowski, A. Real-time investigations of Pt(111) surface transformations in sulfuric acid solutions, *J. Am. Chem. Soc.* **2010**, *132*, 14036–14038.
42. Heinen, M.; Jusys, Z.; Behm, R. J. Ethanol, acetaldehyde and acetic acid adsorption/electrooxidation on a Pt thin film electrode under continuous electrolyte flow: An in situ ATR-FTIRS flow cell study, *J. Phys. Chem. C* **2010**, *114*, 9850–9864.
43. Kutz, R. B.; Braunschweig, B.; Mukherjee, P.; Dlott, D. D.; Wieckowski, A. Study of ethanol electrooxidation in alkaline electrolytes with isotope labels and sum-frequency generation, *J. Phys. Chem. Lett.* **2011**, *2*, 2236–2240.
44. Abd-El-Latif, A. A.; Mostafa, E.; Huxter, S.; Attard, G.; Baltruschat, H. Electrooxidation of ethanol at polycrystalline and platinum stepped single crystals: A study by differential electrochemical mass spectrometry, *Electrochim. Acta* **2010**, *55*, 7951–7960.
45. Gomes, J. F.; Busson, B.; Tadjeddine, A.; Tremiliosi-Filho, G. Ethanol electro-oxidation over Pt(*h k l*): Comparative study on the reaction intermediates probed by FTIR and SFG spectroscopies, *Electrochim. Acta* **2008**, *53*, 6899–6905.
46. Lai, S. C. S.; Koper, M. T. M. Electro-oxidation of ethanol and acetaldehyde on platinum single-crystal electrodes, *Faraday Discussions* **2008**, *140*, 399–416.
47. Iwasita, T.; Hoster, H.; John-Anacker, A.; Lin, W. F.; Vielstich, W. Methanol oxidation on PtRu electrodes. Influence of surface structure and Pt-Ru atom distribution, *Langmuir* **1999**, *16*, 522–529.
48. Chen, Y.-X.; Heinen, M.; Jusys, Z.; Behm, R. J. Bridge-bonded formate: Active intermediate or spectator species in formic acid oxidation on a Pt film electrode? *Langmuir* **2006**, *22*, 10399–10408.
49. Severson, M. W.; Stuhlmann, C.; Villegas, I.; Weaver, M. J. Dipole-dipole coupling effects upon infrared spectroscopy of compressed electrochemical adlayers: Application to the Pt(111)/CO system, *J. Chem. Phys.* **1995**, *103*, 9832–9843.
50. Souza-Garcia, J.; Herrero, E.; Feliu, J. M. Breaking the C–C bond in the ethanol oxidation reaction on platinum electrodes: Effect of steps and ruthenium adatoms, *ChemPhysChem* **2010**, *11*, 1391–1394.
51. Psofogiannakis, G.; St-Amant, A.; Ternan, M. Methane oxidation mechanism on Pt(111): A cluster model DFT study, *J. Phys. Chem. B* **2006**, *110*, 24593–24605.

52. Engelhardt, K.; Rumpel, A.; Walter, J.; Dombrowski, J.; Kulozik, U.; Braunschweig, B.; Peukert, W. Protein adsorption at the electrified air-water interface: Implications on foam stability, *Langmuir* **2012**, *28*, 7780–7787.
53. Wang, J.; Buck, S. M.; Chen, Z. Sum frequency generation vibrational spectroscopy studies on protein adsorption, *J. Phys. Chem. B* **2002**, *106*, 11666–11672.
54. Shen, Y. R.; Ostroverkhov, V. Sum-frequency vibrational spectroscopy on water interfaces: Polar orientation of water molecules at interfaces, *Chem. Rev.* **2006**, *106*, 1140–1154.
55. Braunschweig, B.; Eissner, S.; Daum, W. Molecular structure of a mineral/water interface: Effects of surface NanoRoughness of $\alpha$-$Al_2O_3$(0001), *J. Phys. Chem. C* **2008**, *112*, 1751–1754.
56. Wang, J.; Even, M. A.; Chen, X.; Schmaier, A. H.; Waite, J. H.; Chen, Z. Detection of amide I signals of interfacial proteins in situ using SFG, *J. Am. Chem. Soc.* **2003**, *125*, 9914–9915.
57. Murayama, K.; Tomida, M. Heat-induced secondary structure and conformation change of bovine serum albumin investigated by Fourier transform infrared spectroscopy, *Biochemistry* **2004**, *43*, 11526–11532.
58. Lin, V. J. C.; Koenig, J. L. Raman studies of bovine serum albumin, *Biopolymers* **1976**, *15*, 203–218.
59. Lu, J. R.; Su, T. J.; Penfold, J. Adsorption of serum albumins at the air/water interface, *Langmuir* **1999**, *15*, 6975–6983.

Part Two

# Raman Spectroscopy

# Chapter 5

# Surface-Enhanced Resonance Raman Scattering (SERRS) Studies of Electron-Transfer Redox-Active Protein Attached to Thiol-Modified Metal: Case of Cytochrome *c*

**Agata Królikowska**
Department of Chemistry, Warsaw University, Warsaw, Poland

Studies of redox proteins adsorbed on electrodes have gained increasing attention in recent years. In this area detailed knowledge of the mechanism and dynamics of the interfacial processes, particularly heterogeneous electron-transfer (ET) reaction is essential. Coating metals with a thin layer of organic film allows preventing denaturation of biomolecules and retaining their biological activity. Self-assembled monolayers (SAMs) of ω-substituted alkanethiols are ideal linkage monolayers for protein–enzyme binding since they are firmly bound to the surface via chemisorption. Interchain interactions provide a high degree of order and a terminal group allows controlling the surface properties. Cytochrome *c* (cyt *c*) has been demonstrated to be redox active in many systems involving gold and silver coated with alkanethiol SAMs. Identification of the mechanism of ET promoted by alkanethiol monolayers for cytochrome *c*, which is a simple redox protein, can allow recognition of the parameters governing interfacial charge-transfer (CT) processes in general and

---

*Vibrational Spectroscopy at Electrified Interfaces*, First Edition.
Edited by Andrzej Wieckowski, Carol Korzeniewski, and Björn Braunschweig.

provide the basis for understanding more complex systems. The SERRS technique, combining surface-enhanced Raman scattering (SERS) with the resonance Raman (RR) spectroscopy, gives the opportunity to probe both the structure of the linkage alkanethiol monolayer and the attached cyt *c*, as well as to recognize protein redox form, spin configuration, and iron coordination state. Time-resolved measurements provide an insight into the dynamics of the interfacial electron transfer.

In this chapter the potential of the electrochemical EC-SERRS spectroscopy technique for probing ET properties of the surface-confined redox biomolecules in the systems mimicking biological interfaces is presented. The first part is thus dedicated to the principles of SERS technique and methodology of electrochemical EC-SERRS. The second part is introducing the cyt *c* and the concept of the biomolecules tethering to the ω-substituted alkanethiols SAMs on silver/gold.

The third part describes the methodology of stationary potential and time-resolved EC-SERRS studies of cyt *c* attached to alkanethiol-coated metal electrodes and the information provided by a spectroelectrochemical experiment. It also summarizes the obtained results dealing with the redox behavior and ET kinetics for the mammalian (horse heart) heme protein, cytochrome *c*, attached to thiol linkage SAMs on metal substrates. Special attention is paid to the electric field effects at the metal–SAM–cyt *c* interface on the immobilized protein.

## 5.1 PRINCIPLES AND METHODOLOGY OF SERS AND EC-SERRS TECHNIQUES

### 5.1.1 Surface-Enhanced Raman Scattering

#### *Historical Background*

In 1974 the first report of the observation of unexpectedly intense Raman scattering in the vicinity of a nanostructured metal surface was published by Fleischman et al. [1], who studied adsorption of pyridine on roughened Ag substrate. The nature of this phenomenon was initially not fully understood and ascribed to the increased metal surface area (due to roughening procedure), resulting in a higher pyridine surface concentration in comparison to the smooth electrode. The enhancement factors were determined independently by Jeanmarie and Van Duyne [2] and Albrecht and Creighton [3] to reach up to $10^5$ for pyridine molecule and were related to surface properties of a roughened electrode. Mechanism responsible for this giant enhancement of Raman cross sections is known generally as surface-enhanced Raman scattering (SERS). During the next over 30 years the field of SERS has grown drastically and proven to be a powerful analytical tool for sensitive and selective studies of adsorbates on noble metal nanostructures [4–7].

#### *Mechanism of Surface Enhancement*

The intensity of Raman scattering is proportional to the square of the induced dipole moment (μ), which in turn is dependent on the molecular polarizability (α) and the amplitude of the electric field ($E$). Hence, an increase in the magnitude of $\alpha$ and/or $E$ will result in enhancement of the Raman scattering. Thus, the enhancement factor of the SERS spectrum can be understood as a product of two contributions: chemical and electromagnetic enhancement. This approach provides two classes of models explaining the origin of surface enhancement: electromagnetic (EM) model and chemical (charge transfer, CT) model, which will be now briefly discussed.

**Electromagnetic Enhancement** When an electromagnetic wave illuminates a metal surface, the magnitude of the field at the surface is changed with respect to this observed in the far field. If the surface is rough, the wave may excite localized surface plasmons (LSPs) on the surface. Surface plasmons (SPs) are collective oscillations of free electrons that propagate along the surface of a metal when it is in contact with a dielectric interface. Localized character means their confinement to metallic nanostructure resulting in enhancement of the electromagnetic field in the vicinity of the surface [8]. The size, shape, and material of the nanoscale roughness features must be taken into account to understand the electromagnetic enhancement [9]. These factors determine the resonant frequency of the conduction electrons in a metallic nanostructure. When the frequency of the radiation incident on metallic nanostructure meets this characteristic frequency, the electric field drives the conduction electrons into collective oscillation: excitation of LSPs outcomes at first in a selective absorption and scattering of the electromagnetic resonant radiation and next in the generation of a large electric field at the surface roughness [10]. It provides an enhancement of the Raman signal of a factor up to $10^4$ [with respect to normal Raman (NR) intensity] [11]. For noble metal substrates, when approximating the roughness feature with an isolated sphere model, an expression for the electric field magnitude ($E$) at the surface of a sphere can be given by [12]

$$E \propto E_i^2 \left| \frac{\varepsilon_m - \varepsilon_0}{\varepsilon_m - 2\varepsilon_0} \right|^2 \tag{5.1}$$

where $E_i$ is the incident field magnitude, $\varepsilon_m$ is the wavelength-dependent dielectric constant of the metal, and $\varepsilon_0$ is the dielectric constant of the sphere surroundings. When $\varepsilon_m = -2\varepsilon_0$, the magnitude of $E$ becomes very large. Moreover, the imaginary part of the metal dielectric constant (corresponding to light absorption) must be zero. This relation can be achieved for silver, gold, and copper at certain wavelengths in the visible and near-infrared (IR) region. This explains why the SERS effect for Au and Cu is observed only for red and near-IR lines. Basically, for these metals the localized surface plasmon resonance (LSPR) condition is fulfilled around 600 nm, while for silver it is met for the whole visible light range. EM model of SERS effect reveals the resonance nature of the phenomenon and a long-range character of the effect and is responsible for enhancement of the signal by a factor of $10^4$.

**Chemical Enhancement** The additional effect contributing to SERS is chemical enhancement, also called the first-layer effect. Its origin can be ascribed to specific interactions, that is, electronic coupling between molecule and metal, leading to formation of a molecule–metal complex, followed by an increase of the Raman cross section of the adsorbed molecule in this complex. One approach states it relies on resonance Raman (RR) effect, basing on coincidence of incident photon energy and electronic transition in the molecule, which results in significant increase of signal intensity. The RR effect becomes operative due to shifted and broadened electronic levels in the adsorbed molecule (compared to the free one) or due to a new electronic transition in the metal–molecule system.

There are two possibilities of a CT mechanism [13]:

- Electron transfer from the Fermi level ($E_F$) of metal to the nearly located charge-transfer level ($E_{CT}$), composed of the excited electronic state of the adsorbed molecule and the LUMO (lowest unoccupied molecular orbital) level of metal
- Electron transfer from the adsorbed molecule energetic level ($E_M$), which involves the transfer from the HOMO (highest occupied molecular orbital) level of metal–molecule complex to the metal Fermi level ($E_F$)

In photon-driven charge-transfer theory (PDCT) proposed by Otto [14] the four steps can be distinguished (for the molecule approximately neutral in a ground state):

1. Photon annihilation, excitation of the electron (surface plasmon creation)
2. Tunneling of the electron into the electron affinity level of adsorbate molecule
3. Tunneling of the electron back from electron affinity level (with changed normal coordinates of some internal molecular vibrations) to the metal
4. Recombination of electron–hole pair followed by Raman photon emission

Alternatively, an analogous round-trip of a hole involving the hole affinity level for anionic species (charged with approximately one unit in a ground state) can be imagined.

This dynamical charge transfer is useful in order to explain extremely large surface enhancement factors in the case of single-molecule SERS [15]. In the study by Otto [14], atomic-scale surface roughness is discussed as a requirement necessary for the chemical surface enhancement in terms of PDCT theory. In general, the chemical enhancement is considered to provide enhancement factors on the order of $10–10^2$ [16]. The charge-transfer mechanism explains the potential dependence of the SERS spectrum (shift of the maximum intensity as a potential function for different excitation lines) [17]. This is a consequence of the fact that the polarization of the electrode shifts the Fermi level of the metal, and thus the energy of the laser must be tuned to reestablish the resonance condition.

The contribution of both electromagnetic and chemical enhancement to the overall SERS intensity can be expressed approximately as

$$I_{\mathrm{SERS}} \propto G_{\mathrm{EM}} \sum_{\rho,\sigma} \left|(\alpha_{\rho\sigma})_{nm}\right|^2 \tag{5.2}$$

where $G_{\mathrm{EM}}$ is the electromagnetic enhancement factor for the electromagnetic fields of both incident and scattered light, while the sum term $(\alpha_{\rho\sigma})_{nm}$ is the molecular-specific, frequency-dependent polarizability. These two terms include all the mechanisms discussed here of the electromagnetic and chemical contributions to the enhancement.

The magnitude of the overall enhancement factor (EF) can be determined experimentally using the following expression [18]:

$$\mathrm{EF} = \frac{[I_{\mathrm{SERS}} / N_{\mathrm{surf}}]}{[I_{\mathrm{NRS}} / N_{\mathrm{vol}}]} \tag{5.3}$$

where $I_{\mathrm{SERS}}$ and $I_{\mathrm{NRS}}$ are, respectively, intensities of the surface-enhanced Raman and normal Raman Stokes bands measured at a single excitation wavelength, and $N_{\mathrm{surf}}$ and $N_{\mathrm{vol}}$ are, respectively, the number of molecules bound to the enhancing metallic substrate and in the excitation volume.

**Electromagnetic Surface Selection Rule** The electric field at the metal surface is *strongly anisotropic*, which means that the component normal to the surface is significantly larger than the tangential one, and, therefore, estimation of the average orientation of adsorbed molecules versus the metal surface from the SERS spectrum is possible [19]. Raman modes with the *polarizability component normal to the surface* will exhibit *the strongest enhancement*.

#### *SERS Active Substrates*

The most common metals used as substrates for SERS are coinage metals, that is, Ag, Au, and Cu, as they provide large surface enhancement [20]. Even for Ag, Au, and Cu supports there is another limitation: Surface morphology with a roughness scale of 50–200 nm is prerequisite for high surface enhancement [20]. These submicroscopic dimensions can be fabricated by various preparation procedures:

- Electrochemical roughening with use of oxidation–reduction cycles [21]
- Chemical etching [22]
- Metal film coatings [23]
- Chemical reduction of metal salts (colloids preparation) [24]
- Formation of ordered nanostructured surfaces by means of electron, ion, or light beam lithography and scanning probe microscopy (SPM) lithography [20]

### 5.1.2 Electrochemical SERS (EC-SERS) Technique

To have an insight into the redox and ET properties of the redox-active proteins with SERS spectroscopy, a spectroelectrochemical experiment must be carried out. Performing electrochemical SERS (abbreviated further as EC-SERS) measurements and hence studying an effect of varying potential effect at molecular level requires in situ SERS measurements under potential control. Varying potential applied to the electrode (SERS-active substrate) can affect both the chemical and physical enhancement of the SERS signal, as it influences the position of the Fermi level of metal and the dielectric constant of the electrolyte. Moreover, all these features make EC-SERS a greatly more complicated system than conventional SERS, that is, taken ex situ and/or at open-circuit potential. Therefore, a few aspects of the EC-SERS, that is, properties of the electrochemical double layer, polarity of the surface, effect of potential on SERS enhancement, effect of electrolyte and solvent, design of EC-SERS cell, and choice of a substrate will be briefly discussed within this section.

### *Electrical Double Layer of EC-SERS System and Effect of Polarity of Surface*

The key region when interfacial electrochemistry is discussed is the electrode–electrolyte interface [25]. This important region, where charge transfer occurs and gradients in chemical and electrical potentials arise, driving the electrochemical reactions, is called the electric double layer (EDL). Across the EDL the potential drop over the compact and diffuse layer results in a huge, highly localized electric field, which can reach even a value of $10^9$ V m$^{-1}$ [26]. It is generally accepted that mainly electromagnetic field enhancement contributes to the amplification of the Raman signal in SERS. Therefore, while doing an EC-SERS experiment one has to keep in mind that for this case two types of electric fields are present in the system:

- *Alternating EM field* Used for the sample illumination and excited at nanostructured substrate
- *Static electrochemical field (EC field)* Coexisting in the electrochemical system, mostly across the EDL

A schematic diagram showing the overall distribution of these two electromagnetic electric field contributions is presented in Figure 5.1 [27].

In the same figure, the effect of the surface polarity is also illustrated. In Figure 5.1*a*, when the potential of the electrode is positive with respect to the potential of zero charge (pzc), the surface attracts the negatively charged oxygen of the water

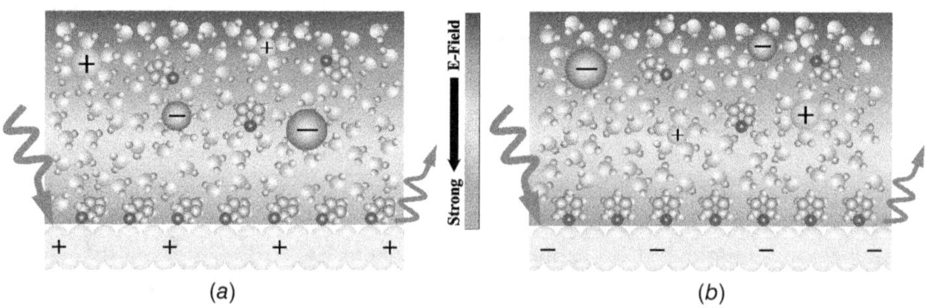


**Figure 5.1** Schematic diagrams of the electrochemical interfaces exhibiting SERS, showing the coexistence of electromagnetic field and the electric field at the electrode potentials: (*a*) positive or (*b*) negative to the pzc (potential of zero charge) [27].

molecules (assuming an aqueous electrolyte solution), however, if some other anionic species are present in the electrolyte, they may move toward the surface and repel the surface water. In Figure 5.1*a*, pyridine is shown as an example, which tends to interact with the positively charged electrode via both the π orbital and the lone-pair orbital on the nitrogen atom, which results in a tilted configuration of the molecule. When the potential is varied to the more negative value than pzc, binding of water via hydrogen atoms will be favored and other cations will approach the surface, as shown in Figure 5.1*b*. For the pyridine this will result in preferential adsorption exclusively via lone-pair orbtital of nitrogen atom. This shows that for pyridine, upon the change of the potential from positive to negative, the orientation of the molecules changes from tilted to vertical. Application of the strongly negative potential to the electrode may result in gradual weakening of its interaction with the surface, proceeding from chemisorption to physisorption and eventual desorption of the molecules.

This latter example shows nicely an effect of the potential-induced changes within the adsorbate, which will be reflected in the SERS spectrum. The changes in molecular orientation can be followed by monitoring the variation of SERS bands relative intensities employing surface selection rules, while the strength of the chemical interactions will be reflected in the shifts of the positions of the SERS bands. This shows that for the EC-SERS the chemical enhancement contribution becomes more significant than for the regular SERS experiment when the potential is constant.

#### *Effect of Electrode Potential on Surface Enhancement in EC-SERS System*

Moreover, the intensity of SERS bands, meaning the magnitude of the surface enhancement, is also dependent on the applied potential. This may indisputably relate to the potential-provoked changes of the electrode surface coverage. A linear dependence of the SERS intensity on the surface coverage was obtained for the 2-amino 5-nitro-pyridine (ANP) adsorbed on the roughened silver [28]. Nevertheless, silver is very often contaminated with oxide–hydroxide species, which questions the accuracy of determination of surface coverage of an adsorbate of a choice. Stolberg et al. [29] proposed that the lack of the relationship between SERS intensities and surface concentration for pyridine adsorbed on roughened gold occurs, since the data obtained from a rough surface (SERS intensities) is compared with the data obtained from a smooth surface (surface coverage). Their further studies of potential profiles of SERS spectra of pyrazine on smooth gold, employing a procedure of calculating the SERS intensity on a smooth surface from the data obtained for a rough surface, revealed that the changes of the SERS intensity of the ring-breathing mode simply track the coverage variations up to abourt two-thirds of a monolayer [30].

However, even when working in a potential window corresponding to undisturbed coverage of the adsorbate, still the SERS intensity is dependent on the applied potential. Moreover, the behavior of different modes is not the same, that is, their

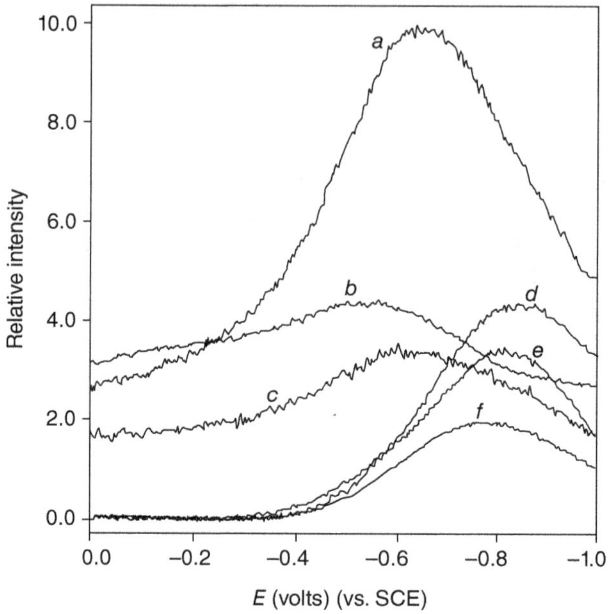


**Figure 5.2** Potential dependence of the SERS intensity ($\lambda_{exc}$ = 514.5 nm) for modes of pyridine adsorbed on silver: *a*, 1006 cm$^{-1}$; *b*, 1035 cm$^{-1}$; *c*, 3056 cm$^{-1}$; *d*, 1215 cm$^{-1}$; *e*, 1594 cm$^{-1}$; and *f*, 623 cm$^{-1}$. Potential was varied (at a rate of ~1 V s$^{-1}$) starting from 0 V (vs. SCE) and each curve is an averaged signal of 10 potential scans [2].

maximum intensity is reached for different potential values. This can be easily seen in Figure 5.2, presenting the discussed variations of the SERS intensity for different modes of pyridine adsorbed on silver. Bands marked with *a*, *b*, and *c* (see the figure caption for band positions) are detectable at the entire studied potential range and exhibit maximum intensity at less negative potential values than bands *d*, *e*, and *f*. The latter become active at strongly negative potential and reach maximum intensity at around −0.8 V (versus saturated calomel electrode; SCE) [2]. Assignment of the pyridine SERS-active vibrations to normal modes and comparison with normal Raman spectrum of liquid neat pyridine [31] suggest that chemical enhancement is involved in the change in the relative Raman intensities with the electrode potential [32]. The reason that the magnitude of the chemical enhancement changes with the potential for a given system (metal–adsorbate and at fixed excitation wavelength) is that energy of the Fermi level is affected by the applied potential. In Figure 5.3 the scheme illustrating the response of the photon-driven charge transfer to the varying potential at the metal–solution interphase of the EC-SERS system is given. On the left the upshift or the downshift of the Fermi level in the metal electrode with the varying potential is shown. The HOMO and LUMO orbitals of the adsorbed molecule are shown in the center. The $\Psi_g(V_i)$ on the right denotes the molecular adsorption ground state, whose energy is potential dependent, while $\Psi_e$(CT) stands for a photon-driven CT excited state formed from a filled level of the metal to the LUMO

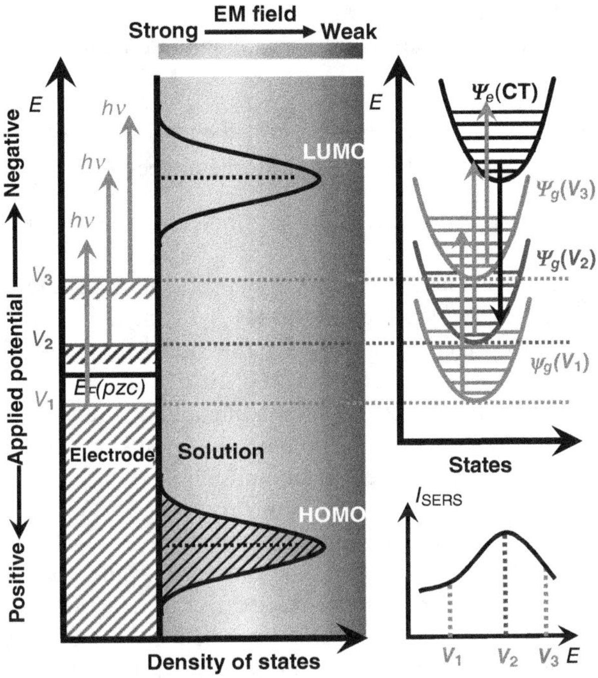


**Figure 5.3** Schematic diagrams of the photon-driven charge transfer (PDCT) from a metal electrode to adsorbed molecule in the EC-SERS system. (*Left*) The conceptual model of the energy levels changed with the electrode potential in the CT process. (*Right*) Relevant energy states involved in the electronic levels and the vibrational levels in the CT process. (*Right bottom*) The corresponding SERS intensity-potential profile. $V_i$ denotes the applied potential [27].

of the adsorbate. Assuming that the energy position of the CT state is independent on the potential, the appearance of the SERS intensity-potential profile seen in the right bottom of Figure 5.3 can be interpreted in terms of a changed resonance condition. Let one compare the difference between energy levels within the metal (on the left) and electronic states of the adsorbate (center) and relevant vibrational levels of combined system (upper right) involved in CT energy states at a given potential with the energy of the exciting photon. At the potential $V_1$ the quantum of excitation energy ($h\nu$) is to small too create the CT states on the interface; therefore, the changes of the SERS intensity are mainly due to the chemical bonding effect. When the potential is changed to $V_2$, the upshift of the Fermi level energy leads to the matching of the excitation energy of $h\nu$ and the charge-transfer energy, which results in a significant enhancement of the SERS signal of the relevant vibrational modes due to resonance-like scattering. Change of the potential to more negative value $V_3$ results in the further upshift of the Fermi lever, whose position no longer matches the resonance condition. Consequently, the decrease of the enhancement by CT states to the SERS intensity for this potential is observed (see the potential-dependent SERS intensity profile shown at right bottom of Fig. 5.3).

Depending on whether the CT takes place from metal to molecule (as in Fig. 5.3) or in the opposite direction, the resonance condition can be given by the following formulas:

$$E[\Psi_e(\text{CT})]-[E_{F(\text{pzc})}+eV_i]=h\nu \quad \text{for the metal} \rightarrow \text{molecule CT} \tag{5.4}$$

$$[E_{F(\text{pzc})}+eV_i]-E[\Psi_g(V_i)]=h\nu \quad \text{for the molecule} \rightarrow \text{metal CT} \tag{5.5}$$

which clearly show the resonance nature and potential dependence of the PDCT enhancement.

Detailed potential- and laser-wavelength-dependent SERS studies of the behavior of spectral intensities for pyridine on silver [17], taking also into account voltage-dependent coverage [33] and further density functional theory (DFT) calculations [32], demonstrated that for this system an EC-SERS potential dependence is quite complex. It is a combination of chemical bonding, charge transfer, and active sites (complexes of the molecule with adsorbed metal adatoms or a few atoms of metal clusters, often charged, responsible for the substantial enhancement of the SERS signal), which all contribute to the total chemical effect. At the early stage of SERS measurements, it was believed that the potential does not affect the EM mechanism of enhancement and that the changes in the intensity profile not exhibiting strong chemical enhancement reflect only the surface coverage changes [34]. What can be undoubtedly stated is that the applied potential will influence the SPR resonance frequency ($\omega_p$), which can be predicted with the use of the following expression:

$$\omega_p=\sqrt{\frac{4\pi ne^2}{m\varepsilon_0}} \tag{5.6}$$

where $e$ and $m$ are the charge and mass of the electron, $\varepsilon_0$ is the vacuum permittivity, and $n$ is the number density of free electrons. The latter quantity is the one responsible for the change of the SPR frequency with potential. When potential becomes negative (with respect to the pzc), it leads to an increased polarizability due to extra electrons in the conductance band, resulting in larger enhancement. Hence, an application of a negative potential to the electrode will increase and upshift (in terms of frequency) the plasmon resonance band, whereas a positive potential will damp and downshift it. Still, according to the surface selection rules [19], only the chemical effect can change the intensity of the vibrational modes belonging to the same irreducible representation, while it is not possible for a purely electromagnetic effect.

### Solvent and Electrolyte-Related Effects in EC-SERS Experiment

To make the solvent in contact with the adsorbate–metal interface electrically conductive, the presence of an electrolyte is necessary. Its type and concentration may also have a strong effect on EC-SERS conditions. First of all, the dimensions of the EDL (electrical double layer) are concentration dependent. High ionic strength may also screen the coulombic attraction of the proteins exhibiting asymmetric charge distribution, which may be the key factor for the attachment of the redox-active

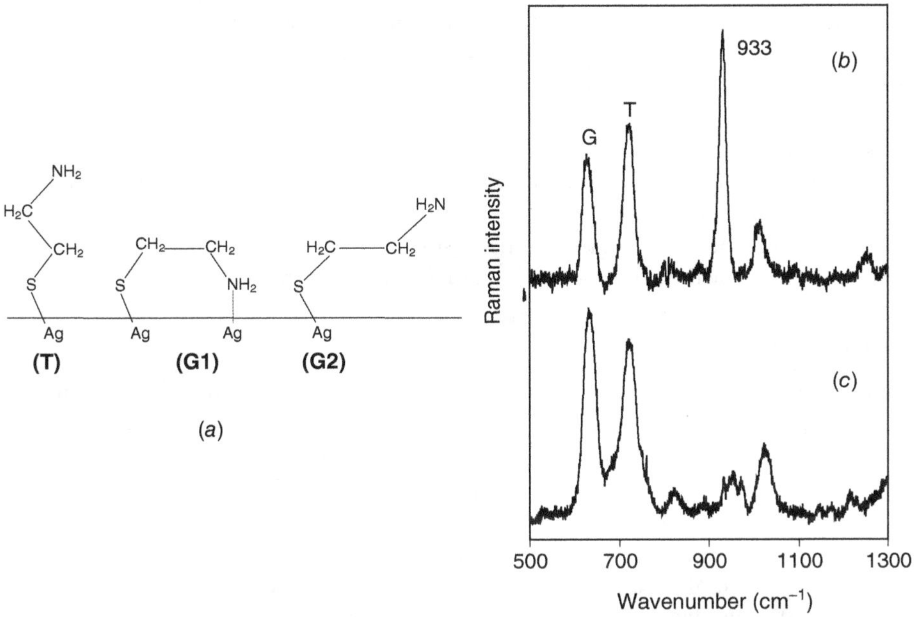


**Figure 5.4** (*a*) Trans (T) and two gauche conformers [(G1) and (G2)] of CYS molecule on Ag surface; SERS spectra of CYS on Ag soaked in (*b*) 1 M $LiClO_4$ and (*c*) water [36].

molecules that cannot interact directly with the metal substrate. Some ions composing the electrolyte may also bind to the metal surface, and, in a case of specific adsorption, it will result in a shift of the SPR band [35]. Coadsorption or competitive adsorption of electrolyte ions in addition to the adsorbate is also possible. SERS studies of cysteamine (CYS) on Ag showed that at OCP (open-circuit potential) the type of electrolyte influences considerably the conformation of adsorbed CYS molecules (see Fig. 5.4*a* for schematic representations of the conformers of CYS molecule on silver) [36]. Hence, halides additionally stabilized the *gauche* (*G*) conformer by the formation of hydrogen bonds with $NH_2$ groups in addition to the direct N· ·Ag bond, while the perchlorate stabilized *trans* (*T*) conformer and there was a spectral evidence of its coadsorption on the Ag surface. SERS spectra of CYS soaked in 1 M $LiClO_4$ and water are compared, respectively, in Figures 5.4*b*, *c*. An increased intensity ratio of the C–S stretching vibrational modes of the *T/G* conformers (marked in the spectra with the respective symbols) corresponding to their relative surface concentration and additional band due to $ClO_4^-$ stretching mode (around 930 $cm^{-1}$) can be seen after soaking in the perchlorate solution. This shows that $ClO_4^-$ anion competes with the amino groups for adsorption sites, not occupied by sulfur and imposes conformational change of the CYS molecules, already under OCP potentials. Varying the charging state of the surface will additionally complicate the conditions, and special emphasis should be thus put to distinguish between the intensity changes corresponding to conformational changes exclusively due to applied potential and due to ion coadsorption.

The type of solvent may considerably change the properties of the EC-SERS system. The most common approach is an exchange of water with some nonaqueous solvent: organic or ionic liquid one [37–39]. The advantages are the expansion of potential window and suppression of hydrogen evolution at negative potentials. However, again, the surface physico-chemical surface properties are changed and as the refractive index increases (compared to water), the red shift of the surface plasmon resonance frequency should be expected. The severe limitation of the method is that spectral features of the adsorbate cannot be easily studied as they may be overwhelmed by the intense scattering signal from the solvent molecules [40].

### Design of EC-SERS Cell

To obtain both a reliable, strong SERS signal and precise control of the electrochemical experiment, a design of the EC-SERS should be thoughtful. As mostly employed electrodes used for EC-SERS are opaque, a setup of the cell suited for a back-scattering Raman system will be presented.

For the aqueous electrolyte, the setup is quite simple and involves a judicious placement of the three-electrode system in order to allow the electrochemical reaction to run and simultaneously assure the minimal loss of the SERS signal. The cell body should be built of a chemically inert material, as it is exposed to electrolyte solution. Various types of Teflon are strongly recommended.

Usually, the working electrode (WE) is mounted at the bottom of the cell. This can be either an SERS-active disck metal electrode or metal nanoparticles attached to a flat, conductive metallic (preferably noble metal) substrate (e.g., gold, platinum etc.). Next, an inert counterelectrode (CE), which is typically a Pt wire ring, is placed to form a closed circuit and a reference electrode (RE) is used to control the potential of the working electrode. Usually, an Ag/AgCl or saturated calomel electrode (SCE) is used as RE. To control the placement of the reference electrode relative to the working electrode, a Luggin capillary can be used, which is a glass tube with a fine (capillary) end. It allows probing precisely the applied potential, as due to small dimensions of the capillary tip and its close distance (two times the capillary diameter) to the WE it allows to minimize the ohmic drop across the electrolyte. An optically transparent quartz or glass window should be used as a cover of the EC-SERS cell, the former especially when the ultraviolet (UV) laser is needed for the excitation.

The glass/quartz coating has a multifunctional role:

- Prevents the solution from evaporation under illumination with a laser beam
- Eliminates contamination of the solution and interference of the electrochemical system with the ambient atmosphere (e.g., oxygen introduction)
- Isolates solution in the cell from the Raman microscope objective to prevent etching

The overall scheme of the cell used by the Ren group can be seen in Figure 5.5*a* [41]. The design of the cell allows also purging with gas and/or serving as a flow cell. This is especially important when performing the experiment in nonaqueous

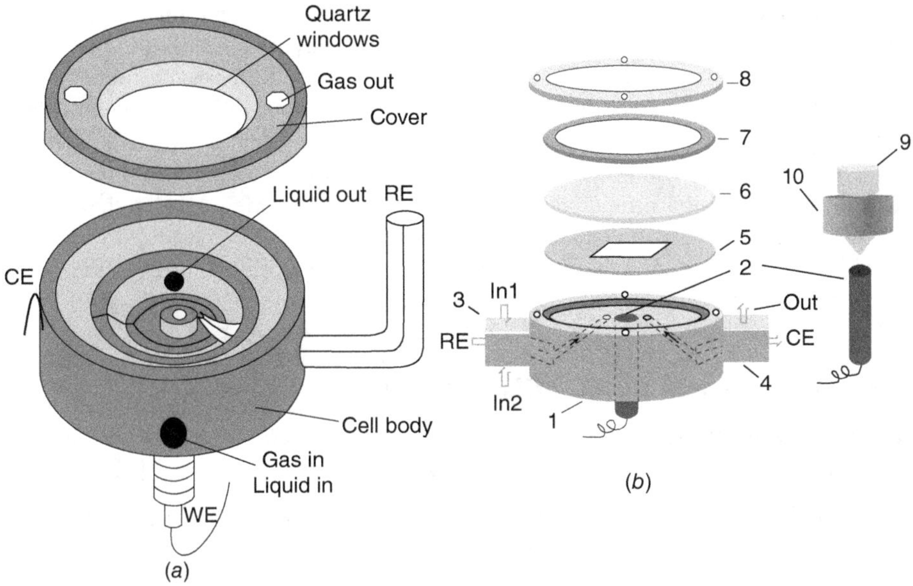


**Figure 5.5** (*a*) Sketch of the typical EC-SERS [41]; (*b*) Design of the flow EC-SERS cell [45]: (1) cell body, (2) WE, (3) four-way connector for inlet capillaries and RE, (4) three-way connector for outlet capillaries and CE, (5) Teflon membrane spacer, (6) quartz optical window, (7) silicone rubber O-ring, (8) stainless steel cover, (9) incident laser beam, (10) microscope objective.

system, for which the O-ring design should be employed, providing water and gas tightness. In this case, a water-free reference electrode is recommended to be used [42, 43]. An exemplary sketch of the flow cell, originally developed by Chen et al. [44] for EC-IRRAS (electrochemical infrared reflection absorption spectroscopy) and further adapted by Zhang et al. [45] for EC-SERS is given in Figure 5.5*b*. A thin Au foil (thickness of 50 μm) was used as a CE in this case instead of a metal wire. A thin Teflon film spacer of variable thickness was used to adjust the thickness of the electrolyte layer between the WE and the quartz window. Such design allows performance of the in situ EC-SERS studies under continuous flow conditions, with well-defined mass transport to/from the electrode. This in turn allows:

- Time-resolved spectroelectrochemical measurements upon sudden exchange of the electrolyte under potential control
- Spectroelectrochemical adsorption/stripping measurements of nonvolatile species
- Modification of the electrode by chemical/electrochemical adsorption/deposition in the flow cell prior to the measurements, aiming at a direct evaluation of different electrocatalysts

The latter cell design provides additional advantage of the efficient coupling of the EC-SERS cell and Raman instrument, due to working with only a thin layer of the electrolyte, adjustable by the employed spacer thickness. The presence of a quartz window and layer of the electrolyte on a optical pathway reduces the solid angle of the objective and hence the efficiency of the light collection is deteriorated. The effect of the electrolyte layer thickness on the intensity of the collected Raman signal is huge. The change of the solution layer size from 0.2 to 2 mm results in a decrease of the scattered light from 89% of the maximum intensity to only 35% [46]. Introduction of a corrosion-protective glass cover results in a 50% loss of the signal. Therefore, alternatively to insulating with a glass, Tian and Ren proposed wrapping the objective with a very thin and highly transparent polyvinyl chloride (PVC) film, which greatly reduces the Raman signal loss, which is now only 10% [46].

### *Substrate Selection for EC-SERS System*

**General Requirements for SERS Electrodes** There are a few criteria that must be met for the substrate to name it ideal for SERS spectroscopy [47, 48]:

1. It should exhibit *high SERS activity* and therefore provide *high sensitivity*. By tailoring the plasmonic properties of the substrate, guided by the size of the nanostructures (more than 50 nm) and interparticle spacing (less than 10 nm), one can tune the LSPR frequency of the substrate to match the incident laser frequency in order to maximize the enhancement.
2. It should provide *a laterally uniform surface enhancement*, so that the deviation in enhancement factor across the whole surface is less than 20%, which requires a relatively ordered arrangement of the nanoparticles on the substrate.
3. It should be characterized by *a good stability and reproducibility*. Long storage time should remain the enhancement effect. The deviation in the enhancement should not be higher than 20% when comparing the substrates from different batches, prepared by the same protocol.
4. *The cleanness of the substrate* should be high enough to allow also the studies of the weak adsorbates or even unknown samples.

Regrettably, it is still very hard to meet all of the features listed above at once. Therefore, usually a choice of the SERS substrate is a compromise, oriented toward a specific application purpose. Hence, for the needs of a quantitative SERS analysis, a uniform and reproducible surface enhancement is the priority, while in a trace analysis the maximum enhancement is the main concern. For the studies of the biomolecules, which is the case described here, purity and a high enhancement of the substrate are the key factors for a decent assignment of the bands, keeping in mind a complexity of the examined biosystems. For the EC-SERS system, independently on the nature of the studies, elimination of any surface impurities that may interfere with the detection of potential-dependent redox species is crucial. This can be achieved in different ways, that is, potential cycling, hydrogen evolution, use of a strong adsorbate, and its subsequent desorption or electrochemical oxidation–reductive desorption.

**EC-SERS Substrates** For the purpose of the EC-SERS measurements, different strategies can be employed to fabricate working electrodes, serving at the same time as a SERS-active substrate. Among them the most popular ones are:

**(a)** *Electrochemical oxidation and reduction cycle(s)* (ORC),

**(b)** *Assembly of metallic nanoparticles (NPs) on the electrically conductive substrate*, that is, GC (glassy carbon), ITO (tin-doped indium oxide), Au, Pt, Pd

**(c)** *Deposition of the core–shell NPs on an electric conductive substrate*, (i.e., GC, ITO, Au, Pt, Pd)

**(d)** *Templated SERS substrates, with ordered arrays of metallic nanoparticles*

The two latter methods of SERS-active substrate will be only briefly discussed, while more attention will be paid to the former two.

The most common metals used as substrates for SERS are coinage metals, that is, Ag, Au, and Cu as they provide large surface enhancement. For silver, the SPR condition is met for the whole visible region, while for gold and copper in the visible region the surface enhancement is observed only for the red lines. Between the latter two metals, gold is employed preferably, especially for the EC-SERS as use of silver introduces additional obstacles to the electrochemical part of the experiment. Superior oxidation resistance, high chemical resistance, and large potential window make gold a preferable WE material for electrochemical techniques. However, silver still remains the most frequent (EC)-SERS substrate, as it provides the strongest enhancement and allows to work using numerous excitation wavelengths in the visible range. Even for Ag and Au supports, there is another limitation: Surface morphology with a roughness scale of 50–200 nm is the prerequisite for high surface enhancement.

Therefore, some procedures to produce nanostructured electrodes for the purpose of EC-SERS-active electrode preparations are required. Among them, electrochemical roughening is the easiest one and can be applied to the bulk electrode. Typically, a metal rod is sealed into an inert, tubular Teflon sheath forming a disk electrode. Before each use, its surface should be mechanically polished with alumina powders down to 0.3 μm, rinsed with ultrapure water, and sonicated to removing alumina residuals. Electrochemical cleaning/hydrogen evolution can be used in addition to removing strong adsorbates and contaminations. This pretreatment should result in improved reproducibility of the electrode EC-SERS performance. Several ORC sequences result in anodic dissolution of the metal, whose atoms are next deposited during the cathodic cycle, forming the rough features on the electrode surface. The most frequently employed oxidation and reduction methods are the step and linear (in time) change of the potential (from the potential corresponding to metal oxidation to the one characteristic to ion reduction and the reverse). This latter method employs the cyclic voltammetry (CV) technique, by which the roughness of the substrate can be controlled by the applied potential range, number of cycles, and sweep rate. The procedures employed successfully in our lab for the silver and gold ORC roughening are given below [49, 50].

***ORC Roughening of Silver*** The polycrystalline silver electrodes are roughened by oxidation–reduction cycling in a 0.1-M KCl solution in a separate cell. Five cycles in which the potential is changed from +0.3 to –0.3 V (versus an Ag/AgCl 1-M $KCl_{aq}$ electrode, being RE), at a sweep rate of 5 mV $s^{-1}$ are applied. The process is performed in a three-electrode system, with the Ag as WE and Pt as CE. The procedure is always terminated in the negative vortex potential (a potential corresponding to reduction peak maximum should be held for 30 s) to provide the fullest possible reduction of the oxidized surface. The color of the electrode after the whole procedure should be milky-brown for the best SERS activity. Roughened silver substrates should be rinsed thoroughly with the ultrapure water just after the roughening procedure and stored for further use, but an immediate adsorption of the analyte is strongly recommended.

***ORC Roughening of Gold*** Prior to SERS measurements polycrystalline gold electrodes are polarized by means of CV technique in the potential range from –0.5 to 1.2 V, at a scan rate of 50 mV $s^{-1}$. The number of the cycles is around 20–30 cycles for each electrode and a good judgment for the termination of the roughening procedure is the brown (but not black) appearance of the gold. In a final cycle a potential close to –0.2 V should be held 30 s to ensure a reduced state of the electrode. All potentials are reported with respect to the Ag/AgCl 1-M $KCl_{aq}$ electrode, which is employed as a reference electrode. A Pt grid is used as counterelectrode. Like for the silver, after careful rinsing with deionized water, the gold SERS-active surface should be exposed to the adsorbate or it can be stored for some time in ultrapure water. Typical scanning electron microscopy (SEM) images of the morphology of silver and gold obtained as described above with ORC roughening procedures are presented in Figure 5.6*a*. The size of the metallic nanostructures ranges from 100 to 300 nm. However, it can be seen that there is some size dispersion across the electrode, and it is expected to change from one electrode to another. This will result, respectively, in somewhat nonuniform distribution of the surface enhancement factor and worsened reproducibility. Elemental analysis deduced from energy dispersive spectroscopy (EDS) spectra taken for such ORC-roughened electrodes (Fig. 5.6*b*) shows that, although the electrodes are rinsed with water, there is still some chlorine contribution at the surface. Its exact quantification is impossible due to strong overlapping of the Cl peaks with the ones indicative of the presence of Ag and Au; however, the existence of the chlorine is slightly more pronounced for the silver substrate.

Chlorine is believed to exist in a form of chloride ions, which stabilize positively charged silver clusters, formed during the roughening procedure [52]. It was proposed that SERS-active sites at rough silver surfaces are introduced by the formation of silver $Ag_4^+$ clusters [53]. From two known stable forms of $Ag_4^+$ clusters—linear and pyramidal—this second was proposed as more relevant to observe SERS modes [53]. Moreover, it was evidenced for silver colloids that $Cl^-$ ions stabilize $Ag_n^+$ complexes on the silver surface [54]. In this roughening procedure, 0.1 M KCl is employed as a supporting electrolyte. In spite of extensive rinsing of roughened substrate with water, $Cl^-$ ions preadsorbed during ORC pretreatment remain on the

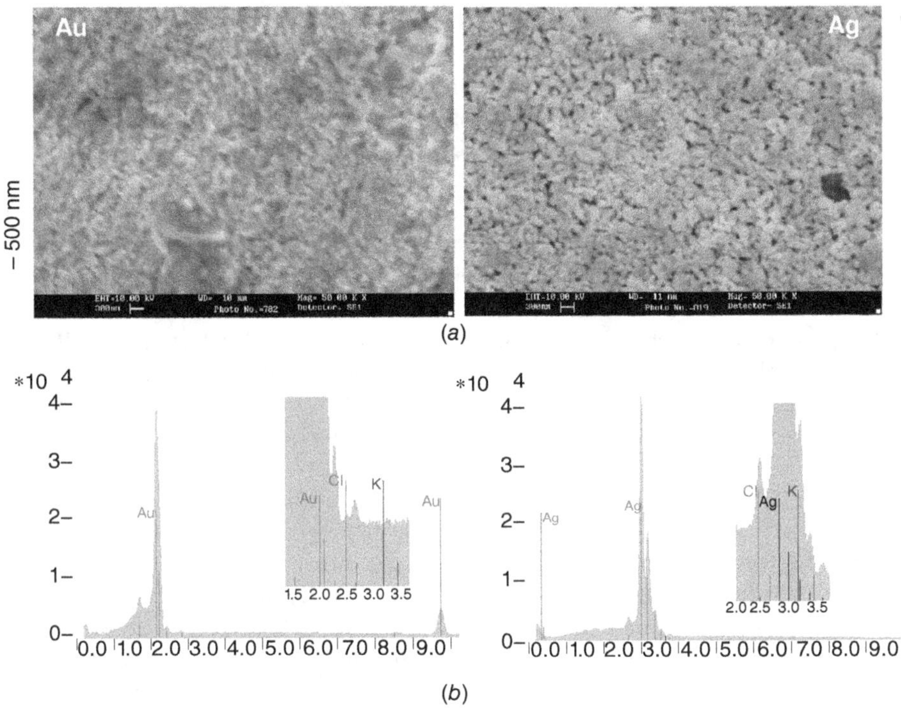


**Figure 5.6** (*a*) SEM images of ORC-roughened gold and silver and (*b*) EDS spectra of ORC-roughened gold and silver [51].

surface even after chemisorption of a strong adsorbate, as evidenced by SERS spectra of CYS on Ag [55].

***Chemical Synthesis of Ag and Au NPs*** As shown above, the ORC-roughened electrodes may suffer from irreproducibility and nonuniform surface enhancement spatial distribution; however, they still exhibit high SERS activity. Anyhow, chemical reduction of metal salts (colloids preparation) is an alternative method for the preparation of SERS-active substrate. Citrate reduction of silver nitrate ($AgNO_3$) [56] and chloroauric acid ($HAuCl_4$) [57] and $AgNO_3$ reduced with hydroxylamine ($NH_2OH$) [58] or sodium borohydride ($NaBH_4$) [24] can serve as the SERS-active colloid examples. Depending on the method and experimental conditions, nanoparticles of different size can be fabricated. The disadvantage of the colloids is their instability, resulting in aggregation of the nanoparticles upon time, leading to the loss of their SERS activity due to too large of a size.

If one needs to tune the plasmon properties of the silver/gold nanoparticles (e.g., in order to use the given laser wavelength, resonant with the adsorbate electronic transitions), there are numerous methods in the literature describing synthesis of such nanoparticles. They rely mostly on size- and shape-controlled synthesis of the metal nanoparticles, which geometry deviates from the spherical, typical for simple chemical metal salt reduction, proceeding under a thermodynamically controlled regime, favoring isotropic growth. Among them, the routes to produce anisotropic nanoparticles, that is, nanoscale rods, disks, triangular prisms, multipods, cubes, and hollow metal spheres can be found [59–62]. The size and growth of the solid anisotropic nanoparticles can be regulated by different routes: capping agent guiding, seed-mediated growth, light irradiated, and microwave- and ultrasound-assisted synthesis. In the case of hollow metal spheres, mostly the template-based methods are employed. As an example a galvanic replacement of cobalt template spheres with the gold shell can be given [63]. The outer and inner diameter, and hence the size of the particle and thickness of the shell of the hollow gold nanospheres (HGNs) can be controlled via experimental conditions: amount of the used gold ions, citrate (capping agent) and $NaBH_4$ (reducing excessive $Co^{2+}$), and rate and order of the reagents addition. Playing with the geometrical properties of such fabricated HGNs, their surface plasmon band absorption can be tuned between 550 and 820 nm [64]. The UV–VIS spectra of suspensions of three different batches (with the observed colors of the solutions containing NPs given in the legend) of Au HGNs prepared according to this procedure in our laboratory are shown in Figure 5.7*a*. In Figure 5.7*b* the TEM image of the "dark blue" Au HGNs NPs is shown, which proves that large FWHM (full width at half-maximum) visible in the VIS spectrum corresponding to this synthesis reflects somewhat broad distribution of the NPs size. The next step prior to the EC-SERS experiment is the deposition of the NPs from suspension onto the conductive support. This can be ITO, GC, or nonroughened metal substrate (hence, SERS inactive), like gold. A solid support requires thorough cleaning or polishing before NPs assembly. The deposition is spontaneous: placing a droplet of NP suspensions and leaving them to dry in air or assisted by a vacuum pump [27].

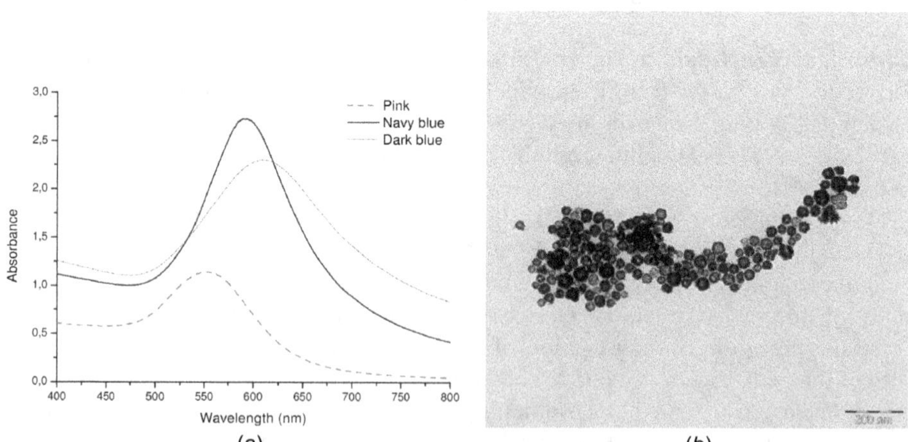


**Figure 5.7** (*a*) VIS spectra of gold HGNs obtained in our lab by cobalt galvanic replacement (according to the protocol given in [64]) and (*b*) representative TEM image of "dark blue" NPs.

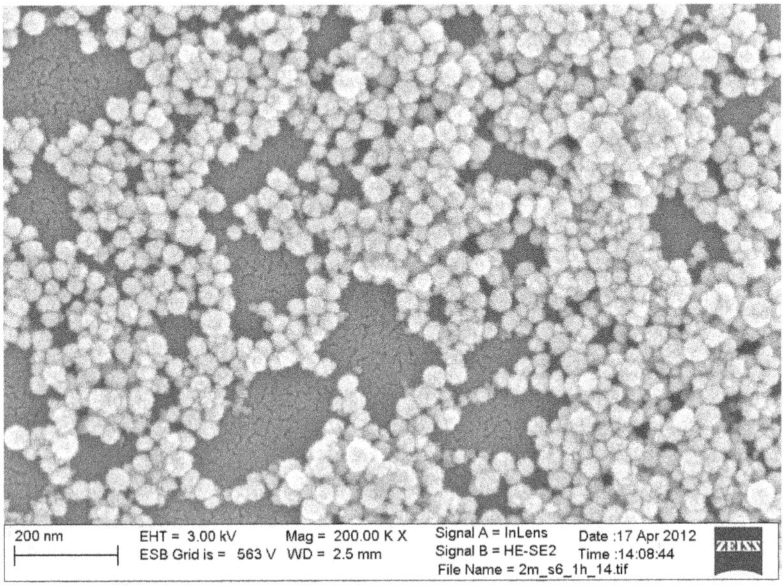


**Figure 5.8** SEM image of the "pink" (see the legend in Figure 5.7*a*) Au HGNs after their immobilization on a solid substrate, with the use of APTMS (according to the protocol by Hajdukova et al. [65]), 24 h of dipping in NPs solution.

In order to improve the deposition efficiency the suspension containing NPs can be centrifuged before dropping. The other approach is coating the substrate with the functional groups capable of bonding the colloidal metal. In Figure 5.8 the SEM image fabricated in our lab citrate capped gold HGNs, utilizing their electrostatic attraction by $NH_3^+$ groups of the substrate, provided by 3-aminopropyl- trimethoxysilane (APTMS) is presented. It can be seen that HGNs demonstrate a much narrower size distribution than those prepared with ORC roughening (compare Figs. 5.6*a* and 5.8).

***Core–Shell NPs Formation for Catalytic Metals SERS Activity*** The other metals that are attractive for EC-SERS studies are those that exhibit catalytical activity such as Pt or Pd. Unfortunately, typically NPs of the Pt group or the Fe group transition metals exhibit relatively weak surface enhancement, especially when compared to their Ag and Au counterparts. Their electrochemical roughening is not trivial, as these metals tend to form a compact oxide layer, which prevents further surface oxidation. Thus, formation of the particles of the size capable of SERS enhancement is nearly impossible.

An excellent solution to overcome this problem was proposed by Zhang and co-workers, who introduced the concept of borrowing SERS intensity by coating Au nanoparticles with a thin layer of transition metal [45]. For such composed core–shell NPs, the SERS activity arises from the gold core, while the chemical properties are determined by the transition metal shell. The enhancement factors are up to $10^4$–$10^5$ [66, 67]. The deposition of NPs on a conductive substrate [e.g., glassy carbon (GC)] is performed by simple drop coating of the colloidal suspension, followed by drying. NPs prepared according to this protocol were successfully employed for time-resolved, potential-controlled EC-SERS studies of CO chemisorption at the Au core–Pt shell NPs, with the use of a spectroelectrochemical flow cell [45].

***Template-Assisted Fabrication of Arrays of SERS-Active NPs*** Metallic NP assemblies on conductive substrates described above provide sufficient uniformity of the size and shape of the nanostructures. However, when the controlled NPs interspacing is also a priority, the template-assisted methods are required to supply the ordered arrays of NPs. Among the numerous template methods of SERS-active substrates fabrication of two of the most frequently employed are *anodic aluminum oxide (AAO) films* and *nanosphere lithography (NSL)*. In the AOO method, the SERS-active metal is electrodeposited into the matrix of a porous AAO membrane. In a final step, the well-ordered metal NP array is exposed upon chemical etching of the AAO matrix [20]. Resulting nanorods exhibit high SERS activity and high ordering, but from the point of view of an EC-SERS experiment, their severe limitation is the difficulty of establishing the electric contact to control the potential. The NSL method involves deposition on the conductive substrate of the monodispersed suspension of the polystyrene or $SiO_2$ nanospheres, which self-assemble, producing a highly ordered surface pattern [68]. Next, a thin layer of the SERS-active metal is deposited on the polymer mask. In the case of the physical vapor deposition (PVD) of the metal, already this metal film over nanosphere (MFON) can serve as an SERS substrate [69]. Spheres can also be removed to form the periodic particle array, which in case of PVD form triangular footprints [70]. Varying the angle between the normal to the substrate and deposited material beam (sputtered metal) [71], metal type, and/or thickness and size of the spheres [72], precise control over the shape, size, and interparticle spacing is achievable. This in turn allows tuning the plasmonic properties of the substrate. The alternative approach to PVD is electrodeposition of the metal film on the nanosphere mask [73]. Lifting off the mask leaves an SERS-active, thin metallic film, consisting of a regular hexagonal array of uniform segment sphere voids.

## 5.2 ATTACHMENT OF CYTOCHROME *c* TO ALKANETHIOL-COATED ELECTRODE PRIOR TO ET STUDIES

### 5.2.1 Strategies for Establishment of Communication between Electrode and Redox Biomolecule

Electrode surface can be considered as a source or sink of electrons (charge) employed to drive electrochemical reactions. Electron transfer (ET) is a key step of almost every type of electrochemical reactions. Redox enzymes and proteins are capable of electron exchange with suitably prepared electrode.

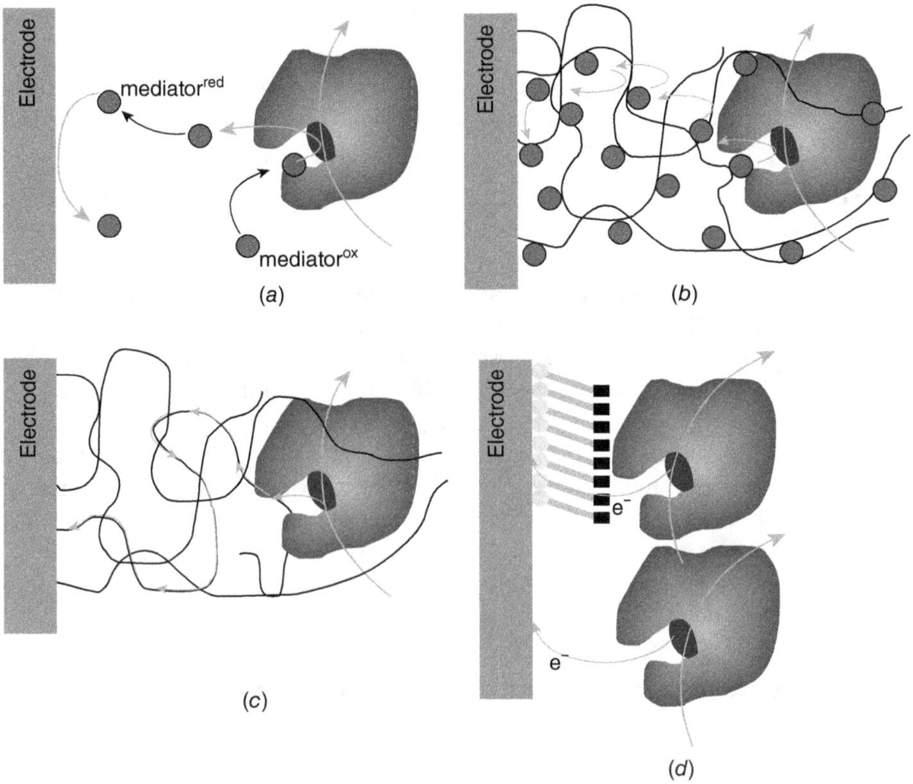


**Figure 5.9** Schematic representation of a possible ET mechanism between biomolecule and electrode [74]: (*a*) shuttle mechanism based on a free-diffusing redox mediator, (*b*) electron hopping within redox polymer matrix (wired biomolecule), (*c*) conducting polymer chain, and (*d*) direct ET—electron tunneling between enzyme–protein and uncoated (*bottom*) or SAM-covered electrode (*top*).

A few strategies can be employed for coupling the redox center with the electrode, which can be categorized according to the ET mechanism (see Fig. 5.9 to view a variety of ET mechanisms) [74]:

- *Shuttle Mechanism.* Based on a free-diffusing redox mediator: Natural enzyme co-substrates were used in so-called first generation of amperometric biosensors, whereas in the second-generation artificial redox mediators, mainly soluble low-molecular-weight metal complexes with reversible ET properties replaced them.
- *Via Sequence of Electron-Hopping Reactions in Redox Polymer.* Electrical wiring of redox-relay modified polymeric hydrogels employed; mediator is tightly retained at the surface, and this improved design also belongs to the second generation of amperometric biosensors.
- *Conducting Polymer as Molecular Cable.* Electron is transferred via conducting polymer chain; in this case again the mediator is fixed at the surface, and

this approach is qualified again as a subtype of second-generation amperometric biosensor.

- *Direct Electron Transfer*. Occurs via a tunneling process between the redox center of the biomolecule and a bare or monolayer-modified electrode; mechanism typical for third generation of amperometric biosensors.

The most attractive feature of direct ET (DET) is avoiding intermediate ET steps *via* self-exchange reactions. The limitation is that DET is only possible with these enzyme/molecules that are immobilized in the first monolayer [75], which is a disadvantage when application as a sensor is considered, since it lowers the sensitivity and the stability of the sensor. In practice, DET is only possible for relatively small enzymes with an easily accessible prosthetic group such as a group of peroxidases, such as cytochrome *c* peroxidase, horseradish peroxidase, or fungal peroxidase [75, 76]. Cytochrome *c* is also an excellent object of ET studies since its heme group is not buried deep in the amino acid matrix. Moreover, cyt *c* structure and redox properties were intensively studied and are very well characterized. Biomolecules directly adsorbed on substrate such as carbon or noble metals undergo at least partial denaturation, leading to electrode fouling and to deterioration of conditions for DET. Chemical modification of the electrode surface prevents denaturation, limiting significantly direct interactions of the protein with the substrate.

## 5.2.2 Structure and Function of Cytochrome *c*

Cytochrome *c* (cyt *c*) belongs to the class of proteins known as cytochromes, which can be defined as electron- or proton-transfer proteins, possessing one or several heme groups. Cytochromes can be subdivided according to their heme iron coordination and heme type. The stucture of cyt *c*, its biological function, and the strategies of attachment to metal surface coated with substituted akanethiols will be characterized below.

### Structure of Cyt c

Cytochromes *c* are electron-transfer proteins, whose heme group or groups are covalently bound to the protein by one or more commonly two-thioether bonds, involving thiol groups of cysteine (Cys) residues. Heme *c* is a complex of iron–protoporhyrin IX (structure presented in Fig. 5.10*a*), where the suffix *proto* points to the presence of two vinyl substituents on the porphyrin ring (see Fig. 5.10*b* for a complete structure of heme). These vinyl side chains are the second constituent of the thioeher link, which is formed in reaction with the thiol groups of two cysteine residues. The single heme iron atom is chelated by four nitrogen atoms of the four pyrrole rings, provided by the porhyrin group. Two additional ligands can bind along the *z* axis. The fifth ligand for *c* type of cytochromes is always a nitrogen provided by histidine (His) residue. Cyt *c* belongs to the class I hemes, according to heme coordination classes of cytochrome *c* proposed by Ambler [77]. Class I includes the low-spin, soluble cytochrome *c* of mitochondria and bacteria, with the heme-

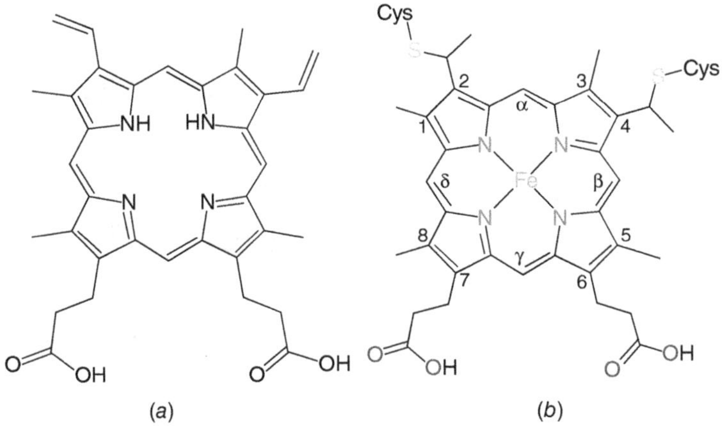


**Figure 5.10** Structure of protoporphyrin IX (*a*) and heme *c* (*b*).

attachment site toward the N terminus and the sixth ligand provided by the methionine [78] residue toward the C terminus. The iron atom is in the ferrous ($Fe^{2+}$) state in the reduced form and in the ferric ($Fe^{3+}$) state in the oxidized form of the cytochrome, which are known as ferro- and ferricytochrome *c*, respectively. The terms low spin and high spin refer to the spin state of the heme iron. Complexes formed according to Hund's rule are known as high spin (HS), whereas these with lower energy orbitals completely filled before population of the upper sets starts are called low spin (LS). For cyt *c* the strength of the ligands is close to the crossover point, thus an equilibrium between low and high spin is attainable.

The subject of this section is a horse heart ferricytochrome *c*, which is a mitochondrial cytochrome *c*. The structure of this protein was extensively studied and it has been thoroughly characterized. It is composed of 104 amino acids and covalently attached heme, water soluble, and its molecular weight is around 12.5 kDa. Its shape is nearly spherical, with the dimesions of 30 × 34 × 30 Å. The solution structure of horse heart ferrocytochrome *c*, deduced from nuclear magnetic resonance (NMR) is presented in Figure 5.11*a* [79]. The general structure of this biomolecule can be described as a one-amino-acid-thick polypeptide layer surrounding a heme prosthetic group. Hydrophobic side chains of these amino acids create the interior of this layer. The next layer consists of main the polypeptide chain, with the polar groups of amino acid residues exposed to the exterior. The secondary structure of reduced horse heart cyt *c* in solution is presented in Figure 5.11*b*. The protein is mostly α-helical (four major helices can be seen); however, according to the IR spectrum the bands characteristic for extended β strand and β turns were found and 13% content of the structure was assigned as unordered [81]. The amino acid sequence of the ferrocytochrome *c* with depicted secondary structure is presented in Figure 5.11*c* [80]. The heme group of cyt *c* is located slightly off center of the biomolecule. It is anchored to the protein by two thiother bonds to the cysteine side chains (Cys-14 and Cys-17). This is quite an unconventional way of the attachment of the heme

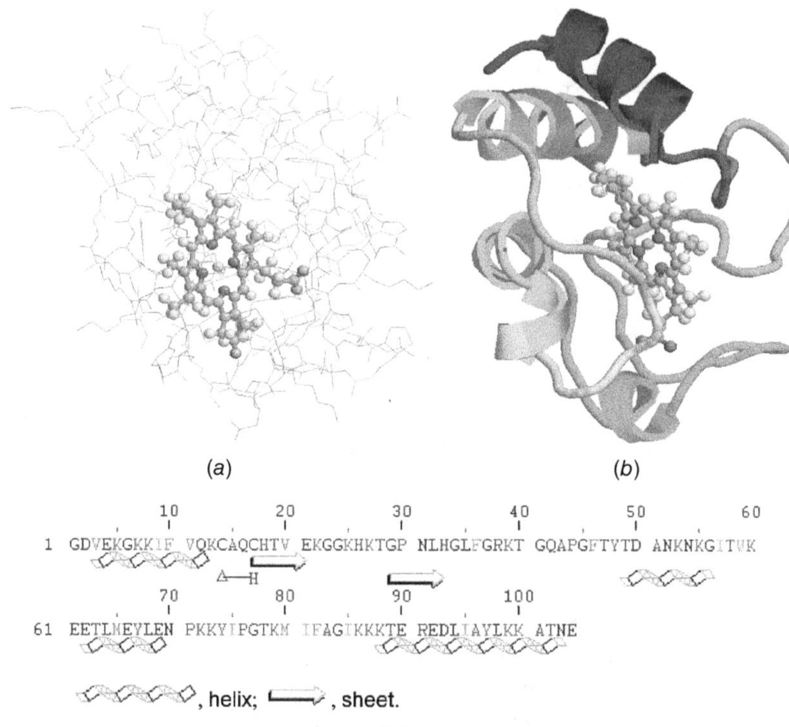


**Figure 5.11** Solution structure of horse heart ferrocytochrome obtained from NMR (1GIW pdb file) [79]: (*a*) amino acid backbone and heme prosthetic group, (*b*) secondary structure of the protein, and (*c*) amino acid sequence of cyt *c* acquired from reversed phase chromatography [80].

prosthetic group, which is typically bound simply by iron ligation, supported by tertiary interactions. The role of *heme covalent attachment* is not completely understood; however, a few possible explanations were presented. Covalent linking can be responsible for the restriction of the heme group orientation, which in the case of exclusive ligation found in other heme proteins may exhibit orientational disorder [82, 83]. XRD (X-ray diffraction) studies showed that two arms of covalent connection can induce nonplanar distortion of heme, maintained by the protein tertiary structure [84]. It was proposed that heme distortion modulates the electron-transfer function through modification of redox potentials of the porphyrin ring and the protein binding properties with redox partner, with the latter also controlled by thioether linkages. Thus, the information about heme oxidation is transferred by these two bonds.

Heme is not only covalently attached but its iron is also ligated. Heme iron is *coordinated by four nitrogen atoms of the porphyrin ring, and the two axial ligands are sulfur from methionine (Met- 80)* and *nitrogen from histidine (His-18). The native configuration* of both reduced and oxidized cyt *c* in aqueous, neutral solution heme iron is the *six coordinated low spin* (referred further in the text as *6cLS*) [85]. The conformation of cyt *c* may change upon binding to the chemically modified metal electrode, which will be further described. The heme group is primarily buried in a protein pocket. However, about 4% of the heme atoms is in contact with the solvent [86]. This is interpreted in terms of exposure of one edge of the heme, making it accessible to the solvent molecules [87].

*Asymmetric charge distribution* across the cyt *c* surface is an intriguing feature. Uncompensated positive charge is provided by a patch of six Lys groups near the exposed heme edge. It plays an important role in cyt *c* binding to biological partners, introduced in the section dealing with the biological function of the protein.

#### Cytochrome c Biological Function

Cytochrome *c* can be found loosely associated with the inner membrane of the mitochondrion. Its role is the transport of electrons in the mitochondrial respiratory chain. The respiratory chain employs the oxidative phosphorylation process, which in general can be described as a metabolic pathway that uses energy released by the oxidation of nutrients to produce adenosine triphosphate (ATP). It includes multistep thermodynamically driven electron transport chains, which employ sets of linked multisubunit protein complexes within mitochondria (in eukaryotes), containing a variety of metal redox cofactors and mobile electron carriers, whose function is shuttling of electrons among former species [88, 89]. The role of electron carriers is performed by small organic molecules (e.g., ubiquiononone/ubihydroquinone) or small proteins (molecular weight of 10–15 kDa) such as cyt *c* or iron–sulfur proteins [88, 89]. Electrons are transferred from electron donors to electron acceptors such as oxygen in redox reactions and released energy is consumed to form ATP. Electron transfers during oxidative phosphorylation are coupled with proton translocation across the inner mitochondrial membrane. Proton movement generates energy in the form of a proton gradient and an electrical potential across the membrane. The flow back of the protons across the membrane lowers this gradient and thus supplies the energy driving the synthetis of ATP by a large enzyme called ATP synthase. This energy is used to convert adenosine diphosphate (ADP) into ATP through a phosphorylation reaction (addidion of a phosphate group $PO_4^{3-}$ to the ADP molecule). The role of cyt *c* is a transfer of the electrons between the so-called complex III and IV. Complex III consists of cytochrome *c* reductase, known also as Q-cytochrome *c* or the cytochrome $bc_1$ complex. Cytochrome *c* oxidase (complex IV) accepts the four electrons transferred by the heme group of cytochrome *c* from complex III.

### 5.2.3 Alkanethiol SAMs as Linkage Monolayers for Biomolecule Immobilization

The subject of this chapter is the characterization of ET properties of cytochrome *c*, immobilized on gold and silver substrates. The most straightforward procedure for binding a biological molecule is physical adsorption at the metal surface, which involves immobilization directly on unmodified metal substrate by hydrophobic or electrostatic interactions. The simplicity of this strategy offers a lack of manipulation with the protein sample, but on the other hand the mechanism of the adsorption is complex [90]. The complexity of the processes coupled with the physical adsorption of the proteins at the metal surface introduces several drawbacks:

- Instability of the protein–metal binding, which results in exchange events or even reversibility of the adsorption process [91]
- Nonspecific adsorption, which introduces large distribution of available orientations or totally random orientation of immobilized protein molecules, which usually blocks their active site or ligand binding site [68]
- Denaturation and loss of biological activity due to high surface energy of a solid–gas interphase [92]

Therefore, a common strategy to attach biomolecules to a metal surface is substrate modification with molecules capable of binding by interactions or reactions with specific moieties present on the biomolecule surface. Selected molecular coating must meet the following criteria to be suitable as linkage monolayer: stability, surface structure uniformity, and relative ease of varying the functionalities. Self-assembled monolayers of ω-terminated alkanethiolates are probably one of the best surface modifications accomplishing these requirements.

### *Concept of Self-Assembly*

Self-assembly can be defined as a process of spontaneous organization of complex, more ordered structures from the primary building units. As a result the atoms, molecules, biomolecules, and the like combine together, forming structures with fewer degrees of freedom. The simplest case of self-assembly are so-called self-assembled monolayers (SAMs).

Typical SAM elements are head group, chain/backbone, and end group (see Fig. 5.12 for details). Every constituent of the monolayer has its own function. SAMs

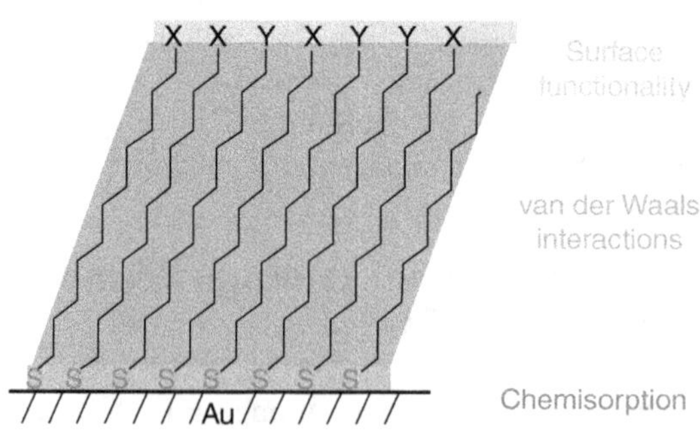


**Figure 5.12** Basic units of the alkanethiol SAM on gold and their function; X and Y represent different end groups.

are formed spontaneously by the adsorption of surfactant molecules on the substrate due to the high and specific affinity of the head group to this substrate. The backbones are responsible for the intermolecular interactions, crucial for the ordering of the monolayer and leading to the formation of closely packed and highly ordered structures. Finally, the end group provides the capability to control surface properties, mainly due to group-specific interactions. Alkanethiols are widely known to form SAMs on silver, gold, and copper substrates [93]. The thiol head group anchors the molecule to the metal surface by metal–sulfur bond, created in a process of chemisorption. The SAM structure is further stabilized by the van der Waals forces occurring between hydrocarbon backbones of neighboring molecules. The terminal group determines the functional properties of the monolayer. It was demonstrated that the end group is responsible for the physico-chemical properties of the monolayer [94, 95]. Hence, the $CH_3$ or $CF_3$ end group makes the SAM hydro- and metalophobic and highly antiadherent, while introduction of the –OH, –COOH, or $NH_2$ groups result in an increased hydrophilicity and strong binding properties toward metal ions and biomolecules. All the building units of the alkanethiol SAMs together with their functions are presented in the Figure 5.12.

The advantages of alkanethiol SAMs as biomolecules linking platforms are listed below [96]:

- Simplicity in forming ordered and stable monolayers
- Introduction of the microenvironment mimicking biological membranes, suitable for biomolecules immobilization
- Surface properties controlled by the SAM functional group
- Minimal amount of the biomolecules (monolayer) necessary for attachment on SAM
- Time stability, enabling performing a series of reliable experiments
- Simplicity in providing data about immobilized biomolecules on molecular level

The limitations of the alkanethiol SAMs-based biomolecule attachment are as follows [96]:

- High sensitivity of immobilized proteins/enzymes to pH, ionic strength, and temperature changes, threatening with a loss of a biological activity
- Some SAMs provide insufficient chemical stability (e.g., risk of oxidation)
- Induced electric field/monolayer desorption may complicate biosensor application
- Highly surface energetic, hydrophobic SAMs surface may accumulate contaminations, blocking the active sites

Although alkanethiol SAMs still provide some drawbacks, the gained benefit still places them as first choice for linking biomolecules to the metal substrate. Therefore, the following sections of this chapter will be dedicated to the (EC)-SERRS studies of the cytochrome *c* attached to alkanethiol SAMs on metal supports.

### *Strategies to Immobilize Cytochrome c on Alkanethiol SAM*

Main interest of this section is ET between cytochrome *c* and alkanethiol-modified metal surface. Therefore, strategies for tailoring the alkanethiol monolayer on Ag/Au with cyt *c* will be discussed. As already mentioned, cyt *c* exhibits asymmetric distribution of charge and possess a patch of positively charged lysine residues. This area of the protein is an excellent binding domain, utilizing electrostatic attraction between protein and anionic docking sites of natural cyt *c* redox partners or, in general, with negatively charged surfaces [97]. Although a Ag electrode surface is positively charged above the pzc (–0.9 V versus Ag/AgCl [98]), this charge is overcompensated by either specific adsorption of anions [99, 100] or the formation of a negatively charged layer provided by the monolayer [101, 102], thus enabling the attachment of cyt *c* due to electrostatic forces (see Fig. 5.13*a* to view electrostatic binding to a SAM–metal interface). The most common cyt *c* linkage monolayers are SAMs carrying carboxylate ($-COO^-$) functional groups [101–103], but sulfonate ($-SO_3^-$) [104, 105] and phosphate ($-PO_3^-$) [106] were also successfully used. Alternative approach for cyt *c* tethering to a SAM-coated electrode is a hydrophobic patch involving amino acid residues 81–85, neighboring with a lysine section and placed above the heme plane, that is, on the side of the axial Met-80 ligand. A possible role of this fragment is stabilization of the complex formed due to driving electrostatic forces between cyt *c* and its redox partners ($bc_1$ complex and CO*x*). Evidence for binding of cyt *c* with *n*-alkanethiol monolayers on Ag by hydrophobic interactions, most likely by the partial penetration of the peptide segment 81–85 into the monolayers was shown with the aid of Raman spectroscopy ([107]; Fig. 5.13*b*).

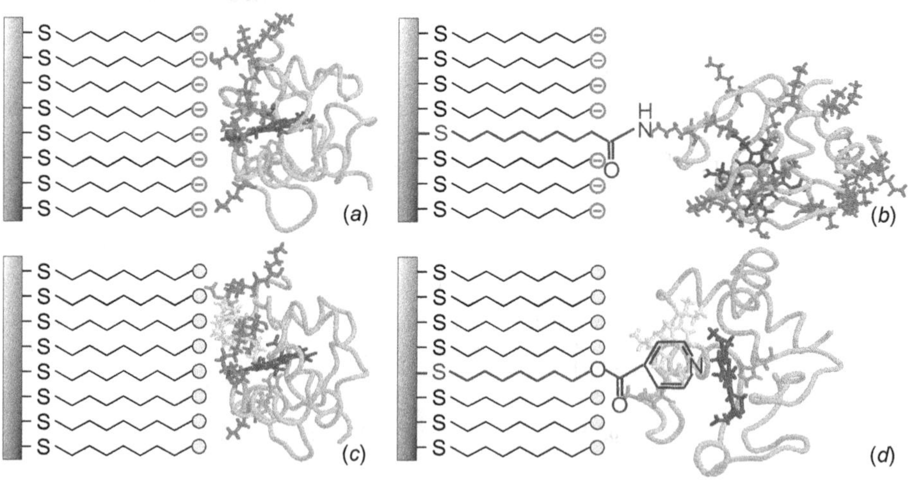


**Figure 5.13** Different modes of cyt *c* immobilization with aid of ω-terminated alkanethiol SAMs [98]: (*a*) anionic termination (e.g., $-COO^-$, $-SO_3^-$, $-PO_3^-$) for electrostatic binding with lysines (blue), (*b*) methyl end groups for hydrophobic binding via the peptide segment 81–85 (yellow), (*c*) carboxylate end groups cross-linked to amino groups of the lysine residues for covalent binding, and (*d*) pyridinyl end groups for coordinative binding to the heme [86].

Cross-linking of the amino groups of the protein and carboxyl groups of the SAM, resulting in the formation of covalent amide bond, is also easily attainable [108] (see Fig. 5.13*c*). Finally, cyt *c* attachment can be achieved by SAM providing a competitive ligand for native heme iron ligands, capable of coordination to the redox center. Monolayers carrying N-ligand end groups, like pirydynyl-functionalized alkanethiols were demonstrated to exchange effectively Met-80 ligand, thus providing direct molecular wiring between redox center and the metal electrode [109, 110] (see Fig. 5.13*d*).

## 5.3 EC–SERRS STUDIES OF CYT *C* ATTACHED TO ALKENETHIOL-COATED ELECTRODES

### 5.3.1 Principles of Surface-Enhanced Resonance Scattering (SERRS) Experiment

Surface enhancement decays strongly with the distance from the surface to the analyzed adsorbed molecules. Therefore, one is not able to observe the bands of the biomolecule attached to thiol-coated surface unless we use the one containing a chromophore such as cyt *c*. This approach overcomes the obstacle of the decaying surface enhancement at a distance provided by a molecular spacer using the excitation wavelength near the electronic transition in the surface-attached molecule and thus combining SERS and resonance Raman (RR) effects into the so-called SERRS (surface-enhanced resonance Raman scattering). SERRS is actually limited to the molecules possessing chromophore, but it is very useful in studies of the chromophoric biomolecules attached to metal by biocompatible spacers [111, 112]. SERRS provides additional $10^3$–$10^4$ enhancement in comparison to a standard SERS experiment. The additional advantages of SERRS are its high sensitivity (down to $10^{-8}$ M) and selectivity (probing only the chromophore group).

### 5.3.2 Resonance Raman Spectrum of Cytochrome *c*

The near UV and visible absorption spectrum of ferrocytochrome contains the Soret band in the near UV and Q bands in the visible region [113]. Assuming idealized $D_{4h}$ symmetry of iron–protoporphyrin IX constituent of heme *c* and ignoring the peripheral constituents, 81 chromophore normal modes can be distinguished [114]. Since the transition dipole moments of hemes lie in the porphyrin plane, resonance enhancement is only expected for the gerade in-plane vibrational modes [115]. Upon Soret band excitation (ca. 413 nm), mainly the totally symmetric $A_{1g}$ modes will gain intensity, due to *A*-term enhancement. When the Q band excitation line (ca. 514.5 nm) is used, the intensity of $A_{1g}$ modes decreases and alternatively the nontotally symmetric modes $B_{1g}$, $A_{2g}$, and $B_{2g}$ gain intensity through the *B*-term enhancement

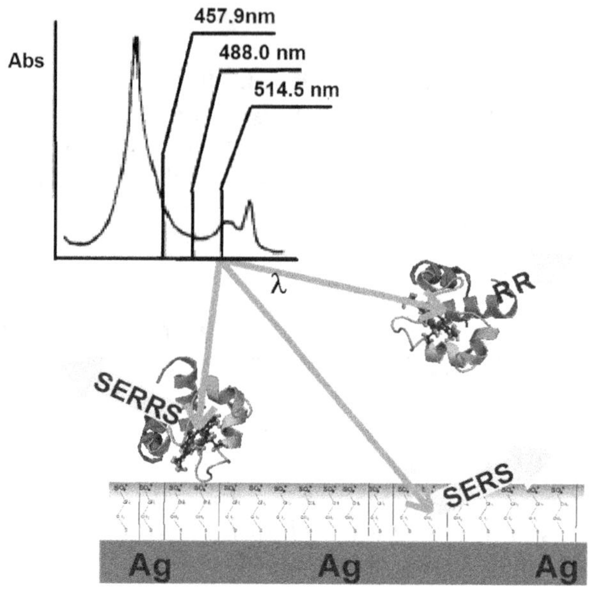


**Figure 5.14** Scheme of contributions to in situ Raman signal of cyt *c* immobilized on alkanethiol SAM-coated Ag. Representation of different scattering mechanisms from different elements of the studied system [51].

mechanism [116, 117]. The *A* and *B* terms refer to formalism of the Albrecht theory of resonance Raman scattering [118]. Frequencies of resonance Raman (RR) marker bands of cyt *c* allow probing the spin configuration, oxidation state, and coordination state of heme iron [117, 119], as they are influenced by the interatomic distances within the heme group, which are different for various electronic structures of the redox center. Measurements performed with excitation wavelength 514.5 nm, pre-resonant with the Q band, increase the contribution of SERS in the SERRS effect (and *applicability of surface selection rules*) in comparison to 413 nm, where RR effect is dominant. Performing in situ SERRS measurements of cyt *c* anchored to alkanethiol SAM, one has to deal with the possibility of different scattering contributions by various components of the examined system in the overall Raman signal (see Fig. 5.14 for a schematic representation). First of all, certainly the SERRS signal of the SAM-attached cyt *c* will be recorded; however, the risk of the RR contribution from the solvated solution protein molecules should be minimized, reducing the protein solution used for immobilization to the micromolar range ($10^{-6}$ M). The other solution is the transfer of the SAM-coated electrode with the attached cyt *c* to a protein-free electrolyte and further performing of the EC-SERRS under these conditions. Typically, in the SERRS spectrum, the SERS signal of the linkage alkanethiol monolayer is also observed, from which the protein-induced structural changes within SAM can be deduced.

### 5.3.3 Choice of Substrate

#### *Gold versus Silver*

As already mentioned, selection of a proper substrate for an SERS experiment, concerning the aim of the measurements, is one of the key factors. In the case of SERRS spectroscopy another restriction is that only metal supports that exhibit SPR in the visible region can be considered. In practice, this limits the choice to the use of Ag and Au substrates. Au supports provide high biocompatibility and easiness of electrochemical studies of redox-active biomolecules in a large potential window [120]. On the other hand, silver is considered as toxic and instable, but it exhibits superior optical properties compared to gold, exhibiting higher surface enhancement and SPR tunable from near UV to infrared [121]. In contrast, wavelength dependence of the SPR for gold imposes use of wavelengths in the red or even infrared region in order to obtain SERS activity, which makes this metal not suitable for combining SERS and RR for chromophores absorbing out of this region. Heme of cyt *c* absorbs unfortunately mostly in the blue and violet region and exhibits only some weaker RR effect at green wavelengths. Therefore, initial SERRS and EC-SERRS spectroscopy investigations of cyt *c* attached to alkanethiol-coated metal were limited to the silver substrates, mostly colloidal or ORC roughened, and its further correlation with the ET properties deduced from electrochemical experiments for the protein immobilized in similar systems but on the flat gold supports. The following sections of this chapter will be dedicated to the description of possible risks resulting from such an approach and presentation of different support architectures allowing better selective enhancement of cyt *c* SERRS modes and more direct analysis of interfacial electron-transfer processes analogous to these on flat gold substrates.

#### *Flat versus Roughened*

A common approach in the initial studies on ET properties of cyt *c* attached to thiol SAM-modified gold was a cyclic voltammetry (CV) experiment in order to obtain kinetics data for cyt immobilized on SAM-covered Au, whereas SERRS, revealing structural information, dealt with the protein anchored to thiol SAMs on Ag. Therefore, one of the topics studied in our lab was a comparison of the electrochemically roughened silver and gold plates with atomically flat Au(111), as substrates for the thiol self-assembly [122]. We coadsorbed two ω-functionalized aliphatic thiols: 2-mercaptoetanosulfonate (MES) and mercaptoundecanol (MUL) on electrochemically roughened gold and silver, expecting to form SAMs capable of selective binding cyt *c* only to the MES component [123]. Next we compared the composition and structure of these two-component SAMs on these two substrates with those formed by analogous protocol but on Au(111) substrates. Quantitative and qualitative examination of the mixed SAMs was performed, respectively, with the reductive desorption technique and scanning tunneling microscopy (STM) imaging on flat Au(111) substrates and with SERS spectroscopy on ORC-roughened Ag and Au.

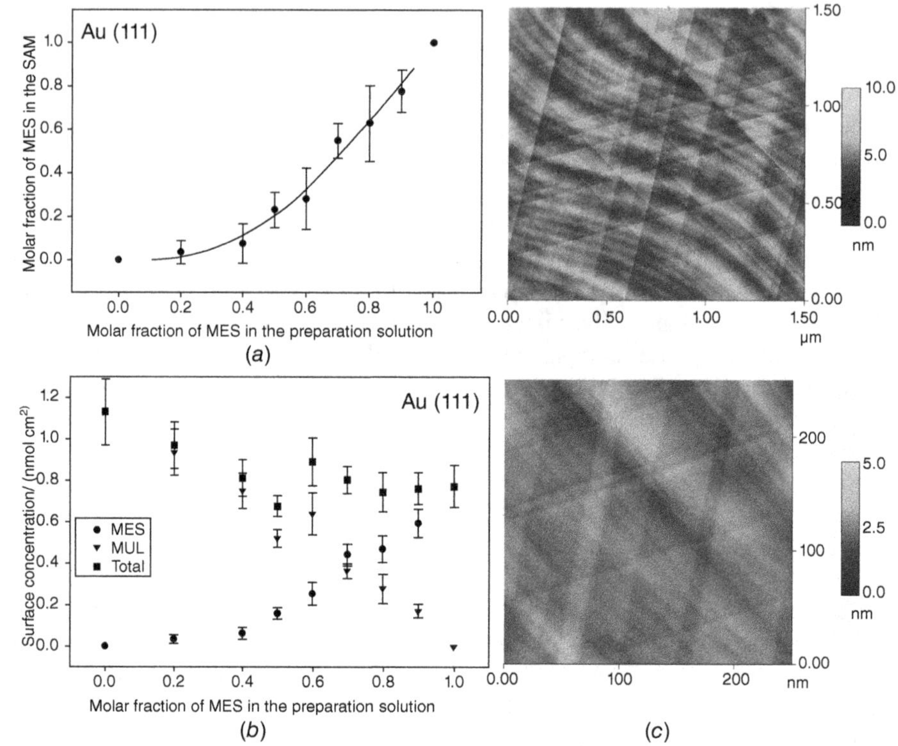


**Figure 5.15** (*a*) Composition of the MES–MUL SAMs determined from the electrochemical desorption data (six SAMs studied for each studied thiol concentration) as a function of molar fraction of MES in solutions used for monolayer deposition. Trend line: fourth-order polynomial fit; (*b*) surface coverage for MES and MUL as a function of molar fraction of MES in solutions used for monolayer deposition; total surface coverage plotted for comparison; both adapted from [122]; (*c*) STM images of Au(111) of different magnification before deposition of a thiol, acquired with 1.5-V bias and current set point of 50 nA [51].

The surface molar fraction of MES determined from thiol reductive desorption on Au(111), plotted as a function of the MES molar fraction in the solution used for SAM formation, is displayed in Figure 5.15*a*. In Figure 5.15*b* the total surface coverage Γ and the respective values for both components of binary SAMs, determined from the area under the reductive desorption peak, are presented. As seen in Figure 5.15*b* the two plots representing the single component surface concentration cross over when the MES molar ratio in the solution used for SAM preparation was about 0.7. This means that at this molar ratio the surface is covered with equal amounts of both components. The same conclusion may be reached from the plot in Figure 5.15*a*. The obtained results pointed at MUL as the preferably adsorbed component in the whole studied bulk (solution) concentration range. According to the electrochemical results for MUL-dominant solutions, MES was adsorbed in fractions significantly lower than its contributions in the bulk or there were no peaks in the CV curves indicating its presence on the surface (for $\chi_{MES} < 0.4$). The morphology of the gold substrate studied for this set of data is shown in two STM images in Figure 5.15*c*. It exhibits terrace-step features characteristic of atomically flat Au(111). The SERS spectra of SAMs grown on ORC-roughened Ag and Au supports from single or two-component thiol solutions of varying molar ratios of MES and MUL were also collected to bring information about the SAM surface composition. The relative intensity ratios of the bands chosen as MES and MUL exclusive markers, plotted as a function of the MES molar fraction in the solution used for the SAM formation on Au- and Ag-roughened substrates are presented in Figures 5.16*a* and *b*, respectively. As here analyzed, bands were assigned to the same vibration (C–S stretching) for the MES and MUL molecules, and we may safely assume the similarity of the Raman scattering cross sections. Hence, the trend deduced from the analysis of the SERS intensities of these bands corresponds to changes of the relative surface composition of the SAM with varying contribution of the components in the self-assembly solution. The results indicate preferential MES adsorption on a rough Ag support. The plateau occurring for MES-dominant solutions ($\chi_{MES} > 0.5$) reaches the band intensity of the one-component MES monolayer, being a perfect illustration of the strong preference of MES adsorption in this case (see Fig. 5.16*b*). Gold is known to provide weaker surface enhancement of the Raman scattering than silver. Therefore, greater errors and lower reproducibility of the spectra are found in the results shown in Figure 5.16*a*, in comparison with the similar plots for the Ag-supported SAMs (Fig. 5.16*b*). In case of gold, MES preferential adsorption cannot be so clearly found, and a more linear trend line fits the experimental data. Differences in mixed thiol composition observed for roughened Ag and Au substrates may arise from slightly different substrate morphology for these two metals, shown already in Figure 5.6*a*. Substrate-dependent structural and conformation properties of the monolayer were already previously reported by our group for the one-component thioglycolic acid (TGA) SAMs on Ag and Au [50]. More interestingly, the reversed preference of adsorption of one thiol from the MES–MUL mixture on the rough support as compared to the flat one [Au(111)] needs some comment. SERS spectra are recorded at the nanostructured surfaces that contain nanoclusters of various sizes. It was reported that SAMs on colloidal Au nanoparticles differ strongly from SAMs on planar Au support, and these differences were interpreted in terms of changes in intermolecular interactions between adsorbed molecules [124]. As proven by IR measurements, the aliphatic chains of long-length thiols adsorbed on nanometer-sized gold clusters are ordered close to the surface but become more disordered at the opposite end [125]. In the case of MUL molecules, gauche defects near the OH terminal groups may cause lateral H bonding among these groups to be less extensive in comparison to the SAMs on the "flat" supports. It is known that the contribution of the hydrogen bonds to the enthalpy of adsorption of ω-terminated thiols is high [126, 127]. Considering that preferential adsorption and phase segregation in thiol SAMs are likely to be the results of a thermodynamically controlled process (at room temperature) [128], one may expect that adsorption of MUL is no longer preferred at the SERS-active supports. Giving this literature background, we can suggest that this phenomenon is related to MES–MES and

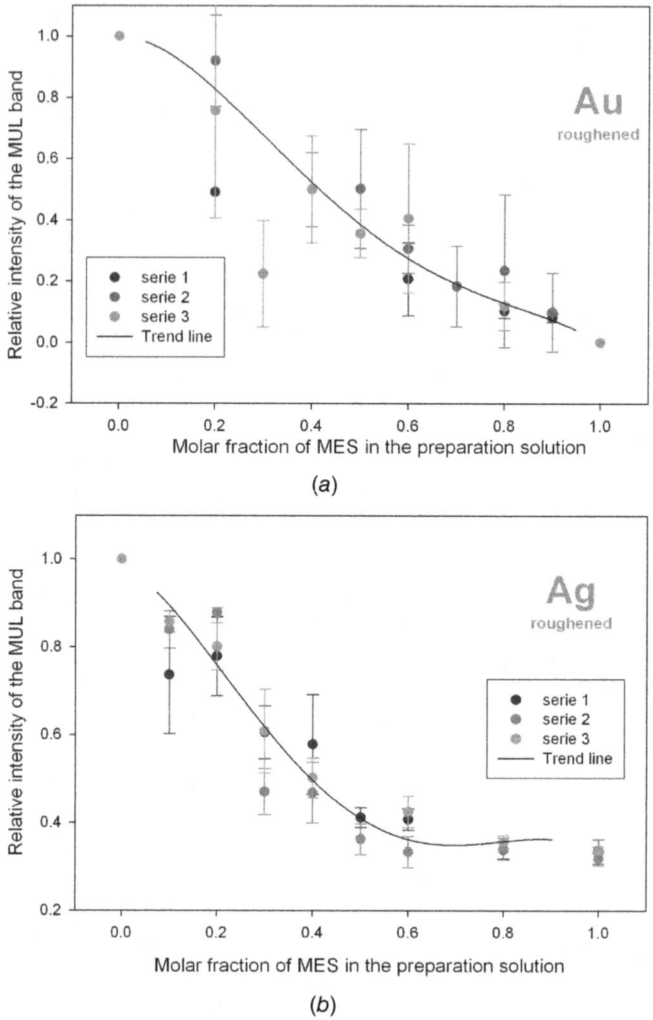


**Figure 5.16** Relative intensities of the SERS MUL marker band, normalized to the sum of the MES and MUL marker plotted as a function of the molar fraction of MES in the solution used for the SAM preparation on ORC-roughened (*a*) Au and (*b*) Ag supports (3 series, 5 points for each sample) [122]. Trend lines: fourth-order polynomial fits.

MUL–MUL interactions and the impact of structure imposed on the SAM by a structured substrate.

This somewhat unexpected finding is extremely important for those combining Raman spectroscopy on rough supports with the electrochemical experiments performed on macroscopically flat metals, aimed at determination of electron-transfer rates for proteins adsorbed on mixed SAMs, among them here discussed cyt *c*. Therefore, researchers should be aware of the possible strong influence of the used substrate (Ag versus Au, smooth versus rough) on the structure and composition of the formed linkage SAM, especially when comparing the results obtained for different substrates, imposed by specific analytical methods (e.g., electrochemistry and SERRS).

### *Performance of SERRS and Electrochemistry on Same Silver Substrate*

The results discussed in the previous section prove that reliable investigation of electronic properties of cyt *c* adsorbed at chemically modified metal support, aiming at their correlation with the structural information provided by SERRS studies, require using electrodes prepared in a manner enabling the performance of SERRS and electrochemistry for the same substrate. Electrochemical measurements for roughened, polycrystalline support would be difficult and ambiguous for several reasons, listed below. First, roughness of the electrode are problematic for cyclic voltammetry, which like all electrochemical techniques is employed preferably on smooth surfaces. Next, electrochemical roughening procedure and previous polishing of the electrode are expected to introduce inhomogeneity, especially in the case of the silver surface [129, 130]. Use of nonsmooth substrate will very likely provide a variety of the available adsorption sites for linkage monolayers, which may hinder the interpretation of the electrochemical data for the attached cyt *c*. Use of silver introduces additional obstacles to the electrochemical experiment, mentioned previously. However, a few years ago the van der Zwan group introduced a new method for preparing Ag electrodes, which allows their simultaneous application as SER(R) S-active platforms and substrate for the performance of the electrochemistry of cyt *c* immobilized on alkanethiolate coatings [131]. A novel polishing procedure involving the aluminum oxide lapping film and a short immersion time of the electrode in the promoter solution (4-mercaptopyridine) was introduced. Such prepared Ag electrodes were found appropriate to perform the direct electrochemistry of solution cyt *c*, providing voltammetric responses essentially identical to those obtained on gold electrodes.

The electrochemical pretreatment of the polished silver was claimed to worsen the voltammetric response of cyt *c* at the Ag/PySH electrode considerably, in terms of smaller current peaks and low signal-to-background level compared with that obtained in the absence of electrochemical pretreatment. Despite this fact, the reports on successfully combined voltammetry and SERRS studies of cyt *c* immobilized on 4-mercaptopyridine [132] and 11-mercapto-1-undecanoic acid (MUA) [133] modified the roughened silver electrode, polished to a mirrorlike appearance prior to an oxidation reduction cycle (ORC) roughening procedure have also appeared.

For the cyt *c* on polished prior roughening Ag electrodes coated with MUA, the nearly ideal Nernstian behavior of the adsorbed protein at pH 7 was derived from the treatment of EC-SERRS data [131]. In fact, the formal potential value obtained from the potential-controlled EC-SERRS experiment for this substrate is in fine agreement with the values reported for native cyt *c* adsorbed on gold electrodes chemically modified with the same carboxylic-terminated alkanethiol (MUA) [134, 135] and altered by only 10 mV in comparison for that determined from the CV curve recorded for such treated silver electrodes. Such insignificant potential shift can be attributed to the local heating of the system upon irradiation with laser beam during an SERRS experiment. The absence of a strong voltammetric response characteristic of the native redox couple cyt $c^{3+}$/cyt $c^{2+}$ upon increase of the pH of the working solution from 7.0 to 9.0 is in nice correlation with the deviation from the ideal Nernstian behavior expected for a one-electron couple deduced from EC-SERRS spectra. The complementary of the results provided by Zwan and co-workers [132, 133] by cyclic voltammetry and Raman spectroscopy should be emphasized. This is a remarkable outcome, which highlights the advantages of operating with these two techniques simultaneously.

### Hybrid Ag–Au Systems

The plasmonic properties of the metallic substrate can be furthermore tuned by changing the surface morphology and the dielectric constant of the surrounding medium. The red shift of the SPR peak for metal NPs is expected for increasing diameter and aspect ratio, defined as the ratio between the longer and the shorter axis of the particle. Further red shift is predicted for the increasing dielectric constant of the surrounding medium [136]. Due to the interparticle plasmonic coupling, aggregated metal NPs exhibit typically higher field enhancement than isolated nanostructures. According to the rule of a thumb mentioned above, the SPR position of the NP aggregates is also considerably red shifted, compared to isolated ones [137, 138]. Keeping in mind these features of the SERS-active nanostructures, the Hilderbrandt group developed in 2009 novel hybrid Au–Ag supports, employing the transfer of plasmon resonance excitation from the Ag to the Au coating, possibly due to the presence of the thin dielectric layer between the two metals [139]. The General design of this device is given in the scheme in Figure 5.17. The more detailed structure of the hybrid Ag–Au system employed for cyt *c* studies is as follows. The bottom layer consists of a mechanically polished and electrochemically roughened Ag electrode on which subsequently an amino-1-undecanethiol (AUT) SAM was deposited. Next, the adsorbed $AuCl_4^-$ ions were adsorbed at the SAM-

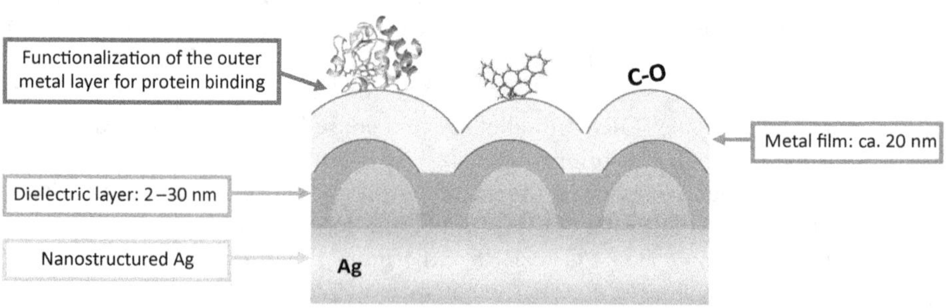


**Figure 5.17** Scheme of the setup of the hybrid Ag–Au substrate designed for different analyte SER(R)S studies [140].

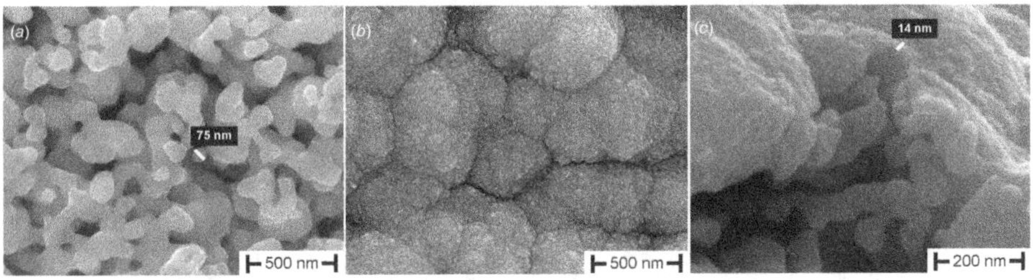


**Figure 5.18** SEM top views of different regions of SERS-active hybrid Ag–Au supports: (*a*) electrochemically roughened Ag, (*b*) Au overlayer on Ag, and (*c*) SEM side view of the Ag–Au junction [139].

modified Ag electrode and electrochemically reduced to form a gold coating. In a final step to guarantee the biocompatibility of the substrate, the mixed SAM of 11-mercapto-undecanoic acid (MUA) and 11-mercapto-1-unodecanol (MUL) was formed on the top gold layer. SEM images showing the morphology of the rough silver bottom layer, gold overlayer, and a section through the hybrid-layered Ag–Au electrode are shown, respectively, in Figures 5.18*a*, *b*, *c* [139].

Yeast cyt *c* was chosen as a test molecule to study the EC-SERRS and electrochemical performance of this novel device [139]. It was demonstrated that this hybrid Ag–Au support is suitable for SE(R)RS spectroscopy in the entire visible spectral region, and the surface enhancement is similar to this for pure Ag substrates. Furthermore, its electrochemical properties are basically the same as those of a pure Au electrode. These results show that the device combines the optical properties of Ag with the chemical advantages of Au. Successful transfer of plasmonic properties of the Ag system to upper gold layer secures the high SER(R)S sensitivity, while displaying the electrochemical stability and higher biocompatibility of gold. Stationary and time-resolved SERRS studies of interfacial ET processes were successfully performed for cyt *c* on hybrid Ag–$SAM_1$–Au–$SAM_2$ ($SAM_1$ and $SAM_2$ were, respectively, –$NH_2$ and –COOH terminated) by Sezer et al. [141].

## 5.3.4 Information Derived from (EC)-SERRS Spectra of Cytochrome *c* Attached to Thiol-Modified Metal

### *Heme Structure*

Similar to RR spectroscopy, the SERRS spectra of cyt *c* attached to Ag (or Ag–Au hybrid structures) provide information on the electronic state of the heme. Hence, the oxidation state, spin configuration, and ligation pattern on the redox center iron atom can be deduced from the SERRS data. In Figure 5.19 the SERRS spectra excited with the 413-nm line, resonant with Soret band, showing the marker bands of iron oxidation state are presented.

The band labeling given in the figure are further discussed according to the complete assignment of the RR spectra via enzymatic reconstitution with isotopically

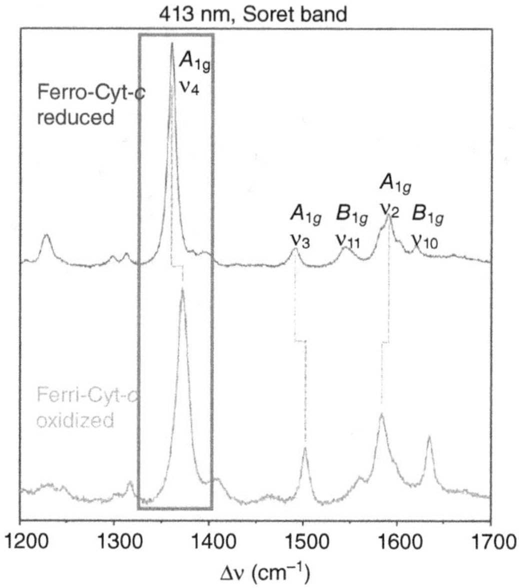


**Figure 5.19** SERRS (excited with 413 nm) spectra of the reduced and oxidized form of cyt *c* attached to alkanethiol-coated Ag [140].

labeled hemes [117]. EC-SERRS spectra collected for cyt *c* adsorbed electrostatically at pH 7.0 of mercaptoethanosulfonate [123] SAM on electrochemically roughened Ag at a gradually varying potential, in a range corresponding to protein reduction–oxidation, are shown in Figure 5.20 [105]. Preservation of protein electroactivity upon electrostatic immobilization on MES SAM on Ag is evidenced by the presented spectra. Application of the potential of −500 mV (versus Ag–AgCl) allows one to safely assume that adsorbed cyt *c* is completely reduced [108, 135]. Markers in the SERRS spectrum taken at this potential confirm total reduction of cyt *c*.

At a potential of 0 mV and higher striking changes become noticeable in the cyt *c* SERRS spectrum, confirming the oxidation of the protein. SERRS spectra shown in Figure 5.20 were excited with the 514.5-nm line, and it can be seen that this line is preresonant with the Q-band results in quite a different intensity pattern this observed for the Soret band excitation (413 nm). However, the position of the bands is conserved. This feature of the signal dependence on the energy of the incident light is typical for the spectra with contribution from RR enhancement. More straightforward comparison of the RR spectra of cyt *c* excited at 413 and 514.5 nm is given in Figure 5.21.

In Figure 5.20, the bands sensitive to the heme ligation pattern are also marked. The native conformational state of cyt *c* is six-coordinated low spin (6cLS) with His-18 and Met-80 as axial ligands, denoted as $B_1$. Exposure of cyt *c* to too strong an electrostatic field or some other factors may lead to conformational conversion from $B_1$ to nonnative $B_2$ spin states of heme iron: five-coordinated high spin (5cHS) due to a loss of the Met-80 native axial ligand or the 6cLS/HS bis-His due to occu-

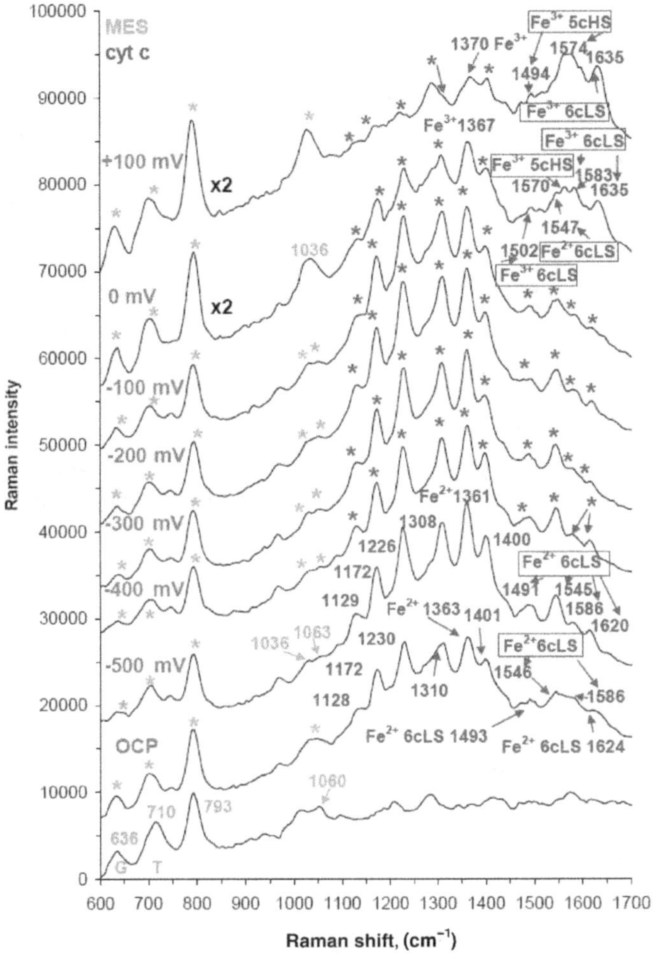


**Figure 5.20** EC-SERRS of cyt *c* electrostatically adsorbed on MES SAM on roughened Ag at pH = 7.0 ($\lambda_{exc}$ = 514.5 nm). (*Bottom*) SERS spectrum of MES SAM grown for 2 h from 10 mM MES aqueous solution. (*Above*) EC-SERRS spectra taken at OCP and at applied potential of MES SAM transferred to $10^{-6}$ M cyt *c* solution of pH = 7.0 [105].

pancy of the vacant position left by Met-80 and His-26 or the His-33 residue [135]. It can be seen in the Figure 5.20 that potential-induced oxidation of cyt *c* attached to MES-coated Ag is concomitant with the appearance of a portion of the $Fe^{3+}$ $B_2$ 5cHS species, in addition to $B_1$ 6cLS ferric ion. Therefore, not only is the information on the oxidation state of the heme iron but also on the coordination state, spin configuration, and type of heme pocket ligation is provided by the SERRS spectra. An exemplary set of SERRS spectra corresponding to altered heme ligation pattern is shown in Figure 5.22. The oxidation state (Red and Ox denote, respectively, reduced and oxidize), coordination type, and spin configuration are given schematically in

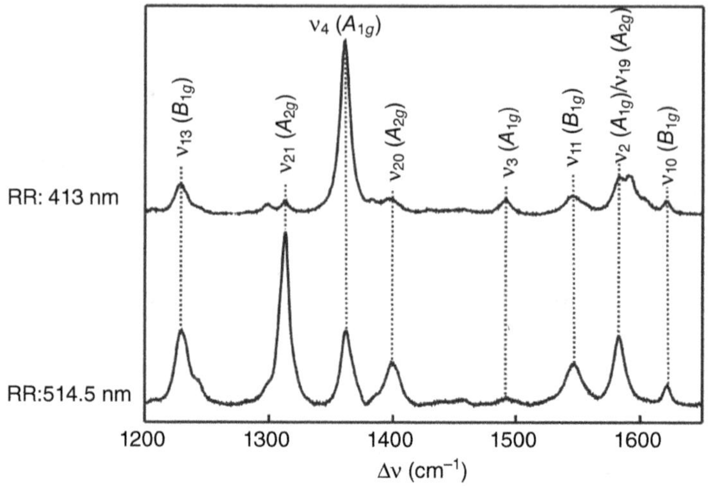


**Figure 5.21** RR spectra of ferrocytochrome *c* taken with Soret resonant (*top*) and Q preresonant (*bottom*) excitation lines [115].

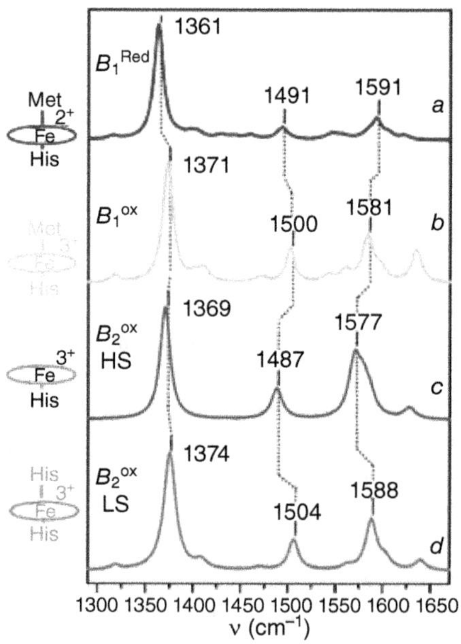


**Figure 5.22** SERRS spectra of different oxidation, spin, and coordination states of cyt *c* immobilized on a COO⁻ SAM measured with 413-nm excitation: (*a*) $B_1$[6cLS]Red, (*b*) $B_1$[6cLS]Ox, (*c*) $B_2$[5cHS]Ox, and (*d*) $B_2$[6cLS]Ox [98].

the legend for Figure 5.22. For a more detailed discussion of RR and SERRS cyt *c* marker bands, especially of nonnative states, the reader is referred to the studies by Hildebrandt's group [142, 143].

### Redox Group Orientation

Strong contribution of electromagnetic enhancement through the surface plasmon resonance effect for 514.5-nm excitation extends applicability of surface selection rules in the interpretation of the SERRS spectrum of cyt *c*. Thus, intensification of particular SERRS bands is accomplished via the individual components of the scattering tensor depending on the direction of the electric field vector and the orientation of the heme plane with respect to the metal surface. When idealized $D_{4h}$ symmetry of the porhyrin group is assumed Raman-active in-plane, $A_{1g}$, $B_{1g}$, $A_{2g}$, and $B_{2g}$ modes will be enhanced when each mode involves a polarization component perpendicular to the surface [19, 144]. This condition is fulfilled irrespective of the symmetry labels when the heme plane is oriented perpendicular to the metal surface, while for parallel alignment of the heme plane with respect to the surface only the intensity of the $A_{1g}$ modes will be selectively enhanced [145]. The reason is that both totally symmetric modes (e.g., $A_{1g}$ mode $\nu_4$) and nontotally symmetric modes (e.g., $B_{1g}$ mode $\nu_{10}$) involve movement along the $x$ and $y$ axes (in the plane of heme group), while the totally symmetric ($A_{1g}$) modes involve also motion along the $z$ direction [113, 146]. Hence, the variations of the relative intensity of the $A_{1g}/B_{1g}$ modes provide information about the dynamics of protein orientation [147]. The supplied information shows only the trend in the changes of the heme plane orientation; however, the absolute orientation cannot be accurately determined. Increase of the $B_{1g}/A_{1g}$ ratio, interpreted in terms of the surface selections rules, implies reorientation of the plane of the heme ring to an increased angular position relative to the metal surface plane. Selecting the modes for the relative intensity analysis, one has to keep in mind that this comparison makes sense only when comparing the species with the same coordination and spin configuration as the heme iron. In Figure 5.23, the changes in $B_{1g}/A_{1g}$ relative intensity ratio for $\nu_{10}$ and $\nu_4$ bands corresponding to "straightening" of the heme plane upon reduction of the redox center are shown. These orientation changes, concomitant to redox-linked changes, can be explained in terms of adapting the more perpendicular orientation, which is considered as the one facilitating ET between the cyt *c* and the metal electrode. Strongly resonant Soret-band excited SERRS of cyt *c*, taken at 413 nm, provide the highest sensitivity for monitoring the redox-linked structural changes but are insensitive to the protein reorientation discussed. In contrast, SERRS spectra excited with substantially less resonant Q-band at 514 nm provide much weaker surface enhancement but are sensitive both to the relative orientation of the heme plane and to the redox state and the heme-pocket structure.

### SER(R)S Studies of Alkanethiol SAM-Induced Effects on Linked Cyt c

This chapter is devoted to the ET properties of cyt *c* attached to ω-terminated alkanethiol SAMs on Ag and Au supports, studied with SERRS and EC-SERRS spectroscopy.

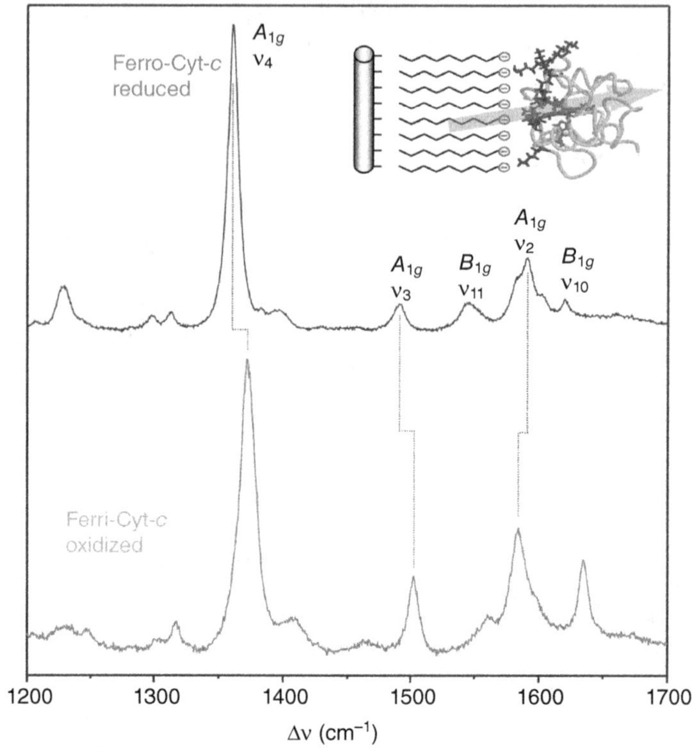


**Figure 5.23** SERRS spectra collected with 514.5-nm excitation (Q-band) for reduced and oxidized form of cyt *c*, exhibiting changed $B_{1g}/A_{1g}$ intensity ratio, corresponding to more perpendicular orientation of heme with respect to the metal surface for ferricytochrome *c* [140].

Therefore, in this section a closer look will be taken at the issue of alkanethiol ω-functionalized coatings on metal acting as linkage monolayers to bind cyt *c* and the risks carried by direct interaction of the biomolecule with the metal substrate or unwilling effects on native cyt *c* structure, induced by the presence of the alkanethiol SAM. SAM structural properties anticipated to influence the ET properties of the anchored cyt *c* will be also shortly discussed.

**SAM Defects of Various Origin** Although alkanethiol SAMs on Au and Ag are known to form monolayers of high order and quality, they are not free of defects. A scheme presenting an overview of reasons for the possible SAM defects is shown in Figure 5.24. Most of them are easy to minimize or eliminate by careful control of the experimental procedure (cleanness of the substrate, purity of solvents and reagents). Molecular defects can also appear in the SAM when the degree of molecular organization is poor: aliphatic chains not fully extended or tilt angle nonuniform across the monolayer [148]. Another important class of the defects is coverage defects [149], which arise from the lack of a monolayer. They can be divided into two main classes:

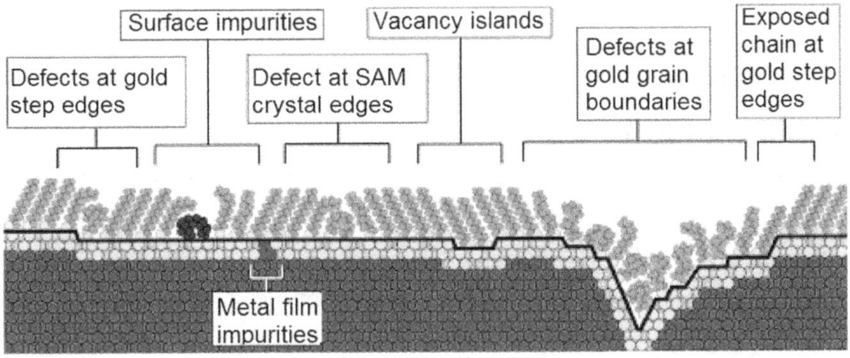


**Figure 5.24** Schematic illustration of the potential factors leading to defects of the SAM formed on polycrystalline substrate. A dark line at the metal–SAM interface is to guide the eye to follow the changing substrate topography [93].

- Single-Site Defect/Pinhole. A void site where the substrate is exposed to solution, whose size can vary from one atom to $>10^3$ gold atoms
- Collapsed Site. When substrate–solution spacing lies below thickness of the full monolayer, thus allowing molecules to physisorb

Elimination of these defects would be crucial to studying the intrinsic mechanism of the ET of cyt *c* across the metal–alkanethiol interphase. As for the defects-rich SAMs, the electron transfer was found to occur primarily through the defects [123].

**Influence of SAM Structural Imperfections on Attached Cyt *c* Properties** The coverage effect is very often related to the solvent used for the self-assembly. Our group studied this effect for cyt *c* immobilized at dicarboxyl-tailored SAMs on ORC-roughened Ag, provided by mercaptosuccinic acid (MSA, also called thiomalic acid). SERRS experiments confirmed successful electrostatic attachment of cyt *c* on MSA SAMs on silver formed from an aqueous solution at neutral pH [150]. However, a negative potential has to be applied to obtain native protein structure. In the case of covalent bonding, native 6cLS configuration is preserved. Although MSA is water soluble as other short ω-terminated thiols, it is a common strategy to dissolve it in ethanol, for preparation of the linkage monolayer [78], which encouraged us to investigate the influence of the solvent on the monolayer structure. From the changed SERS pattern of the MSA SAM on Ag, grown for a shorter time and from a more dilute MSA ethanolic solution, compared to that typical for an MSA self-assembly aqueous solution [151] (3 h for 1 mM MSA versus 48 h for 10 mM MSA solution), we deduced lower surface coverage of MSA on silver for the former. More interestingly, for MSA monolayers on Ag formed for a short time from diluted ethanolic solution, undesirable oxidation of the iron occurs in the case of electrostatic immobilization, as evidenced by SERRS spectrum of cyt *c* attached under these conditions [150], very likely due to incomplete blocking of the metal surface by MSA. The structure of the protein layer is improved in the case of covalent attachment of the cyt *c* and application of extra negative charge to the electrode. A more compact SAM structure, achieved by application of longer time and higher concentration of a thiol self-assembly solution also results in binding of cyt *c* in its native form of heme pocket (reduced 6cLS).

**Alkanethiol SAMs as Insulating Barriers for Electron-Transfer Chain Length Dependence** A densely packed alkanethiol SAM may act as a barrier between the electrode surface and the electroactive species in solution. A distinctive feature of the EDL structure representative for such an insulating barrier is a shift of both IHP (inner Helmholtz plane) and OHP (outer Helmholtz plane) away from the metal–solution interface The increased distance of the planes from the electrode surface results in a linear potential drop within the aliphatic chain layer [152]. The ET rate is strongly dependent on the thickness of the insulating layer (exponential variation with alkyl chain length). Electron transfer in this case is expected to proceed via electron tunneling through the SAM insulator, which reduces the absolute ET rate to a level at which diffusion limitations are either greatly diminished or totally eliminated [153]. An additional advantage is that the occurrence of the ET takes place with the redox couple from a few to tens of Angstroms from the metal surface, which decreases the participation of specific adsorption, double-layer corrections, and image charge effects in comparison to measurements for bare electrodes. Thus, SAMs of derivatized alkanethiols on gold can be considered useful tunneling barriers, allowing ultrafast heterogeneous electron tunneling measurements. A more detailed description of the number of methylene groups on the ET mechanism for the attached cyt *c* case will be given in further sections of this chapter.

**Effect of Electric Field Provided by Alkanethiol SAM** A potential drop across the electrode–SAM–cyt *c*–solution interface can be estimated using a simple electrostatic model. The model considering the interfacial potential changes for electrodes coated with monolayers of electroactive molecules was originally proposed by Smith and White [154] and adapted for the system with the protein adsorbed on SAM-coated electrodes by Murgida and Hildebrandt [98, 135] (see Fig. 5.25).

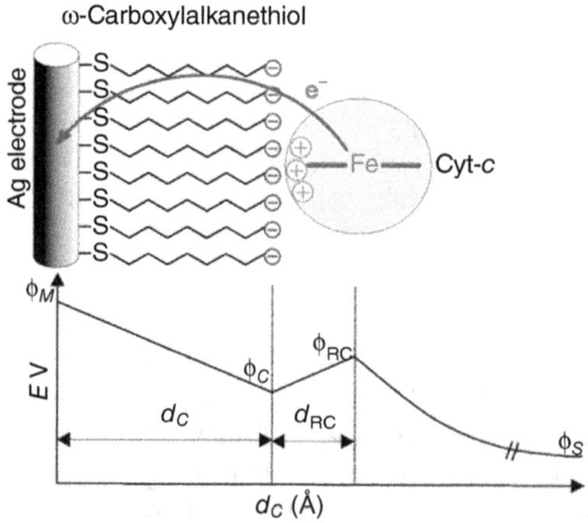


**Figure 5.25** Schematic representation of the potential distribution at the Ag–SAM–cyt *c*–solution interface (see the text for explanation of symbols) [98].

The main assumption of this model is the presence of the net charge densities on the metal ($\sigma_M$), at the carboxylate–protein interface ($\sigma_C$), and in the redox center ($\sigma_{RC}$), which are compensated by the charge density in the bulk solution ($\sigma_S$), to fulfill the electroneutrality condition:

$$\sigma_M + \sigma_C + \sigma_{RC} + \sigma_S = 0 \tag{5.7}$$

Positive and negative charges of amino acid side chains inside the cyt *c* are assumed to cancel, so the protein matrix and the monolayer are treated as dielectric continua. Consequently, in these two regions the potential profile exhibits linear variation with potential. Beyond the plane of the redox center, the potential decays exponentially, consistent with the Gouy–Chapman theory. Such inhomogeneous charge distribution is also observed within the lipid bilayers. The potential drop at the redox site $E_{RC}$ decreases with SAM thickness $d_C$ [98]. The electric field strength $E_F$ at the protein binding site is determined by following parameters:

$$E_F(d_C) = \frac{\varepsilon_0 \varepsilon_S \kappa E_{RC} - \sigma_C - \sigma_{RC}}{\varepsilon_0 \varepsilon_C} \tag{5.8}$$

where $\varepsilon_S$ and $\varepsilon_C$ denote, respectively, the dielectric constants of the solution and the SAM, and $\kappa$ stands for the inverse Debye length. For carboxylate-terminated SAMs the electric field strength at the cyt *c* binding site was estimated to be on the order of $10^9$ (V m$^{-1}$) [98], whose value is similar to that predicted for biological membranes in the proximity of charged lipid head groups [155]. Mimicking important features of the lipid bilayers (potential distribution, electric field strength) and close structural resemblance (hydrophobic core and the charged terminal groups) implies that alkanethiolate SAM-coated electrodes can also be considered as simple models for biomembranes. Higher field strengths are predicted for phosphate and sulfate SAMs since their $\sigma_C$ is noticeably larger, whereas the field strength is distinctly reduced in the case of hydrophobic head groups, which provide $\sigma_C = 0$ C m$^{-2}$. Thus, it can be expected that the magnitude of the electric field dependent on the terminal group of the linkage SAM and its distance from the metal surface will have an effect on the coordination pattern of the heme pocket. Indeed, a cleavage of Fe–(Met-80) bond deduced from the SERRS spectra (5cHS markers) was reported for cyt *c* anchored to carboxylate-terminated alkanethiolate [HS-$(CH_2)_n$-COOH] SAMs on Ag [135]. It was partially assisted with the occupation of the free sixth coordination site by a water molecule or by a new ligand, most likely His-33, to give a $B_2$ six-coordinated high-spin (6cHS) or a six-coordinated low-spin (6cLS) configuration. For 11 and more ethylene groups in the alkyl chain, there was no contribution from 5cHS and other substate forms (only native protein denoted as $B_1$ in 6cLS His/Met state was detected), while for $n < 11$ the content of new conformational states (represented as $B_2$) increased with the alkyl chain length decrease. Relative contribution of 5cHS in the sum of $B_1$ and $B_2$ states satisfied the same relationship; however, for $n = 6$ the 5cHS conformation was the only nonnative substate, while for $n = 3$ the 6cLS appeared and for $n = 2$ in addition 5cLS was present. These structural changes were claimed to be induced by the electric field at the anionic binding site of the monolayer. Obviously, the strength of the electric field will be larger for a shorter distance between the electrode and the SAM binding site, which is responsible for the observed transitional changes. The postulated mechanism of these is that electrostatic binding salt bridges formed between lysine residues of the cyt *c* binding site and the anionic groups of the linkage monolayer support the interfacial electric field and perturb the structure of the heme pocket, which eventually results in dissociation of the Met-80 bond from the heme iron [98]. Certainly, the strength of the electric field provided by the alkanethiol linker and its end group will also affect the ET properties of the attached cyt *c*, but a closer look at these aspects with SERRS spectroscopy will be given in following sections.

**Influence of Monolayer Composition on Protein Orientation and Its Effect on ET Properties** SERRS measurement for cyt *c* attached electrostatically to mixed SAMs of 2- mercaptoethanosulfonate [123] and 2-mercaptoethanol (MEL), formed through co-adsorption on electrochemically roughened Ag were conducted by our group [156]. Spectra were taken both in situ (in the presence of excessive protein) and ex situ (after removal from cyt *c* solution and rinsing with buffer) and for two cyt *c* concentrations of $10^{-4}$ M and $10^{-6}$ M at pH 7.0. The concentration in a range of micromoles is satisfactory to get a complete monolayer and sufficient to obtain a SERRS signal exclusively for adsorbed cyt *c*, but a part of our studies was related to that reported in the literature of a similar system of gold substrate, where cyt *c* concentration in a range of $10^{-4}$ M was used [157]. Moreover, this range of concentration is often used for electrochemical studies of cyt *c*, which is what spiked our interest to examine the properties of protein adsorbed under these conditions. We were aware that especially for in situ experiment a contribution of the RR signal of the solution protein to the SERRS spectrum cannot be totally neglected. On the other hand, the intensity pattern we observed was changed for SERRS experiments with $10^{-4}$ M cyt *c* in comparison to a volumetric sample (data not shown here), and the relative intensities in the SERRS spectra of cyt *c* were affected by the monolayer composition. For these two reasons we are convinced that spectra taken in situ for $10^{-4}$ M were dominated by signals of SAM-bound protein.

The SERRS spectra collected ex situ for cyt immobilized electrostatically using $1 \times 10^{-4}$ M aqueous solution of pH 7 on binary MES/MEL SAMs are shown in Figure 5.26. Alteration of heme orientation, following changes of the monolayer composition was deduced from the dependence of the intensity ratio of $B_{1g}$ to $A_{1g}$ bands of cyt *c* on SAM composition, reproduced in Figure 5.27*a*. A lower molar fraction of MES in solution resulted in more perpendicular orientation of heme with respect to the metal surface (indicated by increasing the $B_{1g}$ to $A_{1g}$ ratio), which should facilitate the electron transfer. The conclusions made for the silver surface correlate very well with the results obtained for gold coated with mixed MES/MEL SAM by Imabayashi et al. [157], who observed increasing $k_{ET}$ (ET rate) on the SAMs, with the increasing contribution of the MEL component (see Fig. 5.27*b* to compare with the trend they observed). Hence, too high a negative charge provided by linkage monolayer can deteriorate ET kinetics of tethered cyt *c*. Similar findings for the in situ experiment with use of $10^{-6}$ M cyt *c* solution prove that control of the

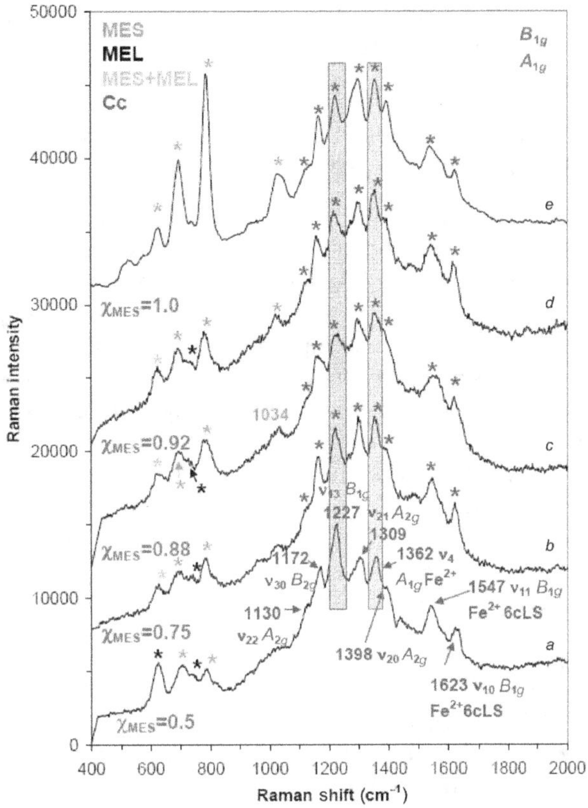


**Figure 5.26** Ex situ SERRS spectra ($\lambda_{exc}$ = 514.5 nm) of $10^{-4}$ M cyt *c* of pH 7 anchored electrostatically to mixed MES–MEL SAMs formed from ethanolic solutions ($c_{total}$ = 1 mM) with varying MES mole fraction in solution ($\chi_{MES}^{sol}$) on roughened Ag at pH = 7.0. Light blue and light green bars mark the $B_{1g}$ and $A_{1g}$ modes [156].

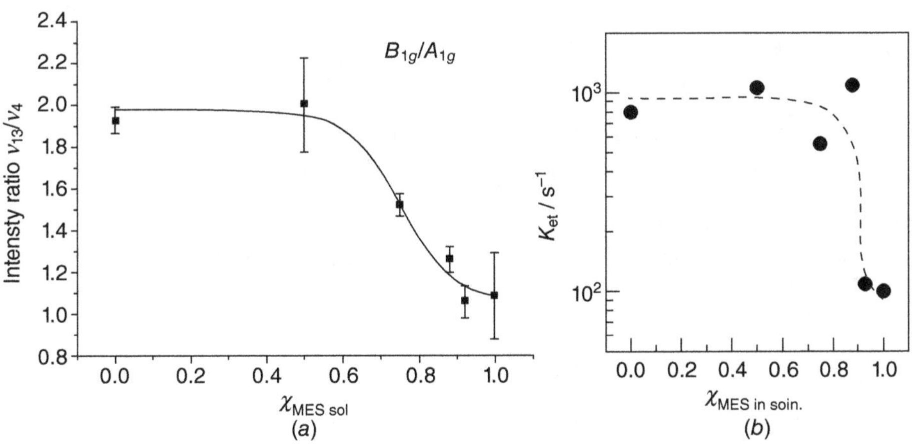


**Figure 5.27** (*a*) Intensity ratio of $\nu_{13}(B_{1g})$ to $\nu_4(A_{1g})$ SERRS modes of cyt *c* vs. MES mole fraction in solution ($\chi_{MES}^{sol}$) used for preparation of mixed MES–MEL SAMs on roughened Ag (ex situ experiment with use of $10^{-4}$ cyt *c* solution) [156]. (*b*) ET rate ($k_{ET}$) of cyt *c* vs. MES mole fraction in solution used to form mixed MES–MEL SAMs on Au(111) [157].

surface composition of the linkage monolayer allows to govern the redox center orientation and hence the ET process.

### *Effect of Excessive Amount of Bulk Protein on Heme Electronic Structure*

Moreover, during of the SERRS experiments described in the previous section, a comparison of ex situ and in situ SERRS spectra of cyt *c* ($10^{-4}$ M solution) attached to MES/MEL-coated Ag showed differences of the heme state [156]. Oxidation of iron and the conversion of native six-coordinated low spin (6cLS) to five-coordinated high spin (5cHS) under in situ conditions was ascribed by our group, respectively, to the presence of surface $Ag_4^+$ species and/or sandwiching protein multilayer by silver clusters. Changes within the heme electronic structure were eliminated during in situ experiments with $10^{-6}$ M cyt *c* solution. These results show the importance of protein multilayer formation for the properties of the electroactive first layer. Demonstration of serious differences between the heme properties (oxidation and configuration state, coordination state of iron) of adsorbed cyt *c* in the presence and absence of bulk protein is an important result for the researchers working with the cyt *c* immobilized on alkanethiolate SAMs, as many electrochemical experiments are conducted under one of these two regimes.

## 5.3.5 Results of SERRS Studies of Redox and ET Properties of Cyt *c* Tethered to Alkanethiol-Coated Metal Substrates

Discussing the EC-SERRS studies on cyt *c* attached to the biomimetic interphase, here metal (Ag or Ag–Au hybrids) coated with functionalized alkanethiolate SAMs, one must distinguish two types of the EC-SERRS experiments, which can be performed and provide different information on a studied system. These are:

- Stationary potential-dependent SERRS spectroscopy
- Time-resolved potential-jump SERRS spectroscopy

Within this section their principles, supplied information, and primarily insight provided by their means on ET properties of cyt *c* attached to coated electrodes will be presented.

### *Stationary Potential SERRS Measurements*

This type of SERRS measurements involves the collection of SERRS spectra, measured at different stationary potentials. As already discussed above, the SERRS spectrum is sensitive to structural and/or electronic properties of the immobilized protein. Therefore, the following information can be deduced from stationary potential SERRS:

- Heme structure
- Conformational and redox equilibria
- Orientational distribution

An example of the potential induced, studies examining a change of the protein redox state of orientational distribution of the heme group, derived from $B_{1g}$ to $A_{1g}$ SERRS modes, the intensity ratio was already given above (see Fig. 5.23).

Nevertheless, probably the most interesting topic to study with stationary potential EC-SERRS is probing the conformational and redox equilibria of cyt *c* attached to coated metal electrodes and studies of the effect of alkanethiolate SAM structural properties on them. Such an SERRS examination relies usually on spectroelectrochemical titrations, which results in quantification of unknown relative contribution of a given redox and/or conformational form of cyt *c* heme, presumed from SERRS spectra, as a function of the potential. To this end, the careful selection of the SERRS marker bands and cautious fitting the experimental data to extract the spectral components indicative of a given redox and/or conformational state are the key steps. And example of the titration curves obtained by Hildebrant's group [140] showing the potential-dependent relative concentration of oxidized (blue curve) and reduced cyt *c* (red curve) and corresponding stationary EC-SERRS spectra, displaying the details of the fitting procedure are presented in Figure 5.28.

When plotting potential versus log [Red]/[Ox], one can validate the Nernstian behavior of the studied system, and the apparent redox potential of the bound cyt *c* ($E_{app}$) can be determined from the interception. An exemplary set of data showing both analyzed stationary potential EC-SERRS spectra and obtained Nernst plot are shown in Figure 5.29. More systematic SERRS study on how the thickness of the linkage SAM, achieved by use of the ω-carboxyl-terminated alkanethiols $C_n$-SAMs with varying *n*, being the number of methylene groups [general formula:

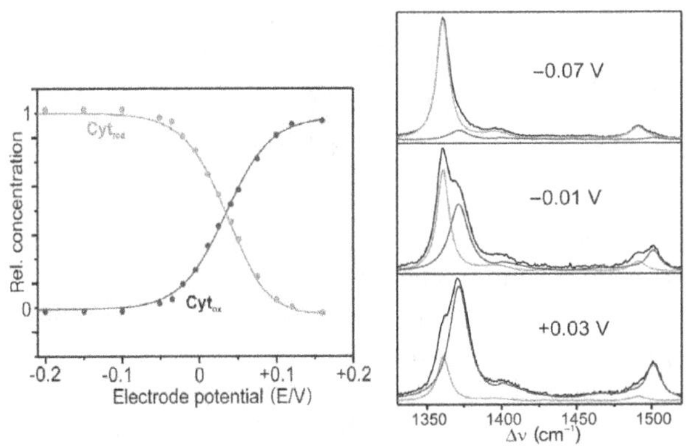


**Figure 5.28** (*Left*) Redox titration curves obtained bases on SERRS spectra of cyt *c* shown in the right panel. (*Right*) Experimental SERRS spectra (excited with 413 nm) of cyt *c* adsorbed on coated electrodes at different potentials. The titration curves and component spectra of the various redox state species are given in different colors: blue, $cyt_{Ox}$; red, $cyt_{Red}$ [140].

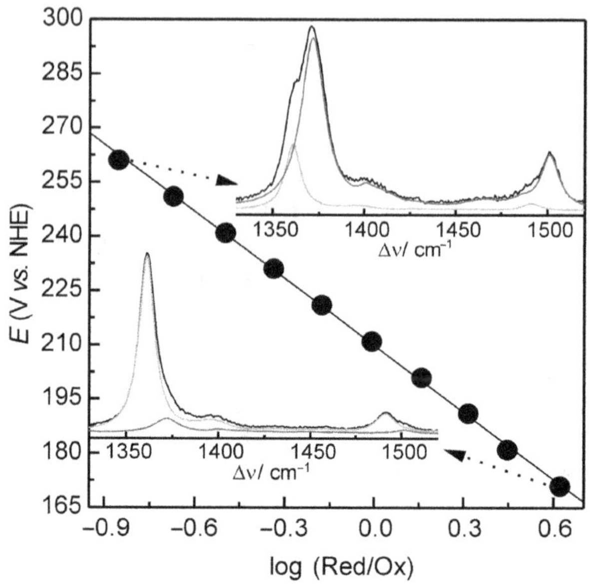


**Figure 5.29** Nernst plot obtained from the quantitative treatment of the SERRS spectra (excited with 413 nm) of cyt *c* on alkanethiol SAM coated Ag, as a function of the electrode potential. The component spectra of the various redox state species are given with solid ($cyt_{Red}$) and dotted ($cyt_{Ox}$) lines [158].

SH-($CH_2$)COOH], on Ag affects the determined apparent redox potentials ($E_{app}$) of bound electrostatically cyt *c* was performed by Murgida and Hildebrandt [135]. The value of $E_{app}$ for $B_1$ species displayed a steady downshift from 244 mV (versus NHE; normal hydrogen electrode) for $n$ = 1 to 202 mV for $n$ = 15. When comparing a potential drop across a monolayer, corresponding to the potential difference given by $E_{app} - E^0$ and further denoted as $E_{RC}$ to potential value for cyt *c* in solution ($E_s^0 = 257$ mV versus NHE [85]), one obtains

$$E_{RC} = E - E_S^0 + \frac{RT}{nF}\ln\left(\frac{C_{B_1(\text{red})}}{C_{B_1(\text{ox})}}\right) \tag{5.9}$$

where $R$, $T$, $n$, and $F$ denote gas constant, temperature, number of electrons involved in ET, and Faraday constant, respectively. It was shown that both $E_{RC}$ and $E_{app}$ decrease nonlinearly with the length of the SAM [135]. It was also demonstrated that these shifts reflect only the distance-dependent interfacial potential drop, which can be quantitatively described by a simple electrostatic model presented already above, dealing with the effect of electric field provided by the alkanethiol SAM. In contrast to electrostatic adsorption, covalently tethered cyt *c* ($B_1$) exhibits a nonideal electrochemical behavior [159]. Most possibly, the cross-linking reactions engage various lysine residues leading to a distribution of orientations that may result in electrochemically less active or inactive species.

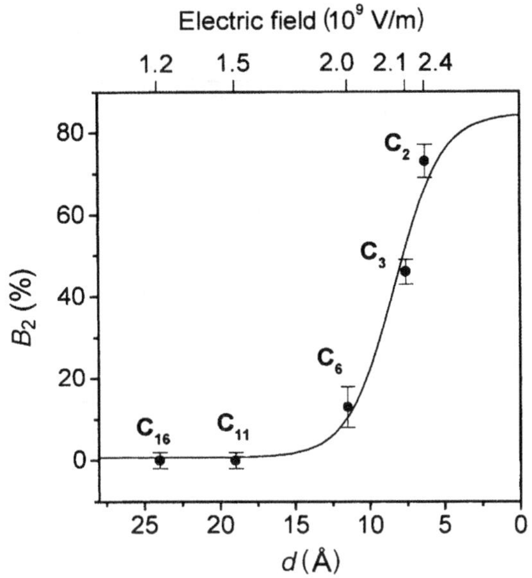


**Figure 5.30** Relative concentration of the $B_2$ state of ferric cyt $c$ as a function of the SH–$C_n$–COOH chain length and the electric field strength, calculated assuming electrostatic model [135].

EC-SERRS gives also insight into conformational equilibrium of cyt $c$ attached to different ω-substituted alkanethiol SAMs. The conformational changes involve mostly the appearance of nonnative $B_2$ states, which involves the loss of the heme axial Met-80 ligand, leading to an equilibrium between a five-coordinated and a six-coordinated species in which this axial coordination site remains vacant or is occupied by a histidine (His-33 or His-26). It was evidenced that both type of terminal group and length of aliphatic strongly influence the content of the $B_2$ species.

On carboxyl-terminated $C_n$-SAMs, the total amount of $B_2$ species measured at $E_{app}$ increases from 0% for $n = 15$ and 10 up to 75% for $n = 1$ (see Fig. 5.30). This finding can be understood on the basis of an electrostatic model that assumes a linear variation of the potential across the SAM and up to the redox site and a Gouy–Chapman distribution for the region beyond the redox site [135].

Assuming a constant ionic strength, the magnitude of the electric field sensed by the adsorbed cyt $c$ increases with the shift of the electrode potential relative to the potential of zero charge, increased charge density on the SAM surface (dependent on SAM terminal group), and shortening of the thiol aliphatic chain. Depending on these parameters, the magnitude of the field may vary between $10^8$ and $3 \times 10^9$ V m$^{-1}$. The plot in Figure 5.30 shows that conformational $B_2/B_1$ equilibrium sensitively depends on the electric field experienced by the adsorbed cyt $c$. Even larger portions of oxidized $B_2$ species than these for short carboxylate-terminated SAMs are found on phosphate and sulfate-terminated monolayers [160], which is in good agreement with a predicted electric field strength influence.

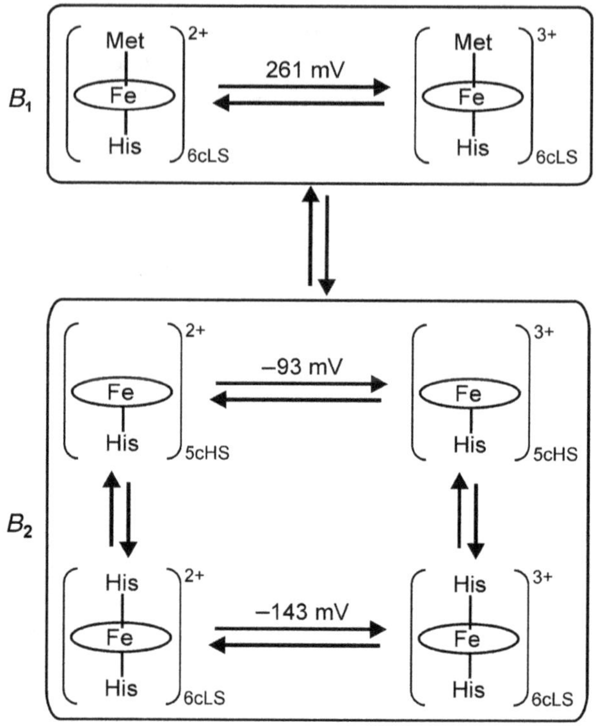


**Figure 5.31** Schematic representation of the redox and conformational equilibria of cyt *c* in electrostatic complexes [119, 135]. Reprinted from [158].

Moreover, redox and structural equilibrium of cyt *c* in electrostatic complexes was determined by Hilderbrandt's group from stationary potential EC-SERRS spectra [119, 135]. Certainly, it employed again the component analysis, to determine quantitatively potential-dependent equilibrium of the reduced and oxidized $B_1$ and $B_2$ forms. The results of this analysis shown in Figure 5.31 demonstrate that these conformational changes are reversible, which indicates they do not involve a protein denaturation. Due to considerable structural changes within the heme pocket and the ligand exchange, the redox potentials of the $B_2$ species are downshifted up to about 400 mV with respect to the native 6cLS $B_1$ conformation. Interestingly, the reduced form of the immobilized protein shifts the equilibrium nearly quantitatively toward the $B_1$ form, independent of the chain length of the alkanethiol SAM linker. The transition to the $B_2$ species, accompanied by the strong downshift of the redox potential, occurs also when cyt *c* binds to liposomes of negatively charged phospholipids, in particular at low protein–lipid ratios equivalent to high local electric fields [161]. Accordingly, it is likely that the $B_2$ state may also be formed when cyt *c* binds to the inner mitochondrial membrane, including its main lipid component, the anionic cardiolipin. Under physiological conditions, this conformational transition would be related to a changed protein function, since due to the low redox potential, state $B_2$ cannot be any more reduced by complex III, eliminating the cyt *c* ability to serve as an electron carrier. On the other hand, loss of the Met-80 ligand results in strongly increased peroxidase activity [162], which may account for the cyt *c* catalytic properties toward peroxidation of cardiolipinand and the resultant increased permeability of the inner mitochondrial membrane [163]. This process is believed to be functional for the release of cytochrome *c* to the cytosol, where the protein may be involved in caspase-dependent apoptosis [164]. As the local electric field promotes the conformational conversion to the $B_2$ state, exhibiting very low redox potential and increased peroxidase activity, it may be the factor triggering the switch of cyt *c* from redox to peroxidase function under physiological conditions.

### *Time-Resolved Potential-Jump SERRS Spectroscopy*

To investigate the dynamics of the heme group of the cyt *c* during an interfacial ET process and provide kinetic data, it is essential to probe within a time scale of this process. Therefore, the use a time-resolved EC-SERRS spectroscopy (TR EC-SERRS) is necessary to examine the ET dynamics. A pioneering work in this field for the TR EC-SERRS studies of ET reaction of cyt *c* attached to silver electrodes was done by Hildebrandt, who introduced the concept of the time-resolved potential-jump (EC) SERRS spectroscopy [99]. In this approach, the redox-linked processes are provoked by a rapid potential jump and further monitored by SERRS spectroscopy after a delay time $\delta$ for the probe interval $\Delta t$. The synchronization is possible due to a mechanical chopper, triggering the potential jump via a photodiode and hence gating the exciting continuous-wave (CW) laser beam. After the probe event, the potential is reset to its initial value ($E_i$). The following relaxation processes that restore thermodynamic equilibrium at the final potential ($E_f$) are eventually probed by SERRS. Restoration of disturbed initial equilibrium allows a continuous repetition of the sequence of potential jumps and probe events. Use of a CW laser, together with a rotating electrode as SERRS substrate, enables the avoidance of unwanted photo-induced degradation and desorption processes of sensitive biological material (cyt *c*), quite likely when pulsed laser excitation and stationary electrodes are used. The scheme showing the principles of the TR SERRS experiment, SERRS spectra corresponding to the potential jump and the analyzed data after varying delay times are shown in Figure 5.32. Original setup was further modified by the Hildebrandt and Murgida group in two aspects [165]. First, a mechanical chopper was replaced with an electronically controlled intensity modulator equipped with a pulse amplifier, which allowed fine-tuning of $\Delta t$ down to the nanosecond time scale. Second, synchronization and timing of the potential jumps, delay times $\delta$, and measuring intervals were achieved with the use of a multichannel pulse delay generator. Further improvement by Kranich et al. [166] involving use of two consecutive laser intensity modulators resulted in obtaining a response time of 20 ns.

Summarizing, in the TR EC-SERRS experiment, potential jumps of variable height and duration are applied to trigger the ET reaction. Next, the SERRS spectra are taken after variable delay times following each jump. This provides an insight into:

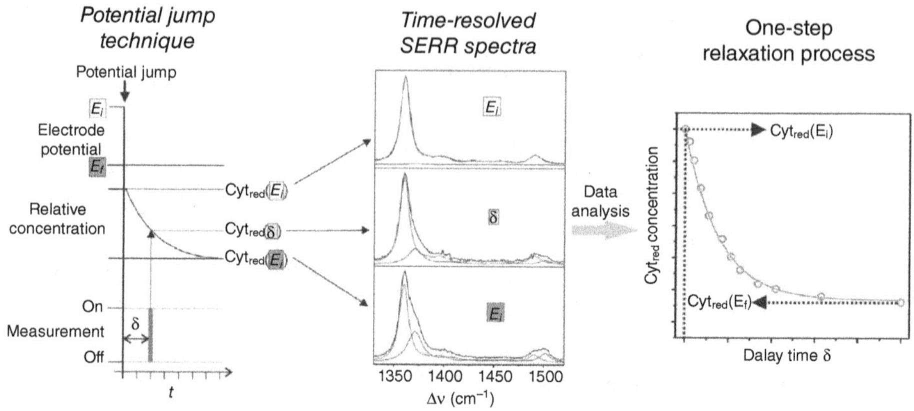


**Figure 5.32** Scheme of the time-resolved SERRS experiment for a potential jump from reductive to oxidizing potential [167]. (*Left*) Temporal relationship between potential jump, concentration profile, and data accumulation. (*Middle*) SERRS spectra taken at the initial potential $E_i$ (*top*), the final potential $E_f$ (*bottom*), and after a delay time δ following the potential jump from $E_i$ to $E_f$. The red and blue lines refer to the component spectra of the reduced and oxidized cyt *c*, respectively. (*Right*) Results of the analysis showing the relaxation of the reduced cyt *c* following the potential jump.

- ET process kinetics
- Protein dynamics
- Redox center reorientation

Depending whether one is interested in redox and conformational equilibria or heme group reorientation disentangling, one should use, respectively, the Soret band (413 nm) or the Q-band (514.5 nm) excited TR SERRS in order to get the required information. The fitting procedure of the individual species to the measured spectra involves obligatory background subtraction and treatment by single-band (514.5 nm) or component analysis (413 nm) [168]. The time evolutions of the SERRS marker band of the oxidized state of cyt *c* (measured with the 413 nm) and $B_{1g}$ to $A_{1g}$ modes of ferric cyt *c* (measured with 514.5 nm), monitoring, respectively, the kinetics of protein reduction and reorientation (as potential jump was from $E_i$ = 250 mV to $E_f$ = 150 mV versus NHE) for cyt *c* adsorbed electrostatically at carboxyl-terminated $C_{15}$-SAM- and $C_5$-SAM-coated silver are presented in Figures 5.33*a*, *b*. At SAMs of mercaptocarboxylic containing 15 methylene groups in the alkyl chain ($C_{15}$-SAM), potential jump-induced protein reorientation, as probed by TR SERRS experiments with Q-band excitation, is much faster than electron tunneling (compare main and inset curves in Fig. 5.33*a*).

When decreasing the number of methylene groups, the rate of protein reorientation decreases until it approaches the rate of electron tunneling, which takes place for mercaptohexanoic acid ($C_5$-SAM), as can be seen in Figure 5.33*b*. Therefore, two different regimes for the interfacial electron transfer of the attached cyt *c* should be distinguished:

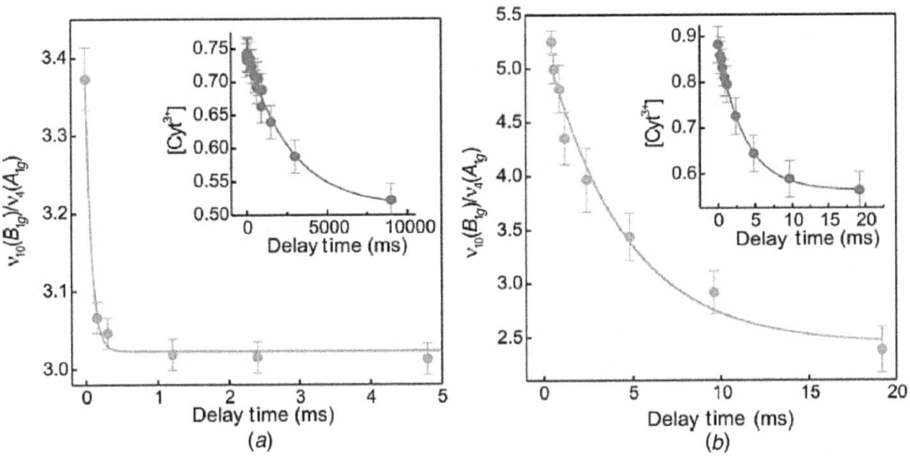


**Figure 5.33** Time evolution of the SERRS $\nu_{10}(B_{1g})/\nu_4(A_{1g})$ intensity ratio for ferric cyt adsorbed at pH = 7.0 on (*a*) $C_{15}$ and (*b*) $C_5$ COOH terminated SAMs after a potential jump from 250 to 150 mV (vs. NHE). SERRS spectra were excited with 514.5 nm [166]. The insets show the kinetics of cyt reduction on the same SAMs as monitored by TR SERRS with 413 nm excitation.

- At long distances, for carboxyl-terminated alkanethiol SAMs with 10 or more $CH_2$ groups, ET is controlled exclusively by electron tunneling.
- At distances shorter than for $n$ = 10, orientational changes of the tethtered protein appear to be rate limiting.

The more precise analysis of the derived TR SERRS spectra allows extracting the kinetic data, when one assumes a correct one-step relaxation process, according to the experimental conditions and results. Usually, the initial and final potential values correspond to the heme redox process, and next the spectral signatures indicative of individual species should be distinguished and analyzed. Hence, for example, when only $B_1$ species were detected in the SERRS spectra, the one-step relaxation being the ET of the native state $B_1$ of the adsorbed cyt *c* can be examined. Then one obtains [165]

$$\frac{\Delta[B_{1\text{red}}]_{\delta'}}{\Delta[B_{1\text{red}}]_0} = \exp\left(\frac{-t}{\tau}\right) \tag{5.10}$$

with $\Delta[B_{1\text{red}}]_\delta$ and $\Delta[B_{1\text{red}}]_0$ denoting concentration differences of the reduced form of $B_1$ with respect to the equilibrium values at the final potential $E_f$ for $\delta' = \delta + \Delta t/2$ and $t = 0$, respectively. The $\tau$ (sometimes referred to as $k_{\text{relax}}$) stands for the relaxation constant. If the final potential was set equal to the redox potential so that the relative equilibrium concentration at $E_f$ of $B_{1\text{red}}$ is 0.5 and the ET rate constants ($k_{\text{ET}}$) for oxidation and reduction are identical, then one acquires:

$$\tau = \frac{1}{2k_{\text{ET}}} \tag{5.11}$$

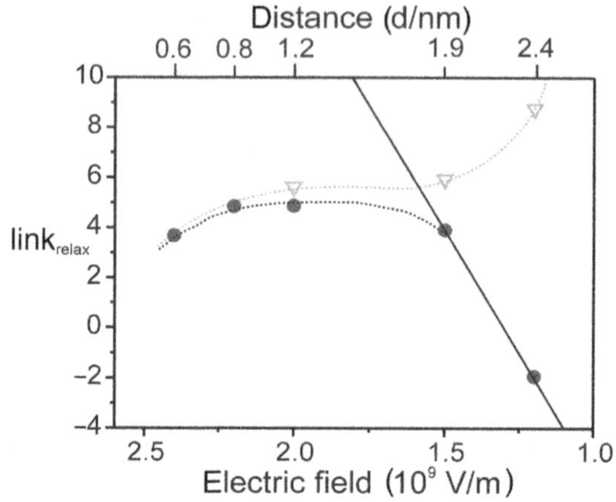


**Figure 5.34** Rate constants for reorientation (triangles) and reduction (circles) of cyt *c* immobilized on Ag electrodes coated with COOH-terminated SAMs of different chain lengths, determined by TR SERRS spectroscopy [86]. The upper and bottom axis indicate, respectively, the distance from the electrode surface and the electric field strength at the SAM–cyt *c* interface, as estimated on the basis of an electrostatic model. The straight line represents the exponential distance dependence of electron tunneling, extrapolated from rate constants determined for $C_{15}$-SAM and $C_{10}$-SAM. The dotted lines are to guide the eyes. Reprinted from [167].

Thus, when plotting $\ln(\Delta[B_{1\text{red}}]_{t=\delta}/\Delta[B_{1\text{red}}]_{t=0})$ as a function of time, the relaxation constant, hence the related $k_{ET}$, can be readily determined.

When plotting rates of reorientation and reduction (determined with TR SERRS) for cyt *c* immobilized at Ag modified with carboxyl-terminated alkanethiol SAMS of varying chain length (see Fig. 5.34, upper axis), one can see again more clearly that at a short distance from the electrode, the orientation changes of the immobilized protein slow down dramatically and become rate limiting, while for the thick SAMs the measured ET rates are determined by electron tunneling probabilities [167]. Again, applying the simple electrostatic model, the distance dependence of the interfacial electric field can be predicted, and, as can be seen from relaxation constants dependence (see Fig. 5.34, bottom axis), the interfacial ET process of immobilized cyt *c* may be alternatively classified in terms of a low-field (long distances) and a high-field (short distances) regime (upper axis).

An expanded view of the apparent ET rates for the $B_1$ state of cyt *c* determined with the help of TR SERRS for coatings different from the carboxyl-terminated SAMs, shown in Figure 5.35, yields a uniform picture of the heterogeneous redox reaction controlled by electron tunneling at long distances (low field) and by electric-field-dependent reorientation at short distances (high field). For $SO_4$–Ag and $PO_3$–SAM coatings, which provide a considerably increased amount of $B_2$ species, the apparent ET rates are much lower than for COOH-terminated coatings at comparable distances. What is worth emphasizing is that the overpotential and temperature dependence TR SERRS measurements for COOH-terminated $C_{15}$-SAM on silver are in good agreement with the Marcus theory for long-range electron tunneling [169].

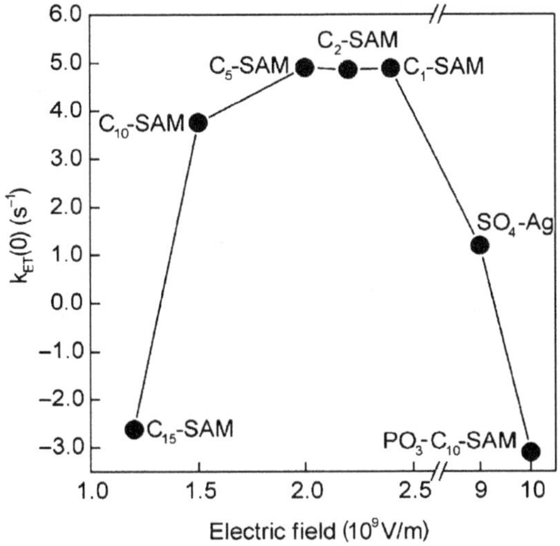


**Figure 5.35** Apparent rate constant of ET at zero driving force as a function of the electric field strength for cyt *c* adsorbed on Ag electrodes with different coatings. SAM denotes the alkanethiol linker with a different functionalization, while $C_n$ refers to the carboxyl-terminated SAMs [158].

The derived reorganization energy from these studies is distinctly lower than that for cyt *c* in solution [170], demonstrating a strongly reduced contribution of the solvent reorganization for the immobilized protein [169]. However, when the substrate was changed from Ag to hybrid Au–Ag (see Section 5.3.3 for details), much weaker overpotential dependence was found [171], providing physically meaningless low reorganization energy values. In an attempt to resolve this discrepancy, it was proposed that also under low-field regime there is an electric field effect on the free-energy term controlling the tunneling rate. Under the assumption of a local electric field strength being proportional to the difference between actual electrode potential and metallic system-specific pzc, an empirical linear correction was included in a free-energy term of the Marcus equation, resulting in electric-field-dependent retardation of electron tunneling. This allowed a reasonable explanation of the inconsistency of ET rates overpotential dependencies on Ag and Au–Ag hybrid observed with TR SERRS [171]. Metal-specific overpotential dependencies and modified Marcus equation describing their behavior are shown in Figure 5.36.

## *Conclusions*

Gathering all the information provided by stationary potential and time-resolved SERRS measurements of cyt *c* for Ag electrodes coated with COOH-terminated alkanethiol SAMs summarized in the preceding sections, one can derive the simplified scheme of the ET reaction between cyt *c* and the corresponding oxidase embedded into the mitochondrial membrane, given in Figure 5.37. The first step is the

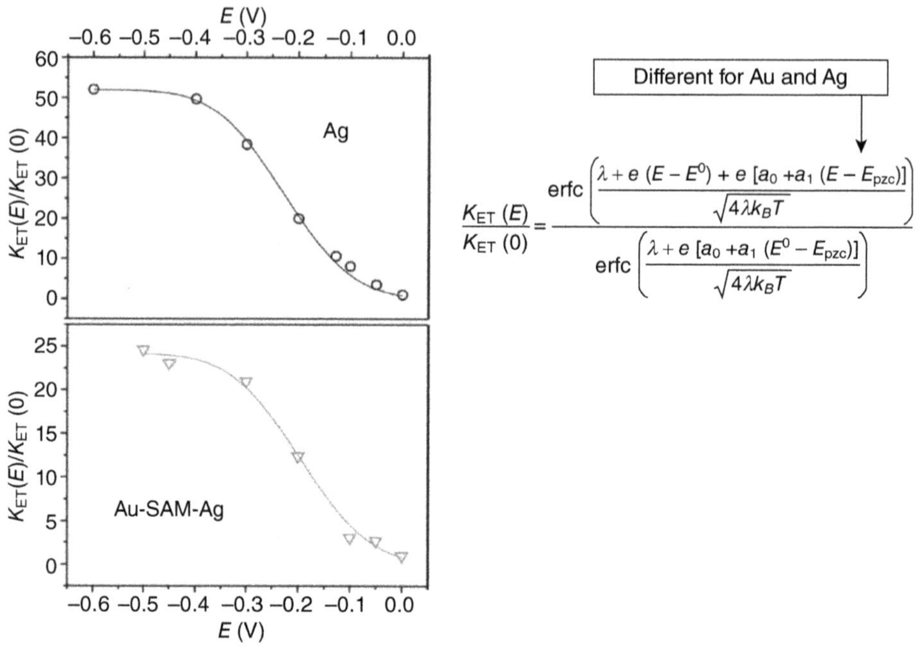


**Figure 5.36** Overpotential dependencies of electron tunneling rates for cyt *c* attached to COOH-terminated $C_{15}$ SAM on Ag (*top left*) and hybrid Au–Ag (*bottom left*). (*Right panel*) Modified Marcus equation including metal-specific electric-field-dependent retardation of electron tunneling [171].

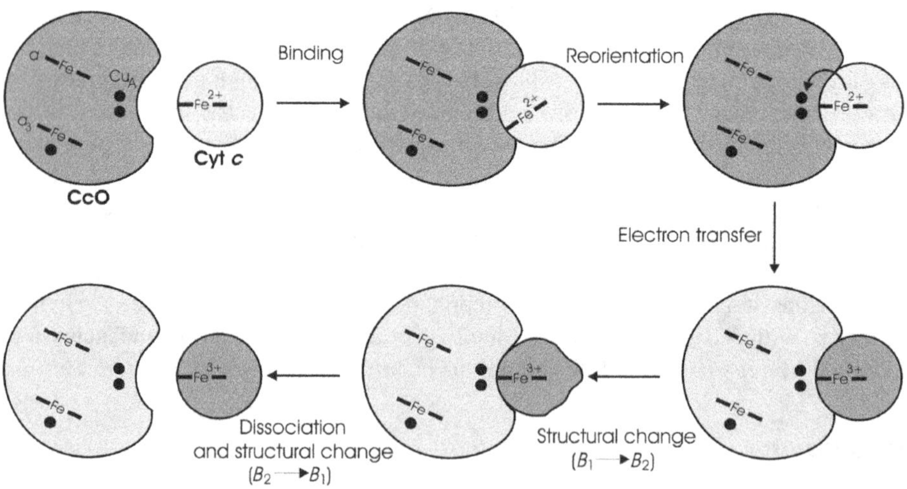


**Figure 5.37** Schematic representation of the proposed mechanism of electron-transfer reaction between cyt *c* and cytochrome *c* oxidase (CcO) [115].

formation of an electrostatic complex via the positively charged area of cyt *c* around the partially exposed redox center and the negatively charged domain of the oxidase. This electrostatic complex is not necessarily optimized for the electron transfer from cyt *c* to the $Cu_A$ (binuclear copper center), the primary electron acceptor of the enzyme. Hence, the second step of the mechanism may involve the reorganization of the electrostatic complex (reorientation of the heme redox center). At a sufficiently low local electric field, this reorganization will proceed rapidly, resulting in efficient electron tunneling. The charge-transfer reaction is fast despite the long distance and extremely low driving force (close to 0 eV) due to the considerably decreased reorganization energy in the electrostatic complex, as suggested by TR SERRS results. The close to zero driving force for the forward redox reaction implies the high probability of efficiently proceeding, competing back electron transfer. On the other hand, oxidation of cyt *c* at low electric fields may trigger a fast conformational transition to the $B_2$ form, whose redox potential is lowered enough to prevent re-reduction of the carrier. The terminal step would be the dissociation of cyt *c* from the binding site, which returns rapidly to the $B_1$ state in a released form in solution. This mechanism would secure an efficient and unidirectional interprotein reaction.

## 5.4 SUMMARY

In this chapter (EC)-SERRS spectroscopy was presented as a powerful tool for the study of redox-active proteins immobilized on metallic substrates. The information provided is limited not only to Faradaic processes, as for traditional electrochemical methods, but it additionally offers simultaneous insight into structural and orientational parameters, also under a dynamic range (time-resolved SERRS technique). Presented are the results for the cytochrome *c* (cyt *c*), being a model redox protein, demonstrated an ability of EC-SERRS technique to monitor protein redox activity, solve the heterogeneous interfacial electron transfer (ET), and indicate the parameters governing the ET process. For cyt *c* attached to chemically modified metal supports, the strength of the electric field at the interface is the key parameter controlling the ET properties. Its second function is switching the protein function from the redox to the peroxidase one, the latter being one of the key events preceding apoptosis. Wishing to study the electrical performance of the metal supports with different immobilized redox biomolecules, one has to carefully choose a SER(R) S-active substrate, metal type, biomimicking coating of the metal support, and adjust the spectroelectrochemical experiment conditions to the particular needs, utilizing the flexibility of the EC-SERRS spectroscopy technique.

## REFERENCES


1. Fleischmann, M.; Hendra, P.; McQuillan, A. Raman spectra of pyridine adsorbed at a silver electrode, *Chem. Phys. Lett.* **1974**, *26*, 163.
2. Jeanmarie, D. I.; Van Duyne, R. P. Surface Raman spectroelectrochemistry: Part I. Heterocyclic, aromatic, and aliphatic amines adsorbed on the anodized silver electrode, *J. Electroanal. Chem.* **1977**, *84*, 1.

3. Albrecht, M. G.; Creighton, J. A. Anomalously intense Raman spectra of pyridine at a silver electrode, *J. Am. Chem. Soc.* **1977**, *99*, 5215.
4. Moskovits, M. Surface-enhanced spectroscopy, *Rev. Mod. Phys.* **1985**, *57*, 783.
5. Kneipp, K.; Kneipp, H.; Itzkan, I.; Dasari, R. R.; Feld, M. S. Surface-enhanced Raman scattering: A new tool for biomedical spectroscopy, *Curr. Sci.* **1999**, *77*, 915.
6. Stuart, D. A.; Yonzon, C. R.; Zhang, X. Y.; Lyandres, O.; Shah, N. C. Glucose sensing using near-infrared surface-enhanced Raman spectroscopy: Gold surfaces, 10-day stability, and improved accuracy, *Anal. Chem.* **2005**, *77*, 4013.
7. Kneipp, K.; Kneipp, H.; Itzkan, I.; Dasari, R. R.; Feld, M. S. Surface-enhanced Raman scattering and biophysics, *J. Phys.: Condens. Matter* **2002**, *14*, R597.
8. Hutter, E.; Fendler, J. H. Exploitation of localized surface plasmon resonance, *Adv. Mater.* **2004**, *16*, 1685.
9. Underwood, S.; Mulvaney, P. Effect of the solution refractive index on the color of gold colloids, *Langmuir* **1994**, *10*, 3427.
10. Hao, E.; Schatz, G. C. Electromagnetic fields around silver nanoparticles and dimers, *J. Chem. Phys.* **2004**, *120*, 357.
11. Haynes, C. L.; McFarland, A. D.; Van Duyne, R. P. Surface-enhanced Raman spectroscopy, *Anal. Chem. (A-pages)* **2005**, *77*, 338A.
12. Schatz, G. C.; Van Duyne, R. P. *Handbook of Vibrational Spectroscopy*, J. M. Chalmers and P. R. Griffiths, Eds., Wiley: New York, **2002**, 759.
13. Arenas, J. F.; Castro, J. L.; Otero, J. C.; Marcos, J. I. SERS of dicarboxylic acids on silver sols, *J. Raman Spectrosc.* **1995**, *26*, 637.
14. Otto, A. Theory of first layer and single molecule surface enhanced Raman scattering (SERS), *Phys. Status Solidi (A)* **2001**, *188*, 1455.
15. Michaels, A. M.; Nirmal, M.; Brus, L. E. Surface enhanced Raman spectroscopy of individual rhodamine 6G molecules on large Ag nanocrystals, *J. Am. Chem. Soc.* **1999**, *121*, 9932.
16. Campion, A.; Kambhampati, P. Surface-enhanced Raman scattering, *Chem. Soc. Rev.* **1998**, *27*, 241.
17. Lombardi, J. R.; Birke, R. L.; Sanchez, L. A.; Bernard, I.; Sun, S. C. The effect of molecular structure on voltage induced shifts of charge transfer excitation in surface enhanced Raman scattering, *Chem. Phys. Lett.* **1984**, *104*, 240.
18. McFarland, A. D.; Young, M. A.; Dieringer, J. A.; Van Duyne, R. P. Wavelength-scanned surface-enhanced Raman excitation spectroscopy, *J. Phys. Chem. B* **2005**, *109*, 11279.
19. Moskovits, M.; Suh, J. S. Surface selection rules for surface-enhanced Raman spectroscopy: Calculations and application to the surface-enhanced Raman spectrum of phthalazine on silver, *J. Phys. Chem.* **1984**, *88*, 5526.
20. Tian, Z.-Q.; Ren, B.; Wu, D.-Y. Surface-enhanced Raman scattering: From noble to transition metals and from rough surfaces to ordered nanostructures, *J. Phys. Chem. B* **2002**, *106*, 9463.
21. Chase, B.; Parkinson, B. A study of the wavelength and potential dependence of surface enhanced Raman scattering on copper, silver, and gold electrodes, *J. Phys. Chem.* **1991**, *95*, 7810.
22. Chen, L.-Y.; Yu, J.-S.; Fujita, T.; Chen, M.-W. Nanoporous copper with tunable nanoporosity for SERS applications, *Adv. Funct. Mater.* **2009**, *19*, 1.
23. Aroca, R.; Jennings, C.; Kovac, G. J.; Loutfy, R. O.; Vincett, P. S. Surface-enhanced Raman sçattering of Langmuir-Blodgett monolayers of phthalocyanine by indium and silver island films, *J. Phys. Chem.* **1985**, *89*, 4051.
24. Creighton, J. A.; Blatchford, C. G.; Albrecht, M. G. Plasma resonance enhancement of Raman scattering by pyridine adsorbed on silver or gold sol particles of size comparable to the excitation wavelength, *J. Chem. Soc.* **1979**, *75*, 790.
25. Brett, C. M. A.; Brett, A. M. O. *Electrochemistry. Principles, Methods and Applications*, Oxford: Oxford University Press, **1998**, Chapter 3.
26. Oklejas, V.; Sjostrom, C.; Harris, J. M. SERS detection of the vibrational stark effect from nitrile-terminated SAMs to probe electric fields in the diffuse double-layer, *J. Am. Chem. Soc.* **2002**, *124*, 2408.
27. Wu, D-Y.; Li, J.-F.; Ren, B.; Tian, Z.-Q. Electrochemical surface-enhanced Raman spectroscopy of nanostructures, *Chem. Soc. Rev.* **2008**, *37*, 1025.

28. Foresti, M. L.; Funtikov, A. M.; Guidelli, R.; Muniz-Miranda, M. Relation between surface enhanced Raman intensity and surface concentration for 2-amino 5-nitro pyridine adsorbed on roughened polycrystalline silver electrodes, *J. Electroanal. Chem.* **1994**, *367*, 223.
29. Stolberg, L.; Lipkowski, J.; Irish, D. E. An examination of the relationship between surface enhanced Raman scattering (SERS) intensities and surface concentration for pyridine adsorbed at the polycrystalline gold/aqueous solution interface, *J. Electroanal. Chem.* **1991**, *300*, 563.
30. Brolo, A. G.; Irish, D. E.; Szymanski, G.; Lipkowski, J. Relationship between SERS intensity and both surface coverage and morphology for pyrazine adsorbed on a polycrystalline gold electrode, *Langmuir* **1998**, *14*, 517.
31. Partal Urena, F.; Fernández Gómez, M.; López González, J. J.; Martinez Torres, E. A new insight into the vibrational analysis of pyridine, *Spectrochim. Acta, Part A* **2003**, *59*, 2815.
32. Wu, D-Y.; Liu, X.-M.; Duan, S.; Xu, X.; Ren, B.; Lin, S.-H.; Tian, Z.-Q. Chemical enhancement effects in sers spectra: A quantum chemical study of pyridine interacting with copper, silver, gold and platinum metals, *J. Phys. Chem. C* **2008**, *112*, 4195.
33. Kudelski, A.; Bukowska, J.; Jackowska, K. Relative SERS enhancement factors for pyridine adsorbed on a silver electrode. The chemical effect in SERS as a product of charge-transfer and active-site mechanisms, *J. Raman Spectrosc.* **1994**, *25*, 153.
34. Brolo, A. G.; Irish, D. E.; Smith, B. D. Applications of surface enhanced Raman scattering to the study of metal-adsorbate interactions, *J. Mol. Struct.* **1997**, *405*, 29.
35. Gordon II, J. G.; Ernst, S. Surface plasmons as a probe of the electrochemical interface, *Surf. Sci.* **1980**, *101*, 499.
36. Michota, A.; Kudelski, A.; Bukowska, J. Influence of electrolytes on the structure of cysteamine monolayer on silver studied by surface-enhanced Raman scattering, *J. Raman Spectrosc.* **2001**, *32*, 345.
37. Pemberton, J. E. Surface enhanced Raman scattering at Ag electrodes in non-aqueous pyridine solutions, *Chem. Phys. Lett.* **1985**, *115*, 321.
38. Santos Jr., V. O.; Alves, M. B.; Carvalho, M. S.; Suarez, P. A. Z.; Rubim, J. C. Surface-enhanced Raman scattering at the silver electrode/Ionic liquid ($BMIPF_6$) interface, *J. Phys. Chem. B* **2006**, *110*, 20379.
39. Yao, J.-L.; Yuan, Y.-X.; Gu, R.-A. Negative role of triphenylphosphine in the inhibition of benzotriazole at the Cu surface studied by surface-enhanced Raman spectroscopy, *J. Electroanal. Chem.* **2004**, *573*, 255.
40. Cao, P.; Sun, Y.; Gu, R. Investigations of chemisorption and reaction at non-aqueous electrochemical interfaces by in situ surface-enhanced Raman spectroscopy, *J. Raman Spectrosc.* **2005**, *36*, 725.
41. Ren, B.; Yao, J. L.; She, C. X.; Huang, Q. J.; Tian, Z. Q. Surface Raman spectroscopy on transition metal surfaces, *Internet J. Vib. Spectrosc.* (www.ijvs.com) **2000**, *4*, 4.
42. Gritnzer, G.; Kuta, J. Recommendations on reporting electrode potentials in nonaqueous solvents, *Pure Appl. Chem.* **1984**, *6*, 461.
43. Bott, A. W. Practical problems in voltammetry 3: Reference electrodes for voltammetry, *Curr. Sep* **1992**, *14*, 64.
44. Chen, Y. X.; Heinen, M.; Jusys, Z.; Behm, R. B. Kinetics and mechanism of the electrooxidation of formic acid—Spectroelectrochemical studies in a flow cell, *Angew. Chem. Int. Ed.* **2006**, *45*, 981.
45. Zhang, P.; Chen, Y.-X.; Cai, J.; Liang, S.-Zi; Li, J.-F.; Wang, A.; Ren, B.; Tian, Z.-Q. An electrochemical in situ surface-enhanced Raman spectroscopic study of carbon monoxide chemisorption at a gold core–platinum shell nanoparticle electrode with a flow cel, *J. Phys. Chem. C* **2009**, *113*, 17518.
46. Tian, Z.-Q.; Ren, B. Adsorption and reaction at electrochemical interfaces as probed by surface-enhanced Raman spectroscopy, *Annu. Rev. Phys. Chem.* **2004**, *55*, 197.
47. Natan, M. J. Concluding remarks surface enhanced Raman scattering, *Faraday Discuss.* **2006**, *132*, 321.
48. Lin, X.-M.; Cui, Y.; Xu, Y.-H.; Ren, B.; Tian, Z.-Q. Surface-enhanced Raman spectroscopy: Substrate-related issues, *Anal. Bioanal. Chem.* **2009**, *394*, 1729.
49. Kudelski, A.; Bukowska, J. Charge-transfer contribution to surface-enhanced Raman scattering and surface-enhanced resonance Raman scattering of dyes at silver and gold electrodes, *Chem. Phys. Lett.* **1996**, *253*, 246.

50. Królikowska, A.; Kudelski, A.; Michota, A.; Bukowska, J. SERS studies on the structure of thioglycolic acid monolayers on silver and gold, *Surf. Sci.* **2003**, *532–535*, 227.
51. Donten, M. L. M. Sc. *Thesis: Investigation of the structure of binary thiol monolayers designed for immobilization of cytochrome c by electrochemical, STM and Raman spectroscopy techniques, Department of Chemistry, Warsaw University, Warsaw*, **2007**.
52. Ershov, B. G.; Janata, E.; Henglein, A. Growth of silver particles in aqueous solution: long-lived "magic" clusters and ionic strength effects, *J. Phys. Chem.* **1993**, *97*, 339.
53. Roy, D.; Furtak, T. E. Vibrational characteristics of silver clusters in surface-enhanced Raman scattering, *Phys. Rev. B* **1986**, *34*, 5111.
54. Muniz-Miranda, M.; Sbrana, G. Evidence for surface $Ag^+$ complex formation by an anion-induced effect in the SER spectra of phthalazine adsorbed on silver sols, *J. Raman Spectrosc.* **1996**, *27*, 105.
55. Michota, A.; Kudelski, A.; Bukowska, J. Chemisorption of cysteamine on silver studied by surface-enhanced Raman scattering, *Langmuir* **2000**, *16*, 10236.
56. Lee, P. C.; Meisel, D. Adsorption and surface-enhanced Raman of dyes on silver and gold sols, *J. Phys. Chem.* **1982**, *86*, 3391.
57. Frens, G. Controlled nucleation for the regulation of the particle size in monodisperse gold suspensions, *Nature Phys. Sci.* **1973**, *241*, 20.
58. Leopold, N.; Lendl, B. A new method for fast preparation of highly surface-enhanced Raman scattering (SERS) active silver colloids at room temperature by reduction of silver nitrate with hydroxylamine hydrochloride, *J. Phys. Chem. B* **2003**, *107*, 5723.
59. Xiao, J.; Qi, L. Surfactant-assisted, shape-controlled synthesis of gold nanocrystals, *Nanoscale* **2011**, *3*, 1383.
60. Jones, M. R.; Osberg, K. D.; Macfarlane, R. J.; Langille, M. R.; Mirkin, A. Templated techniques for the synthesis and assembly of plasmonic nanostructures, *Chem. Rev.* **2011**, *111*, 3736.
61. Hu, J.; Chen, M.; Fanga, X.; Wu, L. Fabrication and application of inorganic hollow spheres, *Chem. Soc. Rev.* **2011**, *40*, 5472.
62. Tao, A. R.; Habas, S.; Yang, P. Shape control of colloidal metal nanocrystals, *Small* **2008**, *4*, 310.
63. Liang, H.-P.; Wan, L.-J.; Bai, C.-L.; Jiang, L. Gold hollow nanospheres: Tunable surface plasmon resonance controlled by interior-cavity sizes, *J. Phys. Chem. B* **2005**, *109*, 7795.
64. Schwartzberg, A. M.; Olson, T. Y.; Talley, C. E.; Zhang, J. Z. Synthesis, characterization, and tunable optical properties of hollow gold nanospheres, *J. Phys. Chem. B* **2006**, *110*, 19935.
65. Hajdukova, N.; Prochazka, M.; Stepanek, J.; Spirkova, M. Chemically reduced and laser-ablated gold nanoparticles immobilized to silanized glass plates: Preparation, characterization and SERS spectral testing, *Colloids Surf., A* **2007**, *301*, 264.
66. Tian, Z.-Q.; Ren, B.; Li, J. F.; Yang, Z. L. Expanding generality of surface-enhanced Raman spectroscopy with borrowing SERS activity strategy, *Chem. Commun.* **2007**, *3514*.
67. Tian, Z.-Q.; Yang, Z. L.; Ren, B.; Li, J. F.; Zhang, Y.; Lin, X. F.; Hu, J. W.; Wu, D. Y. Surface-enhanced Raman scattering from transition metals with special surface morphology and nanoparticle shape, *Faraday Discuss.* **2006**, *132*, 159.
68. Ferretti, S.; Paynter, S.; Russell, D. A.; Sapsfor, K. E.; Richardson, D. J. Self-assembled monolayers: A versatile tool for the formulation of bio-surfaces, *Trends Anal. Chem.* **2000**, *19*, 530.
69. Dick, L. A.; McFarland, A. D.; Haynes, C. L.; Van Duyne, R. P. Metal film over nanosphere (MFON) electrodes for surface-enhanced Raman spectroscopy (SERS): Improvements in surface nanostructure stability and suppression of irreversible loss, *J. Phys. Chem. B* **2006**, *106*, 853.
70. Willets, K. A.; Van Duyne, R. P. Localized surface plasmon resonance spectroscopy and sensing, *Annu. Rev. Phys. Chem.* **2007**, *58*, 267.
71. Haynes, C. L.; McFarland, A. D.; Smith, M. T.; Hulteen, J. C.; Van Duyne, R. P. Angle-resolved nanosphere lithography: Manipulation of nanoparticle size, shape, and interparticle spacing, *J. Phys. Chem. B* **2002**, *106*, 1898.
72. Lin, W.-C.; Liao, L.-S.; Chen, Y.-H.; Chang, H.-C.; Tsai, D. P.; Chiang, H.-P. Size dependence of nanoparticle-SERS enhancement from silver film over nanosphere (AgFON) substrate, *Plasmonics* **2010**, *6*, 201.

73. MAHAJAN, S.; ABDELSALAM, M.; SUGAWARA, Y.; CINTRA, S.; RUSSELL, A. E.; BAUMBERG, J. J.; BARTLETT, P. N. Tuning plasmons on nano-structured substrates for NIR-SERS, *Phys. Chem. Chem. Phys.* **2007**, *9*, 104.
74. SCHUHMANN, W. Amperometric enzyme biosensors based on optimised electron-transfer pathways and non-manual immobilisation procedures, *Rev. Mol. Biotechnol.* **2002**, *82*, 425.
75. ZIMMERMANN, H.; LINDGREN, A.; SCHUHMANN, W.; GORTON, L. Anisotropic orientation of horseradish peroxidase by reconstitution on a thiol-modified gold electrode, *Chem. Eur. J.* **2000**, *6*, 592.
76. HABERMÜLLER, K.; MOSBACH, M.; SCHUHMANN, W. Electron-transfer mechanisms in amperometric biosensors, *Fresenius J. Anal. Chem.* **2000**, *366*, 560.
77. AMBLER, R. P. Sequence variability in bacterial cytochromes *c*, *Biochim. Biophys. Acta* **1991**, *1058*, 42.
78. PASQUINET, E.; BOUVIER, C.; THERY-MERLAND, F.; HAIRAULT, L.; LEBRET, B.; MÉTHIVIER, C.; PRADIER, C. M. Synthesis and adsorption on gold surfaces of a functionalized thiol: Elaboration and test of a new nitroaromatic gas sensor, *J. Colloid Interface Sci.* **2004**, *272*, 21.
79. BANCI, L.; BERTINI, I.; HUBER, J. G.; SPYROULIAS, G. A.; TURANO, P. Solution structure of reduced horse heart cytochrome *c*, *J. Biol. Inorg. Chem.* **1999**, *4*, 21.
80. XU, W.; ZHOU, H.; REGNIER, F. E. Regio-specific adsorption of cytochrome *c* on negatively charged surfaces, *Anal. Chem.* **2003**, *75*, 1931.
81. CALVERT, J. F.; HILL, J. L.; DONG, A. Redox-dependent conformational changes are common structural features of cytochrome *c* from various species, *Arch. Biochem. Biophys.* **1997**, *346*, 287.
82. LAMAR, G. N.; DAVIS, N. L.; PARISH, D. W.; SMITH, K. M. Heme orientational disorder in reconstituted and native sperm whale myoglobin: Proton nuclear magnetic resonance characterizations by heme methyl deuterium labeling in the met-cyano protein, *J. Mol. Biol.* **1983**, *168*, 887.
83. MCLACHLAN, S. J.; LAMAR, G. N.; BURNS, P. D.; SMITH, K. M.; LANGRY, K. C. $^{1}$H-NMR assignments and the dynamics of interconversion of the isomeric forms of cytochrome *b*5 in solution, *Biochim. Biophys. Acta* **1986**, *874*, 274.
84. HOBBS, J. D.; SHELNUTT, J. A. Conserved nonplanar heme distortions in cytochromes *c*, *J. Protein Chem.* **1995**, *14*, 19.
85. SCOTT, R. A.; MAUK, A. G. *Cytochrome c—A Multidisciplinary Approach*, University Science Books: Sausalito, **1995**.
86. BANCI, L.; BERTINI, I.; GRAY, H. B.; LUCHINAT, C.; REDDIG, T. Solution structure of oxidized horse heart cytochrome *c*, *Biochemistry* **1997**, *36*, 9867.
87. KOPPENOL, W. H.; MARGOLIASH, E. The asymmetric distribution of charges on the surface of horse cytochrome *c*. Functional implications, *J. Biol. Chem.* **1982**, *257*, 4426.
88. COWAN, J. A. *Inorganic Biochemistry: An Introduction*, Wiley-VCH: New York, **1996**, Chapter 5.
89. GRAY, H. B.; ELLIS, W. R. Electron transfer, in *Bioinorganic Chemistry*, I. BERTINI, H. B. GRAY, S. J. LIPPARD, and J. S. VALENTINE, Eds., University Science Books: Mill Valley, CA, **1994**, pp. 315–363.
90. MRKISCH, M.; WHITESIDES, G. M. Using self-assembled monolayers to understand the interactions of man-made surfaces with proteins and cells, *Annu. Rev. Biophys. Biomol. Struct.* **1996**, *25*, 55.
91. LÖFÄS, S.; JOHNSSON, B.; EDSTROM, Å.; HANSSON, A.; LINDQUIST, G.; MULLER HILLGREN, R. M.; Stigh, L. Methods for site controlled coupling to carboxymethyldextran surfaces in surface plasmon resonance sensors, *Biosens. Bioelectron.* **1995**, *10*, 813.
92. SMULEVICH, G.; SPIRO, T. G. Surface enhanced Raman spectroscopic evidence that adsorption on silver particles can denature heme proteins, *J. Phys. Chem.* **1985**, *89*, 5168.
93. LOVE, J.; ESTROFF, L. A.; KRIEBEL, J. K.; NUZZO, R. G.; WHITESIDES, G. M. Self-assembled monolayers of thiolates on metals as a form of nanotechnology, *Chem. Rev.* **2005**, *105*, 1103.
94. BAIN, C. D.; TROUGHTON, E. B.; TAO, Y.-T.; EVALL, J.; WHITESIDES, G. M.; NUZZO, R. G. Formation of monolayer films by the spontaneous assembly of organic thiols from solution onto gold, *J. Am. Chem. Soc.* **1989**, *111*, 321.
95. CHEN, D.; LI, J. Interfacial design and functionization on metal electrodes through self-assembled monolayers, *Surf. Sci. Rep.* **2006**, *61*, 445.
96. CHAKI, N. K.; Vijayamohanan, K. Self-assembled monolayers as a tunable platform for biosensor applications, *Biosens. Bioelectron.* **2002**, *17*, 1.

97. Döpner, S.; Hildebrandt, P.; Rosell, F. I.; Mauk, A. G.; von Walter, M.; Buse, G.; Soulimane, T. The structural and functional role of lysine residues in the binding domain of cytochrome *c* in the electron transfer to cytochrome *c* oxidase, *Eur. J. Biochem.* **1999**, *261*, 379.
98. Murgida, D. H.; Hildebrandt, P. Electron-transfer processes of cytochrome *c* at interfaces. New insights by surface-enhanced resonance Raman spectroscopy, *Acc. Chem. Res.* **2004**, *37*, 854.
99. Wackerbarth, H.; Klar, U.; Günther, W.; Hildebrandt, P. Novel time-resolved surface-enhanced (resonance) Raman spectroscopic technique for studying the dynamics of interfacial processes: Application to the electron transfer reaction of cytochrome *c* at a silver electrode, *Appl. Spectrosc.* **1999**, *53*, 283.
100. Wackerbarth, H.; Hildebrandt, P. Redox and conformational equilibria and dynamics of cytochrome *c* at high electric fields, *Chem. Phys. Chem.* **2003**, *4*, 714.
101. Tarlov, M. J.; Bowden, E. F. Electron-transfer reaction of cytochrome *c* adsorbed on carboxylic acid terminated alkanethiol monolayer electrodes, *J. Am. Chem. Soc.* **1991**, *113*, 1847.
102. Sagara, T.; Kubo, Y.; Hiraishi, K. Estimation of the orientation of heme in cytochrome *c* immobilized on a carboxylate-terminated alkanethiol monolayer on a Au electrode by the use of electroreflectance spectroscopy with polarized light incidence, *J. Phys. Chem. B* **2006**, *110*, 16550.
103. Jiang, X.; Ataka, K.; Heberle, J. Influence of the molecular structure of carboxyl-terminated self-assembled monolayer on the electron transfer of cytochrome *c* adsorbed on an Au electrode: In situ observation by surface-enhanced infrared absorption spectroscopy, *J. Phys. Chem. C.* **2008**, *112*, 813.
104. Du, Y.-Z.; Saavedra, S. S. Molecular orientation distributions in protein films. V. cytochrome *c* adsorbed to a sulfonate-terminated, self-assembled monolayer, *Langmuir* **2003**, *19*, 6443.
105. Królikowska, A.; Bukowska, J. Surface-enhanced resonance Raman spectroscopic characterization of cytochrome *c* immobilized on 2-mercaptoethanesulfonate monolayers on silver, *J. Raman Spectrosc.* **2010**, *41*, 1621.
106. Quan, L. T. M.Sc. *Thesis: Electrochemistry of cytochrome c on self-assembled phosphonic acid terminated alkanethiols on gold, Department of Chemistry, North Carolina State University, Raleigh, NC*, **2004**.
107. Rivas, L.; Murgida, D. H.; Hildebrandt, P. Conformational and redox equilibria and dynamics of cytochrome *c* immobilized on electrodes via hydrophobic interactions, *J. Phys. Chem. B* **2002**, *106*, 4823.
108. Dick, L. A.; Haes, A. J.; Van Duyne, R. P. Distance and orientation dependence of heterogeneous electron transfer: A surface-enhanced resonance Raman scattering study of cytochrome *c* bound to carboxylic acid terminated alkanethiols adsorbed on silver electrodes, *J. Phys. Chem. B* **2000**, *104*, 11752.
109. Wei, J. J.; Liu, H. Y.; Khoshtariya, D. E.; Yamamoto, H.; Dick, A.; Waldeck, D. H. Electron-transfer dynamics of cytochrome *c*: A change in the reaction mechanism with distance, *Angew. Chem. Int. Ed.* **2002**, *41*, 4700.
110. Murgida, D. H.; Hildebrandt, P.; Wei, J.; He, Y. F.; Liu, H.; Waldeck, D. H. Surface-enhanced resonance Raman spectroscopic and electrochemical study of cytochrome *c* bound on electrodes through coordination with pyridinyl-terminated self-assembled monolayers, *J. Phys. Chem. B* **2004**, *108*, 2261.
111. Di Paolo, R. E.; Rivas, L.; Murgida, D.; Hildebrandt, P. Redox equilibria of cytochrome C-3 immobilised on self-assembled monolayers coated silver electrodes, *Phys. Scr.* **2005**, *T118*, 225.
112. Todorovic, S.; Verissimo, A.; Wisitruangsakul, N.; Zebger, I.; Hildebrandt, P.; Pereira, M. M.; Teixeira, M.; Murgida, D. H. SERR-spectroelectrochemical study of a $cbb_3$ oxygen reductase in a biomimetic construct, *J. Phys. Chem. B* **2008**, *112*, 16952.
113. Spiro, T. G.; *Biological applications of Raman Spectroscopy, Vol. 3, RR Spectra of Heme and Metalloproteins*, Wiley: New York, **1988**.
114. Spiro, T. G.; Strekas, T. C. Resonance Raman spectra of heme proteins. Effects of oxidation and spin state, *J. Am. Chem. Soc.* **1974**, *96*, 338.
115. Murgida, D. H.; Hildebrandt, P. Disentangling interfacial redox processes of proteins by SERR spectroscopy, *Chem. Soc. Rev.* **2008**, *37*, 937.
116. Desbois, A. Resonance Raman spectroscopy of *c*-type cytochromes, *Biochimie* **1994**, *76*, 693.

117. Hu, S. Z.; Morris, I. K.; Singh, J. P.; Smith, K. M.; Spiro, T. G. Complete assignment of cytochrome *c* resonance Raman spectra via enzymic reconstitution with isotopically labeled hemes, *J. Am. Chem. Soc.* **1993**, *115*, 12446.
118. Tang, J.; Albrecht, A. C. H. *Raman Spectroscopy*, H. Szymanski, Ed., Plenum: New York, **1970**, p. 33.
119. Wackerbarth, H.; Murgida, D. H.; Oellerich, S.; Döpner, S.; Rivas, L.; Hildebrandt, P. J. Dynamics and mechanism of the electron transfer process of cytochrome *c* probed by resonance Raman and surface enhanced resonance Raman spectroscopy, *J. Mol. Struct.* **2001**, *563–564*, 51.
120. Leger, C.; Bertrand, P. Direct electrochemistry of redox enzymes as a tool for mechanistic studies, *Chem. Rev.* **2008**, *108*, 2379.
121. Le Ru, E. C.; Etchegoin, P. G. *Principles of Surface-Enhanced Raman Spectroscopy and Related Plasmonic Effects*, Elsevier: Oxford, UK, **2009**.
122. Donten, M. L.; Królikowska, A.; Bukowska, J. Structure and composition of binary monolayers self-assembled from sodium 2-mercaptoetanosulfonate and mercaptoundecanol mixed solutions on silver and gold supports, *Phys. Chem. Chem. Phys.* **2009**, *11*, 3390.
123. Benítez, G.; Vericat, C.; Tanco, S.; Remes Lenicov, F.; Castez, M. F.; Vela, M. E.; Salvarezza, R. C. Role of surface heterogeneity and molecular interactions in the charge-transfer process through self-assembled thiolate monolayers on Au(111), *Langmuir* **2004**, *20*, 5030.
124. Boal, A. K.; Rotello, V. M. Intra- and intermonolayer hydrogen bonding in amide-functionalized alkanethiol self-assembled monolayers on gold nanoparticles, *Langmuir* **2000**, *16*, 9527.
125. Hostetler, M. J.; Stokes, J. J.; Murray, R. W. Infrared spectroscopy of three-dimensional self-assembled monolayers: *N*-alkanethiolate monolayers on gold cluster compounds, *Langmuir* **1996**, *12*, 3604.
126. Cooper, E.; Leggett, G. J. Influence of tail-group hydrogen bonding on the stabilities of self-assembled monolayers of alkylthiols on gold, *Langmuir* **1999**, *15*, 1024.
127. Valiokas, R.; Ostblom, M.; Svedhem, S.; Svensson, S. C. T.; Liedberg, B. Thermal stability of self-assembled monolayers: Influence of lateral hydrogen bonding, *J. Phys. Chem. B* **2002**, *106*, 10401.
128. Chen, S.; Li, L.; Boozer, Ch. L.; Jiang, S. Controlled chemical and structural properties of mixed self-assembled monolayers of alkanethiols on Au(111), *Langmuir* **2000**, *16*, 9287.
129. Taylor, C. E.; Garvey, S. D.; Pemberton, J. E. Carbon contamination at silver surfaces: Surface preparation procedures evaluated by Raman spectroscopy and X-ray photoelectron spectroscopy, *Anal. Chem.* **1996**, *68*, 2401.
130. Sauer, G.; Nickel, U.; Schneider, S. Preparation of SERS-active silver film electrodes via electrocrystallization of silver, *J. Raman Spectrosc.* **2000**, *31*, 359.
131. Millo, D.; Ranieri, A.; Koot, W.; Gooijer, C.; van der Zwan, G. Towards combined electrochemistry and surface-enhanced resonance Raman of heme proteins: Improvement of diffusion electrochemistry of cytochrome *c* at silver electrodes chemically modified with 4-mercaptopyridine, *Anal. Chem.* **2006**, *78*, 5622.
132. Millo, D.; Bonifacio, A.; Ranieri, A.; Borsari, M.; Gooijer, C.; van der Zwan, G. Voltammetric and surface-enhanced resonance Raman spectroscopic characterization of cytochrome *c* adsorbed on a 4-mercaptopyridine monolayer on silver electrodes, *Langmuir* **2007**, *23*, 4340.
133. Millo, D.; Bonifacio, A.; Ranieri, A.; Borsari, M.; Gooijer, C.; van der Zwan, G. pH-induced changes in adsorbed cytochrome *c*. Voltammetric and surface-enhanced resonance Raman characterization performed simultaneously at chemically modified silver electrodes, *Langmuir* **2007**, *23*, 9898.
134. Clark, R. A.; Bowden, E. F. Voltammetric peak broadening for cytochrome *c*/alkanethiolate monolayer structures: Dispersion of formal potentials, *Langmuir* **1997**, *13*, 559.
135. Murgida, D. H.; Hildebrandt, P. Heterogeneous electron transfer of cytochrome *c* on coated silver electrodes. Electric field effects on structure and redox potential, *J. Phys. Chem. B* **2001**, *105*, 1578.
136. Kelly, K. L.; Coronado, E.; Zhao, L. L.; Schatz, G. C. The optical properties of metal nanoparticles: The influence of size, shape and dielectric environment, *J. Phys. Chem. B* **2003**, *107*, 668.
137. Lal, S.; Grady, N. K.; Kundu, J.; Levin, C. S.; Lassiter, J. B.; Halas, N. J. Tailoring plasmonic substrates for surface enhanced spectroscopies, *Chem. Soc. Rev.* **2008**, *37*, 898.

138. PRODAN, E.; RADLOFF, C.; HALAS, N. J.; NORDLANDER, P. A hybridization model for the plasmon response of complex nanostructures, *Science* **2003**, *302*, 419.
139. FENG, J.-J.; GERNERT, U.; SEZER, M.; KUHLMANN, U.; MURGIDA, D. H.; DAVID, C.; RICHTER, M.; KNORR, A.; HILDEBRANDT, P.; WEIDINGER, I. M. Novel Au–Ag hybrid device for electrochemical SE(R)R spectroscopy in a wide potential and spectral range, *Nano Lett.* **2009**, *9*, 298.
140. HILDEBRANDT, P. Conference proceedings: Electric field effects on the protein and electron transfer dynamics of immobilized heme proteins-vibrational spectroscopic approaches, in *Seeing Enzymes in Action*, Lorentz Center: Leiden, Netherlands, **2010**.
141. SEZER, M.; FENG, J.-J.; LY, H. K.; SHEN, Y.; NAKANISHI, T.; KUHLMANN, U.; HILDEBRANDT, P.; MÖHWALD, H.; WEIDINGER, I. M. Multi-layer electron transfer across nanostructured Ag-SAM-Au-SAM junctions probed by surface enhanced Raman spectroscopy, *Phys. Chem. Chem. Phys.* **2010**, *12*, 9822.
142. OELLERICH, S.; WACKERBARTH, H.; HILDEBRANDT, P. Spectroscopic characterization of nonnative conformational states of cytochrome *c*, *J. Phys. Chem. B* **2002**, *106*, 6566.
143. HILDEBRANDT, P.; STOCKBURGER, M. Surface-enhanced resonance Raman spectroscopy of cytochrome *c* at room and low temperature, *J. Phys. Chem.* **1986**, *90*, 6017.
144. MOSKOVITS, M. Surface selection rules, *J. Chem. Phys.* **1982**, *77*, 4408.
145. CREIGHTON, J. A. *Spectroscopy of Surfaces*, R. J. H. CLARK, R. E. HESTER, Eds., Wiley: New York, **1988**, p. 37.
146. CARTLING, B. Intermediate and stable redox states of cytochrome *c* studied by low temperature resonance Raman spectroscopy, *Biophys. J.* **1983**, *43*, 191.
147. KEATING, C. D.; KOVALESKI, K. M.; NATAN, M. J. Protein: colloid conjugates for surface enhanced Raman scattering: Stability and control of protein orientation, *J. Phys. Chem. B* **1998**, *102*, 9404.
148. ULMAN, A. Formation and structure of self-assembled monolayers, *Chem. Rev.* **1996**, *96*, 1533.
149. LEE, L. Y. S.; LENNOX, R. B. Ferrocenylalkylthiolate labeling of defects in alkylthiol self-assembled monolayers on gold, *Phys. Chem. Chem. Phys.* **2007**, *9*, 1013.
150. KRÓLIKOWSKA, A.; BUKOWSKA, J. Self-assembled monolayers of mercaptosuccinic acid monolayers on silver and gold surfaces designed for protein binding. Part II: vibrational spectroscopy studies on cytochrome *c* immobilization, *J. Raman Spectrosc.* **2007**, *38*, 943.
151. KRÓLIKOWSKA, A.; BUKOWSKA, J. Self-assembled monolayers of mercaptosuccinic acid on silver and gold surfaces designed for protein binding. Part I: Structure of the monolayer, *J. Raman Spectrosc.* **2007**, *38*, 936.
152. FINKLEA, H. O. *Electroanalytical Chemistry*, A. J. BARD and I. RUBINSTEIN, Eds., Marcel Dekker: New York, **1996**, p. 109.
153. BECKA, A. M.; MILLER, C. J. Electrochemistry at ω-hydroxy thiol coated electrodes. 3. Voltage independence of the electron tunneling barrier and measurements of redox kinetics at large overpotentials, *J. Phys. Chem.* **1992**, *96*, 2657.
154. SMITH, C. P.; WHITE, H. S. Theory of the interfacial potential distribution and reversible voltammetric response of electrodes coated with electroactive molecular films, *Anal. Chem.* **1992**, *64*, 2398.
155. CLARKE, R. J. The dipole potential of phospholipid membranes and methods for its detection, *Adv. Colloid Interface Sci.* **2001**, *89*, 263.
156. KRÓLIKOWSKA, A.; BUKOWSKA, J. SERRS studies of cyt c adsorbed on mixed SAMs of mercaptoethanesulfonate and mercaptoethanol on Ag: influence of monolayer composition on protein orientation, *J. Raman Spectrosc.* **in preparation**.
157. IMABAYASHI, S.-I.; MITA, T.; KAKIUCHI, T. Effect of the electrostatic interaction on the redox reaction of positively charged cytochrome *c* adsorbed on the negatively charged surfaces of acid-terminated alkanethiol monolayers on a Au(111) electrode, *Langmuir* **2005**, *21*, 1470.
158. MURGIDA, D. H.; HILDEBRANDT, P. Redox and redox-coupled processes of heme proteins and enzymes at electrochemical interfaces, *Phys. Chem. Chem. Phys.* **2005**, *7*, 3773.
159. MURGIDA, D. H.; HILDEBRANDT, P. The heterogeneous electron transfer of cytochrome *c* adsorbed on Ag electrodes coated with ω-carboxyl alkanethiols. A surface enhanced resonance Raman spectroscopic study, *J. Mol. Struct.* **2001**, *565*, 97.
160. RIVAS, L.; MARTI, M.; SMITH, A.; MURGIDA, D. H.; HILDEBRANDT, P. Unpublished data.

161. Oellerich, S.; Lecomte, S.; Paternostre, M.; Heimburg, T.; Hildebrandt, P. Peripheral and integral binding of cytochrome *c* to phospholipids vesicles, *J. Phys. Chem. B* **2004**, *108*, 3871.
162. Basova, L. V.; Kurnikov, I. V.; Wang, L.; Ritov, V. B.; Belikova, N. A.; Vlasova, I. I.; Pacheco, A. A.; Winnica, D. E.; Peterson, J.; Bayir, H.; Waldeck, D. H.; Kegan, V. E. Cardiolipin switch in mitochondria: Shutting off the reduction of cytochrome *c* and turning on the peroxidase activity, *Biochemistry* **2007**, *46*, 3423.
163. Kagan, V. E.; Tyurin, V. A.; Jiang, J.; Tyurina, Y. Y.; Ritov, V. B.; Amoscato, A. A.; Osipov, A. N.; Belikova, N. A.; Kapralov, A. A.; Kini, V.; Irina Vlasova, I.; Zhao, Q.; Zou, M.; Peter, D.; Dimitry Svistunenko, A.; Kurnikov, I. V.; Borisenko, G. G. Cytochrome *c* acts as a cardiolipin oxygenase required for release of proapoptotic factors, *Nat. Chem. Biol.* **2005**, *1*, 223.
164. Jiang, X.; Wang, X. Cytochrome *c* mediated apoptosis, *Annu Rev. Biochem.* **2004**, *73*, 87.
165. Murgida, D. H.; Hildebrandt, P. Proton-coupled electron transfer of cytochrome *c*, *J. Am. Chem. Soc.* **2001**, *123*, 4062.
166. Kranich, A.; Ly, H. K.; Hildebrandt, P.; Murgida, D. H. Direct observation of the gating step in protein electron transfer: Electric-field-controlled protein dynamics, *J. Am. Chem. Soc.* **2008**, *130*, 9844.
167. Ly, H. K.; Sezer, M.; Wisitruangsakul, N.; Feng, J. J.; Kranich, A.; Millo, D.; Weidinger, I. M. Surface-enhanced vibrational spectroscopy for probing transient interactions of proteins with biomimetic interfaces: Electric field effects on structure, dynamics and function of cytochrome *c*, *FEBS J.* **2011**, *278*, 1382.
168. Dopner, S.; Hildebrandt, P.; Mauk, A. G.; Lenk, H.; Stempfle, W. Analysis of vibrational spectra of multicomponent systems. Application to pH-dependent resonance Raman spectra of ferricytochrome *c*, *Spectrochim. Acta A* **1996**, *52*, 573.
169. Murgida, D. H.; Hildebrandt, P. Electrostatic-field dependent activation energies modulate electron transfer of cytochrome *c*, *J. Phys. Chem. B* **2002**, *106*, 12814.
170. Cheng, J.; Terrettaz, S.; Blankman, J. I.; Muller, C. J.; Dangi, B.; Guiles, R. D. Electrochemical comparison of heme-proteins by insulated electrode voltammetry, *Isr. J. Chem.* **1997**, *37*, 259.
171. Ly, H. K.; Wisitruangsakul, N.; Sezer, M.; Feng, J. J.; Kranich, A.; Weidinger, I.; Zebger, I.; Murgida, D. H.; Hildebrandt, P. Electric-field effects on the interfacial electron transfer and protein dynamics of cytochrome *c*, *J. Electroanal. Chem.* **2011**, *660*, 367.

# Chapter 6

# Depolarization of Surface-Enhanced Raman Scattering Photons from a Small Number of Molecules on Metal Surfaces

**Fumika Nagasawa,[1] Mai Takase,[2] Hideki Nabika,[3] and Kei Murakoshi[1]**
[1]Department of Chemistry, Faculty of Science, Hokkaido University, Sapporo, Japan
[2]Catalysis Research Center, Hokkaido University, Sapporo, Japan
[3]Department of Material and Biological Chemistry, Faculty of Science, Yamagata University, Yamagata, Japan

## 6.1 INTRODUCTION

Understanding molecular processes at electrified interfaces is a key issue for the development of various electrochemical technologies such as catalysis, photocatalysis, photoenergy conversion, and other renewable energy processes. To achieve this goal, we need to understand both the vibrational and the electronic structures of the molecules at the interface [1]. Raman spectroscopy is well-suited for this problem because vibrational characteristics of molecules can be measured by shifts in visible to near-infrared photons that are transmitted to/from the electrode

*Vibrational Spectroscopy at Electrified Interfaces*, First Edition.
Edited by Andrzej Wieckowski, Carol Korzeniewski, and Björn Braunschweig.

surface through most electrolyte solutions [2]. The electronic structure of molecules can also be analyzed from the observed Raman modes and their intensities, such as the Raman polarizability tensor, which defines the Raman scattering intensity and the vibrational mode selectivity, reflect the response of molecular polarization under light illumination [3, 4]. When Raman scattering occurs via resonances accompanied by electronic excitation, the origin of the resonance often gives information on the potential energy and the special distribution of electronic states of target molecules at interfaces [2, 5, 6]. The interaction between molecules and surfaces is the most fundamental way to understand electrochemical properties. As a consequence of the interaction, specific vibrational and electronic structures of molecules at interfaces emerge that are not normally observed in the free molecules. Raman spectra thus provide useful information both on vibrational and electronic structures of molecules at interfaces.

One drawback of Raman spectroscopy is that the cross-section of Raman scattering is very small. For most molecules, it therefore requires a relatively strong excitation intensity (>1 mW $\mu m^{-2}$) with long exposure times (>$10^2$ s) to obtain vibrational spectra with acceptable signal-to-noise (S/N) ratio. This disadvantage can be overcome by enhanced Raman resonances by localized electromagnetic fields at metal surfaces, that is, surface-enhanced Raman scattering (SERS) [2, 5, 6]. Due to the interaction of molecules with localized electromagnetic fields produced by plasmons in metallic nanostructures under light illumination, Raman scattering may be enhanced by factors of $10^3$ to $10^{10}$ times that observed without the metal nanostructures. SERS thus could provide vibrational information on very small numbers of molecules in situ. The detection and characterization of molecules at less than monolayer coverage on electrochemical interfaces have been proposed for more than 40 years [7–12]. Detailed analyses of SERS spectra are difficult, particularly with molecular specificity. A very limited set of molecules, such as pyridines and certain dyes, provide ultra-high sensitivity at the single molecule level [13–15]. Other molecules, such as water, have been observed at electrochemical interfaces only under very specific conditions [16–24]. The absence of SERS for various molecules suggests an incomplete understanding of the SERS mechanism. After clarifying the origin of the molecular-specificity of SERS, we may then use it as an in situ, ultra-high sensitivity monitoring method for electrochemical interfaces.

To better understand SERS, a detailed spectral analysis should be carried out using a system with a well-defined metal nanostructure and a small number of molecules. If the orientation of a single molecule is defined relative to the polarization of the highly localized electromagnetic field on the metal surfaces, Raman scattering modes could be assigned rigorously. Molecular orbital theory is used to assign the vibrational and electronic structures of the adsorbed molecules [25–28]. Recent techniques have been reported for the creation of SERS metal nanostructures in order to control the modes of the electromagnetic field [29–31]. In addition, methods have been developed to introduce molecules into these electromagnetic fields in well-controlled ways for the purpose of single-molecule detection [13–15]. A remaining issue for well-defined SERS observations is polarization measurements. According

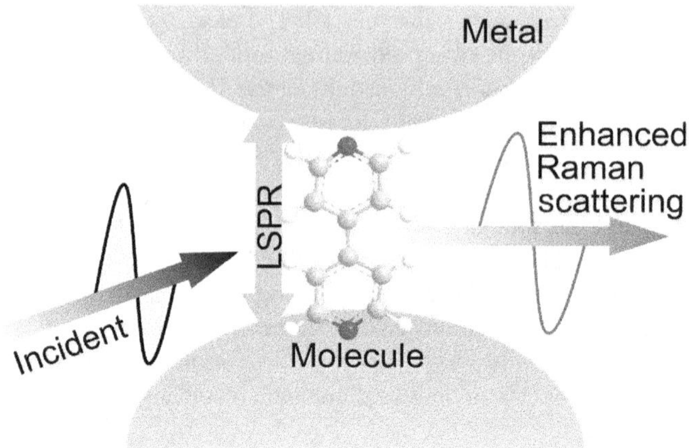


**Figure 6.1** Illustrated surface-enhanced Raman scattering process from a single molecule at metal nanogap.

to the twofold electromagnetic enhancement theory of SERS [2, 5, 6, 32, 33] in which enhancement occurs via polarizations by the field for both excitation and scattering, the polarization behavior is mainly defined by the polarization anisotropy of the localized electromagnetic field (Fig. 6.1). Thus one should expect that SERS photons would be highly polarized, reflecting the anisotropy of the electromagnetic field. However, SERS enhancement also has contribution from the electronic excitation-induced dipole. The SERS polarization thus may reflect contributions from the electronic structure of the system that also provide additional enhancement of the Raman scattering. These additional polarization effects could be monitored as SERS depolarization.

In this chapter, the depolarization of SERS photons is summarized by giving several examples from previous reports. A method for observing polarized SERS is also discussed. After examining SERS depolarization for various vibrational modes, the application to electrified interfaces is considered. Finally, potential applications of spectroelectrochemical methods for monitoring electrochemical processes at a single molecule level are discussed.

## 6.2 BRIEF HISTORY OF POLARIZED SERS MEASUREMENTS

SERS is the consequence of the interaction of localized plasmon resonances with adsorbed molecules. Thus, the most significant contribution to polarization is due to the anisotropic electromagnetic fields having energies that are matched with the excitation and the scattering process. It has been demonstrated that polarized SERS reveals information on both the incident and Raman scattering polarizations [34–46]. Pioneering work on polarized SERS was carried out by Ushioda et al. in early 1990 [34]. The scattering-angle dependence of the SERS intensity on the incident and scattered polarization was monitored using pyridine molecules adsorbed on a roughened Ag surface. The angular dependence and depolarization ratio was not explained by electromagnetic theory. They considered a model in which the scattering dipoles are embedded in an effective dielectric layer. Depolarized SERS has been also reported for chemically synthesized metal nanocolloids [37], tips used in scanning probe microscopies [40, 43], and roughened metal surfaces [34, 47] in which the possible origin of the depolarization is attributed to optical properties of metal structures. Details on the polarization effect were investigated using an isolated, asymmetric metallic nanoparticle trimer that shows elliptically polarized SERS (Fig. 6.2) [44]. The trimer was characterized by scanning electron microscopy, and generalized Mie theory was used to explain the optical response that created the wavelength-dependent polarization pattern. Polarization modulation was also very sensitive to the interparticle distances. These results demonstrated the importance of the optical properties of the nanostructures on the SERS depolarization behavior.

To avoid complexity of the metallic nanostructures, polarized SERS measurements were carried out using arrays of highly uniform gold nanoparticles prepared by electron beam lithography (Fig. 6.3) [41]. SERS depolarization on a prolate nanoparticle was analyzed via the optical properties of the plasmon resonances. Polarization characteristics of the incident and the scattering photons at the metal structure qualitatively explained the SERS depolarization. On the other hand, molecules adsorbed on the same nanostructure show distinct depolarization. This molecular-dependent depolarization implies a possible effect of both the electronic resonance and the molecular orientation on the intensity of SERS photons having specific polarization directions.

The documented observation of SERS depolarization reveals optical properties of the metallic nanostructures, and the contribution of adsorbed molecules to the depolarization is also suggested. However, difficulties in the precise placement of molecules in the localized magnetic field have made it unclear as to the origins of the molecule-dependent depolarization, even when using well-defined metallic nanostructures.

## 6.3 METHOD FOR SERS POLARIZATION MEASUREMENT

For the observation of polarized SERS from a small number of molecules at a single metal nanodimer, we used a home-made polarized Raman microprobe spectrometer modified for near-infrared (NIR) laser light ($\lambda_{ex}$=785 nm). Photons having parallel

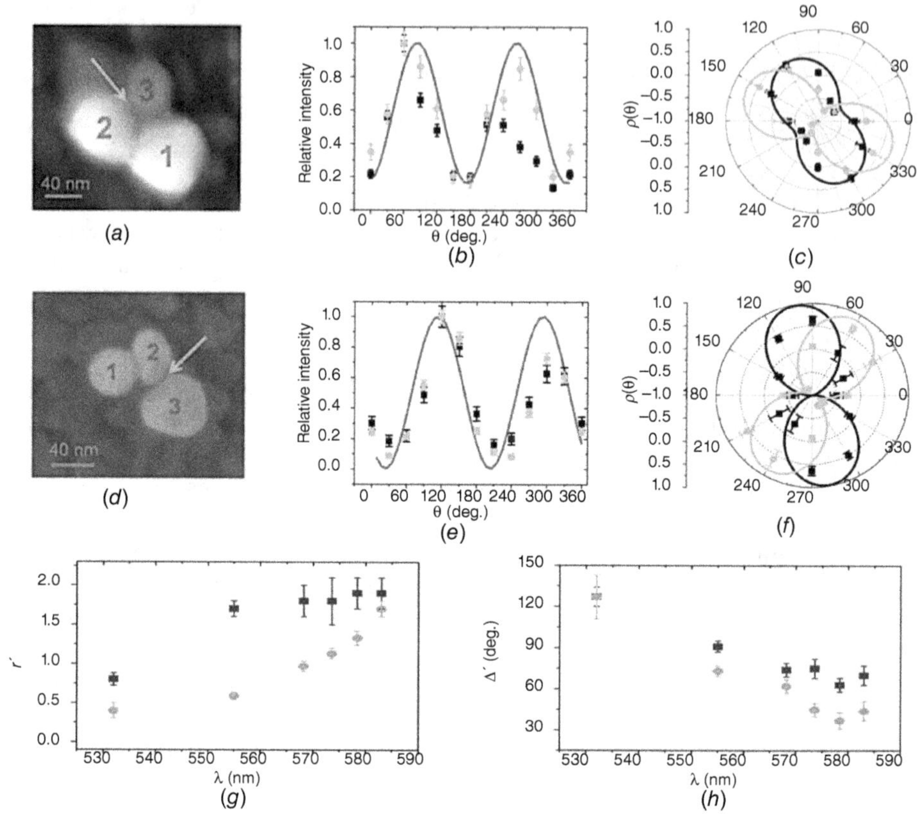

**Figure 6.2** Polarization response of a nanoparticle trimer. (*a*) Scanning electron microscopy (SEM) image of a trimer. A white arrow indicates the position of the molecule that leads to the best agreement between experiment and calculation. (*b*) Normalized Raman intensity at 555 nm (black squares) and 583 nm (gray circles) as a function of the angle of rotation of the incident polarization. The intensities at both wavelengths show approximately the same profile, but the maximal intensity is observed at 75°, which does not match any pair of nanoparticles in the trimer. The black line is the result of a calculation assuming that the molecule is situated at the junction with a gap of 1 nm marked with a white arrow in the SEM image, and the corresponding geometrical parameters from the SEM image $R_1 = 44$ nm, $R_2 = 35$ nm, and $R_3 = 28$ nm as the input. (*c*) Depolarization ratio (ρ) measured at 555 nm (black squares) and 583 nm (gray circles). Depolarization profiles are wavelength dependent in this case and are aligned differently than the intensity profiles. The black and gray lines show the result of calculations at the two wavelengths, assuming that the molecule is situated at the junction marked with the white arrow in the SEM image. (*d*) SEM image of a second trimer. A red arrow indicates the position of the molecule that leads to the best agreement between experiment and calculation. (*e*) Normalized Raman scattering intensity at 555 nm (black squares) and 583 nm (gray circles) as a function of the angle of rotation of the incident polarization. As in (*b*), the intensity profile does not peak along the direction of the axis connecting particles 2 and 3. (*f*) Depolarization ratio (ρ) measured at 555 nm (black squares) and 583 nm (gray circles). As in (c), depolarization profiles are wavelength dependent and are aligned differently than the intensity profiles. The black and gray lines show the result of calculations at the two wavelengths, assuming that the molecule is situated at the junction marked with a white arrow in the SEM image. (*g*) Wavelength dependence of the parameter $r'$ (representing the ratio between $g$ tensor elements), calculated from measurements on the trimer in (*a*) (black) and (*d*) (gray). (*h*) Wavelength dependence of the parameter (representing the phase difference between $\Delta'$ tensor elements), calculated from measurements on the trimer in (*a*) (black) and (*d*) (after Ref. [44]).

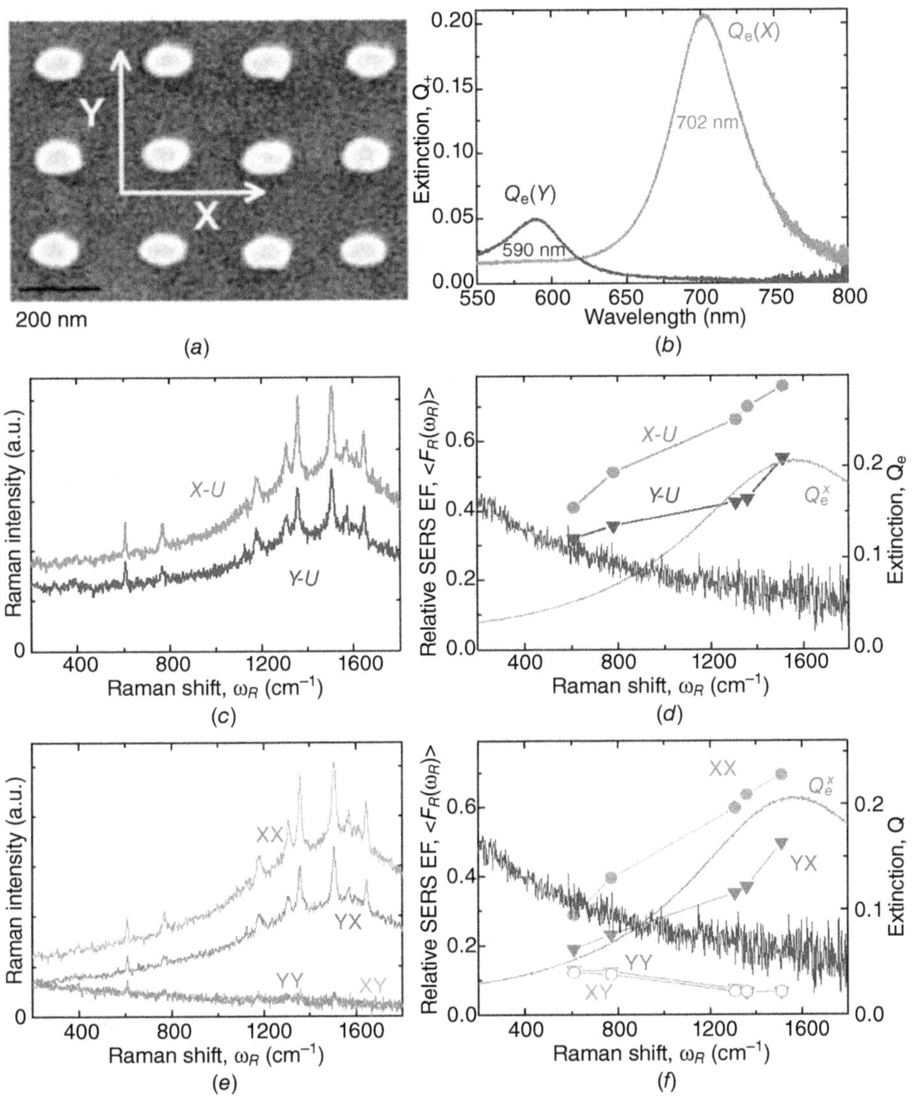


**Figure 6.3** (*a*) Representative SEM image of gold prolate nanoparticles (NPs). Height is 50 nm and dimensions in the plane are 140 nm (*X*) by 90 nm (*Y*). (*b*) Extinction spectrum for *X*- and *Y*-polarized excitation. (*c*) Corresponding SERS spectra at 633-nm excitation (with unpolarized detection) after dipping the array in a $10^{-5}$ M Rhodamine 6G solution. (*d*) Comparison between the relative average SERS enhancement factors at Raman frequency $\omega_R$ $F_R(\omega_R)$ (symbols) of the SERS peaks in (*c*) and the extinction profile $Q_e(\omega_R)$ (lines) in part (*b*) for each excitation polarization. Enhancement factor is the SERS spectrum enhancement relative to the normal Raman spectrum at the same Raman shift $\omega_R$. (*e*) SERS spectra for the four excitation–detection (ED = *XX*, *XY*, *YX*, *YY*) polarization configurations. (*f*) Comparison of the relative SERS EFs in part (*e*) with the extinction profiles of part *b* (after Ref. [41]).

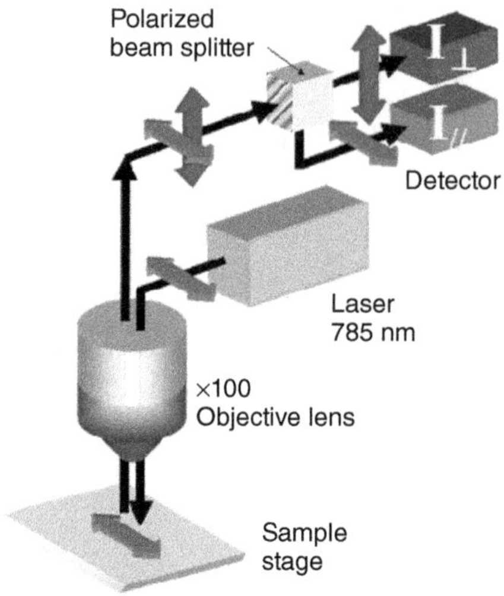


**Figure 6.4** Schematic presentation of a polarized Raman microprobe spectrometer.

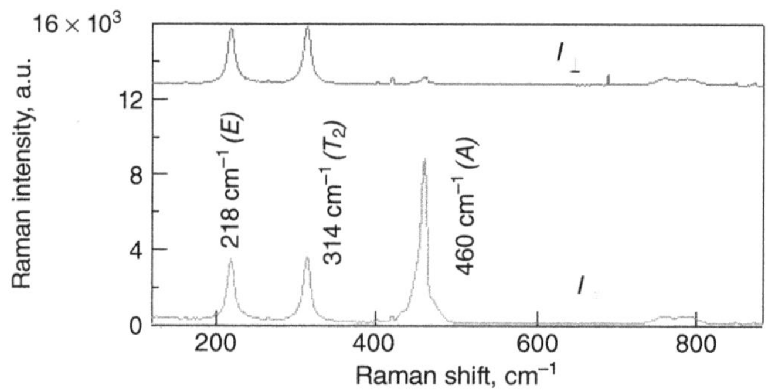


**Figure 6.5** Polarized Raman spectra of $CCl_4$ observed by the polarized Raman microprobe spectroscopy; $I_{ex}$ = 10 mW, $t_{ex}$ = 10 s. The spectral response of the CCD element is not calibrated.

and perpendicular polarization were collected simultaneously in a backscattering configuration (Fig. 6.4).

Polarized spectra were acquired with a multichannel detection system (PIXIS 1024BR; Princeton Instruments) consisting of a 1024 × 256 pixel array charge-coupled device (CCD) (back illuminated, deep depletion type) cooled to –25°C. Spectral images of both polarizations were simultaneously projected onto upper and lower areas (26 × 26 $mm^2$) of the array. The instrument's response was corrected with spectra of pure $CCl_4$, which has a 3/4 depolarization ratio for nontotally symmetric modes [218 $cm^{-1}$ (*E*), 314 $cm^{-1}$ (*T*2)] with linearly polarized incident light (Fig. 6.5).

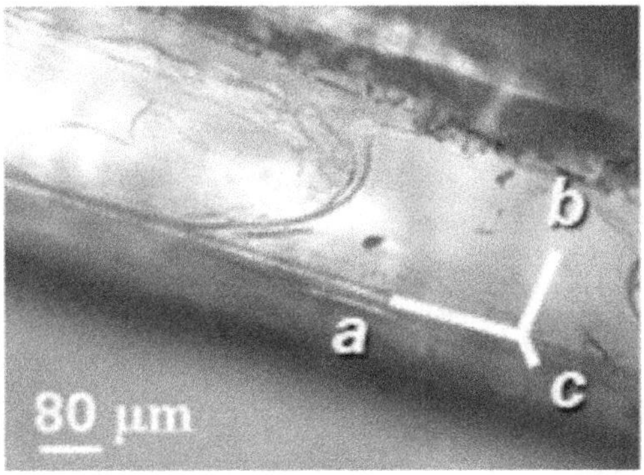


**Figure 6.6** Photograph of a 4,4′-bipyridine single crystal.

The expanded NIR beam was focused onto the sample with a 100× water-immersion objective lens having a numerical aperture of 1.0. The estimated spot size of irradiation was 1 μm, with powers ranging from 10 μW to 20 mW. All Raman measurements were carried out in aqueous solutions containing 1 μM to 1 mM 4,4′-bipyridine (reagent grade, Wako Co. Ltd.).

For Raman band assignments, conventional polarized Raman measurements were carried out on a 15 mM 4,4′-bipyridine aqueous solution. The parallel polarized scattering configuration showed intense peaks at 1010 and 1298 $cm^{-1}$ and weak peaks at 764, 873, and 1229 $cm^{-1}$ (Fig. 6.7*a*). For the perpendicular polarized configuration, scattering was not observed (Fig. 6.7*a*), indicating that the peaks assigned to totally symmetric *a* modes of 4,4′-bipyridine molecules [13] have a depolarization ratio of zero in aqueous solution. Raman spectra (Fig. 6.7*c,d*) for a 4,4′-bipyridine single crystal (Fig. 6.6) were acquired in air for excitation polarization directions relative to the crystal axis as determined by X-ray diffraction. For the parallel polarization configuration for X-excitation and X-scattering directions, shown in the inset of Figure 6.7, relatively strong peaks at 762, 1000, and 1301 $cm^{-1}$ and weak peaks at 880, 1075, and 1230 $cm^{-1}$ were observed (Fig. 6.1*c*). For the perpendicular polarization configuration for *X*-excitation and *Y*-scattering directions, additional peaks at 862, 1042, 1215, and 1325 $cm^{-1}$ were observed (Fig. 6.1*d*). The polarization dependence of the single crystal spectra indicates that the peaks observed at the parallel and the perpendicular configurations are assigned to totally symmetric *a* modes and nontotally symmetric $b_1$ modes, respectively [48, 49]. The assignments were also supported by density functional theory (DFT) calculations using Gaussian 03, revision E.01 at the B3LYP level, and with 6-31G** basis sets. Calculated modes are shown in the slides in Figure 6.8; 770 and 870 $cm^{-1}$ are assigned to the out-of-plane ring and C–H deformations (Fig. 6.8*a,b*); 1240 $cm^{-1}$ is assigned to the in-plane C–H bending mode (Fig. 6.8*c*); and 1300 $cm^{-1}$ is assigned to the inter-ring stretching mode (Fig. 6.8*d*). Although the wavenumbers of the experimentally observed Raman modes differ from those obtained by calculation, agreement both

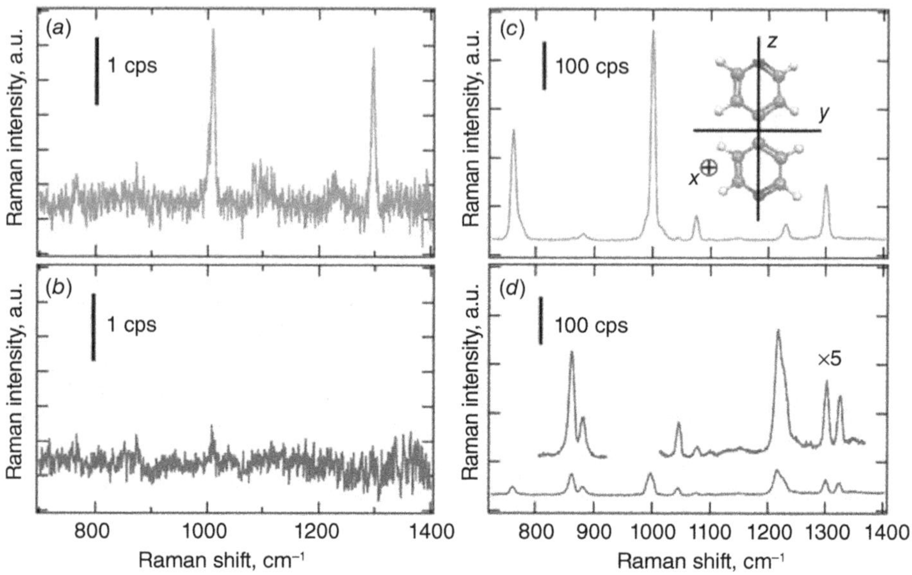


**Figure 6.7** Polarized Raman spectra of 15 mM 4,4′-bipyridine aqueous solution with the (*a*) parallel and (*b*) perpendicular scattering configurations ($I_{ex}$ = 20 mW, $t_{ex}$ = 100 s), 4,4′-bipyridine single crystal with the parallel polarization configuration on *X*-excitation and *X*-scattering directions, (*c*) and the perpendicular polarization configuration on *X*-excitation and *Y*-scattering directions, and (*d*) $I_{ex}$ = 20 mW, $t_{ex}$ = 10 s) (after Ref. [46]).

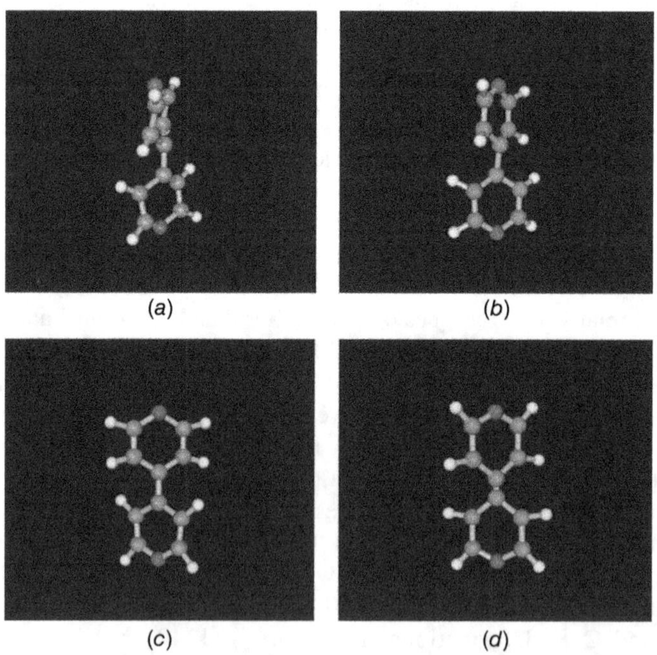


**Figure 6.8** Slides of Raman modes at (*a*) 770 cm$^{-1}$ assigned to the out-of-plane ring deformation, (*b*) 870 cm$^{-1}$ assigned to the out-of-plane C–H deformation, (*c*) 1230 cm$^{-1}$ assigned to the C–H bending, and (*d*) 1300 cm$^{-1}$ assigned to the interring stretching.

in the number of modes and the depolarization behavior supports the validity of the mode assignments presented here.

## 6.4 DEPOLARIZATION BEHAVIOR OF SERS PHOTONS

We have developed a Ag nanodimer structure in order to observe SERS from a small number of molecules at the nanogap. The optical properties of the nanodimer were optimized for confocal NIR Raman observation using 785 nm ($E_{ex}$=0.58 eV) laser excitation. The adsorption isotherm of 4,4′-bipyridine indicates that the coverage on the metal surface can be controlled from 10% to full coverage (>100%) when the solution concentration is varied from 1 μM to 1 mM [50, 51]. We have previously demonstrated that the Ag nanodimer structure enables observation of single-molecule SERS at the gap in a bi-analyte aqueous solution containing bipyridine isomers [52]. In this section, the polarization dependence of SERS from 4,4′-bipyridine at the gap of the Ag dimer is discussed with respect to the optical properties of the Ag nanostructure and the 4,4′-bipyridine electronic structure.

Ag dimer arrays were prepared by angle-resolved nanosphere lithography (AR-NSL) using repeated vapor depositions onto a polystyrene (PS) particle (Polysciences Inc., diameters = 350 nm) monolayer on a glass substrate (Fig. 6.9*a*). The aqueous PS particle suspension was concentrated to 8 and 10 wt%, and then diluted with ethanol (60 and 50%, respectively). In addition to the previously documented drop-coating method to prepare PS monolayers, we used another method based on liquid–gas interfaces in order to improve the quality of the array [46, 52–56]. In this technique, the PS solutions (20 μL) were dropped on the convex surface of a watch glass immersed in Milli-Q water. The thin water layer on the glass surface creates uniform spreading of PS beads at the air–water interface. The change in the surface tension of the water around the PS layer results in the formation of well-ordered monolayers with few defects, dislocations, and vacancies at the liquid–water interface. The high quality monolayer was then lifted off with a clean glass substrate and

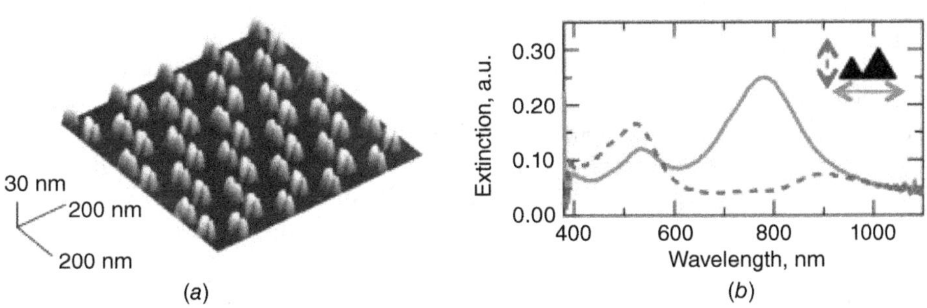


**Figure 6.9** AFM image and polarized extinction spectrum of Ag nanodimer (after Ref. [46]).

the Ag was deposited. After the first Ag deposition, a second deposition was performed at a different angle. Then the PS mask was removed by sonication in Milli-Q water for 10–30 s. The extinction spectrum in the visible to near-infrared region of the Ag dimer array was recorded utilizing the multi-channel spectrometer. The structure of the dimer on the glass substrate was imaged in air by atomic force microscopy (Fig. 6.9*a*). It consists of a triangle with one 50 nm side, and two 100 nm sides; the height is 25–30 nm. The gap sizes are estimated as follows. From the extinction spectra in Fig. 6.9*b*, the polarized excitation along the long axis leads to the appearance of a peak at 800 nm. This agrees well with the theoretical spectra of "tip-to-side" Ag triangular dimers, whose distances are a few nanometers. So the AR-NSL dimers have a gap distance of a few nanometers. The 1 μm excitation spot size covers approximately 9–10 dimers. The stability of the metal structure was checked by a polarized Rayleigh scattering light measurement. It was found that the polarization properties were unchanged before and after light irradiation.

Intense SERS was observed when the substrate was immersed in an aqueous solution containing 1 mM 4,4′-bipyridine. Polarized SERS measurements of the Ag dimers for excitation parallel to the dimer axis exhibited well-defined polarization behavior (Fig. 6.10). The SERS spectrum for the parallel polarization configuration along the long dimer axis excitation showed intense SERS peaks at 770, 871, 1016, 1074, 1231, and 1296 $cm^{-1}$ (Fig. 6.10, *upper*). These peaks can be assigned to the totally symmetric *a* modes. They were not observed in scattering at the perpendicular configuration to the long dimer axis (Fig. 6.10, *lower*). For excitation perpendicular to the dimer axis, scattering was not observed at either the parallel or the perpendicular configuration. This difference between the excitation polarization directions demonstrates the significant contribution of the localized electromagnetic field at the dimer gap.

The system showed very unique behavior, whereby the polarization of the scattering was dependent upon the SERS active site and the concentration of the solution. Figure 6.11 depicts polarized SERS spectra observed in 1 μM and 1 mM 4,4′-bipyridine solutions at distinct SERS active sites for excitation parallel to the long dimer axis. At site A, perpendicular polarized scattering was not observed, as in the case shown in Fig. 6.10. When the concentration was increased to 1 mM at the same site, there was still no perpendicular scattering (Fig. 6.11*b*). At site B, relatively weak perpendicular scattering was observed both in 1 μM and 1 mM solutions (Figs. 6.11*c*,*d*). The perpendicular scattering became much more apparent at site C (Figs. 6.11*e*,*f*). For 1 μM solutions (Fig. 6.11*e*), perpendicular polarized scattering was observed at site C with comparable intensity to the parallel polarization. For 1 mM solutions (Fig. 6.11*f*), perpendicular polarized scattering was weak as in the case of site B. It should be noted that all of the observations in Fig. 6.11 were carried out before and after confirmation that no SERS signal was observed with excitation polarization perpendicular to the long dimer axis at the same site.

The SERS spectral features were also dependent on the molecular orientation. Differences in the relative intensities between the bands and the wavenumbers should reflect the orientation of molecules relative to the direction of the localized electromagnetic field at the dimer gap. Anisotropy of the highly localized electro-

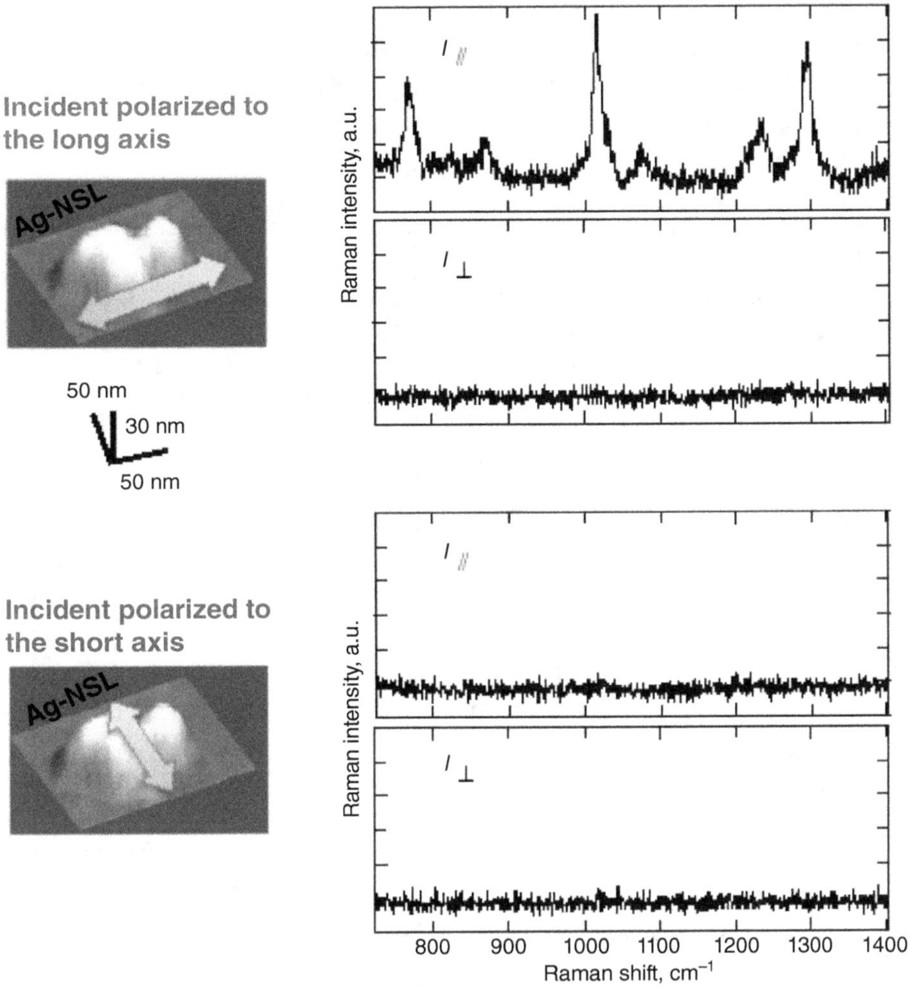


**Figure 6.10** AFM image of a single Ag nanodimer and SERS spectra observed in a 1-mM 4,4′-bipyridine aqueous solution at the parallel and the perpendicular polarization configuration with the parallel excitation to the dimer axis (*upper*), and the parallel and perpendicular polarization configuration with the perpendicular excitation to the dimer axis (*lower*); $I_{ex}$ = 50 mW, $t_{ex}$ = 1 s.

magnetic field at the gap leads to a sensitive response in the intensity of the SERS band [24–26]. To analyze the molecular orientation on the surface, the spectral features were compared with those predicted by DFT. SERS spectra at site A exhibited relatively intense bands at 770 and 871 cm$^{-1}$ that are assigned to the out-of-plane ring and C–H deformation modes, respectively. These features were reproduced in DFT calculations for polarization in the *X* direction for both excitation and scattering (middle of Figs. 6.11*a*,*b*). Relatively strong out-of-plane modes versus the in-plane modes could be attributable to the "flat" adsorption of the 4,4′-bipyridine rings on

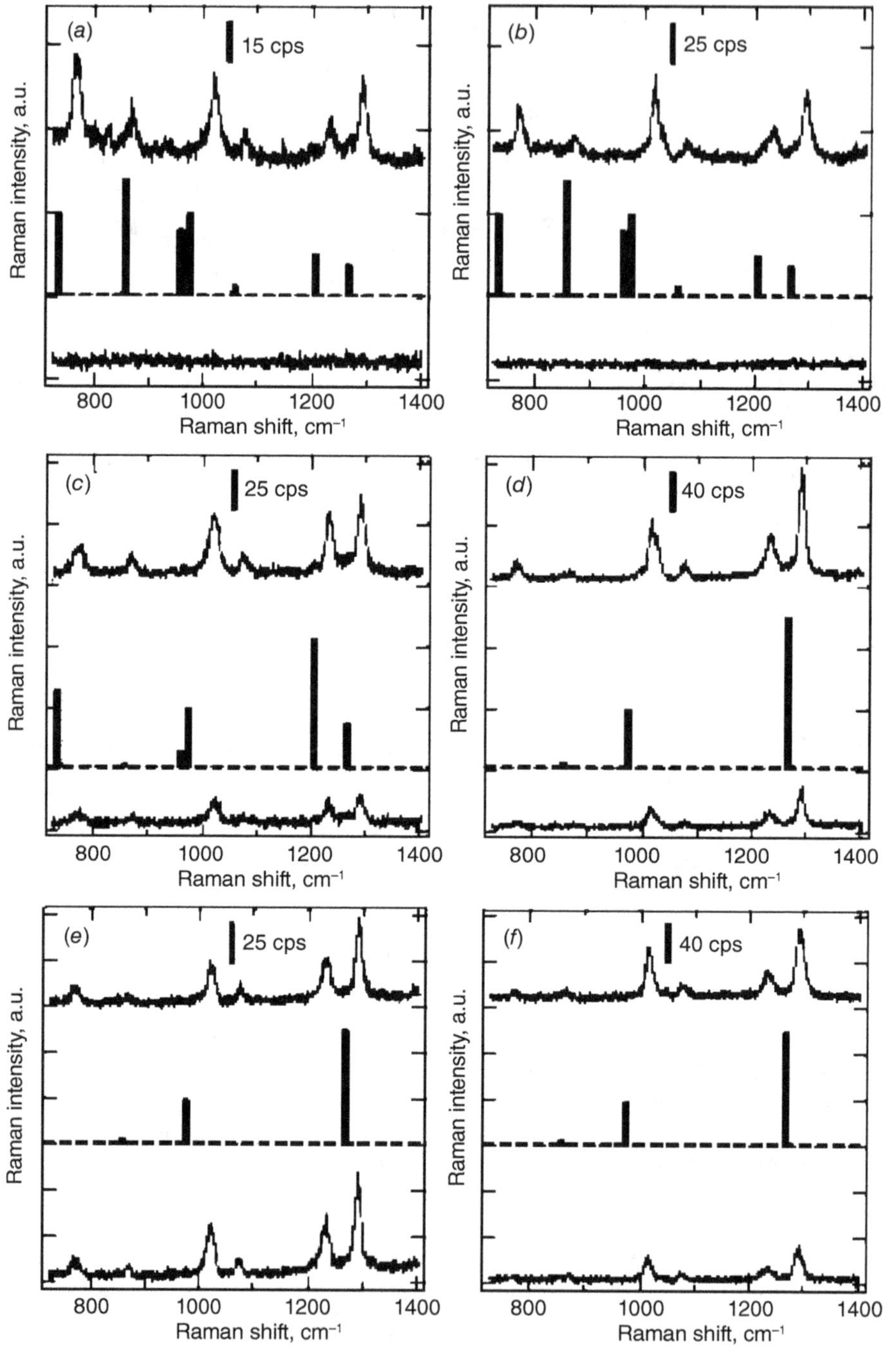


**Figure 6.11** SERS spectra observed in 1 mM (*a*, *c*, *e*) and 1 mM (*b*, *d*, *f*) 4,4′-bipyridine aqueous solutions observed at site A (*a*, *b*), site B (*c*, *d*), and site C (*e*, *f*) at the parallel (*upper*) and perpendicular (*lower*) polarization configuration with the excitation parallel to the dimer axis ($I_{ex}$ = 10 mW, $t_{ex}$ = 1 s); DFT calculation (*middle*) for the polarization of *X*-excitation and *X*-scattering (*a*, *b*), *Y*-excitation and *Y*-scattering (*c*, *d*), *Z*-excitation and *Z*-scattering directions (*d*, *e*, *f*) (after Ref. [46]).

the Ag surface. A localized electromagnetic field perpendicular to the bipyridine rings on the surface may enhance the relative intensity of the band. At site B, the SERS spectrum observed at 1 μM showed a sharp increase in the relative intensity of the band at 1240 $cm^{-1}$ (Fig. 6.11*c*). At 1 mM, the band at 1296 $cm^{-1}$ was observed (Fig. 6.11*d*). DFT calculations suggest that the band at 1240 $cm^{-1}$ assigned to the in-plane C–H bending mode becomes intense for the “flat” orientation in which the $Y$ axis of the molecule is normal to the metal surface. An intense band observed at 1300 $cm^{-1}$, assigned to the inter-ring stretching mode, could be attributed to the “standing” orientation in which the molecular $Z$ axis is normal to the surface. The features at site C were comparable to that of site B for 1 mM solutions. The band analysis thus indicates the site dependence of the molecular orientation in this system. In addition to this observation, the depolarization behavior clearly demonstrates that the orientation of the molecules at the gap affected the polarization of the scattering process.

The polarization dependence of SERS has been discussed with respect to the individual tensors for the excitation and scattering fields. To explain the present observation of depolarized scattering photons, a selective contribution to the scattering process should be considered. One of the possible contributions is the two different resonance modes seen in localized surface plasmon resonance spectroscopy relative to the incident and scattered light [45]. The extinction of the Ag dimer at the longer wavelengths (>850 nm) for perpendicular polarization contributes to the resonance at perpendicular polarization of the Raman scattering around 1300 $cm^{-1}$. This process, however, cannot explain the polarization dependence of the molecular orientation induced by changing the solution concentration at the same site [41]. Another possibility is that the scattering depolarization observed in the present system could be induced by the resonance of the scattering photons with the localized electronic states caused by the specific adsorption. In the present system, bipyridine molecules adsorbed on the metal surface tend to have the “standing” orientation when the concentration is 10–50 μM due to the coordination ability of the nitrogen atom and the intermolecular interaction. Specific adsorption of molecules could lead to the formation of localized electronic states at the molecule–metal interface. The contribution of localized resonance states to the SERS process has been considered as an excitation process involving charge-transfer (CT) contribution, the so-called chemical enhancement, leading to the observation of nontotally symmetric modes [10, 12, 57, 58]. Although the bands in the SERS spectra observed here were fully assigned to totally symmetric modes, characteristics of the SERS spectral features such as relatively intense bands at 1229 and 1298 $cm^{-1}$, compared with those in bulk solution and with those of the single crystal, could imply a CT contribution. To demonstrate the contribution of the resonance of the localized states to the depolarization, further detailed analysis would be required for a well-controlled single-molecule SERS.

The electrochemical potential of the Au dimer also can be utilized to control the molecular orientation. Figure 6.12 depicts SERS spectra observed in a 1 mM 4,4′-bipyridine aqueous solution under electrochemical potential control. As the potential was changed from +500 mV to 0 mV and −500 mV versus Ag/AgCl, the

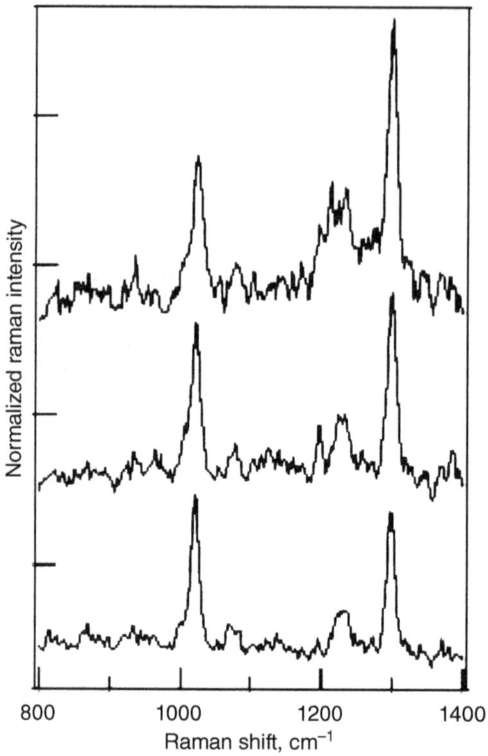


**Figure 6.12** SERS spectra observed in 1 mM 4,4′-bipyridine and 0.1 M $Na_2SO_4$ aqueous solutions at the Ag dimer array electrode under electrochemical potential control. Raman band intensities were normalized at 1020 cm$^{-1}$. From upper to lower, the potential of the electrode was controlled at +500, 0, and −500 mV versus Ag/AgCl.

relative intensity of the 1300 cm$^{-1}$ band becomes weaker compared with that of the 1240 cm$^{-1}$ band. This spectral difference at distinct potentials indicates that the "standing" orientation of 4,4′-bipyridine at +500 mV changes to the "flat" orientation as the electrochemical potential becomes negative. The "standing" spectrum at +500 mV also shows the difference in the bands at 1240 and 1300 cm$^{-1}$, compared with the "standing" spectra of the system in the absence of the electrochemical potential (Figs. 6.11*e,f*). The relatively strong intensity of these bands may also suggest changes in the vibrational and electronic structure of the molecules by the strong interaction with the electrode at positive potential. Similar effects of CT could be also considered. Polarized SERS measurements are also under way using the electrochemical system to obtain insights into the electronic structure of the system. Depolarization is also expected if additional polarization occurs due to the induced dipole of the electronic excitation.

## 6.5 CONCLUSION

Current progress in the experimental observation of polarized SERS has been summarized. Several examples were introduced along with the experimental methods of

Raman measurements. The orientation of molecules on surfaces can be analyzed by comparison of SERS spectra with those obtained by conventional polarized Raman spectra from a homogeneous aqueous solution as well as from a single crystal. Several systems show depolarized SERS that is not solely attributed to the optical properties of the metallic nanostructure. Continuous polarized Raman monitoring on the same metal structure indicates that the molecular orientation affects the polarization characteristics of SERS photons. This result is similar to the previous observation of molecule-dependent depolarization ratios on highly uniform prolate gold nanoparticles prepared by electron beam lithography. Possible contributions to the depolarization could be polarization of electronic excitation. Recently, we reported SERS for a single 4,4′-bipyridine molecule under conditions of in situ current flow in a nanogap, by using nanofabricated mechanically controllable break junction electrodes [59]. In this system, nontotally symmetric $b_1$ and $b_2$ modes of 4,4′-bipyridine were strongly enhanced. The contribution of molecule-to-metal or metal-to-molecule CT excitation was discussed as a means to modify the Raman polarizability. The modulation of the selection rules for optical absorption processes at a highly localized electromagnetic field should also be considered for such nanostructures [60, 61]. To clarify the detailed characteristics of the additional polarization of SERS photons, the analysis of plasmonic circular dichroism should be performed as well. These observations would be indispensable for a better understanding of the molecular specificity of SERS. After these investigations, we may begin in situ SERS monitoring of electrochemical interfaces [62].

## ACKNOWLEDGMENTS

The authors are grateful to Mr. Ichiro Chiba (Photon Design Co Ltd.) for construction of the polarized Raman microscope. This work was partially supported by the Grants-in-Aid for scientific research 18750001, from the Ministry of Education, Science and Culture, Japan. Particularly, support from the Priority Area "Strong Photon-Molecule Coupling Fields (No. 470)" is acknowledged.

## REFERENCES

1. Bockris, J. O. M.; Khan, S. U. M. *Surface Electrochemistry: A Molecular Level Approach*, Springer: New York, **1984**, pp. 37–50.
2. Birke, R. L.; Lombardi, J. R. In *Spectroelectrochemistry*, R. J. Gale Ed., Plenum: New York, **1988**, pp. 263–348.
3. Chang, R. K.; Laube, B. L. In *Critical Review in Solid State and Materials Science*. CRC Press: New York, **1984**, pp. 1–73.
4. Rhee, H.; Choi, J.-H.; Cho, M. *Acc. Chem. Res.* **2010**, *43*, 1527–1536.
5. Burstein, E.; Lundqvist, S.; Mills, D. L. In *Surface Enhanced Raman Scattering*; Chang, R. K.; Furtak, T. E., Eds.; Plenum Press: New York, **1982**, pp. 67–89.
6. Moskovits, M. *Rev. Mod. Phy.* **1985**, *57*, 783–826.
7. Jeanmaire, D. L.; Van Duyne, R. P. *J. Electroanal. Chem. Interfacial Electrochem.* **1977**, *84*, 1–20.
8. Bindemann, R.; Schwabe, R.; Haensel, T. *Phys. Stat. Solidi B Basic Res.* **1978**, *87*, 169–177.

9. Gersten, J. I.; Birke, R. L.; Lombardi, J. R. *Phys. Rev. Lett.* **1979**, *43*, 147–150.
10. Lombardi, J. R.; Birke, R. L. *J. Phy. Chem. C* **2008**, *112*, 5605–5617.
11. Lombardi, J. R.; Birke, R. L. *Acc. Chem. Res.* **2009**, *42*, 734–742.
12. Lombardi, J. R.; Birke, R. L.; Haran, G. *J. Phys. Chem. C* **2011**, *115*, 4540–4545.
13. Sawai, Y.; Takimoto, B.; Nabika, H.; Ajito, K.; Murakoshi, K. *J. Am. Chem. Soc.* **2007**, *129*, 1658–1662.
14. Blackie, E.; Le Ru, E. C.; Meyer, M.; Timmer, M.; Burkett, B.; Northcote, P.; Etchegoin, P. G. *Phys. Chem. Chem. Phys.* **2008**, *10*, 4147–4153.
15. Dieringer, J. A.; Wustholz, K. L.; Masiello, D. J.; Camden, J. P.; Kleinman, S. L.; Schatz, G. C.; Van Duyne, R. P. *J. Am. Chem. Soc.* **2008**, *131*, 849–854.
16. Chen, T. T.; Smith, K. E.; Owen, J. F.; Chang, R. K. *Chem. Phys. Lett.* **1984**, *108*, 32–38.
17. Sanchez, L. A.; Lombardi, J. R.; Birke, L. *Chem. Phys. Lett.* **1984**, *108*, 45–50.
18. Kung, H. P.; Chen, T. T. *Chem. Phys. Lett.* **1986**, *130*, 311–315.
19. Chase, D. B.; Parkinson, B. A. *Appl. Spec.* **1988**, *42*, 1186–18187.
20. Mrozek, I.; Bornemann, T.; Otto, A. *Surf. Sci.* **1988**, *204*, 358–373.
21. Gui, J.; Devine, T. M. *Surf. Sci.* **1989**, *224*, 525–542.
22. Kwon, M.-Y.; Lee, B.-W.; Kim, J.-J. *Chem. Phys. Lett.* **1989**, *162*, 238–242.
23. Tian, Z. Q.; Lin, T. Q. *J. Electroanal. Chem.* **1989**, *265*, 277–282.
24. Ambrosio, R. C.; Gewirth, A. A. *Anal. Chem.* **2010**, *82*, 1305–1310.
25. Li, S.; Pedano, M. L.; Chang, S.-H.; Mirkin, C. A.; Schatz, G. C. *Nano Lett.* **2010**, *10*, 1722–1727.
26. Masiello, D. J.; Schatz, G. C. *J. Chem. Phys.* **2010**, *132*, 064102/1–064102/8.
27. McMahon, J. M.; Li, S.; Ausman, L. K.; Schatz, G. C. *J. Phy. Chem. C* **2011**, *116*, 1627–1637.
28. Ausman, L. K.; Li, S.; Schatz, G. C. *J. Phys. Chem. C* **2012**, *116*, 17318–17327.
29. Wang, H.; Brandl, D. W.; Nordlande, P.; Halas, N. J. *Acc. Chem. Res.* **2007**, *40*, 53–62.
30. Mukherjee, S.; Sobhani, H.; Lassiter, J. B.; Bardhan, R.; Nordlander, P.; Halas, N. J. *Nano Lett.* **2010**.
31. Nabika, H.; Takase, M.; Nagasawa, F.; Murakoshi, K. *J. Phys. Chem. Lett.* **2010**, *1*, 2470–2487.
32. Itoh, T.; Yoshida, K.; Biju, V.; Kikkawa, Y.; Ishikawa, M.; Ozaki, Y. *Phys. Rev. B* **2007**, *76*, 085405 /1–085405/5.
33. Itoh, T.; Sujith, A.; Ozaki, Y. *Front. Mol. Spectrosc.* **2009**, 289–319.
34. Sano, H.; Mizutani, G.; Ushioda, S. *Phys. Rev. B* **1993**, *47*, 13773–13781.
35. Bosnick, K. A.; Jiang, J.; Brus, L. E. *J. Phys. Chem. B* **2002**, *106*, 8096–8099.
36. Jiang, J.; Bosnick, K.; Maillard, M.; Brus, L. *J. Phys. Chem. B* **2003**, *107*, 9964–9972.
37. Xu, H.; Kaell, M. *ChemPhysChem* **2003**, *4*, 1001–1005.
38. Etchegoin, P. G.; Galloway, C.; Le Ru, E. C. *Phys. Chem. Chem. Phys.* **2006**, *8*, 2624–2628.
39. Shegai, T. O.; Haran, G. *J. Phys. Chem. B* **2006**, *110*, 2459–2461.
40. Gucciardi, P. G.; Lopes, M.; Déturche, R.; Julien, C.; Barchiesi, D.; Chapelle, M. L. D. L. *Nanotechnology* **2008**, *19*, 215702.
41. Le Ru, E. C.; Grand, J.; Felidj, N.; Aubard, J.; Lelvi, G.; Hohenau, A.; Krenn, J. R.; Blackie, E.; Etchegoin, P. G. *J. Phys. Chem. C* **2008**, *112*, 8117–8121.
42. Le Ru, E. C.; Meyer, M.; Blackie, E.; Etchegoin, P. G. *J. Raman Spectrosc.* **2008**, *39*, 1127–1134.
43. Motohashi, M.; Hayazawa, N.; Tarun, A.; Kawata, S. *J. Appl. Phys.* **2008**, *103*, 031309.
44. Shegai, T.; Li, Z.; Dadosh, T.; Zhang, Z.; Xu, H.; Haran, G. *Proc. Natl. Acad. Sci.* **2008**, *105*, 16448–16453.
45. Li, Z.; Shegai, T.; Haran, G.; Xu, H. *ACS Nano* **2009**, *3*, 637–642.
46. Nagasawa, F. T., Mai; Nabika, Hideki; Murakoshi, Kei *Chem. Commun.* **2011**, *47*, 4514–4516.
47. Mrozek, I.; Otto, A. *J. Electron Spectrosc. Relat. Phenom.* **1987**, *45*, 143–152.
48. Wandlowski, T.; Ataka, K.; Mayer, D. *Langmuir* **2002**, *18*, 4331–4341.
49. Topacli, A.; Akyuz, S. *Spectrochim. Acta, Part A* **1995**, *51A*, 633–641.
50. Uosaki, K.; Hill, H. A. O. *J. Electroanal. Chem. Interfacial Electrochem.* **1981**, *122*, 321–326.

51. Yang, H.; Sun, Y.; Ji, J.; Song, W.; Zhu, X.; Yao, Y.; Zhang, Z. *Corros. Sci.* **2008**, *50*, 3160–3167.
52. Sawai, Y.; Takimoto, B.; Nabika, H.; Ajito, K.; Murakoshi, K. *J. Am. Chem. Soc.* **2007**, *129*, 1658–1662.
53. Nabika, H.; Sasaki, A.; Takimoto, B.; Sawai, Y.; He, S.; Murakoshi, K. *J. Am. Chem. Soc.* **2005**, *127*, 16786–16787.
54. Takimoto, B.; Sawai, Y.; Nabika, H.; Ajito, K.; Murakoshi, K. *Trans. Mater. Res. Soc. Jpn.* **2005**, *30*, 561–564.
55. Sawai, Y.; Takimoto, B.; Nabika, H.; Murakoshi, K. *Can. J. Anal. Sci. Spectros.* **2007**, *52*, 142–149
56. Takase, M.; Sawai, Y.; Nabika, H.; Murakoshi, K. *Trans. Mater. Res. Soc. Jpn.* **2007**, *32*, 409–412.
57. Lombardi, J. R.; Birke, R. L.; Lu, T.; Xu, J. *J. Chem. Phys.* **1986**, *84*, 4174–4180.
58. Lombardi, J. R.; Birke, R. L. *Acc. Chem. Res.* **2009**, *42*, 734–742.
59. Konishi, T.; Kiguchi, M.; Takase, M.; Nagasawa, F.; Nabika, H.; Ikeda, K.; Uosaki, K.; Ueno, K.; Misawa, H.; Murakoshi, K. *J. Am. Chem. Soc.* **2013**, *135*, 1009–1014
60. Jain, P.; Ghosh, D.; Baer, R.; Rabani, E.; Paul, A. *Proc. Natl. Acad. Sci.* **2012**, *109*, 8016–8019.
61. Takase, M.; Ajiki, H.; Mizumoto, Y.; Komeda, K.; Nara, M.; Nabika, H.; Yasuda, S.; Ishihara, H.; Murakoshi, K. *Nat. Photonics*, **in press**.
62. Cortés, E.; Etchegoin, P. G.; Le Ru, E. C.; Fainstein, A.; Vela, M. E.; Salvarezza, R. C. *J. Am. Chem. Soc.* **2013**, *135*, 2809–2815.

Part Three

# IRRAS Spectroscopy (Including PM–IRRAS)

# Chapter 7

# DFT and In Situ Infrared Studies on Adsorption and Oxidation of Glycine, L-Alanine, and L-Serine on Gold Electrodes

**Andrea P. Sandoval,[1]* José Manuel Orts,[2] Antonio Rodes,[2] and Juan M. Feliu[2]**
[1]Departamento de Quimica, Facultad de Ciencias, Universidad Nacional de Colombia, Bogotá, Colombia
[2]Instituto de Electroquímica, Universidad de Alicante, Alicante, Spain

## 7.1 INTRODUCTION

Naturally occuring amino acids are chemical compounds of great importance in different scientific fields [1–9], especially in those related to life sciences. One interesting aspect of the behavior of amino acids is their adsorption on different surfaces. The combination in relatively small molecules of different functional groups, of different polarity and hydrophilicity, together with the coexistence of positive and negative centers of charge (in the zwitterionic form) makes very attractive the study of amino acids, which can be considered as good molecular probes for testing the effect of introducing structural and functional changes on their behavior, in particular for adsorption. Moreover, their intrinsic chirality (with the exception

*Present address: Departamento de Química, Facultad de Ciencias, Universidad Nacional de Colombia, Cra. 30# 45-03, Edificio 451, Bogotá (Colombia)

*Vibrational Spectroscopy at Electrified Interfaces*, First Edition.
Edited by Andrzej Wieckowski, Carol Korzeniewski, and Björn Braunschweig.

of glycine) can be used for adding chirality to intrinsically nonchiral surfaces [10]. The study of the adsorption of amino acids, both from theoretical and experimental points of view, is the first step in order to understand the way in which proteins interact with surfaces. In particular, the study of the adsorption of amino acids on gold, because of the good properties of this metal and its biocompatibility, is of relevance for technological applications, such as in bioelectronics [2, 3], the development of biosensors [4–6], medical devices [7], as well as for the understanding of the processes related with surface fouling [8–9].

A number of reports exist regarding the adsorption and electrochemical behavior of a number of amino acids on platinum [11–21] and gold [21–27] electrodes, including single crystals, both for acid and alkaline media. Either reversible or irreversible adsorption behavior has been reported depending on the amino acid, the electrode surface, the electrode potential, and the solution pH. The study of the behavior of several amino acids on platinum electrodes shows as common points the reversible, anionlike adsorption at low electrode potentials, linked to the presence of the carboxylate group, and the detection of $CO_2$ as their main oxidation product [11, 15–18, 20]. In alkaline solutions it was found that irreversible dissociative adsorption takes place [12, 21], forming adsorbed cyanide that was spectroscopically detected using in situ external reflection infrared [15–18, 20] or surface-enhanced Raman spectroscopy (SERS) [21]. Xiao et al. [21] compared the adsorption behavior of glycine, L-threonine, and L-serine in alkaline solutions and found that the threshold potential for the dissociation of these amino acids on platinum to produce cyanide adlayers increases in the order glycine < L-threonine < L-serine. Another adsorbate, CO, that results from dissociative adsorption has been reported for L-serine both on Pt(111) [16] and Pt(100) [18, 20] electrodes. On the latter surface, adsorbed NO and cyanate species were also detected during the oxidation of glycine [17].

Several investigations on the electrochemical behavior of amino acids on polycrystalline gold [21–23, 26] have been reported and also a few dealing with well-defined substrates such as gold single crystals [24, 25, 27]. Xiao et al. [21] extended their SERS study on the adsorption and electrooxidation behavior of glycine, L-serine, and L-threonine in alkaline media to polycrystalline gold electrodes, concluding that the formation of cyanide on gold surfaces requires the electrooxidation of the amino acids. Xiao's group also studied the electrochemical behavior of glycine in alkaline media on Au(111) and polycrystalline gold electrodes, using in situ infrared spectroscopy and the electrochemical quartz crystal microbalance (EQCM) [25]. A voltammetric and alternating-current (AC) impedance study of the adsorption of L-serine, L-tyrosine, and L-hystidine on polycrystalline gold electrodes [22] using cyclic voltammetry and AC impedance measurements found that the adsorption energy decreases in the order L-tyrosine ≥ L-hystidine > L-serine, while the maximum coverage has the opposite behavior. In the case of asparagine [26], a strong adsorption of the amino acid on gold electrodes was observed, concluding from infrared spectroscopy measurements that asparagine is anchored to the surface by the oxygen atoms of the carboxylate group and of the amide group. Finally, a spectroelectrochemical study by Li et al. [24] indicates that in neutral solutions L-phenylalanine adsorbs on Au(111) electrodes in the neutral form at potentials below the potential of zero charge, whereas the zwitterionic form, linked to the surface by the carboxylate group, predominates at more positive potentials [24].

Our group has recently studied the behavior of glycine on Au thin-layer and single-crystal electrodes by combining electrochemical measurements, in situ infrared reflection spectroscopy, and density functional theory (DFT). We found that glycine adsorbs reversibly on the gold surface through the oxygen atoms of the carboxylate group in an asymmetric bridge bidentate configuration, and there is a significant co-adsorption of perchlorate anions that compensates the positive charge density due to the ammonium groups of adsorbed zwitterionic glycine [27]. The formation of adsorbed cyanide and dissolved carbon dioxide were detected during the electrooxidation of glycine at high enough potentials [27].

It is interesting to compare the results of the aforementioned studies on the adsorption and oxidation of amino acids at platinum and gold electrodes with those carried out under ultrahigh vacuum (UHV) conditions and with DFT studies for these systems. In this way, glycine adsorption at the metal–vacuum interface has been studied for the (111) and (100) orientations of copper [28]. Upon adsorption, glycine undergoes deprotonation, yielding glycinate as the stable adsorbate under UHV conditions. DFT studies for these systems [29–32] concluded, in agreement with experiments, that the glycinate adsorbs on these surfaces in a bidentate, bridge configuration, at low coverages, with bonding through the two oxygen atoms of the carboxylate. At high coverage, the nitrogen atom, which is placed in a "top" adsorption site, seems to form an additional bond to the metal. We are not aware of similar studies for gold surfaces. Other theoretical studies reveal that the zwitterionic configurations of amino acids are the stable species in solution due to strong interactions with the solvent [33, 34]. Some theoretical studies of the adsorption of L-alanine on copper surfaces have been reported [35–37]. The adsorbed zwitterion of L-alanine was found to be less stable than the neutral form on Pd(111) surfaces [38]. Only one study has been reported on the adsorption of L-serine on metals, dealing with the neutral form [39].

The possibility of interconversion between the neutral and zwitterionic form of the amino acids must be taken into account when studying these systems. It is well known that the attempts to optimize the zwitterions structures in the gas phase led to the nonzwitterionic form. In the absence of water, geometry optimization of zwitterionic L-alanine on Ni(111) [40] resulted in adsorbed neutral L-alanine that could bond to the nickel surface either by the N atom or by the oxygen of the carbonyl group. However, with the addition of a water molecule, the zwitterionic form was stable enough to adsorb through one of the oxygen atoms of the carboxylate. The interconversion between the neutral and zwitterionic forms of amino acids implies the transfer of a proton from the ammonium to the carboxylate group, through the formation of a cyclic transition state with intramolecular hydrogen bonds. Theoretical studies of the microsolvation of amino acids show that the addition of some water molecules can stabilize the zwitterionic form. The minimum number of water molecules needed depends on the identity of the amino acid and also on the details of the calculation model (functional, basis sets, etc.). For instance, seven water molecules are needed to stabilize the zwitterion with respect to the neutral L-alanine. If continuum solvent models are additionally included in the calculations, only two or three explicit water molecules are needed [41]. The effect of water molecules on the stabilization of the zwitterionic form has also been studied for L-serine [42]. At the metal–solution electrochemical interface, the situation is even more complex because of the interaction with the surface metal atoms and with the interfacial water, whose dipole orientation is strongly affected by the electric field. In this chapter we present some results that provide insight on the effect of the interactions with the metal surface, a water molecule and a external electric field, on the geometry and vibrational frequencies of adsorbed zwitterions of the three amino acids studied.

The experimental investigations on the nature and adsorption configuration of adsorbates at electrode surfaces are based in most cases on the use of in situ vibrational spectroscopies (including infrared and Raman). Two main approaches can be adopted for the infrared characterization of the metal–solution interfase [43–45]. A first possibility is to use the external reflection approach in a thin-layer configuration with either thin-film or bulk electrodes, including single crystals. On the other hand, the use of internal reflection spectroscopy with a Kretschmann configuration requires working with a thin metal film deposited onto the surface of a high refraction index window but has several significant advantages over external reflection spectroscopy, such as a higher signal-to-noise ratio, low electric resistance, minimal interference by species in the solution bulk, and enhancement of the absorption by some orders of magnitude due to the surface enhanced infrared absorption (SEIRA) effect [43, 46]. This makes SEIRA spectroscopy under attenuated total reflection conditions (ATR-SEIRAS) one of the most powerful techniques not only for the in situ detection of adsorbed species at low coverages (including intermediates) but also for detecting changes in the infrared absorption by the first molecular layers (composed of solvent as well as other species) in the immediate vicinity of the electrode surface. As a drawback, the need of using thin-layer electrodes means that the control of surface orientation and order is not as straightforward as in single-crystal work. However, it is possible to obtain thin-layer surfaces with preferential (111) orientation by carrying out an electrochemical annealing of gold thin films deposited at low deposition rates, in solutions containing anions that adsorb specifically such as sulfate [45,47] or acetate [48]. The latter procedure was chosen for improving the surface order of gold thin films thermally evaporated on a silicon prism, allowing the in situ spectroelectrochemical study of amino acids on (111)-oriented Au electrodes using ATR.

This chapter reports a comparative spectroelectrochemical study of the behavior of the three amino acids with the simpler structures: glycine, L-alanine, and L-serine. Following a previous work devoted to the study of glycine adsorption at gold electrodes [27], we focus here on the electrochemical behavior of L-alanine and L-serine, presenting their adsorption on Au(111) and Au(100) single-crystal and quasi-Au(111) thin-film electrodes in acidic media. Cyclic voltammetry (CV) and in situ infrared spectroscopy results, both with internal and external reflection configurations, will be discussed and compared with the previously published results obtained with glycine [27]. In order to better support the assignment of experimental vibrational frequencies, the optimized adsorbate geometries and theoretical harmonic vibrational frequencies of adsorbed L-alanine and L-serine species are reported.

## 7.2 MODEL AND COMPUTATIONAL DETAILS

A theoretical study of the adsorbed glycine, L-alanine, and L-serine species was carried out for supporting the assignment of experimental bands recorded in the in situ vibrational spectroscopy experiments. The supermolecule approach was chosen where molecular adsorbates are put in contact with a metal cluster representing the metallic surface. Despite the limitations of this approach (e.g., for obtaining accurate adsorption energies), it has been widely used for calculating adsorption geometries and vibrational frequencies of adsorbates that correlate well with experimental results. This is a consequence of the local character of the vibrational frequencies, which are not very sensitive to the size of the metal cluster.

Our calculations have been obtained with a $Au_{31}$ cluster modeling the unreconstructed Au(111) surface. This choice is based on two facts: the first one is that the gold thin-film electrodes used in this work have a marked (111) orientation, with wide (in the 20- to 30-nm range) flat ordered atomic domains, and a voltammetric response very close to that of the Au(111) single-crystal surface. The second reason is that previous experimental and theoretical studies with metal surfaces of low Miller indices show that the frequencies of adsorbed carboxylate anions do not depend significantly on the crystallographic orientation of the metal surface [49–51]. The metal atoms in the $Au_{31}$(111) cluster are arranged in two layers of 19 and 12 atoms. The size of the metal cluster is big enough to prevent, during a full optimization of the adsorbate geometry, the bonding of the adsorbate species to the metal atoms in the cluster border. The geometry of the metal cluster was kept fixed, with the gold nuclei located at their positions in the truncated crystal and the same distances between neighbouring gold atoms as in the bulk metal (0.28837 nm) [52].

The calculations were carried out for one adsorbed zwitterion ($NH_3^+$–CHX–$COO^-$ with X = H, $CH_3$, and $CH_2OH$), as this is the expected form of the amino acids in aqueous solutions (as opposed to the neutral form, $NH_2$–CHX–COOH, which is the stable one in the gas phase). Calculations were carried out both in the absence and in the presence of one added water molecule, in order to understand the effect of water–zwitterion interactions (in particular hydrogen bonding) on the equilibrium geometry and vibrational frequencies. Another significant variable affecting the adsorption behavior under electrochemical conditions is the electric field present in the metal–solution electrified interface. In order to qualitatively assess its effects, calculations were also carried out with an external electric field in the normal direction to the metal surface of 0.01 atomic units (equivalent to $5.14 \times 10^9$ V $m^{-1}$), such as to generate a positive surface charge density in the cluster side facing the adsorbate. This corresponds to experiments where the electrode potential is positive to the potential of zero charge (PZC). These two ingredients in the model account (at least partially) for the importance of solvation, by screening the positive charge in the ammonium group, and applied potential, by removing negative charge density from the oxygen atoms of the carboxylate group.

A full optimization of the geometry of the adsorbate system was carried out using the B3LYP functional as implemented in the Gaussian 03 code [53]. The B3LYP functional combines the three-parameter hybrid exchange functional of

Becke [54] with the Lee–Yang–Parr correlation functional [55]. This functional, in combination with the 6-31+G(d) [56–58] basis set for the C, O, and H atoms and the LANL2DZ [59] effective core potential and associated double-zeta basis set for describing the gold metal atoms, has been used to estimate theoretical values of harmonic frequencies that coincide remarkably well with the experimental values measured for adsorbed acetate on Cu [60, 61], Ag [49], and Au [50] surfaces. A good agreement between theoretical and experimental frequency values was also found for some acetate derivatives, such as fluoroacetates [51], glyoxylate [62], and glycolate [63] anions adsorbed on gold surfaces. All frequency values are given without scaling. Assignments of the calculated frequencies are based on the visualization of the vibrational normal modes using Molden [64].

## 7.3 EXPERIMENTAL SECTION

Working solutions of perchloric acid were prepared from concentrated acid (Merck Suprapur, Darmstadt) and ultrapure water (18.2 MΩcm, Veolia Water, Paris). Glycine, L-alanine, or L-serine (>99%, Sigma-Aldrich, St. Louis, MO) was added to the perchloric acid solution to obtain the desired concentrations. Solutions were deaerated with Ar (N50, Air Liquide, Paris). Solutions in deuterated water were prepared with deuterium oxide (99.9 atom %D, Aldrich), which was used as received.

All the experiments were performed in glass cells using a large-area gold wire spiral as the counterelectrode and a reversible hydrogen electrode (RHE) as the reference electrode in a separate vessel connected to the main cell body through a Luggin capillary. The cells used in spectroelectrochemical experiments [44, 65] are equipped with a prismatic window beveled at 60°. This prism was made from Si or $CaF_2$ for the internal and external reflection experiments, respectively. All the spectra are presented in absorbance units (a.u.). They were obtained with a resolution of 8 $cm^{-1}$ and are referred to the single-beam spectrum collected at 0.10 V in the same solution. In external reflection experiments, the spectra collected in the potential region where the amino acids show reversible adsorption/desorption were obtained from 10 sets of 100 interferograms, which were accumulated alternately at the reference and sample potentials using the so-called SNIFTIR (subtractively normalized interfacial Fourier transform infrared) technique. A single set of 100 interferograms was collected in the ATR-SEIRAS experiments.

The working electrodes used in the electrochemical and in situ external reflection infrared spectroscopy experiments were gold single crystals with diameters around 2.0 and 4.5 mm, respectively. They were prepared from a high-purity gold wire (99,9998% Alfa-Aesar, Ward Hill, MA) following Clavilier's method [66], which was first developed for platinum electrodes [67]. Prior to each experiment, the electrodes were heated in a gas–oxygen flame, cooled down in air, and protected with a droplet of ultrapure water.

In the internal reflection infrared spectroscopy experiments a 25-nm-thick gold thin- film thermally evaporated on a silicon prism was used as the working electrode. Deposition was carried out in the vacuum chamber of a PVD75 coating system (Kurt

J. Lesker, Clairton, PA) equipped with a turbomolecular pump. Before deposition, pressure was lowered at ca. $10^{-6}$ Torr. The deposition rate, which was fixed at 0.006 nm $s^{-1}$, and the thin-film thickness were monitored with a quartz crystal microbalance. Once assembled the spectroelectrochemical cell, the gold film was electrochemically annealed by cycling the electrode potential between 0.10 and 1.20 V RHE in a 0.1 M $HClO_4$ + 10 mM $CH_3COONa$ solution at 20 mV $s^{-1}$ for 1–2 h [48]. Then the solution was replaced by an acetate-free perchloric acid solution.

## 7.4 RESULTS AND DISCUSSION

### 7.4.1 Espectroelectrochemical Results

Figures 7.1 and 7.2 show voltammetric profiles for Au(100) and Au(111) electrodes in glycine, L-alanine, or L-serine containing perchloric acid solutions in the low-potential region. The existence of a similar adsorption/desorption behavior in the double-layer potential region can be deduced from the curves reported in Figure 7.1 for these simple amino acids. Namely, an increase of the voltammetric charge is observed on both electrode surfaces in the presence of the amino acid with respect to that measured in the amino-acid-free perchloric acid solution. Note that in the case of L-serine, the upper potential limit in the double-layer region is restricted to 0.90 V because at higher potentials its irreversible oxidation takes place, as will be discussed below. The voltammetric curves recorded for amino acid concentrations up to 50 mM clearly show that the voltammetric charge between 0.10 and 1.10 V steadily increases when increasing the amino acid concentration (not shown). At the same time, it can be observed that the charge under the peak for the lifting of the surface reconstruction, which appears at 0.85 V for the Au(100) electrode in the amino-acid-free perchloric acid solution [68, 69], increases with the amino acid concentration while shifting toward less positive potentials. For the Au(111) electrode (Fig. 7.1*b*), for which it has also been reported the existence of a potential-dependent lifting of the reconstructed surface obtained after flame annealing [68, 69], the shift toward lower potentials of the voltammetric peak associated to this process in the presence of these amino acids is less evident than in the case of the Au(100) electrode, but is still observed. The similarity of the features within the double-layer zone suggests a pattern of adsorption that could be common to these three amino acids. Also, as the peak for the lifting of the reconstruction is shifted in almost the same extent for the three amino acids, it can be concluded that the adsorption of these amino acids makes more stable the nonreconstructed $(1 \times 1)$ surface phase.

The voltammograms recorded for the Au(100) and Au(111) electrodes in the potential zone for the oxidation of the studied amino acids show significant differences, showing that this reaction is a surface-sensitive process on gold electrodes that could be promoted by the surface oxides. Figure 7.2 allows the comparison of the voltammetric curves recorded up to 1.70 V for 10 mM glycine, L-alanine, or L-serine solutions with Au(100) and Au(111) electrodes. It can be observed that glycine oxidation at the Au(100) electrode, for which a reversible preoxidation of

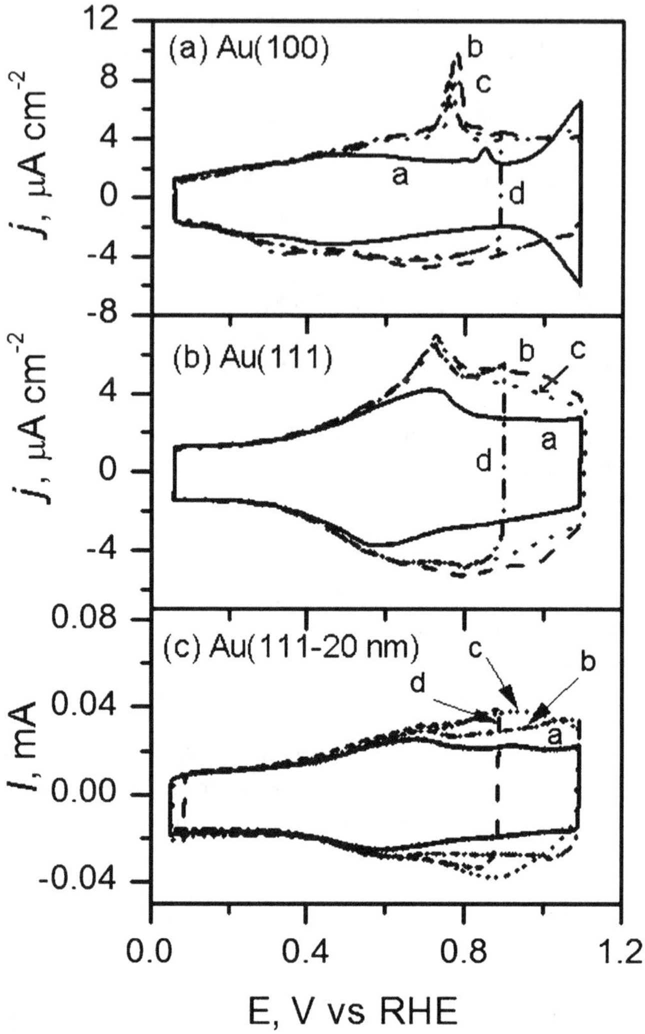


**Figure 7.1** Cyclic voltammograms within the double-layer zone for (*a*) Au(100), (*b*) Au(111) single-crystal, and (*c*) thin-film gold electrodes in 0.1 M $HClO_4$ before (*a*, solid line) and after the addition of 10 mM of *b* (dashed line) glycine, *c* (dotted line) L-alanine, and *d* (dashed-dotted line) L-serine. Scan rate = 50 mV $s^{-1}$.

the surface occurs between 1.0 and 1.25 V, takes place at lower potentials than at Au(111) electrodes. In the case of L-alanine, its oxidation on Au(100) electrodes starts at 1.1 V, that is, at potential less positive than the onset of the oxidation of the gold surface, and takes place in two voltammetric peaks at ca. 1.35 and 1.50 V. The voltammetric profile corresponding to L-alanine oxidation on Au(111) shows oxidation currents in the potential region above the onset of surface oxidation, with two peaks appearing at ca. 1.40 and 1.60 V. The relative intensity of the peak at the lower

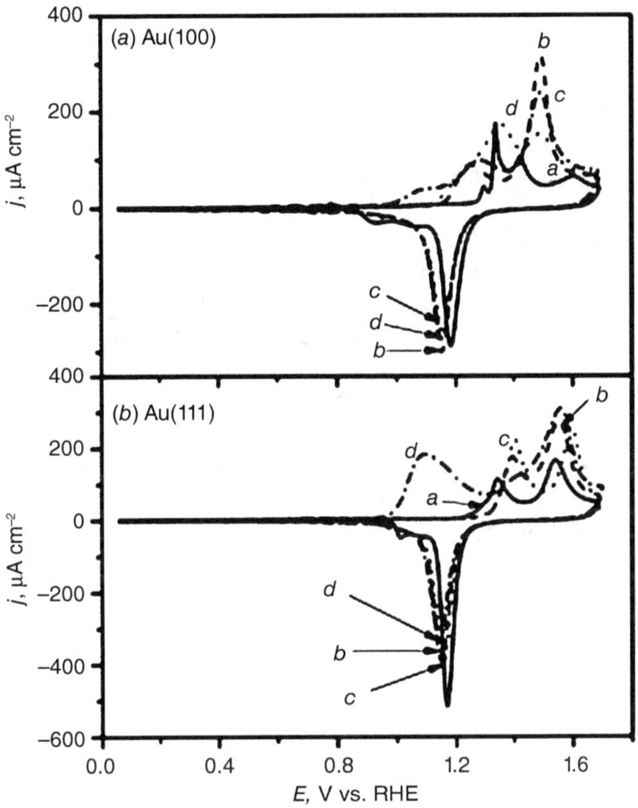


**Figure 7.2** Cyclic voltammograms for (*a*) Au(100) and (*b*) Au(111) electrodes in 0.1 M $HClO_4$ before (*a*, —) and after the addition of 10 mM of (*b*) (- - -) glycine, (*c*) (. . . .) L-alanine and (*d*) (- · · -) L-serine. Scan rate = 50 mV $s^{-1}$.

potential depends on L-alanine concentration. As shown in Figure 7.2, the oxidation of L-serine starts below 1.0 V, that is, at potentials much lower than for the other two amino acids studied. This behavior is probably a consequence of the presence of the $-CH_2-OH$ group that is more easily oxidizable than the methyl group, thus making easier the decarboxilation of the adsorbed L-serine molecules. This point will be discussed further in the light of the in situ spectroscopic data obtained in the potential region where the oxidation of the amino acids takes place.

The voltammetric study reported above on Au(111) and Au(100) single-crystal electrodes has been extended to gold thin-film electrodes deposited on silicon substrates. The cyclic voltammograms obtained for these gold thin-film electrodes in perchloric acid solutions in the absence and in the presence of glycine, L-alanine, or L-serine are shown in Figure 7.1*c* for the double-layer region. It can be clearly seen from these curves that the increase of the voltammetric charge related to the presence of each amino acid takes place in the same potential region as in the case of the Au(111) electrode surface (see Fig. 7.1*b*). Potential excursions up to 1.70 V have been avoided in the case of the gold thin-film electrodes since they can decrease their stability.

In order to gain further understanding at the molecular level about the electrode processes taking place at the gold electrodes in the amino-acid-containing solution, in situ infrared spectroscopy has been used to characterize the species present on the electrode during adsorption and oxidation of the corresponding amino acids. The external reflection infrared spectra obtained at 1.10 V in glycine- and L-alanine-containing solutions (spectra *a* and *b*) and at 1.0 V in the L-serine-containing solutions (spectrum *c*) are shown in Figures. 7.3*a*, *b* for the Au(111) and Au(100) single-crystal electrodes, respectively. On the other hand, Figures 7.4*a*, *b* allow the comparison of the ATR-SEIRA spectra collected for gold thin-film electrodes at 0.90 and 1.30 V, respectively. All these potential-difference spectra show bands related to the amino acid adsorption–desorption process for potentials above 0.60 V. In addition to a band at ca. 1610 $cm^{-1}$ related to the uncompensated absorption by the water bending mode, the external reflection spectra in Figure 7.3 show other negative-going bands at ca. 1730 and 1260 $cm^{-1}$ that can be assigned to the carbonyl CO stretch and C–OH bending modes of the cationic amino acid species ($NH_3^+$–R–

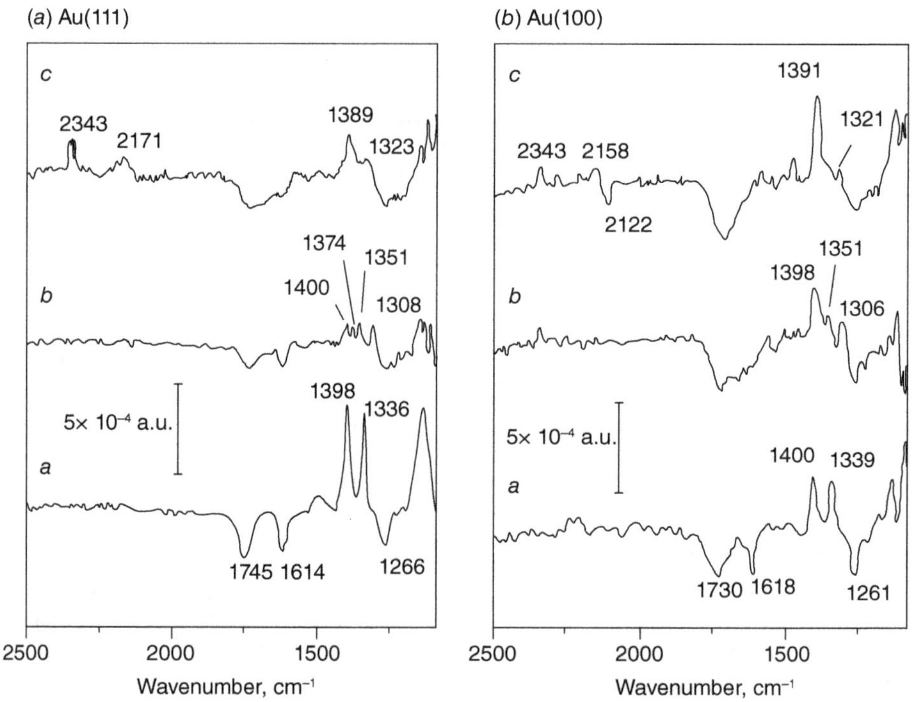


**Figure 7.3** In situ external reflection infrared spectra collected for (*a*) Au(111) and (*b*) Au(100) single-crystal electrodes in 10 mM amino acid + 0.1 M $HClO_4$ solutions. *a*, glycine; *b*, L-alanine, and *c*, L-serine. Sample potentials: 1.10 V (*a* and *b*) and 1.00 V (*c*). The reference spectrum was collected for each solution at 0.10 V. At each potential, 1000 interferograms were co-added.

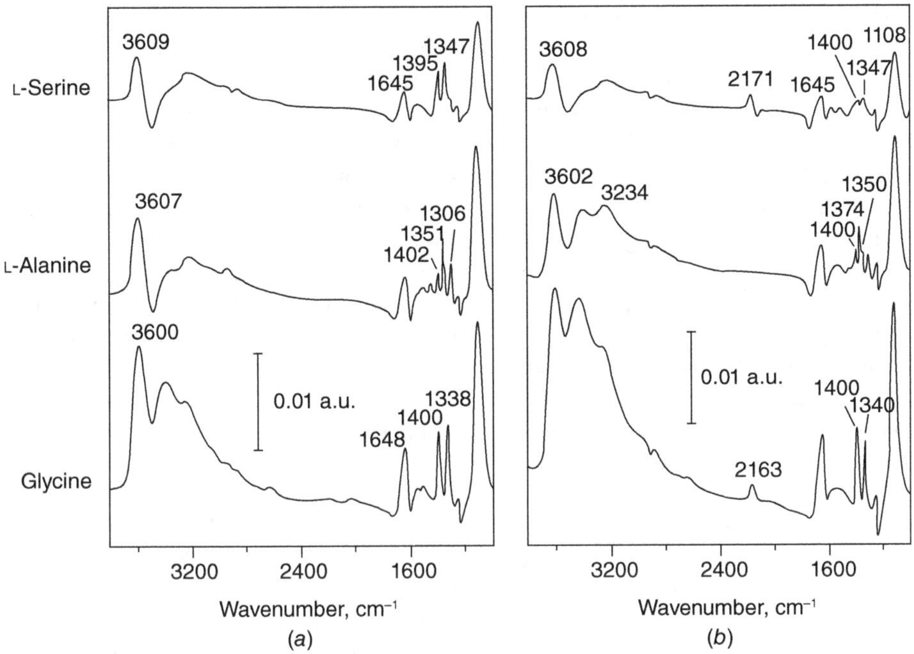


**Figure 7.4** ATR-SEIRA spectra collected for a Au(111) 25-nm thin-film electrode in 10 mM amino acid +0.1 M $HClO_4$ solutions at (*a*) 0.90 V and (*b*) 1.30 V. Reference potential 0.10 V. 100 Interferograms were collected at each potential.

COOH) in solution. The latter are consumed when stepping from the reference to the sample electrode potential. The corresponding bands cannot be appreciated in the internal reflection spectra reported in Figure 7.4 due to the surface specificity of the ATR-SEIRA bands.

Positive-going bands in Figure 7.3 (and some of those observed in Fig. 7.4) correspond to absorption by the adsorbed species formed from the corresponding amino acid. Even if the relatively low signal-to-noise ratio in the external reflection experiments makes somewhat difficult the comparison of the spectra obtained for different electrodes, it can be concluded from the spectra reported in Figures 7.3*a*, *b* that similar bands appear in the spectra obtained for the Au(111) and Au(100) electrodes. With L-alanine, four positive bands at 1400, 1374, 1352, and 1308 $cm^{-1}$ are observed experimentally on the Au(111) electrode (Fig. 7.3*a*). For L-serine, the spectra reported in Figure 7.3*a* for the Au(111) electrode show a band around 1389 $cm^{-1}$, and a shoulder around 1323 $cm^{-1}$. Finally, bands for adsorbed glycine appear at ca. 1400 and 1340 $cm^{-1}$ for both the Au(111) and Au(100) electrodes. All these bands can be observed with a higher signal-to-noise ratio in Figure 7.4 for the gold thin-film electrodes. In a first approximation, the comparison of the observed band frequencies with those of the carboxylate species [50, 51], including glycine [27], suggests that the observed features for adsorbed species correspond to modes involving contributions from the symmetric OCO stretch of the unprotonated amino acid and –CH bending modes [27]. This behavior indicates again that there is a similar pattern of adsorption as it was stated from the voltammetric results. Namely, the three amino acids seem to undergo deprotonation of the carboxylic group in order to bond to the metal through their carboxylate groups. This conclusion holds for both the Au(111) and Au(100) electrodes. The absence in the spectra obtained for the three studied amino acids of any adsorbate signal attributable to the asymmetric stretch of the OCO group, which would be expected to appear between 1500 and 1700 $cm^{-1}$, is also noteworthy. In order to be sure that there is not a band related to the asymmetric stretching of the carboxylate group, which would be expected to appear near the complex zone of OH bending, additional experiments (not shown) were carried out in deuterated water.

At 1.0 V two additional positive bands appear at 2344 and 2171 $cm^{-1}$ in the experimental external reflection spectra obtained in the presence of L-serine (Fig. 7.3*a*, *b*) that can be associated to oxidation products of L-serine, in agreement with the voltammetric results. The first one corresponds to the symmetric stretching of dissolved $CO_2$, which remains trapped in the thin layer of solution between the electrode surface and the calcium fluoride window. Moreover, another band appears at 2171 $cm^{-1}$, which lies in the zone of the stretching of C–N bonds and can be associated to the presence of adsorbed cyanide as a product of L-serine decarboxylation, as it happens on Pt(111) [16, 20]. Note that the spectra reported in Figure 7.3 have been collected in the SNIFTIR mode, with a relatively high number of interferograms being collected at the sample potential. This favors the accumulation of the species formed at the sample potential. When comparing the spectra obtained on the onset of the amino acid oxidation region for glycine, L-alanine, and L-serine (Fig. 7.3), it can be observed that, whereas the formation of adsorbed cyanide and carbon dioxide is detected in the external reflection experiments already at 1.0 V in the case of L-serine, no absorption bands for these species are observed for glycine and L-alanine at 1.10 V. Since none of the oxidation products is detected in these latter cases (even collecting the sample spectra at more positive potentials than in the case of L-serine), it can be confirmed that glycine and L-alanine are more stable toward oxidation than L-serine. In this way, it can be recalled here that the spectra collected in the glycine-containing solution with a Au(111) electrode during a potential scan only show a band for carbon dioxide for potentials above 1.20 V [27]. Under these conditions, no adsorbed cyanide could be detected despite the strong absoption of this species (probably associated with both the attainment of a low cyanide coverage and the low signal-to-noise ratio inherent to the external reflection experiments). On the other hand, adsorbed cyanide is detected at potentials above 1,20 V from the ATR-SEIRA spectra obtained with gold thin-film electrodes in the presence of glycine and L-serine (but not in the presence of L-alanine) taking advantage of the much more favorable sensitivity associated to the SEIRA effect (see Fig. 7.4). On the contrary, the formation of carbon dioxide cannot be confirmed from ATR-SEIRAS experiments due to the absence of surface enhancement in the case of dissolved species. The absence of the cyanide band in the presence of L-alanine means that the $-CH_3$ group stabilizes the molecule, as observed for the same amino acid in the case of platinum single-crystal electrodes [16, 18]. As stated above, the presence of the –OH group in the L-serine molecule seems to facilitate the oxidation of this amino acid.

### 7.4.2 Geometry Optimization and Theoretical Vibrational Frequencies of Adsorbed Glycine, L-Alanine, and L-Serine on $Au_{31}(111)$ Cluster

The optimized geometries obtained for adsorbed L-alanine zwitterion on the $Au_{31}(111)$ cluster are shown in Figure 7.5. Some of the most significant geometrical parameters are summarized in Table 7.1. The optimized geometries of the L-alanine zwitterion correspond, irrespective of the presence of a water molecule and external field, to the bidentate, asymmetric bridge adsorption, with the OCO plane perpendicular to

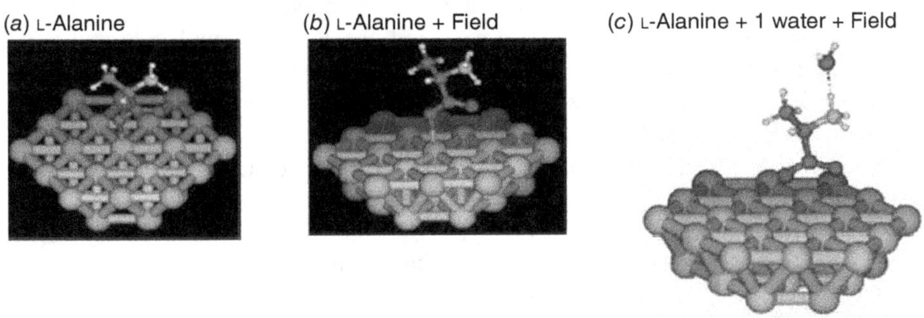


**Figure 7.5** Optimized geometries [B3LYP/6-31+G(d), LANL2DZ] of L-alanine zwitterions on a $Au_{31}(111)$ cluster: (*a*) without electric field, (*b*) in the presence of an external electric field of 0.01 a.u., and (*c*) in the presence of an external electric field of 0.01 a.u. and a water molecule.

**Table 7.1** Selected Geometrical Parameters for Adsorbed L-Alanine Zwitterions on $Au_{31}(111)$ Cluster

| | Ala[a] | Ala + Field[b] | Ala + Water + Field[c] |
|---|---|---|---|
| *d*-Au-$O_1$ (Å) | 2.45 | 2.34 | 2.32 |
| *d*-Au-$O_2$ (Å) | 3.09 | 2.94 | 2.71 |
| θ (Au-Au-$O_1$) (deg) | 86.6 | 87.6 | 91.1 |
| *d*-H ··O(CO) (Å) | 1.63 | 2.00 | 2.70 |
| *d*-O-C (Å) | 1.249 | 1.269 | 1.271 |
| *d*-O-C (Å) | 1.262 | 1.250 | 1.252 |
| α-(OCO) (deg) | 130.7 | 129.1 | 128.8 |

[a]Without electric field.
[b]In the presence of an external electric field of 0.01 a.u.
[c]In the presence of an external electric field of 0.01 a.u. and a water molecule.

the metal surface. The shorter Au–O distance corresponds to the oxygen atom located close to the methyl group. This oxygen atom is slightly tilted from the position just "on top" of the central Au atom in the cluster.

Both the application of electric field and the addition of a water molecule cause significant effects in the adsorbate geometry. The distances between each of the oxygen atoms in the carboxylate group and the surface gold atoms decrease. This effect is more important for the oxygen atom close to the ammonium group, especially for the case with water and field. A significant increase is observed for the distance between the H atom in the ammonium group and the oxygen atom in the carboxylate group. This distance amounts to 1.633 Å for the adsorbed zwitterion, and increases to 2.00 Å upon field application. In both cases, an intramolecular hydrogen bond exists between the ammonium and the carboxylate. This results in the NHCCO atoms lying approximately in the same plane. The increase in the H···O distance is a consequence of the positive surface charge density on the metal surface, which causes a decrease in electronic density in the carboxylate group, and this implies a weakening in the strength of the hydrogen bond.

The H···O distance increases with the addition of a water molecule and field. The water orientation is strongly influenced by the external electric field, which forces the H atoms to point toward the "solution side". This water molecule also interacts strongly with one of the H atoms of the ammonium group, causing this NH bond to be oriented perpendicular to the metal surface, rotated by about 60° with respect to the other two geometries previously shown. As a consequence of this rotation of the ammonium group, there is no longer an intramolecular hydrogen bond, as indicated by the increase of the H··O(CO) distance to 2.70 Å.

The tilt of the O–O line with respect to the surface plane is very small, in particular in the presence of water and electric field. This means that the dynamic dipole corresponding to the OCO asymmetric stretch is essentially parallel to the surface and, according to the surface selection rule that holds in surface infrared (IR) spectra, both for the external [70] and internal [71] reflection configurations should not be observed experimentally, which is in agreement with our in situ spectra.

From these optimized geometries, we can conclude that the interaction of the carboxylate with the gold metal surface is strong enough to stabilise the zwitterionic form of L-alanine and prevent the evolution toward the nonzwitterionic form. The applied field gives an additional stabilization, resulting in shorter distances to the surface metal atoms. In the presence of field, the main effect of adding water molecules is to form water–ammonium hydrogen bonds, which compete with and replace the intramolecular hydrogen bond in the adsorbed L-alanine zwitterions. Neither the applied field nor the water molecule modifies the bonding mode or the orientation of the carboxylate group.

The behavior observed (Fig. 7.6 and Table 7.2) for the geometries of the adsorbed L-serine zwitterion on the $Au_{31}(111)$ cluster is rather similar to that of L-alanine. Overall, the main trends observed upon adding field and water are very similar to those described above for the case of L-alanine. However, some differences can be observed that are related to the additional hydroxyl group present in the L-serine molecule. The electronic density of the lone pairs of the oxygen atom

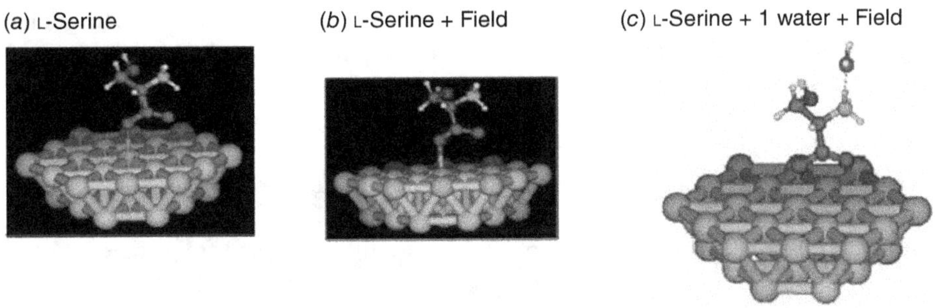


**Figure 7.6** Optimized geometries [B3LYP/6-31+G(d), LANL2DZ] of L-serine zwitterions on a $Au_{31}(111)$ cluster: (*a*) without electric field, (*b*) in the presence of an external electric field of 0.01 a.u., and (*c*) in the presence of an external electric field of 0.01 a.u. and a water molecule.

**Table 7.2** Selected Geometrical Parameters for Adsorbed L-Serine Zwitterions on $Au_{31}(111)$ Cluster

| | Ser[a] | Ser + Field[b] | Ser + Water + Field[c] |
|---|---|---|---|
| *d*-Au-$O_1$ (Å) | 2.444 | 2.357 | 2.334 |
| *d*-Au-$O_2$ (Å) | 2.939 | 2.715 | 2.617 |
| θ (Au-Au-$O_1$) (deg) | 88.8 | 91.0 | 92.7 |
| *d*-H···O(CO) (Å) | 1.731 | 2.045 | 2.289 |
| *d*-H···O(H) (Å) | 2.033 | 2.029 | 2.150 |
| *d*-O-C (Å) | 1.250 | 1.264 | 1.267 |
| *d*-O-C (Å) | 1.258 | 1.254 | 1.256 |
| α-(OCO) (deg) | 131.1 | 129.2 | 128.8 |

[a]Without electric field.
[b]In the presence of an external electric field of 0.01 a.u.
[c]In the presence of an external electric field of 0.01 a.u. and a water molecule.

in this latter group interacts with the positive charge in the ammonium group, and in the absence of water makes the intramolecular hydrogen bond with the carboxylate slightly longer than in the case of L-alanine. Upon addition of the water molecule and considering the electric field, some rotation of the ammonium group is observed, but in a lesser extent than in the case of L-alanine, as a consequence of the more complex interaction between the ammonium moiety and the three (instead of two) oxygen centers.

For glycine, the situation is qualitatively different from that reported above for L-alanine, and L-serine. The interaction with the surface is not enough to stabilize the zwitterions in the absence of water and electric field, and it evolves toward the nonzwitterionic form, which remains adsorbed in a unidentate form in a geometry optimization. In order to stabilize sufficiently the adsorbed glycine zwitterion, the addition of a water molecule plus the external electric field were needed. The geometry and frequencies corresponding to the calculations with glycine+water+field have been reported in a previous study [27].

Besides the effects on geometry, significant changes are also observed in the calculated harmonic vibrational frequencies of the adsorbed zwitterions due to the added water and electric field. We will discuss in the following these effects in the case of L-alanine and L-serine. Calculated frequencies of the adsorbed glycine zwitterion are discussed in Ref. [27]. Table 7.3 shows the calculated frequencies for the

**Table 7.3** Calculated B3LYP/6-31+G(d),LANL2DZ Frequencies for L-Alanine Zwitterion Adsorbed on $Au_{31}(111)$ Cluster and Effect of Adding Water Molecule and External Electric Field of 0.01 a.u.

| Alanine $\upsilon$ ($cm^{-1}$) | Assignment | Ala + Water + Field $\upsilon$ ($cm^{-1}$) | Assignment |
|---|---|---|---|
| – | – | 3779 | Asym str $H_2O$ |
| – | – | 3687 | Sym str $H_2O$ + str N–H (···O) |
| 3544 | Asym str $NH_2$ | 3507 | Asym str $NH_2$ |
| 3467 | Sym str $NH_2$ | 3448 | Sym str $NH_2$ |
| 3161 | Asym str $CH_3$ | 3168 | Asym str $CH_3$ |
| 3129 | Str C–H + asym str $CH_3$ | 3129 | Str C–H + asym str $CH_3$ |
| 3112 | Str C–H + asym str $CH_3$ | 3111 | Str C–H + asym str $CH_3$ |
| 3053 | Sym str $CH_3$ | 3060 | Sym str $CH_3$ |
| 2661 | Str N–H···O | 2954 | Str N–H ·(·· $OH_2$) |
| 1719 | Asym str OCO + bend CCNH | 1742 | Bend $NH_3$ |
| 1683 | Bend $NH_3$ | 1709 | Bend $H_2O$ + bend $NH_3$ |
| 1654 | Bend $NH_3$ | 1694 | Bend $H_2O$ + bend $NH_3$ + asym str OCO |
| 1521 | Bend $CH_3$ | 1619 | Asym str OCO + bend $NH_3$ |
| 1515 | Bend $CH_3$ | 1589 | Bend $NH_3$ + str CC + asym str OCO |
| 1442 | Bend $CH_3$ + Bend $NH_3$ | 1520, 1517 | Bend $CH_3$ |
| 1399 | Sym str OCO + bend $NH_3$; $CH_3$; CH | 1438 | Bend $CH_3$ |
| 1378 | Sym str OCO + bend $NH_3$; $CH_3$; CH | 1398 | Bend $NH_3$; $CH_3$; CH |
| 1344 | Sym str OCO + bend $NH_3$; $CH_3$; CH | 1364 | Sym str OCO + bend $NH_3$; $CH_3$; CH |
| 1300 | Bend $NH_3$; $CH_3$; CH | 1324 | Sym str OCO + bend $NH_3$; $CH_3$; CH |
| 1217 | Bend $NH_3$; $CH_3$; CH | 1227 | Bend $NH_3$; $CH_3$; CH |

adsorbed zwitterion of L-alanine on $Au_{31}(111)$ and also when a water molecule and external electric field are present. The frequencies corresponding to vibrations of the methyl group, appearing in the range 3000–3200 $cm^{-1}$ remain unaffected by the inclusion of water and electric field in the model. At higher wavenumbers, the presence of water and field causes a weakening of the $NH_2$ stretchings. The NH stretching of the hydrogen-bonded H atom in the ammonium group shifts from 2661 to 2954 $cm^{-1}$ when passing from the intramolecular to the intermolecular hydrogen bond.

The effects on the vibrations in the range between 2000 and 1200 $cm^{-1}$ are more complex. The breaking of the intramolecular hydrogen bond affects significantly the contributions to the normal modes, which in a number of cases correspond to combination modes. In particular, the addition of water and fields shifts the asymmetric OCO stretch from 1719 $cm^{-1}$ to values between 1589 and 1694 $cm^{-1}$ for three modes having significant contributions from this stretch. The calculated frequencies of the modes with significant contribution from the symmetric OCO stretch, also undergo red shift with water and field. Besides these effects of red shifting, it must be pointed out that the composition of movements corresponding to the normal modes are significantly altered.

Regarding the vibrational frequencies and assignments for L-serine (Table 7.4), the most important effects arise from the rupture of the intramolecular H bond in the presence of water and external field, affecting in particular the spectral features in the range between 1750 and 1200 $cm^{-1}$.

### *Experimental Band Assignment*

In the spectral range between 1200 and 1500 $cm^{-1}$, the ATR-SEIRA spectra in Figure 7.4 for L-alanine adsorbed on the (111) preferentially oriented thin film electrodes show bands at 1306, 1351, 1374, and 1402 $cm^{-1}$, which compare well with those observed with single-crystal surfaces in external reflection measurements (Fig. 7.3) at 1307, 1352, 1373, and 1400 $cm^{-1}$. As discussed previously, bands in this spectral region of the calculated spectra correspond to combination modes involving the symmetric OCO stretching and various bending modes (see Table 7.2). In the case of L-serine, the improved sensitivity of the ATR-SEIRA spectra in Figure 7.4 allows the detection of adsorbate bands at 1255, 1308, 1347, and 1395 $cm^{-1}$. These frequencies compare well with the theoretical harmonic frequencies of combination bands at around 1292, 1347, 1366, and 1392 $cm^{-1}$. Again, these bands involve several bending movements (see Table 7.4). Besides, the bands at 1347, 1366, and 1392 $cm^{-1}$ also have, as a main contribution, that of the symmetric OCO stretch of the carboxylate group. Finally, it can be recalled here that there is also a good agreement between the frequency of observed experimental bands for adsorbed glycine (see the corresponding spectra in Figs. 7.3 and 7.4) and those for the combination bands appearing in the calculated spectrum for zwitterionic glycine [27].

The comparison of the calculated frequencies for the adsorbed zwitterion of L-alanine with the experiment (see Table 7.5) shows that the theoretical values are

**Table 7.4** Calculated B3LYP/6-31+G(d), LANL2DZ Frequencies for L-serine Zwitterion Adsorbed on $Au_{31}(111)$ Cluster and Effect of Adding Water Molecule and External Electric Field of 0.01 a.u.

| Serine υ ($cm^{-1}$) | Assignment | Ser + Water + Field υ ($cm^{-1}$) | Assignment |
|---|---|---|---|
| – | – | 3778 | Asym str $H_2O$ |
| 3777 | Str OH ($CH_2OH$) | 3749 | Str OH ($CH_2OH$) |
| – | – | 3694 | Sym str $H_2O$ + str ···H–N |
| 3527 | Asym str $NH_2$ | 3482 | Asym str $NH_2$ |
| 3383 | Sym str $NH_2$ | 3386 | Sym str $NH_2$ |
| 3126 | Str CH + asym str $CH_2$ | 3128 | Asym str $CH_2$ |
| 3121 | Asym str $CH_2$ + str CH | 3116 | Str CH |
| 3037 | Sym str $CH_2$ | 3069 | Sym str $CH_2$ |
| 2914 | Str N-H(···OCO) | 2953 | Str N–H···$OH_2$ |
| 1711 | Asym Str. OCO + bend $NH_3$ | 1743 | Bend $NH_3$ |
| 1701 | Bend $NH_3$ | 1713 | Bend $H_2O$ + bend $NH_3$ |
| 1622 | Bend $NH_3$ | 1663 | Bend $H_2O$ + bend $NH_3$ |
| 1534 | Bend $CH_2$ | 1595 | Asym Str. OCO + bend $NH_3$ + bend $CH_2$ |
| 1456 | Bend (COH; $CH_2$; $NH_3$) | 1564 | bend $NH_3$ + Asym str OCO + bend $CH_2$ |
| 1420 | Bend (COH; $CH_2$; $NH_3$; CCH) | 1530 | Bend (sciss) $CH_2$ |
| 1380 | Sym str OCO + Bend (CCH; $CH_2$) | 1448 | Bend (wagging) $CH_2$ + bend C–O–H |
| 1364 | Sym str OCO + Bend (CCH; $CH_2$) | 1392 | Sym str OCO + Bend (CCH; $NH_3$; $CH_2$) |
| 1327 | Sym str OCO + Bend (CCH; $NH_3$; $CH_2$; COH) | 1366 | Sym str OCO + Bend (CCH; $NH_3$; $CH_2$; COH) |
| 1257 | Sym str OCO + Bend (CCH; $NH_3$; $CH_2$; COH) | 1347 | Sym str OCO + Bend (CCH; $NH_3$; $CH_2$; COH) |
| 1229 | Sym str OCO + Bend (CCH; $NH_3$; $CH_2$; COH) | 1292 | Bend ($CH_2$; CCH; COH; $NH_3$) |

slightly higher than the experimental ones. The differences from 17 to 38 $cm^{-1}$ are somewhat higher than the experimental uncertainty, 8 $cm^{-1}$, and amount to relative errors that are below 2.5% for the calculated frequencies. In the same way, the calculated frequencies for L-serine are only slightly higher (up to 39 $cm^{-1}$) than the corresponding experimental ones (see Table 7.5). The maximum relative error (around 3%, for the signal at 1347 $cm^{-1}$, observed at 1308 $cm^{-1}$ in ATR-SEIRAS) is in the range expected for the frequencies calculated at this theory level. The agreement between experimental and calculated frequencies is found to be better in the case of glycine. It is well known that the harmonic frequencies calculated using the

**Table 7.5** Comparison of Experimental and Calculated (B3LYP/6-31+G(d), LANL2DZ) Harmonic Vibrational Frequencies of Zwitterionic L-Alanine and L-Serine Coadsorbed with One Water Molecule on $Au_{31}(111)$ Model Cluster in Presence of External Electric Field of 0.01 a.u.

| | $\nu$ / $cm^{-1}$ | | |
|---|---|---|---|
| | B3LYP | External reflection Au(111) | ATR-SEIRAS |
| *Assignment: Glycine* | | | |
| Sym str OCO + str CC + wagg $CH_2$ | 1402 | 1401 | 1400 |
| Twist $CH_2$ + twist $NH_2$ + str CC + sym str OCO | 1355 | 1339 | 1339 |
| Sym str OCO + str CC + wagg $CH_2$ + bend $NH_3$ | 1348 | | |
| *Assignment: L-Alanine* | | | |
| Bend (umbrella) $CH_3$ + str CC | 1438 | 1400 | 1400 |
| Bend HCN + bend $NH_3$ + bend $CH_3$ | 1398 | 1373 | 1374 |
| Sym str OCO + bend HCN + bend $NH_3$ + bend $CH_3$ | 1364 | 1352 | 1351 |
| Bend HCC + sym str OCO + bend $NH_3$ + bend $CH_3$ | 1324 | 1307 | 1307 |
| *Assignment: L-Serine* | | | |
| Sym str OCO + Bend oop CCH + bend $NH_3$ + bend $CH_2$ | 1392 | 1390 | 1395 |
| Sym str OCO + Bend oop CCH + bend $NH_3$ + bend $CH_2$ + bend COH | 1366 | – | 1347 |
| Sym str OCO + Bend CCH + bend $NH_3$ + bend $CH_2$ + bend COH | 1347 | 1328 | 1308 |
| Bend (twist) $CH_2$ + bend CCH + bend COH + bend $NH_3$ | 1292 | – | 1255 |

B3LYP functional in combination with Pople basis sets similar to the one used by us are slightly higher than the experimental values for gas-phase molecules, typically by 2–3%. This is usually corrected by applying an average scaling factor that has been obtained from the comparison of calculated and experimental frequencies for a set of test molecules. We have not applied any scaling to the calculated frequencies in this chapter, as there is no reported scaling factor for the cases involving adsorption on metals described using LANL2DZ pseudopotentials. We do not expect, however, the scaling factor to differ significantly from those reported for gas-phase calculations. Taking into account this remark, the agreement between experimental and calculated frequencies can be considered as rather satisfactory.

The bands with significant contribution from the asymmetric stretch of the OCO group have calculated frequency values of 1589, 1619, and 1694 $cm^{-1}$ for L-alanine (Table 7.2), 1564 and 1595 $cm^{-1}$ for L-serine (Table 7.4), and 1626 and 1602 $cm^{-1}$ in the case of glycine [27]. No bands assignable to these modes can be observed in the experimental spectra. In this spectral range, the features observed in the ATR-SEIRA spectra for the adsorbed amino acids can be attributed to the changes in the water layers in the immediate vicinity of the electrode (see the discussion below). Taking into account the surface selection rule for external [70] and internal [71] reflection experiments, the latter results indicate that the transition dipole corresponding to the asymmetric OCO stretching has an essentially negligible component in the direction normal to the metal surface. This, together with the observation of the band for the symmetric stretching of the carboxylate group, which is not compatible with a flat (OCO plane parallel to surface) configuration for the adsorbate, is in agreement with the bidentate asymmetrical bridge adsorption geometry suggested by the DFT calculations showing that glycine, L-alanine, and L-serine are bonded to the metal atoms through the two oxygen atoms of the carboxylate group.

The zwitterionic nature of the adsorbates is further confirmed from the observation in the ATR-SEIRA spectra of adsorbate bands related to the protonated ammonium group of the zwitterions. In this way, both for L-alanine and L-serine, adsorbate bands are also seen between 2900 and 3000 $cm^{-1}$, which are assigned to the N–H stretch of the hydrogen atom involved in a hydrogen bond of the ammonium group with water molecules (theoretical frequency value around 2950 $cm^{-1}$ both for L-alanine and L-serine). Finally, it can be mentioned here that, according to the DFT theoretical harmonic frequencies (in parentheses for L-alanine), absorption bands corresponding to the asymmetric stretch of $CH_3$ (3168 $cm^{-1}$) and to combination modes (at 3129 and 3111 $cm^{-1}$) involving the asymmetric stretch of $CH_2$, and the stretch of the CH in the $\alpha$ position should appear for the adsorbed amino acids in the spectral region between 3100 and 3200 $cm^{-1}$. A small absorption feature at 3107 $cm^{-1}$ in some of the ATR-SEIRA spectra reported for adsorbed L-alanine (Fig. 7.4) could be attributed to these vibrational modes. However, the low intensity of this feature, which cannot be clearly distinguished in the spectra for L-serine and overlaps for both amino acids with water absorption bands (see below), make this assignment tentative.

For the three amino acids, a very intense absorption band is observed around 1100 $cm^{-1}$, which is related to the co-adsorbed perchlorate anions. In addition, typical water bands appear both in the O–H stretching (between 3000 and 4000 $cm^{-1}$) and O–H bending (between 1600 and 1700 $cm^{-1}$) regions. These features have frequencies and shapes very similar to those observed for the gold electrode in the amino-acid-free perchloric acid solution [72], thus indicating that they are mainly related to the adsorption/desorption of perchlorate anions at the gold electrode surface. The presence of the latter species would be necessary to neutralize the positive charge of the protonated ammonium group. Different from the cases of glycine and alanine, the intensity of the band of adsorbed perchlorate decreases slightly when the concentration of serine increases (see Fig. 7.4). This behavior has been observed (although with a much more important weakening) with adsorbed carboxylates having neutral functional groups facing the solution side (e.g., acetate [50] or fluoroacetate [51]), as in these cases there is no positive charge density accumulated on the side chains, and this means no need of perchlorate counterions for compensating it. This suggests that these anions compete with perchlorate for the surface adsorption sites, causing the decrease of the absorption band of the latter. In the case of L-serine, it seems that, although some perchlorate anions are necessary to compensate the charge of the protonated ammonium groups, the OH groups in serine could help to stabilize this charge by forming some hydrogen bonds with neighboring ammonium groups. This is in agreement with the DFT optimized geometries that show that the presence of the hydroxyl group of L-serine induces some rotation in the ammonium group, as compared with the case of L-alanine, because of the interaction of the electronic density of the oxygen with the one of the protons in the ammonium, comparable to a hydrogen bond. In this way, for higher concentration of L-serine, more hydrogen bonds involving OH and $NH_3^+$ groups of serine adsorbed zwitterions are expected to be formed in the serine adlayer, and consequently less perchlorate anions would be needed.

## 7.5 CONCLUSIONS

The spectroelectrochemical behavior of glycine, L-alanine, and L-serine has been studied on Au(111) and Au(100) single-crystal electrodes and on preferentially (111)-oriented gold thin-films deposited on Si. Cyclic voltammetry profiles in the double-layer potential region (0.06–1.10 V for glycine and L-alanine, 0.06–0.90 V for L-serine) and in situ infrared spectra in the same potential region indicate that, in these acidic conditions, the three amino acids adsorb reversibly. This behavior is in contrast with that reported for Au electrodes in alkaline media, where dissociative adsorption of the amino acids occurs. From the changes in the voltammetric peak related to the lift of the surface reconstruction, observed in the voltammograms of Au(100) and Au(111) electrodes, it can be concluded that the adsorption of the amino acids favors the nonreconstructed surfaces.

The main experimental infrared bands due to the adsorbed amino acids appear between 1300 and 1400 $cm^{-1}$ and are due to normal modes with significant contributions from the symmetric stretch of the carboxylate group, together with bendings involving $CH_x$ and $NH_3^+$ groups. From the absence in the experimental spectra of any band due to the asymmetric stretch of the carboxylate group, the surface selection rule allows us to conclude that these three amino acids are linked to the metal surface by the two oxygens of the carboxylate group in a bidentate configuration. From the DFT-calculated optimized geometries of the adsorbed zwitterions on cluster models of the Au(111) surface, this seems to be an asymmetric bridge adsorption configuration, with the OCO plane perpendicular to the metal surface, and aligned with the surface atomic dense rows. Additionally, the ATR-SEIRAS measurements have allowed the observation of changes in the vibrational signature of interfacial water due to the co-adsorption with perchlorate anions. The significant amounts of co-adsorbed perchlorate confirm that these amino acids are adsorbed in their zwitterionic forms. Additional experimental evidence for the presence of ammonium charged groups was the observation in the ATR-SEIRAS experiments of adsorbate absorption bands around 2900 $cm^{-1}$, corresponding, on the basis of DFT-calculated vibrational frequencies, to the stretching of the N–H bond involved in hydrogen bonding with water molecules solvating the $NH_4^+$ group.

The voltammetric results in the region of higher potentials show that the electrooxidation of these amino acids is sensitive to the structure of the electrode surface. Moreover, the behavior shown by the peak potentials with concentration and sweep rate, as well as by the oxidation and reduction currents indicate that these are complex processes, probably due to the interaction with the surface oxides, that can affect the adsorption-limited oxidation observed for glycine electrooxidation. The main oxidation product observed by external reflection infrared spectroscopy was $CO_2$. In the case of L-serine, adsorbed cyanide species were also observed. Serine also showed the particularity of being oxidized at potentials less positive than those for other amino acids. This had been previously observed with Pt electrodes and can be related to the easier oxidation of the $-CH_2-OH$ group with respect to the $-CH_3$ group.

The interaction with the metal surface contributes to the stabilization to the zwitterionic form of the adsorbate. This is sufficient for avoiding the evolution toward the neutral form in a geometry optimization in the case of L-alanine and L-serine. In the case of glycine, however, the application of field and the addition of a water molecule are needed in order to stabilize the charged form of the amino acid and avoid transformation to the neutral amino acid during the geometry optimization calculation. The theoretical analysis shows clearly that assigning experimental bands can greatly benefit from the calculation of optimized geometries and theoretical frequencies of the adsorbates, using models that include ingredients such as electric field, water, and the like, which play important roles in the electrochemical interface, in particular for zwitterionic adsorbates. The field and water do not change the bonding mode of the zwitterions of L-alanine and L-serine, which bond to the gold surface through the oxygen atoms of the carboxylate group in a bidentate asymmetric bridge configuration as it was the case for other carboxylate anions, such as acetate, fluoroacetates, bioxalate, glycolate, and glyoxalate. However, they significantly affect the adsorption geometry, in particular of the groups involved in hydrogen bonding, and also modify many of the vibrational frequencies.

From the results reported in this work, it can be concluded that, for the interpretation of the experimental spectra of some adsorbed species, in particular for the assignment of bands to modes, the situation can become nontrivial, and the use of DFT calculations of optimized geometries and vibrational frequencies could be of great help, in particular if the relevant ingredients (such as electric field and water molecules) are included in the model used for such calculations. The insight obtained from this theoretical analysis strongly favors the use of theoretical calculation of adsorbate frequencies, instead of relying on tabulated frequencies for gas-phase, solution, or coordination complex species.

## ACKNOWLEDGMENTS

Financial support from Ministerio de Ciencia e Innovación (projects CTQ2010-16271 and CTQ2009-13142, Fondos FEDER) and Generalitat Valenciana (Prometeo/2009/045, ACOMP/2011/200 and ACOMP/2012/137, Fondos FEDER) is gratefully acknowledged. A.P.S is endebted to the Generalitat Valenciana for the award of a Santiago Grisolía grant.

## REFERENCES

1. Lehninger, A. L.; Nelson, D. L.; Cox, M. M. *Principles of Biochemistry*, 2nd ed., Worth: New York, **1993**.
2. Maruccio, G.; Biasco, A.; Visconti, P.; Bramanti, A.; Pompa, P. P.; Calabi, F.; Cingolani, R.; Rinaldi, R.; Corni, S.; Di Felice, R.; Molinari, E. *Adv. Mater.* **2005**, *17*, 816.
3. Daniel, M. C.; Astruc, D. *Chem. Rev.* **2004**, *104*, 293.
4. Herzog, G.; Arrigan, D. W. M. *Analyst* **2007**, *132*, 615.
5. Pingarrón, J. M.; Yáñez-Sedeño, P.; González-Cortés, A. *Electrochim. Acta* **2008**, *53*, 5848.
6. Mizutani, F. *Sensors Actuators B* **2008**, *130*, 14.
7. Cobley, C. M.; Chen, J.; Cho, E. C.; Wang, L. V.; Xia, Y. *Chem. Soc. Rev.* **2011**, *40*, 44.
8. Kanagaraja, S.; Lundström, I.; Nygren, H.; Tengvall, P. *Biomaterials* **1996**, *17*, 2225.
9. Horvatovich, P.; Govorukhina, N.; Bischoff, R. *Analyst* **2006**, *131*, 1193.
10. Zhao, X. Y.; Yan, H.; Zhao, R. G.; Yang, W. S. *Langmuir* **2002**, *18*, 3910.
11. Horányi, G.; Rizmayer, E. M. *J. Electroanal. Chem.* **1975**, *64*, 15.
12. Horányi, G.; Rizmayer, E. M. *J. Electroanal. Chem.* **1986**, *198*, 393.
13. Marangoni, D. G.; Smith R. S.; Roscoe, S. G. *Can. J. Chem.* **1989**, *67*, 921.
14. Marangoni, D. G.; Wylie, I. G. N.; Roscoe, S. G. *J. Electroanal. Chem.* **1991**, *320*, 269.
15. Huerta, F.; Morallón, E.; Cases, F.; Rodes, A.; Vázquez, J. L.; Aldaz, A. *J. Electroanal. Chem.* **1997**, *421*, 179.
16. Huerta, F.; Morallón, E.; Cases, F.; Rodes, A.; Vázquez, J. L.; Aldaz, A. *J. Electroanal. Chem.* **1997**, *431*, 269.
17. Huerta, F.; Morallón, E.; Vázquez, J. L.; Pérez, J. M.; Aldaz, A. *J. Electroanal. Chem.* **1998**, *445*, 155.
18. Huerta, F.; Morallón, E.; Vázquez, J. L.; Aldaz, A. *J. Electroanal. Chem.* **1999**, *475*, 38.
19. Wright, J. E. I.; Fatih, K.; Brosseau, C. L.; Omanovic, S.; Roscoe, S. G. *J. Electroanal. Chem.* **2003**, *550–551*, 41.
20. Gu, Y. J.; Sun, S. G.; Chen, S. P.; Zhen, C. H.; Zhou, Z. Y. *Langmuir* **2003**, *19*, 9823.
21. Xiao X. Y.; Sun, S. G.; Yao, J. L.; Wu, Q. H.; Tian, Z. Q. *Langmuir* **2002**, *18*, 6274.
22. Slojkowska, R.; Jurkiewicz-Herbich, M. *Coll. Surf. A*, **2001**, *178*, 325.
23. Lori, J. A.; Honawa, T. *Corr. Sci.* **2001**, *43*, 2111.
24. Li, H. Q.; Chen, A.; Roscoe, S. G.; Lipkowski, J. *J. Electroanal. Chem.* **2001**, *500*, 299.
25. Zhen, C. H.; Sun, S. G.; Fan, C. J.; Chen, S. P.; Mao, B. W.; Fan, Y. J. *Electrochim. Acta* **2004**, *49*, 1249.
26. Slojkowska, R.; Palys, B.; Jurkiewicz-Herbich, M. *Electrochim. Acta* **2004**, *49*, 4109.
27. Sandoval A. P.; Orts, J. M.; Rodes, A.; Feliu, J. M. *J. Phys. Chem. C* **2011**, *115*, 16439.
28. Efstathiou, V.; Woodruff, D. P. *Surf. Sci.* **2003**, *531*, 304.
29. Rankin, R. B.; Sholl, D. S. *J. Phys. Chem. B* **2005**, *109*, 16764.
30. Rankin, R. B.; Sholl, D. S. *Surf. Sci.* **2004**, *548*, 301.
31. Mae, K.; Morikawa, K. *Surf. Sci.* **2004**, *553*, L63.
32. Balnkenburg, S.; Schmidt, W. G. *Nanotechnology* **2007**, *18*, 424030.
33. Mullin, J. M.; Gordon, M. S. *J. Phys. Chem. B* **2009**, *113*, 8657.

34. KUMAR, S.; RAI, A. K.; RAI, S. B.; RAI, D. K.; SINGH, A. N.; SINGH, V. B. *J. Molec. Struct.* **2006**, *791*, 23.
35. RANKIN, R. B.; SHOLL, D. S. *J. Chem. Phys.* **2006**, *124*, 074703.
36. RANKIN, R. B.; SHOLL, D. S. *Langmuir* **2006**, *22*, 8096.
37. RANKIN, R. B.; SHOLL, D. S. *Surf. Sci.* **2005**, *574*, L1.
38. JAMES, J. N.; SHOLL, D. S. *J. Mol. Catal.A* **2008**, *281*, 44.
39. HONG, G.; HEINZ, H.; NAIK, R. R.; FARMER, B. L.; PACHTER. R. *ACS Appl. Mater. Interf.* **2009**, *1*, 388.
40. GHIRINGHELLI, L. M.; SCHRAVENDIJK, P.; DELLE SITE, L. *Phys. Rev. B* **2006**, *74*, 035437.
41. MULLIN, J. M.; GORDON, M. S. *J. Phys. Chem. B* **2009**, *113*, 8657.
42. JEON, I. S.; AHN, D. S.; PARK, S. W.; LEE, S.; KIM. B. *Int. J. Quantum Chem.* **2005**, *101*, 55.
43. OSAWA, M. *Bull. Chem. Soc. Jpn.* **1997**, *70*, 2861.
44. RODES, A.; PÉREZ, J. M.; ALDAZ, A. In *Handbook of Fuel Cells. Fundamentals, Technology and Applications*, W., VIELSTICH, H. A., GASTEIGER, and A., LAMM, Eds., Wiley: Chichester, **2003**, pp 191–219.
45. WANDLOWSKI, T.; ATAKA, K.; PRONKIN, S.; DIESING, D. *Electrochim. Acta* **2004**, *49*, 1233.
46. AROCA, R. *Surface Enhanced Vibrational Spectroscopy*, Wiley: Chichester, **2006**.
47. ATAKA, K.; OSAWA, M. *Langmuir* **1998**, *14*, 951.
48. DELGADO, J. M.; ORTS, J. M.; PÉREZ, J. M.; RODES, A. *J. Electroanal. Chem.* **2008**, *617*, 130.
49. DELGADO, J. M.; ORTS, J. M.; RODES, A. *J. Phys. Chem. C* **2007**, *111*, 14476.
50. BERNÁ, A.; DELGADO, J. M.; ORTS, J. M.; RODES, A.; FELIU, J. M. *Electrochim. Acta* **2008**, *53*, 2309.
51. DELGADO, J. M.; BLANCO, R.; ORTS, J. M.; PÉREZ, J. M.; RODES, A. *J. Phys. Chem. C* **2009**, *113*, 989.
52. LIDE, D. R. *CRC Handbook of Chemistry and Physics, 86th ed.*, CRC Press: Boca Raton, FL, **2005**.
53. FRISCH, M. J.; TRUCKS, G. W.; SCHLEGEL, H. B.; SCUSERIA, G. E.; ROBB, M. A.; CHEESEMAN, J. R.; MONTGOMERY Jr., J. A.; VREVEN, T.; KUDIN, K. N.; BURANT, J. C.; MILLAM, J. M.; IYENGAR, S. S.; TOMASI, J.; BARONE, V.; MENNUCCI, B.; COSSI, M.; SCALMANI, G.; REGA, N.; PETERSSON, G. A.; NAKATSUJI, H.; HADA, M.; EHARA, M.; TOYOTA, K.; FUKUDA, R.; HASEGAWA, J.; ISHIDA, M.; NAKAJIMA, T.; HONDA, Y.; KITAO, O.; NAKAI, H.; KLENE, M.; LI, X.; KNOX, J. E.; HRATCHIAN, H. P.; CROSS, J. B.; BAKKEN, V.; ADAMO, C.; JARAMILLO, J.; GOMPERTS, R.; STRATMANN, R. E.; YAZYEV, O.; AUSTIN, A. J.; CAMMI, R.; POMELLI, C.; OCHTERSKI, J. W.; AYALA, P. Y.; MOROKUMA, K.; VOTH, G. A.; SALVADOR, P.; DANNENBERG, J. J., ZAKRZEWSKI, V. G., DAPPRICH, S.; DANIELS, A. D.; STRAIN, M. C.; FARKAS, O.; MALICK, D. K.; RABUCK, A. D.; RAGHAVACHARI, K.; FORESMAN, J. B.; ORTIZ, J. V.; CUI, Q.; BABOUL, A. G.; CLIFFORD, S.; CIOSLOWSKI, J.; STEFANOV, B. B.; LIU, G.; LIASHENKO, A.; PISKORZ, P.; KOMAROMI, I.; MARTIN, R. L.; FOX, D. J.; KEITH, T.; AL-LAHAM, M. A.; PENG, C. Y.; NANAYAKKARA, A.; CHALLACOMBE, M.; GILL, P. M. W.; JOHNSON, B.; CHEN, W.; WONG, M. W.; GONZALEZ C.; POPLE, J. A. *Gaussian 03, Revision D.02*, Gaussian, Inc., Wallingford, CT, **2004**.
54. BECKE, A. D. *J. Chem. Phys.* **1993**, *98*, 5648.
55. LEE, C.; YANG, W.; PARR, R. G. *Phys. Rev. B* **1988**, *37*, 785.
56. HEHRE, W. J.; DITCHFIELD, R.; POPLE, J. A. *J. Chem. Phys.* **1972**, *56*, 2257.
57. HARIHARAN, P. C.; POPLE, J. A. *Theor. Chim. Acta* **1973**, *287*, 213.
58. CLARK, T; CHANDRASERKHAR, J.; SPITZNAGEL, G. W.; SCHLEYER, P. V. R. *J. Comput. Chem.* **1983**, *4*, 294.
59. HAY, P. J.; WADT, W. R. *J. Chem.Phys.* **1985**, *82*, 299.
60. YORK, S. M.; HAQ, S.; KILWAY, K. V.; PHILLIPS, J. M.; LEIBSLE, F. M. *Surf. Sci.* **2003**, *522*, 34.
61. PHILLIPS, J. M.; LEIBSLE, F. M.; HOLDER, A. J.; KEITH, T. *Surf. Sci.* **2003**, *545*, 1.
62. DELGADO, J. M.; BLANCO, R.; ORTS, J. M.; PÉREZ, J. M.; RODES, A. *J. Phys. Chem. C* **2010**, *114*, 12554.
63. DELGADO, J. M.; BLANCO, R.; ORTS, J. M.; PÉREZ, J. M.; RODES, A. *Electrochim. Acta* **2010**, *55*, 2055.
64. SCHAFTENAAR, G.; NOORDIK, J. H. *J. Comput.-Aided Mol. Des.* **2000**, *14*, 123.
65. DELGADO, J. M.; ORTS, J. M.; RODES, A. *Langmuir* **2005**, *21*, 8809.

66. Rodes, A.; Herrero, E.; Feliu, J. M.; Aldaz, A. *J. Chem.Soc. Faraday Trans.* **1996**, *20*, 3769.
67. Clavilier, J.; Armand, D.; Sun, S. G.; Petit, M. *J. Electroanal. Chem.* **1986**, *205*, 267.
68. Kolb, D. M. *Prog. Surf. Sci.* **1996**, *51*, 109.
69. A. Hamelin, *J. Electroanal. Chem.* **1996**, *407*, 1.
70. Greenler, R. G. *J. Chem. Phys.* **1966**, *44*, 310.
71. Osawa, M.; Ataka, K.; Yoshii, K.; Nishikawa, Y. *Appl. Spectrosc.* **1993**, *47*, 1497.
72. Ataka, K.; Yotsuyanagi, T.; Osawa, M. *J. Phys. Chem.* **1996**, *100*, 10664.

# Chapter 8

# Composition, Structure, and Reaction Dynamics at Electrode–Electrolyte Interfaces Using Infrared Spectroscopy

**Angel Cuesta**
Instituto de Química Física "Rocasolano," CSIC, Madrid, Spain

## 8.1 INTRODUCTION

At least since the 1960s, and particularly the 1970s, the need for an adequate description of the composition and structure of the electrode–electrolyte interface, of the intermediates involved in electrochemical reactions, and of electrochemical reaction mechanisms stimulated the search for the direct study of the electrochemical interface, following the path already taken by the research of the gas–solid interface. It was soon realized that optical spectroscopies, which combine high (often submonolayer) sensitivity, high selectivity and, with an adequate experiment design, very short response times, were especially suited for the study of electrochemical interfaces. Ultraviolet–visible (UV–vis) reflectance spectroscopy was popular at that time [1–14], and the discovery of surface-enhanced Raman scattering (SERS) [15–17] opened new perspectives. In 1980, Bewick and co-workers [18] reported the first external reflectance infrared (IR) spectra of the electrode–electrolyte interface, and soon IR spectroscopy became the most popular nonelectrochemical technique for the characterization of the electrochemical double layer. I would like to emphasize

*Vibrational Spectroscopy at Electrified Interfaces*, First Edition.
Edited by Andrzej Wieckowski, Carol Korzeniewski, and Björn Braunschweig.

that, although the first attempts to apply IR spectroscopy to the study of electrochemical interfaces employed the attenuated total reflection (ATR) configuration [5, 19], this was soon abandoned for the external reflection approach chosen by Bewick, who disproved, in a magnificent tour de force, the generalized opinion that it would not work due to the strong absorption of IR radiation by the solvent (usually water) [20]. Only after Osawa and co-workers introduced during the 1990s [21–26] simple methods for the deposition on infrared-transparent substrates with high refraction index (like Si) of thin metal films with a grain size suitable for producing surface enhancement of infrared absorption (SEIRA) [27] has IR spectroscopy in the ATR–SEIRAS configuration become increasingly popular.

Although since Bewick's seminal work some other sophisticated optical spectroscopies, like sum frequency generation (SFG) [28], have been introduced, IR spectroscopy (both in the external reflectance and in the ATR configuration) is still by far the most frequently applied optical technique for the study of the electrode–electrolyte interface. There are three reasons for this:

1. All the metals (the most frequently employed electrode materials) have a very high reflectivity in the IR region, and their contribution to the (always differential) IR reflectance spectra is minimal (at most a featureless, potential-dependent baseline). Therefore, contrary to optical spectroscopies in other spectral ranges (i.e., UV–vis spectroscopy), external reflectance or ATR infrared spectra obtained with bulk metal electrodes or with highly reflecting thin metal films, respectively, are similar to transmission spectra, because the molecules at the interface (both adsorbed and in solution) produce a decrease of the high IR reflectivity and retain their spectral fingerprints, associated with the vibrational modes of their functional groups. This results in a relatively simple interpretation of the IR reflectance spectra. However, if the reflectivity of the electrode surface is low, the optical properties of the substrate must also be taken into account [29], as in the UV–vis spectral range.
2. Compared with SERS, which is limited to the coinage metals, IR spectra can be obtained with any electrode material.
3. Compared with more sophisticated techniques, like SFG, IR spectroscopy requires a simple experimental setup, and relatively low-cost instrumentation is commercially available.

In this chapter, after a theoretical description of IR reflectance spectroscopy, with particular attention to the often neglected case of a moderately or poorly reflecting substrate, the application of IR spectroscopy to the characterization of the electrode–electrolyte interface and to the elucidation of the mechanisms of electrocatalytic reactions is illustrated with recent contributions from my group.

## 8.2 REFLECTANCE SPECTROSCOPY

Obtaining spectra from interfaces necessarily involves working in one of the several reflectance configurations and, hence, understanding the laws that govern the reflection of light at interfaces. This knowledge can be used to optimize the experimental configuration, to optimize the absorption of light at the interface, and to achieve a high signal-to-noise ratio.

Since the signal-to-noise ratio will increase if the number of photons reaching the detector increases, a configuration that maximizes the reflection of IR radiation at the interface will be preferred. Absorption of light by matter is due to transitions between different energy levels and requires the interaction between the oscillating electric field of the radiation and the transition dipole moment, which in the case of IR spectroscopy corresponds to variations of molecular electric dipole moments during molecular vibrations. The higher the intensity of the radiation (given by the mean square electric field strength, $\langle E^2 \rangle$) and the transition dipole moment, the more intense the absorption. Accordingly, and since the magnitude of the transition dipole moment is a property of the corresponding vibrational mode that cannot be manipulated, the configuration used must also maximize $\langle E^2 \rangle$ at the interface.

In the IR region, metals have a very high optical conductivity (i.e., a high electrical conductivity at optical frequencies). As indicated above, and as will be illustrated in more detail below, this simplifies the description of reflection at the electrode–electrolyte interface, and the interpretation of the corresponding IR spectra. Metals are by far the most usual electrode materials, and this can lead many researchers to interpret spectra obtained with electrode surfaces that do not show metallic reflectance with concepts and models developed for metals. During the last decades, an increasing number of IR spectroscopic studies have been devoted to the investigation of electrochemical reactions at nanoparticles deposited on a carbonaceous or metallic substrate or at nanostructured metal surfaces. Even when the nanoparticles have been deposited on a highly reflecting metallic substrate, or when the nanostructured metal surface initially showed a high reflectivity, the resulting surfaces often have a low or moderate reflectance in the IR, and this must be taken into account when analyzing the corresponding spectra.

### 8.2.1 Reflectivity of Interphase: Surface Selection Rule and Its Applicability

The reflectivity of an interface is given by

$$R_{s,p} = \hat{r}_{s,p} \cdot \hat{r}^*_{s,p} = \left|\hat{r}_{s,p}\right|^2 \tag{8.1}$$

where $\hat{r}$ is the Fresnel reflection coefficient, defined as the ratio between the complex amplitudes of the electric field vectors of the reflected and incident waves ($\hat{r} = \hat{E}_r / \hat{E}_i$), $\hat{r}^*$ is the complex conjugate of $\hat{r}$, and the $s$ and $p$ subindexes refer to $s$- and $p$-polarized light, respectively. The Fresnel coefficients of an interface are given by the complex refractive indexes of the incident and reflecting media:

$$\hat{n}(\omega) = n(\omega) + ik(\omega) \tag{8.2}$$

where $n$ is the refraction index, $k$ is the extinction coefficient, and $\omega$ is the frequency of the radiation in hertz.

Maxwell's relation, $\hat{\varepsilon} = \hat{n}^2$, relates the optical constants $n$ and $k$ to a yet more fundamental parameter, the complex dielectric function, which, in the last instance, determines the optical properties of a material, and whose real and imaginary parts ($\hat{\varepsilon} = \varepsilon' + i\varepsilon''$) can be related with the optical constants of the material as follows:

$$\varepsilon' = n^2 - k^2 \tag{8.3}$$

$$\varepsilon'' = 2nk \tag{8.4}$$

Taking into account the relation between the imaginary part of the complex dielectric function and the optical conductivity, $\sigma$, of the material, ($\varepsilon'' = 4\pi\sigma/\omega$), the relation between the latter and the optical constants can be easily obtained:

$$\sigma(\omega) = \frac{2nk\omega}{4\pi} \tag{8.5}$$

where $\sigma(\omega)$ is related with the decay length or, if preferred, the penetration depth, of the radiation in the material. If the conductivity is very high [$\sigma(\omega)/\omega > 10$], the radiation will penetrate in the material only a small fraction of a wavelength, while in the limit $\sigma(\omega) = 0$, the material will be transparent.

When light reflects from a surface, the incident and reflected waves sum up vectorially, yielding a standing-wave electric field in the incident medium. In addition, fulfillment of Fresnel's equations is based on the boundary conditions that the tangential component of the electric field vector, $E_t$, and the normal component of the electric displacement vector, $D_n = \varepsilon \cdot E_n$ (where $E_n$ is the normal component of the electric field vector and $\varepsilon$ is the dielectric constant at the interface) are continuous across the interface [31, 32]. Altogether, these conditions are the basis of the equations yielding the mean square electric field strength at the interface derived by Hansen [33] (Hansen's matrix formulas introduced in ref. [33] contained a number of misprints, some of which were corrected in Refs. [34–36]). Figure 8.1 shows the angular dependence for the Au–air interface of the components of $\langle E^2 \rangle$ (normalized to that of the incident beam, $\langle E_0^2 \rangle$) parallel and perpendicular to the Au surface for two different cases: $h\nu = 0.64$ eV (Fig. 8.1*a*) and $h\nu = 3.5$ eV (Fig. 8.1*b*).

Figure 8.1*a* illustrates the optical behavior of a metal surface in the infrared region, where metals have a high optical conductivity and $R$ close to unity. Under these conditions, the components of the electric field strength of the radiation parallel to the surface suffer a phase change of 180° upon reflection, and the incoming and reflected beams cancel each other (see inset in Fig. 8.1*a*). This constitutes the basis for the well-known surface selection rule, according to which vibrational modes completely parallel to the surface are invisible to IR radiation and molecules adsorbed at the surface of a metal or any other highly reflecting substrate cannot be detected by IR spectroscopy at normal incidence or with *s*-polarized light. On the contrary, the components of the electric field strength of the radiation perpendicular to the surface of the incident and reflected beams are in phase and add to each other (see inset in Fig. 8.1*a*). As a consequence, the mean square electric field strength of the component of the *p*-polarized radiation perpendicular to the surface depends strongly on the angle of incidence, and at angles of incidence between 40° and 80° it is clearly

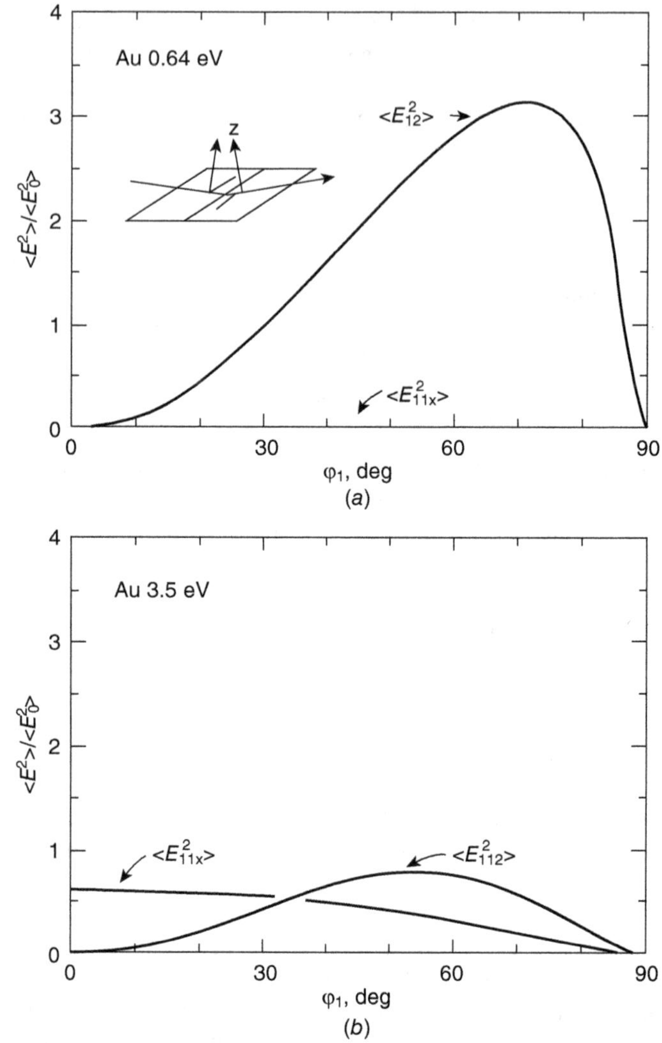


**Figure 8.1** Angular dependence of the electric field strength for the parallel (E11x) and perpendicular (E12) components of *p*-polarized light at a gold surface at (*a*) 0.64 eV (5161.3 $cm^{-1}$; $n_{Au} = 1.09$; $k_{Au} = 13.43$; $\sigma_{Au}/\omega \approx 15$; $R = 97\%$ at normal incidence) and (*b*) 3.5 eV (354.3 nm; $n_{Au} = 1.64$; $k_{Au} = 1.75$; $\sigma_{Au}/\omega \approx 2.87$; $R = 27\%$ at normal incidence). The inset in (*a*) shows the electric field of the incident and reflected beams for *s*-polarized radiation, and the component perpendicular to the surface of the electric field of the incident and reflected beams for *p*-polarized radiation. Reproduced from [30], with kind permission from Springer Science+Business Media B.V.

larger than that of the incident beam. For this reason, high angles of incidence (around 80°) are typically used in IR external reflection spectroscopy at the metal–gas interface. At the electrode–electrolyte interface other factors come into consideration, as we will see below, and lower angles of incidence are used.

The behavior is completely different in the ultraviolet region (Fig. 8.1*b*), where metals have moderate-to-low optical conductivity. As can be observed, the components of the electric field strength of the radiation parallel to the surface vanish only at very high angles of incidence, and the surface selection rule is not effective any more, that is, transitions with their transition dipole moment parallel to the surface can be excited in the ultraviolet region using *s*-polarized light or working at normal incidence. This situation, which is illustrated in Figure 8.1*b* for the case of a specular Au surface at 354 nm, will also apply in the infrared if the reflectivity of the surface is low.

In summary, the IR surface selection rule, often assumed to be of general applicability, applies only to specular surfaces with a high reflectivity, this being true for both the external reflection and the ATR configuration. If the surface reflectivity is low, IR spectra can be obtained with both *p* and *s* polarizations. For example, bands of poly(methylmethacrylate) adsorbed on glassy carbon [37] and of CO adsorbed on particulate Ir deposited on glassy carbon [38] can be observed both with *p*- and with *s*-polarized radiation (actually, the use of unpolarized light would be preferable in these cases because both the *s* and *p* components carry spectral information and the signal-to-noise ratio will be improved). This is true whatever the origin of the low reflectivity of the surface, be it due to the intrinsic low reflectivity of the electrode material (e.g., glassy carbon), to the electrochemical roughening of a metallic surface (platinized platinum or nanoporous metal electrodes), or to the deposition of a film of nanoparticles on a metallic substrate.

## 8.2.2 Band Shape in IR Reflection Spectroscopy

The substrate's reflectivity does not only affect the applicability of the surface selection rule but also influences the shape and the sign of absorption bands in the IR spectra of adsorbates.

In the infrared region, typically over 95% of the light inciding on a metal surface is reflected, and deposition of an IR-absorbing film decreases its reflectance. As a consequence, the absorption bands in external reflectance spectra of adsorbates on metal surfaces, or in ATR spectra of species adsorbed on highly reflecting thin metal films deposited on IR-transparent substrates, are positive, as in the corresponding transmission spectra. Although the band shape is affected by the dielectric properties of the metal and by the angle of incidence (asymmetric band shapes are obtained at grazing angles and for the less reflecting metals), these spectra are relatively easy to interpret.

Complications in the infrared spectra of adsorbed species may arise, however, when moderately reflecting or transparent substrates are used in external reflectance, or in ATR if the metal film deposited on the IR-transparent window is too thin and shows a poor reflectivity. Under these circumstances, positive, derivative-like (bipolar), or negative absorption (antiabsorption) bands may appear, depending on the incidence angle and on the optical constants of the incident medium, the adsorbate layer and the substrate (further complications in the form of artifact bands may arise if, in addition, IR radiation can excite phonon modes of the substrate). For example, antiabsorption bands appear for poly(methylmethacrylate) adsorbed on glassy carbon [37], for ethyl xanthate adsorbed on the semiconducting mineral chalcocite ($Cu_2S$) [39–42], for CO chemisorbed on particulate metals deposited on graphite [43], and for CO and other adsorbates on particulate metals deposited on glassy carbon [44].

Even if the substrate is a highly reflecting metal (external reflection), or if the thin metal film on the IR-transparent window shows a high reflectivity (ATR), a decrease of its reflectance due to the deposition of a particulate metal film may also provoke a transition from positive absorption bands to derivative or antiabsorption bands. A beautiful example is the case of platinized Pt, for which the band of chemisorbed CO was seen to change progressively with increasing thickness of the particulate Pt deposit from positive absorption for thin, highly reflecting deposits to bipolar and, finally, to negative absorption when the platinized Pt coating was so thick that it had a very low reflectance [45]. Essentially the same behavior has been observed for Pt black on Au [46], for Pt electrodeposited on a Pt microelectrode [47], and for carbon-supported Pt nanoparticles on Au [46].

The observation of bipolar or negative absorption bands can be adequately described with Fresnel's equations applied to the simple three-layer model, without having to resort to any special phenomenon outside the bounds of the formalism of classical electromagnetism. Recently [48, 49], we have used a simple yet realistic model for the adsorption of CO on a metal particle, and an advanced effective medium model [50], to calculate quantitatively the small changes of the refractive index of a thin layer of metal particles deposited on a substrate due to the adsorption of CO on the surface of the particles. This was then used to calculate, also quantitatively, the shape and intensity of the bands that should be observed. Figure 8.2*a* illustrates the conventional three-layer model used for the differential reflectance

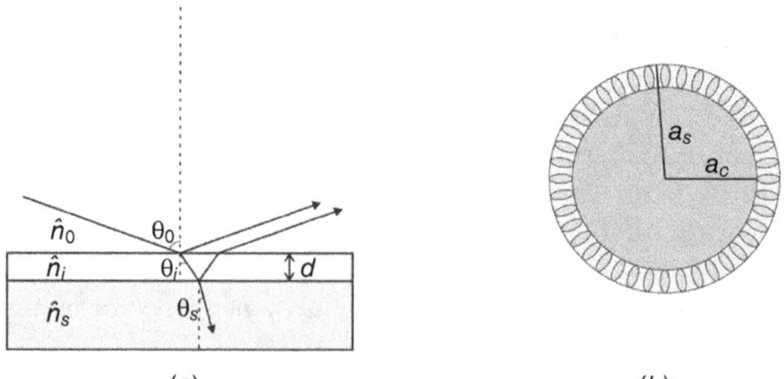


**Figure 8.2** Conventional three-layer model (*a*) used for the differential reflectance calculations and model of a metal particle completely covered by an adsorbate (*b*). Adapted from [49].

calculations, composed of three parallel phases: a transparent medium of complex refractive index $\hat{n}_0$, a thin layer of complex refractive index $\hat{n}_i$ and thickness $d$ composed of metallic particles dispersed into a continuous matrix (air or water), and a polished substrate of complex refractive index $\hat{n}_s$; $\theta_0$ is the angle of incidence at the interface between the transparent medium and the thin layer of nanoparticles, $\theta_i$ is the complex angle of incidence at the interface between the thin layer and the substrate, and $\theta_s$ is the complex refraction angle at the thin layer–substrate interface. The Fresnel relationship,

$$\hat{n}_0 \sin\theta_0 = \hat{n}_i \sin\theta_i = \hat{n}_s \sin\theta_s \tag{8.6}$$

relates these three angles with each other.

Figure 8.2*b* shows the model of a metal particle completely covered by an adsorbate used to describe the adsorption of a compound on the metal particles in the thin layer. Each particle is composed of a metallic core surrounded by an adsorbate shell, and the thin layer is described as a composite of metallic nanospheres (the core in Fig. 8.2*b*), completely coated by an anisotropic layer of the adsorbate and surrounded by ellipsoidal particles of a dielectric matrix (air or water).

The adsorption of a chemical species on the surface of the metal particles will provoke a slight change in the refractive index of the thin layer from $\hat{n}_{i0}$ to $\hat{n}_i = \hat{n}_{i0}(1+\Delta)$, where $\Delta \ll 1$ is the contribution of the adsorbate. Using a thin-layer approximation, that is, $d/\lambda \ll 1$, it was shown in [48] that the differential reflectance can be written as a power series of $\Delta$:

$$\frac{\delta R_{s,p}}{R_{s,p}} = \frac{R_{s,p} - R_{0;s,p}}{R_{0;s,p}} = 2\,\mathrm{Re}\left(\rho_{1;s,p}\Delta\right) + 2\,\mathrm{Re}\left(\rho_{2;s,p}\Delta^2\right) + \left|\rho_{1;s,p}\Delta\right|^2 + \mathrm{O}(\Delta^3) \tag{8.7}$$

where $R_{s,p}$ and $R_{0;s,p}$ are the reflectance in the presence and absence of the adsorbate, respectively, $\rho_{1;s,p}$ and $\rho_{2;s,p}$ are complex coefficients that depend exclusively on $\hat{n}_0$, $\hat{n}_s$, $\theta_0$; and $\hat{n}_{i0}$ (i.e., they are independent of the adsorbate), $\mathrm{O}(\Delta^3)$ is usually infinitesimally small and can be neglected, and the $s$ and $p$ subindexes refer to $s$- and $p$-polarized light, respectively. Thus, the problem reduces to the calculation of $\hat{n}_{i0}$ and $\Delta$, for which an advanced effective medium model [50] based on Bruggeman's effective medium approximation [51] (restricted to particles with a diameter much smaller than the wavelength) was used. The procedure necessary to calculate $\rho_{1;s,p}$, $\rho_{2;s,p}$, and $\Delta$ can be found in Ref. [49].

The parameters that determine the shape of the IR bands of a molecular adlayer on the surface of the metal particles in a thin film deposited on a substrate are: (i) the optical properties of the substrate; (ii) the polarization state of the radiation; (iii) the angle of incidence; (iv) the fill factor, $f$, that is, the fraction of the volume of the thin film of particles that is occupied by the particles; and (v) the percolation threshold, $f_c$, which is the fill factor at which electrical contact between all the metal particles in the thin film is reached.

The parameter $f_c$ is particularly important because surface plasmons appear in its neighborhood and critically affect the shape of the differential reflectance bands due to the formation of a molecular adlayer on the metal particles in the film. The real, $\langle\varepsilon'\rangle$, and imaginary, $\langle\varepsilon''\rangle$, parts of the calculated effective dielectric constant at

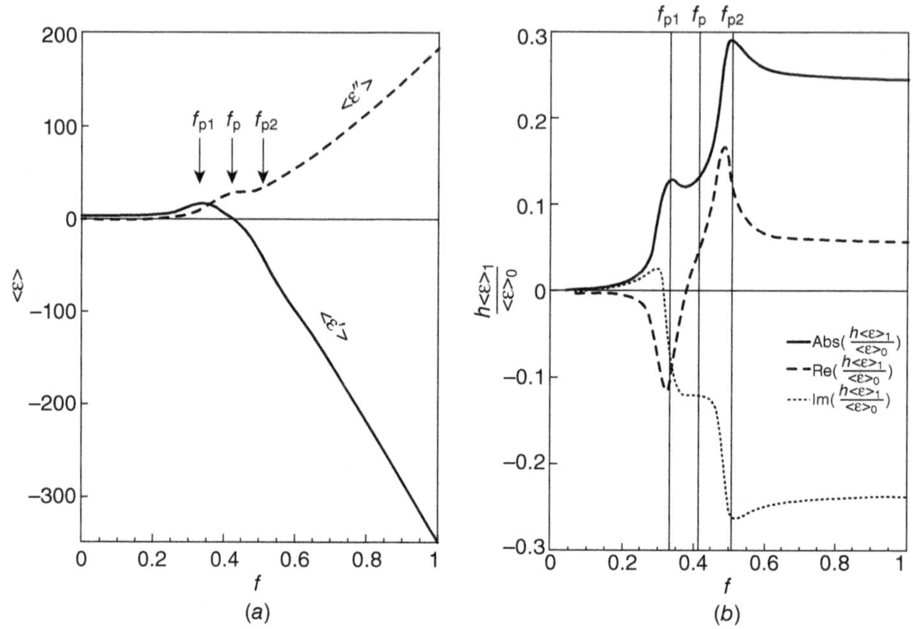


**Figure 8.3** (*a*) Calculated real, ⟨ε′⟩, and imaginary, ⟨ε″⟩, components of the effective dielectric constant, ⟨ε⟩, at 2050 cm$^{-1}$ of a thin layer of spherical Pt particles embedded in water as a function of the fill factor. The arrows indicate plasmons. (*b*) Calculated absolute value (solid line), and real (dashed line) and imaginary (dotted line) components, of the relative change of the effective dielectric constant at 2050 cm$^{-1}$ of a thin layer of Pt particles embedded in water, due to the adsorption of CO, as a function of the fill factor. Adapted from [49].

2050 cm$^{-1}$ of a layer of spherical Pt particles (200 nm in diameter) embedded in water are shown in Figure 8.3(*a*). The percolation threshold was fixed at $f_c = 0.4$, and the optical constants of Pt were taken from Ref. [52]. Three surface plasmons around the percolation threshold, $f_{p1}$, $f_p$, and $f_{p2}$, have been marked with arrows in Figure 8.3(*a*). Parameters $f_{p1}$ and $f_{p2}$ correspond to the lower and upper limit, respectively, of the range of fill factors in which Eq. (8.8) (which gives the effective dielectric constant of the thin film if the particles of the matrix surrounding the metal particles are spherical) does not yield real solutions [49]:

$$c_2\langle\varepsilon\rangle^2 + c_1\langle\varepsilon\rangle + c_0 = 0 \tag{8.8}$$

The coefficients are, for an isotropic adsorbate shell,

$$c_2 = 6 + 2h\left(\frac{\varepsilon_c}{\varepsilon_p} - 1\right) \tag{8.9}$$

$$c_1 = (1-3f)\varepsilon_c(3-2h) - (2-3f)\varepsilon_m \times\left[3 + h\left(\frac{\varepsilon_c}{\varepsilon_p} - 1\right)\right] + 2h\varepsilon_p(1-3f) \tag{8.10}$$

$$c_0 = -\varepsilon_c\varepsilon_m(3-2h) - 2h\varepsilon_p\varepsilon_m \tag{8.11}$$

and for an anisotropic adsorbate shell

$$c_2 = 6 + 2h\left(\frac{\varepsilon_c}{\varepsilon_p} - 1\right) \tag{8.12}$$

$$c_1 = (1-3f)\varepsilon_c(3-2h) - (2-3f)\varepsilon_m \times\left[3 + h\left(\frac{\varepsilon_c}{\varepsilon_p} - 1\right)\right] + 2h\varepsilon_p(1-f) \tag{8.13}$$

$$c_0 = -\varepsilon_c\varepsilon_m(3-2h) - 2h(1-f)\varepsilon_p\varepsilon_m \tag{8.14}$$

where $h = 5\times 10^{-3}$ is the fraction of the volume of the CO-covered Pt particle covered by the CO adlayer, and the $c$, $p$, and $m$ subindexes refer to core, particle, and matrix, respectively.

Parameter $f_p$ is the arithmetic mean of $f_{p1}$ and $f_{p2}$ and corresponds to the fill factor at which $\langle\varepsilon'\rangle = 0$, that is, at which the condition for the appearance of a bulk plasmon is achieved. Part $\langle\varepsilon'\rangle$ has a positive maximum around $f_{p1}$, at which $\langle\varepsilon''\rangle$ starts to increase. $f_p$, where $\langle\varepsilon'\rangle = 0$ and $\langle\varepsilon''\rangle$ reaches a local maximum, appears very close to $f_c = 0.4$. Finally, at $f_{p2}$ $\langle\varepsilon''\rangle$ reaches a local minimum due to the cancellation of the plasmon condition.

The effect of the surface plasmon can be better appreciated if the relative increment of the effective dielectric constant at 2050 $cm^{-1}$ due to the adsorption of a CO adlayer on the surface of Pt particles, $h\langle\varepsilon\rangle_1/\langle\varepsilon\rangle_0$ (where $h = 5\times10^{-3}$ is the fraction of the volume of the CO-covered Pt particle covered by the CO adlayer) is plotted versus the fill factor (Fig. 8.3*b*). The absolute value of $h\langle\varepsilon\rangle_1/\langle\varepsilon\rangle_0$ shows two sharp maxima at $f_{p1}$ and $f_{p2}$ and a minimum close to $f_c \simeq f_p$. The maximum at $f_{p1}$ nearly coincides with the minimum of the real part and is not far from the maximum of the imaginary part, while, on the contrary, the maximum at $f_{p2}$ occurs near the maximum of the real part and nearly coincides with a minimum of the imaginary part.

A particularly simple case is that of very low fill factors, $f \to 0$. Under these conditions, the complex refractive index of the composite layer formed by the metal particles and the dielectric matrix must be very close to that of the latter ($\hat{n}_i \to \hat{n}_0$), that is, the optical properties of the surface will be that of the substrate ($h\langle\varepsilon\rangle_1/\langle\varepsilon\rangle_0 = 0$ in Fig. 8.3*b*). The expression giving the relative change in reflectance due to the adsorption of a molecule on the metal particles [Eq. (8.7)] can be simplified to

$$\frac{\delta R_{s,p}}{R_{s,p}} \cong -2\rho''_{1;s,p}(\omega_0, \hat{n}_i \to \hat{n}_0)\Delta'' \tag{8.15}$$

where $\rho''_{1;s,p}(\omega_0, \hat{n}_i \to \hat{n}_0)$ is the imaginary part of $\rho_{1;s,p}(\hat{n}_i \to \hat{n}_0)$ and depends only on $\hat{n}_0$ and on the angle of incidence; $\omega_0$ is the absorption frequency of the adsorbate; and $\Delta''$ is the imaginary part of $\Delta$. Explicit and relatively concise expressions can be obtained for $\rho_{1;s,p}(\hat{n}_i \to \hat{n}_0)$ for this particular case of a very thin layer of metal particles covering a very small fraction of the substrate surface [48], and the angles of incidence at which absorption, bipolar, and antiabsorption bands will be obtained can be determined. This implies determining, for each angle of incidence, the couples of $\mathrm{Re}(n_s)$ and $\mathrm{Im}(n_s)$ for which $\rho''_{1;s,p}(\omega_0, \hat{n}_i \to \hat{n}_0) = 0$. The curves in Figure 8.4 correspond to the values of $\mathrm{Re}(n_s)$ and $\mathrm{Im}(n_s)$ that satisfy this condition, for

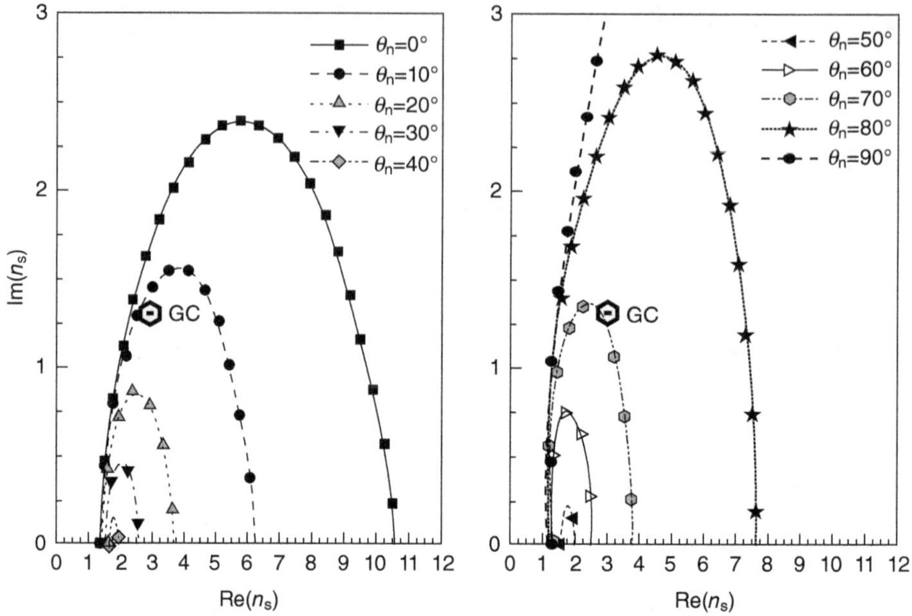


**Figure 8.4** Calculated band inversion condition for a very thin layer of metal particles covering a very small fraction of the substrate surface (i.e., for a very low fill factor), as a function of the complex refractive index of the substrate, for *p*-polarized radiation and several angles of incidence at the water–thin layer interface. Substrates with a complex refractive index located outside/inside the arcs will yield absorption/antiabsorption adsorbate bands. From Ref. [48].

incidence angles from $\theta_0 = 0$ to 90° and *p* polarization. Substrate materials whose complex refractive index lies outside the arcs in Figure 8.4 will produce absorption bands, while those located inside the arcs will yield antiabsorption bands. Thus, a molecule adsorbed on a very thin film of metal particles covering a very low fraction of a glassy carbon substrate will yield absorption bands at angles of incidence between 10° and 70° but antiabsorption bands if $\theta_0 < 10°$ or $\theta_0 > 70°$. On the contrary, if the substrate is a highly reflecting metal surface (for which typically $k > 10$), the substrate complex refractive index lies always outside the arcs, and absorption bands are always obtained for a thin, low–coverage film of metal particles.

Summarizing this section, it can be concluded that for a thin layer of metal particles on a substrate (which can be considered as a metal–insulator composite) the shape of the band will be governed by the optical properties of the substrate at very low fill factors, and remarkable changes in band shape and intensity in differential IR spectra of adsorbates are to be expected in the range of fill factors from $f_{p1}$ to $f_{p2}$ around the percolation threshold, where plasmons can be excited. Taking into account that the SEIRA effect requires the presence of metal islands embedded in an insulator matrix and a low-reflecting substrate that can hold electromagnetic fields parallel to its surface, and that the SEIRA effect is due to the excitation of plasma oscillations in the metal islands, it is by no way surprising that bipolar and antiabsorption bands are always accompanied by an enhancement of the IR signal. Fortunately, the opposite is not true, and the SEIRA effect can occur without necessarily implying the concomitant observation of bipolar or antiabsorption bands. Theoretical descriptions of the SEIRA effect within the frame of an effective medium theory can be found in Refs. [24] and [53].

### 8.2.3 Experimental Configuration: Angles of Incidence

The geometrical configuration of the experimental setup will determine the angle of incidence at the electrode–electrolyte interface and, consequently, the electric field strength at the electrode surface. As indicated above, the higher the electric field strength, the more intense the IR signal, and, accordingly, the angle of incidence has to be optimized in order to obtain the highest possible signal-to-noise ratio. Also to be minimized is the amount of light reflected at the air–window interface since any photon reflected here will either be lost (prismatic or hemicylindrical window) or will reach the detector without having been reflected at the electrode–electrolyte interface (planar window) and will affect the signal-to-noise ratio negatively.

#### *External Reflectance, Flat Fluorite Window*

The amount of light reflected at the air–fluorite interface will be null at normal incidence, but, in this geometry, the electric field strength at the electrode surface will also be null (see above). In principle, the best choice for the incidence angle at the air–window interface would be the Brewster angle ($\theta_B$). At $\theta_B$, the reflected and the refracted beams will be perpendicular to each other:

$$\theta_B + \theta_r = 90° \tag{8.16}$$

$$\theta_B = \arctan\frac{n_2}{n_1} \tag{8.17}$$

where $\theta_r$ is the refraction angle and $n_1$ and $n_2$ are the refraction indexes of the incident (air) and refracting (fluorite) media, respectively. Under these conditions, all the *p*-polarized radiation is transmitted across the interface and (in the case of metal electrodes and if *p*-polarized light is used in order to profit from the surface selection rule) no decrease of the signal-to-noise ratio due to reflection loses should occur. In the case of the air–fluorite interface $\theta_B$ is around 52° ($n_{air} \simeq 1$; $n_{fluorite}$ at 1000 $cm^{-1} \simeq 1.3$), and the refraction angle (which will be identical to the incidence angle at the fluorite–electrolyte interface) when the incidence angle coincides with $\theta_B$ is about 38°, resulting in an angle of incidence at the electrode–electrolyte interface of about 40° at 1000 $cm^{-1}$. As illustrated in Figure 8.1, at these low angles of incidence the electric field strength at the electrode surface is rather low and far from the optimum that would maximize IR absorption at the electrode–electrolyte interface. For this reason, when using a planar fluorite window, it is better to choose a geometry such that the beam, once reflected at the electrode surface, exits across the fluorite–air interface just below the corresponding critical angle, $\theta_c$. Angle $\theta_c$ is defined as the angle of incidence at which, when passing from a medium of higher refraction index to another medium of lower refraction index, $\theta_r = 90°$, and is given by

$$\theta_c = \arcsin \frac{n_2}{n_1} \tag{18.18}$$

At incidence angles above the critical angle, total reflection occurs at the interface (i.e., no radiation is transmitted). The most appropriate incidence angle at the air–fluorite window interface, when working with planar windows and with aqueous electrolytes, is 68°, for which the angle of incidence at the fluorite–water interface is, at 1000 $cm^{-1}$, about 45° (which will coincide with the angle of incidence with which the beam will exit through the fluorite–air interface after having been reflected at the electrode surface, just below the corresponding $\theta_c = 46°$). With this geometry, the angle of incidence at the electrode–electrolyte interface will be, at 1000 $cm^{-1}$, about 50°. Although some *p*-polarized light will be reflected at the air–fluorite interface, its detrimental effect on the signal-to-noise ratio will be compensated by the higher angle of incidence and the corresponding higher field strength at the electrode–electrolyte interface.

### *External Reflectance, Prismatic Fluorite Window*

The signal-to-noise ratio can be considerably improved in the external reflectance configuration if a prismatic window with triangular basal faces is used instead of a planar window. If the IR beam incides on one of the lateral faces of a fluorite prism with triangular basal faces at normal incidence, it will reach the adjacent lateral face with an angle of incidence that will coincide with the angle between the lateral faces involved. In order to achieve the highest possible angles of incidence at the electrode–electrolyte interface, the angle between the lateral faces of the triangular prism must be chosen just below the critical angle of the fluorite–water interface. Good choices are an equilateral triangular fluorite prism or a prism beveled at 65° ($1.5 > n_{water} > 1.2$ and $1.43 > n_{fluorite} > 1.3$ in the spectral range between 4000 and 1000 $cm^{-1}$, in which $90° > \theta_c > 70°$; please note that in a narrow region between about 2860 and about 2950 $cm^{-1}$, the refraction index of water is higher than that of fluorite, and total reflection cannot occur at the fluorite–water interface). At 1250 $cm^{-1}$, and if an equilateral triangular fluorite prism is used, the angle of incidence at the electrode–electrolyte interface is about 64°. The lack of reflection losses at the air–window interface and the higher electric field strength at the electrode surface due to the higher angle of incidence combine to yield a signal-to-noise ratio higher than that achievable using a flat fluorite window.

An alternative to prismatic windows are hemicylindrical windows, with which the incidence is always normal at the air–window interface. These windows are very appropriate if the angle of incidence at the electrode–electrolyte interface is to be changed. It has to be noted, however, that such a window will act as a lens and will displace the position of the focal point of the infrared beam.

### *Attenuated Total Reflectance*

In this configuration the angle of incidence at the window–electrolyte interface must be above the corresponding critical angle, so that total internal reflection occurs. For this reason, high-refractive index materials are the most adequate for the window, Si ($n_{Si} > 3.4$) being the most popular choice. As before, prismatic or hemicylindrical windows are preferred to planar windows in order to minimize reflection losses at the air–Si interface and thus improve the signal-to-noise ratio. Although, at the Si–water interface, $\theta_c \simeq 20°$, equilateral triangular Si prisms and angles of incidence at the Si–water interface of 60° are typically used, because, as with external reflectance, the band intensity increases with increasing incidence angle. Although the highest band intensity would be expected for an angle of incidence of about 80°, this is largely due to the expansion of the beam on the interface along the incidence direction and, for the typical electrode sizes of 1–2 cm in diameter, no significant improvement of the band intensity is achieved when the angle of incidence is increased from 60° to 80°. Using an angle of incidence of 60° or 65° has, in addition, the practical advantage that the same reflectance accessory can be used for external reflectance IR spectroscopy and for ATR–SEIRAS.

## 8.3 EXTERNAL REFLECTANCE INFRARED SPECTROSCOPY FOR STRUCTURAL STUDIES OF CHEMISORBED ADLAYERS

The only advantage of external reflectance over ATR–SEIRAS is the possibility to use single–crystal electrodes and thus study the influence of the substrate's structure on the structure of adsorbate adlayers. A particular case that has been extensively studied, and that can hardly be excluded from a review of IR spectroscopy on electrode surfaces, is that of CO chemisorbed on Pt single-crystal electrodes. The interest of this system derives not only from the ubiquitous presence of adsorbed CO ($CO_{ad}$), a catalytic poison, during the electrooxidation of organic molecules (Pt being the archetypal catalyst), but also from the sensitivity of the C–O bond to the chemical, geometric, and electrostatic environment, which affect the vibrational properties of the CO molecule. Consequently, CO is often used in infrared spectroscopy as a probe of the structure of the electrode surface and/or of the electrochemical double layer.

The first IR studies of CO adsorption on Pt single-crystal electrodes date from the late 1980s [54–58], but the contributions of Weaver and co-workers during the 1990s remain, in my opinion, unsurpassed. They compared IR spectra of CO adsorbed on Pt single-crystal electrodes in aqueous electrolytes with those in ultra-high vacuum (UHV) [58, 59] and nonaqueous solvents [60] in order to understand the effect of the double layer and of the electrolyte composition on the structure and properties of chemisorbed CO, which, for the saturation coverage, were found to be controlled essentially (when not exclusively) by the potential drop across the adlayer. They also studied systematically the effect of CO coverage and electrode potential [61–63], with the aim of understanding dipole–dipole coupling effects, the role of $d\pi$–$2\pi^*$ back-bonding in the geometry of the CO binding, and the influence of co-adsorbed water, anions, or hydrogen on the structure of the CO adlayer. Their conclusion that low-coverage CO adlayers formed by stripping of a saturated adlayer are different to those formed by controlled dosing of CO from the solution, compact islands with a local high CO coverage being formed in the former case, even at rather low total coverage, are often used as an argument in the recurring discussion regarding the mobility of CO chemisorbed on Pt electrodes and on its role in the electro-oxidation of CO. Finally, they pioneered the combination of in situ scanning tunneling microscopy (STM) and Fourier transform infrared spectroscopy (FTIRS) [64] for the characterization of CO and, in general, adsorbate layers on single-crystal metal electrodes, and identified for the first time the (2 × 2)-3CO and $(\sqrt{19} \times \sqrt{19})$ $R23.4°$ structures of CO on Pt(111) electrodes, as well as their corresponding IR spectral fingerprints.

The adsorption of CO on Pt(111) is a good example of how IR spectroscopy can be used to obtain information regarding the structure of adsorbate layers on well-ordered metal electrodes. Figure 8.5 shows series of potential-dependent FTIR spectra of CO adsorbed on a Pt(111) electrode in CO-saturated (Fig. 8.5*a*) and in CO-free (Figs. 8.5(*b*,*c*) 0.1 M $H_2SO_4$. In all the cases, a saturated CO adlayer was initially formed at a dosing potential $E_d = 0.05$ V versus the reversible hydrogen electrode (RHE).

In CO-saturated solutions (Fig. 8.5*a*), an intense band, whose frequency increases from 2066 $cm^{-1}$ at 0.05 V to 2077 $cm^{-1}$ at 0.50 V with a Stark tuning rate of 29 $cm^{-1}$ $V^{-1}$, can be observed over the whole potential range in which the CO adlayer is stable [57, 59, 65]. The intensity and the frequency of this band, which corresponds to CO adsorbed at on-top sites ($CO_L$), starts to decrease at $E > 0.50$ V due to the decrease in the CO coverage as a consequence of the oxidation of the adlayer, which is complete at 0.75 V. In the frequency range below 1900 $cm^{-1}$, and at $E < 0.50$ V, there is a single band whose frequency increases from 1780 $cm^{-1}$ at 0.05 V to 1799 $cm^{-1}$ at 0.45 V, with a Stark tuning rate of 47 $cm^{-1}$ $V^{-1}$ [59, 65], which corresponds to CO adsorbed on threefold-hollow sites ($CO_M$) [64]. At $E = 0.50$ V, the intensity of this band decreases, and a new band, which corresponds to CO adsorbed on twofold bridge sites ($CO_B$), appears at 1843 $cm^{-1}$. The intensity of the $CO_B$ band increases at the expenses of the $CO_M$ band, which has nearly completely disappeared at 0.60 V. Infrared spectra consisting of a band around 2070 $cm^{-1}$ ($CO_L$) and a band around 1780 $cm^{-1}$ ($CO_M$) are characteristic of the (2 × 2)-3CO structure ($\theta_{CO} = 0.75$) on Pt(111), while spectra consisting of a band around 2070 $cm^{-1}$ ($CO_L$) and a band around 1830 $cm^{-1}$ ($CO_B$) are characteristic of the $(\sqrt{19} \times \sqrt{19})$ $R23.4°$ structure ($\theta_{CO} = 0.68$) on Pt(111). The emergence of the $CO_B$ band marks the transition between the two structures, provoked by a decrease in the CO coverage due to partial oxidation of the adlayer, and it is well known that the potential at which this transition takes place is the more negative the higher the density of defects on the Pt(111) surface [66]. The observation of both the band around 1780 $cm^{-1}$ and

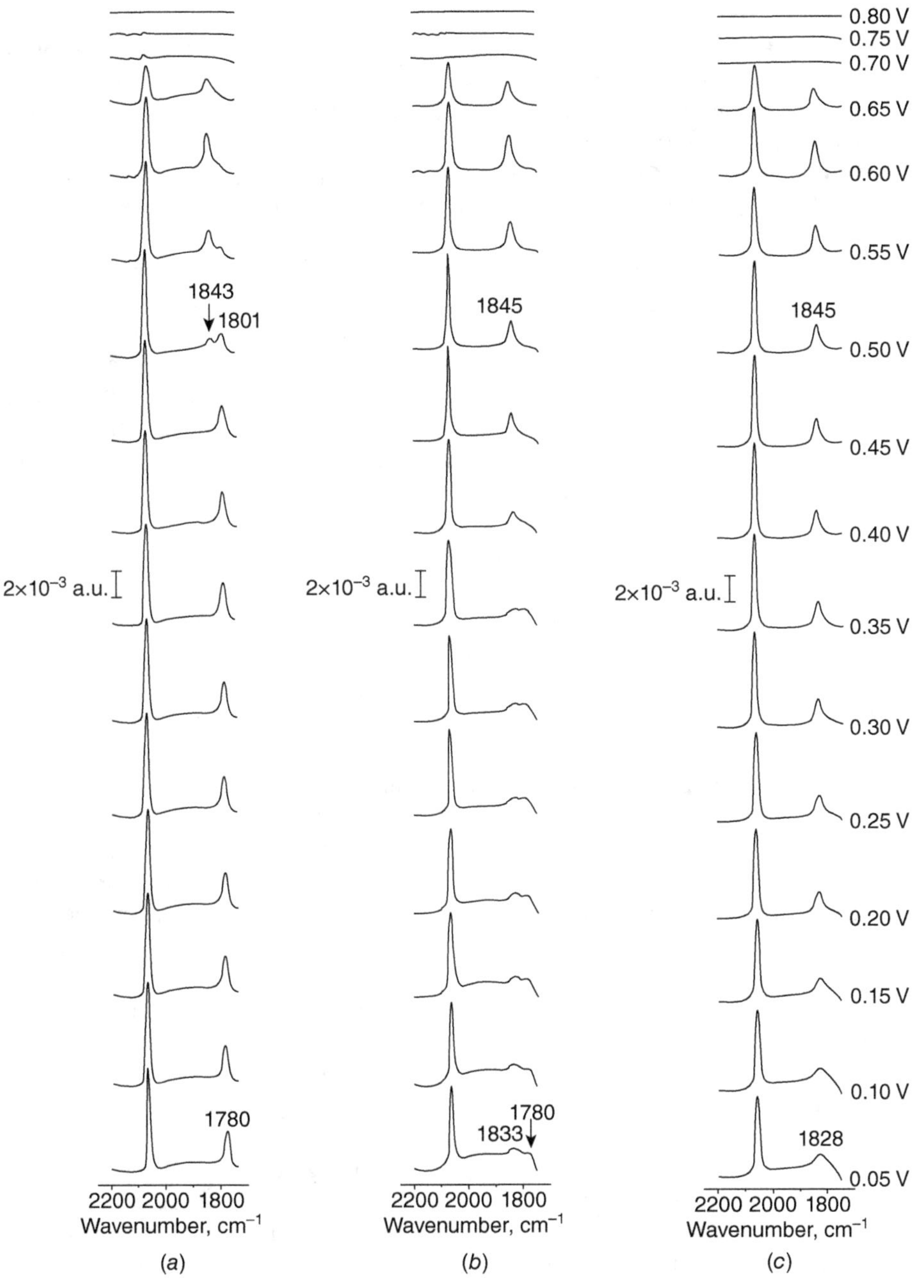


**Figure 8.5** Potential-dependent FTIR spectra of the CO adlayer on a Pt(111) electrode in (*a*) CO-saturated 0.1 M $H_2SO_4$ and after purging the initially CO saturated solution by bubbling $N_2$ through it for (*b*) 20 min and (*c*) 3 h. The reference spectrum was taken at $E_{ref} = 0.80$ V. From Ref. [65].

the band around 1830 $cm^{-1}$ at 0.50 and 0.55 V (Fig. 8.5*a*) suggests that, at CO coverages between 0.68 and 0.75, patches of the $(\sqrt{19} \times \sqrt{19})R23.4°$ and the (2 × 2)-3CO structures coexist on the electrode surface.

The spectra in Figure 8.5*c* were obtained in CO-free solution, after forming the CO adlayer at $E_d$ = 0.05 V versus RHE and after purging the solution with $N_2$ for 3 h. In the absence of CO in the solution, the spectra are characterized by a band around 2060 $cm^{-1}$ ($CO_L$) and a band around 1828 $cm^{-1}$ ($CO_B$). The latter band shows a shoulder at the low-frequency side, which disappears at $E$ = 0.30 V, at which potential an accurate determination of the CO coverage as a function of the potential by electrochemical means showed that $\theta_{CO}$ decreases from 0.68 to 0.67. Similar correlations between the CO coverage, determined by an electrochemical procedure, and the FTIR spectrum of the CO adlayer were found for Pt(100) [67] and for polycrystalline Pt [68]. The spectra in Figure 8.5*a*,*b* suggest that, although the highest possible CO coverage on a Pt(111) electrode is 0.75 ML [corresponding to the (2 × 2)-3CO structure], this coverage can only be attained in CO-saturated solutions. In CO-free solutions the spectra show the characteristic fingerprint of the $(\sqrt{19} \times \sqrt{19})R23.4°$ structure, and the CO coverage is, at most, 0.68 ML. This is confirmed in Figure 8.5*b*, which shows spectra recorded in an initially CO-saturated 0.1-M $H_2SO_4$ solution after purging with $N_2$ during 20 min. Under these conditions, CO could not be completely removed from the solution, and the spectra show a band around 1833 $cm^{-1}$ with a shoulder around 1780 $cm^{-1}$ that disappears at 0.45 V. This supports the suggestion outlined above that, at CO coverages between 0.68 and 0.75, patches of the $(\sqrt{19} \times \sqrt{19})R23.4°$ and the (2 × 2)-3CO structures coexist on the electrode surface.

The higher CO coverage on Pt electrodes in CO-containing solutions, as compared with CO-free ones, was confirmed in a later work [69] in which the CO coverage of a Pt(111) electrode was accurately determined in solutions equilibrated with $N_2$–CO mixtures of different CO partial pressures ($0.005\% \le P_{CO} \le 0.8\%$), using the charge in CO-stripping voltammograms, and the exact thermodynamic double-layer correction [70], based on the CO charge displacement technique developed by the Alicante group [71, 72]. Although IR spectroscopy cannot provide quantitative information regarding the CO coverage, its ability to distinguish between the $(\sqrt{19} \times \sqrt{19})R23.4°$ and the (2 × 2)-3CO structures was used to confirm the coverage data obtained by electrochemical means and to detect qualitatively changes in the CO coverage too small to be detected quantitatively from the CO stripping voltammograms.

Table 8.1 lists the values of the total CO stripping charges, $Q_{total}^{CO\ strippping}$, of the corresponding calculated net Faradaic charges, $Q_{net}^{CO\ oxidation}$, and of the resulting CO coverages in 0.1 M $H_2SO_4$ solutions equilibrated with different $N_2$–CO mixtures, and Figure 8.6 shows the corresponding FTIR spectra at selected CO concentrations. For $P_{CO} \le 0.01\%$, $\theta_{CO}$ = 0.68 ± 0.01, which corresponds to the coverage of the $(\sqrt{19} \times \sqrt{19})R23.4°$ structure and coincides with that obtained in CO-free solutions. According to the data in Table 8.1, a clear increase in $\theta_{CO}$ occurred only at $P_{CO} \ge 0.1\%$, although the observation of a shoulder around 1780 $cm^{-1}$, characteristic of the (2 × 2)-3CO structure in the corresponding FTIR spectrum (Fig. 8.6*b*), suggests that

**Table 8.1** Experimentally Measured Total CO Stripping Charges, $Q_{total}^{CO\ strippping}$,[a] Calculated Net Faradaic Charges for CO Stripping, $Q_{net}^{CO\ oxidation}$,[b] and CO Coverages, $\theta_{CO}$,[c] in 0.1 M $H_2SO_4$ Solutions in Equilibrium with $N_2$–CO Mixtures of Different CO Concentrations

| $P_{CO}$ (%) | $Q_{total}^{CO\ strippping}$ ($\mu$C cm$^{-2}$) | $Q_{net}^{CO\ oxidation}$ ($\mu$C cm$^{-2}$) | $\theta_{CO}$ |
|---|---|---|---|
| **0.005** | 466 ± 3 | 327 ± 6 | 0.68 ± 0.01 |
| **0.01** | 467 ± 2 | 328 ± 5 | 0.68 ± 0.01 |
| **0.1** | 472 ± 2 | 333 ± 5 | 0.69 ± 0.01 |
| **0.2** | 481 ± 1 | 342 ± 5 | 0.71 ± 0.01 |
| **0.3** | 484 ± 3 | 345 ± 6 | 0.72 ± 0.01 |
| **0.4** | 494 ± 3 | 355 ± 6 | 0.74 ± 0.01 |
| **0.6** | 498 ± 3 | 359 ± 6 | 0.75 ± 0.01 |
| **0.7** | 499 ± 4 | 360 ± 6 | 0.75 ± 0.01 |
| **0.8** | 502 ± 9 | 363 ± 10 | 0.76 ± 0.01 |

[a]Obtained by integration of CO stripping voltammograms at 500 mV s$^{-1}$ between 0.10 and 1.00 V.
[b]Calculated using the charge density displaced during the potentiostatic adsorption of CO at 0.10 V, $Q_{initial}^{CO\ adsorption} = 148 \pm 3\ \mu C\ cm^{-2}$, and the charge density obtained by integration of the voltammogram at 500 mV s$^{-1}$ of CO-free Pt(111) between 0.10 and 1.00 V, $Q_{ox}^{Pt} = 287 \pm 4\ \mu C\ cm^{-2}$, to determine the true, exact thermodynamic double-layer correction.
[c]Calculated as $\theta_{CO} = Q_{net}^{CO\ oxidation}/2Q_{Pt(111)}$, where $Q_{Pt(111)} = 240\ \mu C\ cm^{-2}$ is the charge density corresponding to a process in which one electron per Pt atom of the Pt(111) surface flows across the electrode–electrolyte interface.

Adapted from Ref. [69].

a coverage higher than 0.68 is already reached at $P_{CO}$ = 0.01%. The very low CO concentration necessary to increase $\theta_{CO}$ above 0.68 explains the irreproducibility found in the IR spectra of CO on Pt(111) electrodes in "CO-free" solutions in the region below 1900 cm$^{-1}$, where a single band around 1830 cm$^{-1}$ (Fig. 8.5*c*), a band around 1833 cm$^{-1}$ with a shoulder around 1780 cm$^{-1}$ (Fig. 8.5*b*), or a band around 1780 cm$^{-1}$ with a shoulder around 1830 cm$^{-1}$ [66], have been reported, these differences being certainly due to small variations in the CO concentration reached after purging the solution with $N_2$ or Ar.

The quantitative data in Table 8.1 show that $\theta_{CO}$ increases monotonically with increasing $P_{CO}$ and that the highest possible coverage, $\theta_{CO}$ = 0.75 (corresponding to the (2 × 2)-3CO structure), is reached when the CO concentration is only 0.6% of the saturation value, as confirmed by FTIR spectroscopy (Fig. 8.6*d*). Furthermore, for $P_{CO}$ between 0.01 and 0.6%, both the band around 1830 cm$^{-1}$ [characteristic of the $(\sqrt{19} \times \sqrt{19})R23.4°$ structure] and the band around 1780 cm$^{-1}$ [characteristic of the (2 × 2)-3CO structure] are present in the corresponding FTIR spectra, confirming the suggestion above that, at CO coverages intermediate between 0.68 and 0.75, patches with the $(\sqrt{19} \times \sqrt{19})R23.4°$ and the (2 × 2)-3CO structures coexist on the surface of Pt(111) electrodes. This still requires confirmation by in situ STM and/or surface X-ray scattering (SXS).

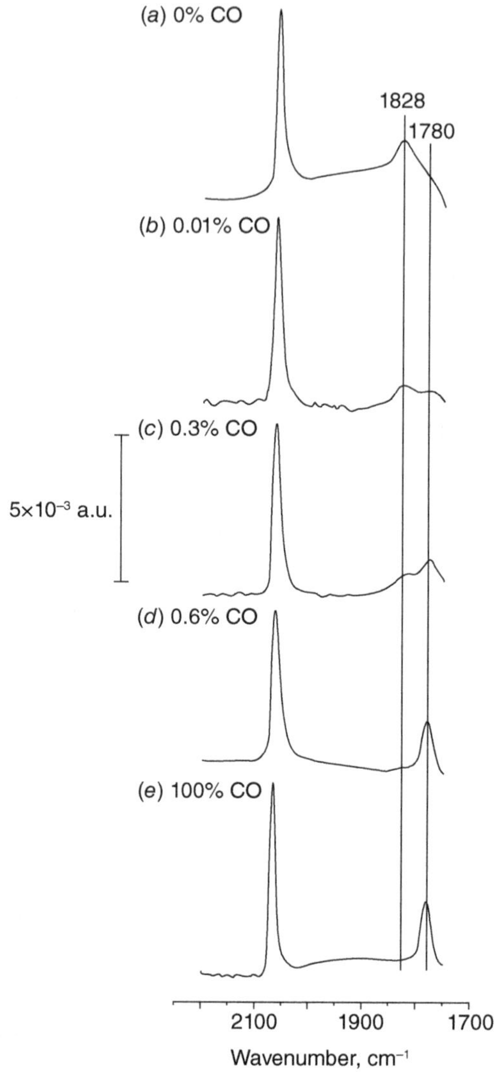


**Figure 8.6** External reflectance FTIR spectra at 0.05 V versus RHE of CO adlayers on a Pt(111) in 0.1 M $H_2SO_4$ solutions in equilibrium with $N_2$ + CO gas mixtures with (*a*) $P_{CO} = 0\%$, (*b*) $P_{CO} = 0.01\%$, (*c*) $P_{CO} = 0.3\%$, (*d*) $P_{CO} = 0.6\%$, and (*e*) $P_{CO} = 100\%$. The reference spectrum was taken at $E_{ref} = 0.80$ V versus RHE.

## 8.4 TIME-RESOLVED ATR–SEIRAS FOR STUDY OF MECHANISM OF ELECTROCATALYTIC REACTIONS

The possibility of identifying the intermediates involved in electrocatalytic reactions, due to the ability of IR spectroscopy to provide chemical information regarding the composition of the electrochemical double layer and about the bonding mode of adsorbates, was certainly one of the reasons that spurred the development of in situ IR spectroscopy for the study of the electrochemical interface and is still today one of the causes of its sustained popularity.

Although it was already widely accepted by the end of the 1970s that the electrooxidation of C1-oxygenated organic molecules (methanol, formaldehyde, and formic acid) proceeds through a dual-path mechanism, one of the paths (usually called the indirect pathway) leading to an electrocatalytic poison [73–77], electrochemists had not reached by then an agreement regarding the nature of the poisoning species. The unambiguous determination of CO as the poison intermediate formed in the indirect pathway was one of the earliest successes of in situ external reflectance IR spectroscopy [78–80]. However, the mass-transport limitations and the high uncompensated electrolyte resistance associated to the thin-layer configuration necessary for in situ external reflectance IR spectroscopy have limited its use in studies of reaction mechanisms and kinetics in electrocatalysis. The development by Osawa and co-workers of ATR–SEIRAS [21–26] has overcome these limitations and has opened the road to IR spectroscopic studies of the mechanisms and kinetics of electrocatalytic reactions. Due to its excellent signal-to-noise ratio, ATR–SEIRAS allows one to monitor these processes because spectra composed of a single interferogram can be obtained. The time resolution is, hence, limited exclusively by the time required by the spectrometer to record an interferogram, typically between some tens and some hundreds of milliseconds, depending on the speed at which the mobile mirror of the spectrometer moves. If the spectrometer allows recording the spectra in the step-scan mode, the time resolution can easily be of some microseconds, but only completely reversible processes can be studied.

One of the most relevant contributions to this area of research by Osawa and co-workers was the observation for the first time during the oxidation of methanol [81], formaldehyde [82], and formic acid [83] on Pt electrodes of bridge-bonded adsorbed formate ($HCOO_{ads}$), which was suggested to be a reactive intermediate in the direct path in all of these three cases. In the particular case of formic acid, Chen et al. [84, 85] argued later that $HCOO_{ads}$ is a site-blocking spectator and that most of the oxidation current is carried by a third path involving the direct dehydrogenation of formic acid via a weakly adsorbed molecular formic acid precursor. However, it has been very recently unambiguously shown [86, 87] that this hypothesis contradicts the experimental results and that $HCOO_{ads}$ must indeed be the reactive intermediate in the direct pathway. In the following, I will illustrate, with recent work from my group on the oxidation of formic acid on metal electrodes, how ATR–SEIRAS can be used to obtain qualitative and quantitative information regarding the mechanism and kinetics of electrocatalytic reactions.

Unveiling the mechanism of the electrooxidation of formic acid on metals involves describing the intermediates and obtaining the kinetic equations for both the direct and the indirect paths. The latter involves the dehydration of formic acid to $CO_{ad}$ and the oxidation of $CO_{ad}$ to $CO_2$, while the former involves the oxidation of HCOOH to $CO_2$ without breaking one of the C–O bonds. Since these two paths run in parallel and start from a common point (HCOOH), there must be another point where the paths separate, that is, there must be an intermediate that is the last

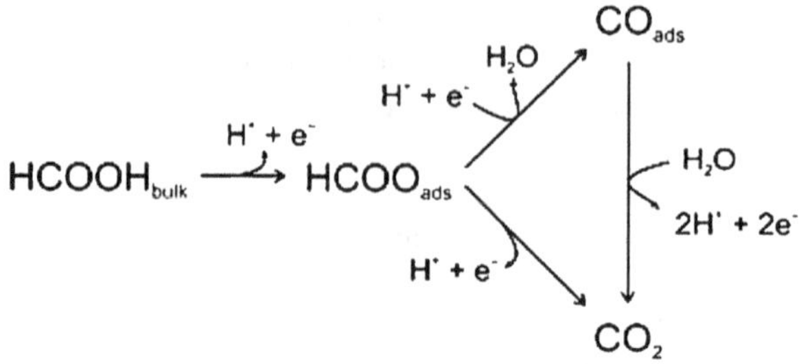


**Figure 8.7** Dual-path mechanism for the electrocatalytic oxidation of formic acid, with bridge-bonded adsorbed formate as the point where the reaction bifurcates into two separate paths.

one shared by the direct and the indirect paths. As we have recently shown [88], and as I will illustrate below, this last intermediate is $HCOO_{ad}$ (Fig. 8.7), which is, hence, the key reaction intermediate, the crossroads where it has to be acted if we wish to design an electrocatalyst with a high activity for formic acid oxidation and a low tendency to CO poisoning.

### 8.4.1 Dehydration of Formic Acid to Adsorbed Carbon Monoxide on Pt Electrodes

According to the reaction mechanism in Figure 8.7, the dehydration of formic acid to $CO_{ad}$ must comprise an electrooxidation step followed by an electroreduction step:

$$HCOOH \rightarrow HCOO_{ad} + H^+ + e \quad \text{(Reaction 1)}$$
$$HCOO_{ad} + H^+ + e \rightarrow CO_{ad} + H_2O \quad \text{(Reaction 2)}$$

If, as we have argued above, $HCOO_{ad}$ is not only the intermediate in the direct path but also in the dehydration of formic acid to $CO_{ad}$, there must be a relationship between the coverage of $HCOO_{ad}$ ($\theta_{formate}$) and the rate at which $CO_{ad}$ is formed ($d\theta_{CO}/dt$).

At negative enough potentials ($E \leq 0.45$ V versus RHE), the current corresponding to HCOOH oxidation on Pt is null or negligible, and the only reaction taking place on the electrode surface is the dehydration of HCOOH to yield $CO_{ad}$. Thus, $\theta_{formate}$ can be assumed to be proportional to the integrated intensity of the $HCOO_{ad}$ band ($I_{HCOO}$), an assumption applicable if the orientation of the adsorbate does not change with coverage [25, 89], and if the adsorbate–adsorbate interaction is weak except at very high coverage [90]. Similarly, $\theta_{CO}$ can be assumed to be proportional to the $CO_L$ band ($I_{CO}$) at low CO coverage [90, 91].

In order to obtain a sufficiently large number of points, instead of a potential-step technique from a potential positive enough as to have a $CO_{ad}$- and $HCOO_{ad}$-free Pt surface to the region below 0.45 V versus RHE, we applied the following strategy: After polarizing the Pt electrode at the potential at which the HCOOH dehydration reaction was to be followed, the recording of a spectral series was started, and then the amount of concentrated HCOOH necessary for reaching a final concentration of

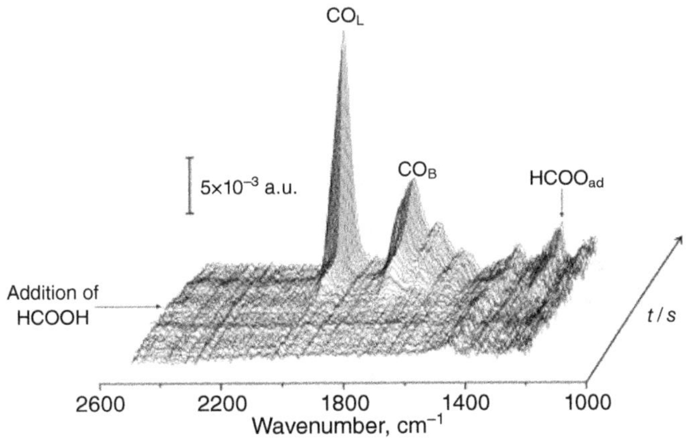


**Figure 8.8** Time-dependent ATR–SEIRA spectra recorded at 0.305 V versus RHE in 0.1 M $HClO_4$ during the dehydration of formic acid on a thin-film Pt electrode deposited on Si. The amount of HCOOH necessary to reach a final concentration of 0.1 M was added at the point indicated by the arrow. Only the spectra recorded during the first 5 s after addition of HCOOH are shown for the sake of clarity. The time interval between spectra is of 140 ms. From Ref. [88].

0.1 M was added with a micropipette to the unstirred working solution. Under these conditions, HCOOH diffuses slowly to the electrode–electrolyte interface until the final concentration value is reached and, if the electroadsorption of $HCOO_{ad}$ is a fast process and the equilibrium is attained instantaneously, the evolution with time of the bands corresponding to $HCOO_{ad}$ and $CO_{ad}$ can be followed and analyzed quantitatively. Figure 8.8 shows a series of time-resolved ATR–SEIRA spectra recorded with an interval of 140 ms at 0.305 V versus RHE. At the point indicated by the arrow, the volume of concentrated HCOOH necessary to reach a final concentration of 0.1 M was added to the solution. As illustrated in Figure 8.9*a*,*b*, both $HCOO_{ad}$ and $CO_{ad}$ appear immediately after the addition of formic acid, but, while the intensity of the $CO_{ad}$ bands increases monotonically until a saturation value, the $HCOO_{ad}$ band increases up to a maximum and then decreases to zero.

The time evolution of the $HCOO_{ad}$ and the $CO_{ad}$ coverages (Figs. 8.9*a*,*b*) is consistent with a reaction mechanism composed of two consecutive unimolecular reactions, in which a reactant (HCOOH) yields a product ($CO_{ad}$) through an intermediate ($HCOO_{ad}$), as indicated in Reactions (1) and (2). If the mechanism composed of Reactions (1) and (2) applies, assuming that the reduction of $HCOO_{ad}$ to $CO_{ad}$ is the rate-determining step (rds), and that it requires at least one additional Pt atom adjacent to those two occupied by $HCOO_{ad}$ [92], the rate equation for the dehydration of formic acid should be

$$\frac{d\theta_{CO}}{dt} = k_{dehyd}\theta_{CO}(1-\theta_{total}) \tag{8.19}$$

where $k_{dehyd}$ is the rate constant for the reduction of $HCOO_{ad}$ to $CO_{ad}$, and $\theta_{total}$ is the total coverage of adsorbed species on the surface. Plots of $dI_{CO}/dt$ versus $I_{HCOO}$ at

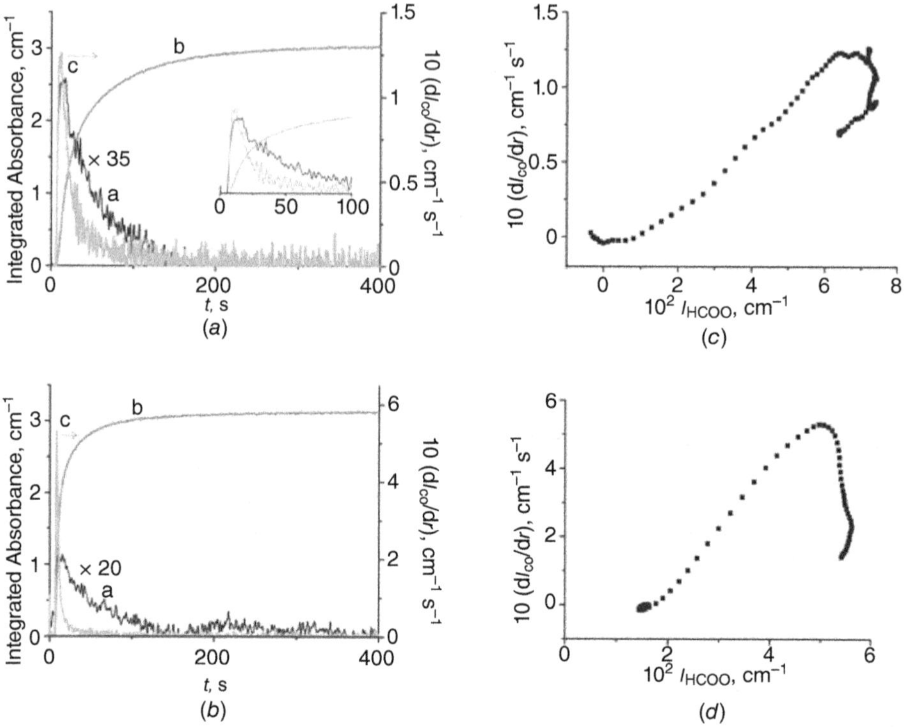


**Figure 8.9** Time dependence (*a*, *b*) of $I_{HCOO}$ (*a*, black), $I_{CO}$ (*b*, dark gray) and $dI_{co}/dt$ (*c*, light gray) obtained from a series of time-resolved ATR–SEIRA spectra recorded during the dehydration of HCOOH on Pt (interval between spectra: 140 ms) and dependence of $dI_{co}/dt$ on $I_{HCOO}$ during the initial stages of the reaction (*c*, *d*) at 0.355 (*a*, *c*) and 0.405 V versus RHE (*b*, *d*). The inset in (*a*) shows the first 100 s in an expanded scale. From Ref. [88].

0.355 and 0.405 V versus RHE (Figs. 8.9*c*,*d*, respectively) show that, initially, when both $\theta_{HCOO}$ and $\theta_{CO}$ (and, hence, $\theta_{total}$) are very low, the rate of formation of $CO_{ad}$ increases linearly with $\theta_{HCOO}$, as expected from Eq. (8.19). The plot deviates from linearity when $\theta_{CO}$ is high enough as to start displacing $HCOO_{ad}$ from the surface, this being a clear indication of the necessity of free sites adjacent to $HCOO_{ad}$ for its reduction to $CO_{ad}$ to proceed. The necessity of unoccupied sites around an $HCOO_{ad}$ is also responsible for the well-known fact that the maximum $\theta_{CO}$ achievable by dehydration of HCOOH is much smaller than that achieved by directly dosing CO. The slope ($k'$) obtained from a linear fit of the experimental data in Figs. 8.9*c*,*d* in the region where $dI_{CO}/dt$ increases linearly with $I_{HCOO}$ is proportional to the rate constant, $k_{dehyd}$, for the reduction of $HCOO_{ad}$ to $CO_{ad}$. A plot of $E$ versus log $k'$ yields a Tafel slope of −128 mV [87, 92]. The only mechanism compatible with all these experimental data is that composed of Reactions (1) and (2) [87, 88].

Assuming that $\theta_{formate}$ can be described by the Langmuir isotherm for reversible electroadsorption, the initial rate of dehydration (i.e., when $\theta_{total} \rightarrow 0$) will be

$$r_{\text{dehyd}} = \frac{d\theta_{\text{CO}}}{dt} = k_{\text{dehyd}} \exp\left(\frac{-(1-\beta_{\text{dehyd}})\eta_{\text{dehyd}}F}{RT}\right) \frac{K_L \exp\left(\frac{\Delta\phi F}{RT}\right) c_{\text{HCOOH}}}{1+K_L \exp\left(\frac{\Delta\phi F}{RT}\right) c_{\text{HCOOH}}} \tag{8.20}$$

where $\beta_{\text{dehyd}}$ and $\eta_{\text{dehyd}}$ are the anodic symmetry factor and the overpotential, respectively, for the reduction of asorbed formate to adsorbed carbon monoxide [Reaction (2)], $\Delta\phi$ depends on the local potential of zero total charge (pztc), $K_L$ is the Langmuir constant of formate adsorption, and $c_{\text{HCOOH}}$ is the concentration of formic acid.

Figure 8.10 shows the potential dependence of the initial rate of HCOOH dehydration predicted by Eq. (8.20) for $k_{\text{dehyd}} = 1\ \text{s}^{-1}$, $\beta_{\text{dehyd}} = 0.5$, $K_L c_{\text{HCOOH}} = 0.1$, 1, and 10, and for $\Delta\phi = 0$ V at $\eta_{\text{dehyd}} = -0.1$, 0, and 0.1 V. As can be seen, the initial rate of dehydration of HCOOH has a symmetric bell shape with a maximum around the pztc, in agreement with the results of Grozovski et al. [93]. The maximum shifts to more negative potentials with increasing $K_L c_{\text{HCOOH}}$, that is, with increasing concentration of formic acid, and the dehydration rate at the maximum is the higher the more negative is the pztc, this explaining the observation by Grozovski et al. [93] that the activity for the dehydration of HCOOH of Pt(111) is negligible since its pztc is very positive (0.37 V versus RHE in 0.1 M $HClO_4$ [94]). Furthermore, this also explains that, for the same density of steps, Pt single-crystal electrodes with (111) terraces and (100) steps are less active than those with (111) terraces and (110) steps (compare Figs. 7B and 9B in Ref. [93]), even if the activation barrier (i.e., $k_{\text{dehyd}}$) is the same, because the pztc of (100)-oriented steps is more positive than that of (110)-oriented steps.

The potential at which the dehydration rate will be maximum can be found by obtaining the derivative of Eq. (8.20) and making it equal to 0. This yields

$$E_{\max} = \left\{2.3\frac{RT}{F}\left[\log\left(\frac{\beta_{\text{dehyd}}}{1-\beta_{\text{dehyd}}}\right) - \frac{1}{2}(\log K_L)\right] + E_{\text{pztc}}\right\} - 2.3\frac{RT}{F}\log c_{\text{HCOOH}} \tag{8.21}$$

where $E_{\text{pztc}}$ is the pztc of the active site. According to Eq. (8.21), a plot of $E_{\max}$ versus $\log c_{\text{HCOOH}}$ must be linear with a slope of −60 mV. Substituting Eq. (8.21) into (8.20) yields an expression for the maximum rate of HCOOH dehydration, which, in logarithmic form, is

$$\log r_{\text{dehyd}}^{\max} = \left\{\log\frac{k_{\text{dehyd}}}{2} - (1-\beta_{\text{dehyd}})F\left[\frac{(E_{\text{pztc}} - E_{\text{eq}})}{2.3RT} + \left[\log\beta_{\text{dehyd}} - \log(1-\beta_{\text{dehyd}}) - \log\sqrt{K_L}\right]\right]\right\} + (1-\beta_{\text{dehyd}})F\log c_{\text{HCOOH}} \tag{8.22}$$

where $E_{\text{eq}}$ is the equilibrium potential of Reaction (2). Equation (8.22) predicts that, if $\beta_{\text{dehyd}} = 0.5$, the maximum rate of HCOOH dehydration should increase by a factor of 10 when $c_{\text{HCOOH}}$ increases by a factor of 100, and that the more negative the pztc, the higher $r_{\text{dehyd}}^{\max}$, in perfect agreement with the activity sequence (110) > (100) > (111) found by Grozovski et al. [93]. Moreover, for a difference in the pztc of about 100 mV, similar to that between (110) and (100) step sites, Eq. (8.22) predicts that the ratio between the corresponding maximum rates of dehydration must be about

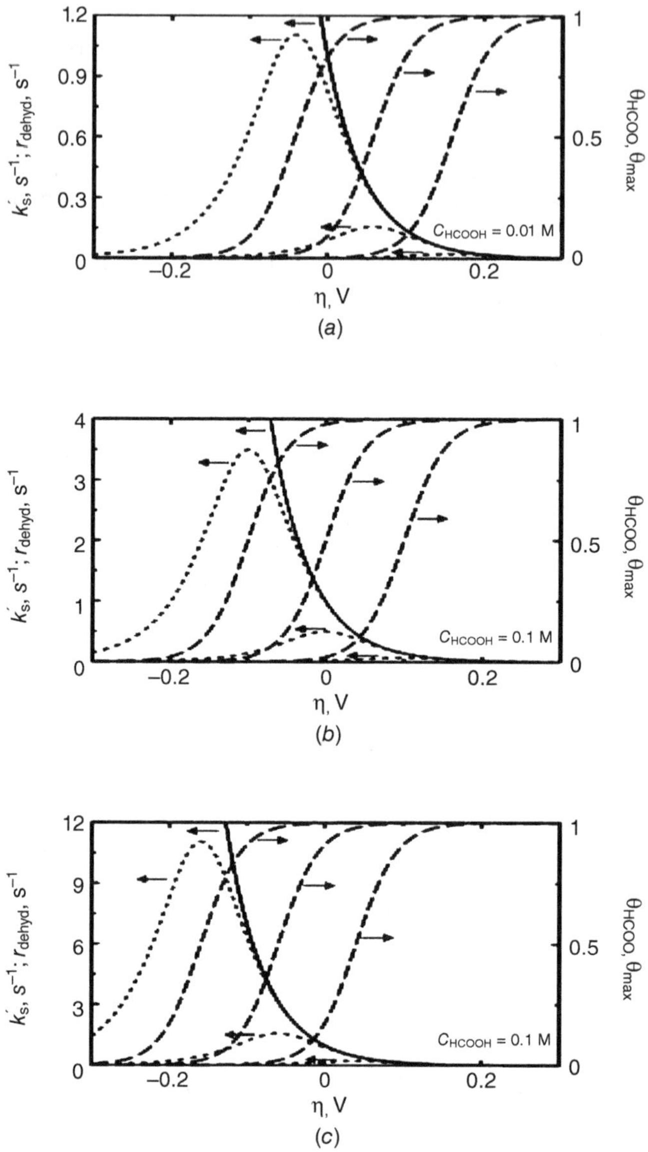


**Figure 8.10** Plot of the rate constant of reduction of $HCOO_{ad}$ to $CO_{ad}$, $k_{dehyd} \exp((-(1-\beta_{dehyd})\eta_{dehyd}F)/RT)$ (solid line), of the formate coverage at Langmuir equilibrium, $\theta_{HCOO}$ (dashed lines), and of their product, which is the rate of dehydration of HCOOH to $CO_{ad}$ (dotted lines), as a function of potential. The curves have been calculated for $k_{dehyd} = 0.1\ s^{-1}$, $K_L = 10$, and $c_{HCOOH} = 0.01$ M (*a*), 0.1 M (*b*), and 1 M (*c*). Three cases have been considered: $\Delta\phi = 0$ at $\eta_{dehyd} = -0.1$ V (left dashed and dotted lines), $\Delta\phi = 0$ at $\eta_{dehyd} = 0$ V (middle dashed and dotted lines), and $\Delta\phi = 0$ at $\eta_{dehyd} = 0.1$ V (right dashed and dotted lines). From Ref. [87].

6.8, in very good agreement with the ratio calculated from the data provided in [93] between the maximum rate of dehydration of HCOOH on Pt(554) and on Pt(544) [please note that these two surfaces have (111)-oriented terraces of the same width, separated by (110)-oriented steps in the former case and by (100)-oriented steps in the latter one].

### 8.4.2 Mechanism of Direct Path of Formic Acid Electrooxidation on Metal Electrodes: Oxidation of Adsorbed Formate to $CO_2$

In the case of Pt, studying the mechanism of the direct path combining spectroscopic and electrochemical measurements is more complicated than unveiling the mechanism of HCOOH dehydration (which is the first stage of the indirect path) because the indirect pathway always contributes to the total oxidation current. An alternative is to use Au electrodes [95], onto which the reaction proceeds exclusively through the reactive pathway [96]. In addition, Au shows only a moderate activity toward formic acid electrooxidation, and obtaining a relation between $\theta_{formate}$ and the oxidation current becomes easier.

As described above, a spectral series and a current transient recorded simultaneously at constant potential, upon addition to the solution of the amount of concentrated HCOOH necessary for reaching a given final concentration, allows to obtain, provided that the equilibrium or the steady state $\theta_{formate}$ is attained instantaneously, a sufficiently large number of points for $i$ versus $\theta_{formate}$ plots, which can then be quantitatively analyzed. Figure 8.11*a* shows a series of time-resolved ATR–SEIRA spectra recorded during the electrooxidation of formic acid on Au at 0.76 V versus RHE following this procedure, with an interval of 140 ms. At the point indicated by the arrow, the volume of concentrated HCOOH necessary to reach a final $c_{HCOOH}$ = 0.1 M was added to the solution. Since, as demonstrated by Beltramo et al. [96], on Au the electrooxidation of HCOOH proceeds exclusively through the reactive pathway, only two bands at 1720 and 1329 $cm^{-1}$, corresponding to HCOOH in the interfacial region and to $HCOO_{ad}$, respectively, emerge in the spectra after addition of HCOOH.

Figure 8.11*b* shows the current transient recorded during the experiment and the time evolution of the integrated intensity of the formic acid band ($I_{HCOOH}$) and of $I_{HCOO}$. It can be clearly observed that, as HCOOH reaches the interface, the oxidation current, $I_{HCOOH}$ and $I_{HCOO}$ increase rapidly and simultaneously to a maximum, after which they all slowly decrease. The decrease is much slower in the case of $I_{HCOO}$. The parallel decrease of the oxidation current and of $c_{HCOOH}$ (obviously $c_{HCOOH}$ is proportional to $I_{HCOOH}$) evidenced in Figure 8.11*b* clearly show that the current decrease is due to transport limitations, as suggested by Osawa et al. [97].

As shown in Figure 8.12, the dependence of the rate of formic acid electrooxidation on Au on $\theta_{formate}$ is quadratic, and the slope of an $i$ versus $\theta^2_{formate}$ plot is potential independent. Under conditions in which the contribution of the indirect path to the total current remains negligible or at least constant ($E > 0.6$ V), the same quadratic dependence between $i$ and $\theta_{formate}$ can be found with Pt [87]. Hence

$$i = k_{direct}\theta^2 \tag{8.23}$$

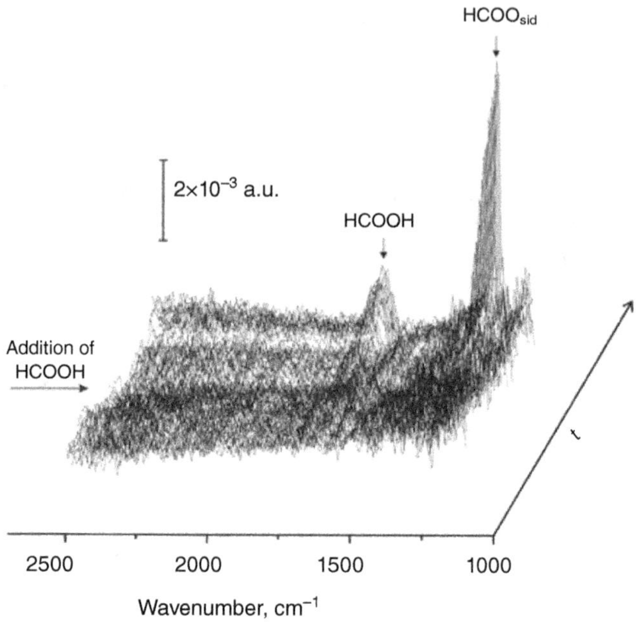


(a)

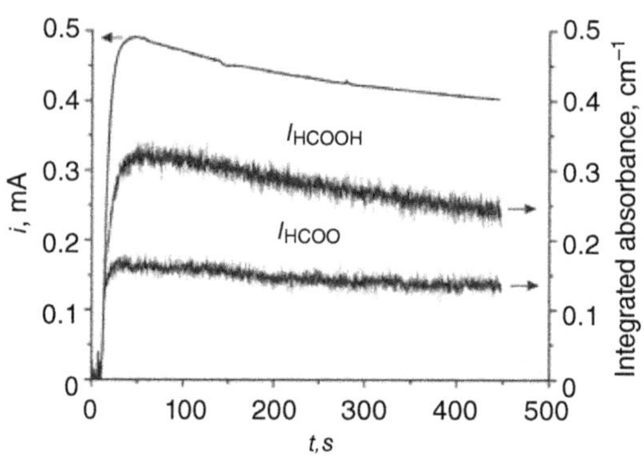


(b)

**Figure 8.11** (*a*) Time-resolved ATR–SEIRA spectra of a gold electrode at 0.76 V versus RHE in 0.1 M $HClO_4$ to which the amount of HCOOH necessary to reach a final concentration of 0.1 M was added at the point indicated by the arrow. A spectrum recorded at 0.76 V before the addition of HCOOH was taken as reference. Time interval between spectra: 140 ms. (*b*) Evolution with time of the oxidation current (*i*, top line), of the integrated absorbance of the formic acid band ($I_{HCOOH}$, middle line) and of the integrated absorbance of the adsorbed formate band ($I_{HCOO}$, bottom line). The decrease of the oxidation current after reaching a maximum is due to mass transport limitations, as indicated by the parallel decrease in $I_{HCOOH}$. From Ref. [95].

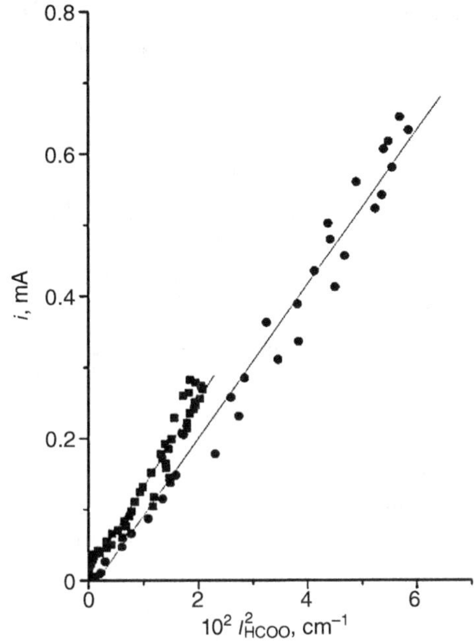


**Figure 8.12** Plot of the oxidation current versus the square of the integrated absorbance of the adsorbed formate band at 0.76 V (squares) and 0.96 V (circles). The lines are linear fits of the experimental data. From Refs. [87] and [95].

This implies that, on both Au and Pt, the rate-determining step (rds) in the oxidation of $HCOO_{ad}$ to $CO_2$ is a purely chemical, same-species bimolecular reaction between adjacent formates, and that this chemical step is necessarily intercalated between the first and second electron transfers. We have suggested that, in the transition state corresponding to this rds step, the two $HCOO_{ad}$ remain upright but lean toward each other until their H atoms are so near that the simultaneous rupture of the two C–H bonds and the formation of the H–H bond ensue [87, 95]. The so-formed $H_2$ molecule is immediately oxidized, even in the case of Au, due to the huge overpotential. We believe that this bimolecular reaction provides a feasible route for the rupture of the C–H bond in $HCOO_{ad}$, requiring a lower activation barrier than that required for the monomolecular rupture of the C–H bond of an isolated $HCOO_{ad}$, which was calculated to be of 1.1–1.2 eV by Wang and Liu [98] and by Neurock et al. [99], and of 0.73 eV by Gao et al. [100]. We expect that new theoretical calculations, which have until now never considered the possibility of a bimolecular mechanism, will provide additional support to our hypothesis.

It has to be noted, however, that a full description of the reaction mechanism, including the observation of a maximum in current versus potential plots, requires the introduction in Eq. (8.23) of an additional factor, $(1 - \theta)^n$ [86, 90, 95, 101–103], which can be attributed to the necessity of free metal atoms adjacent to the couple of reacting adsorbed formates, which would be required for them to rearrange and form the activated complex.

### 8.4.3 First, Common Step in Electrooxidation of Formic Acid on Metals: Formate Electroadsorption

A complete description of the mechanism of the electrocatalytic oxidation of formic acid on metals must include the electroadsorption of formate [Reaction (1)], which is the first step of the reaction, common to both the direct and the indirect pathways. This involves obtaining the adsorption isotherm, that is, the relation between $\theta_{formate}$ and $c_{HCOOH}$. For the reasons given above, such a relationship can be relatively easily obatined in the case of Au electrodes using ATR–SEIRAS.

Crépy et al. [104] were the first to propose that the reversible electroadsorption equilibrium between HCOOH in solution and $HCOO_{ad}$ is the first step in the electrooxidation of HCOOH on gold in acidic media. Langmuir plots in 0.1 M $HClO_4$ (Fig. 8.13) of the form $1/I_{HCOO}$ versus $1/I_{HCOOH}$, obtained from experiments like that in Figure 8.11, show a good linear relationship, suggesting that the adsorption of formate can be described by the (completely reversible) Langmuir electroadsorption isotherm. However, in the case of a Langmuir adsorption equilibrium, a decrease of the slope of the Langmuir plot by a factor of 2154 would be expected when the potential increases by 200 mV, while a decrease by a factor of only 1.3 is observed (Fig. 8.13). The small decrease in the slope of the Langmuir plots with increasing potential (or, equivalently, the necessity of more than 800 mV to reach the saturation coverage) might suggest that the electroadsorption of formate on Au follows a Frumkin isotherm with a large repulsive parameter. It must be noted, however, that a Frumkin isotherm cannot explain the experimentally found linear relationship between $1/I_{HCOO}$ and $1/I_{HCOOH}$, unless we assume that the plot in Figure 8.13 corre-

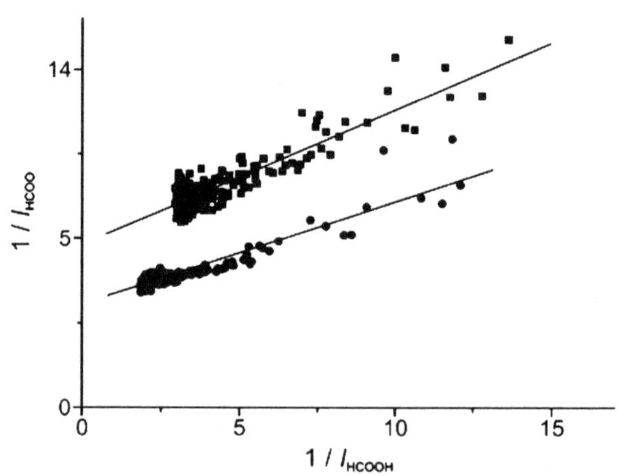


**Figure 8.13** Langmuir plots of formate adsorption on Au at 0.76 V (squares) and 0.96 V (circles). Each plot was obtained from a single experiment like that illustrated in Figure 8.11 in which the amount of HCOOH necessary to reach a final concentration of 0.1 M was added to the electrolyte (0.1 M $HClO_4$). The lines are linear fits of the experimental data. From Refs. [87] and [95].

sponds to the low-coverage limit, under which conditions the Langmuir and the Frumkin isotherms are undistinguishable.

The small decrease of the slope of the $1/I_{HCOO}$ versus $1/I_{HCOOH}$ plot when the potential increases by 200 mV can be accounted for if the adsorption of formate is assumed to be completely irreversible, that is, if adsorbed formate never desorbs as formic acid but is further oxidized [95]. In the steady state, adsorbed formate is oxidized at the same rate at which it is adsorbed, so that its coverage remains constant, and its concentration dependence can be formally described by a Langmuir-like isotherm. The pseudo-Langmuir constant of this steady state would be the ratio between the rate constant of formate adsorption and the rate constant of formate oxidation, while the Langmuir constant of a reversible equilibrium is the ratio between the rate constants of adsorption of a species and of desorption of exactly the same species. This nonequilibrium, steady-state pseudo-Langmuir adsorption can also explain the high Tafel slope values, between 240 and 270 mV, obtained by Crépy et al. [104], by Beltramo et al. [96], and by ourselves [95], but is in disagreement with the observation by Grozovski et al. [105] that adsorbed formate is in equilibrium with formic acid in the solution, this being also consistent with the kinetics derived by us for the dehydration of formic acid on Pt. The nature of the electroadsorption of formate is, hence, a subject that still requires further research.

## 8.5 ATR–SEIRAS FOR STUDY OF ENZYMES IMMOBILIZED ON METAL ELECTRODES

Covalent immobilization of biomolecules on functionalized surfaces has been demonstrated to be a useful tool to develop stable and efficient bioelectronic devices, such as biosensors and biofuel cells [106]. Covalent linkage of biomolecules onto electrodes requires the chemical modification of surfaces with appropriate functional groups or binding motives, a modification that can be achieved by different methods, the most popular being the formation of a thiol self-assembled monolayer (SAM) onto a metallic surface, particularly Au or Pt. A robust alternative, introduced by Savéant and co-workers [107–109], is the electroreduction of aromatic diazonium salts in order to modify carbon electrodes with appropriate functionalities, a method that has been used for covalent immobilization of biomolecules both on carbon [109–111] and Au [112–114] electrodes. Recently, a Au electrode was modified with a mixed monolayer of an aromatic diazonium salt derivative and 6-mercapto-1-hexanol and was used as a platform onto which laccase (a multicopper-containing enzyme that catalyzes the four-electron reduction of $O_2$ to $H_2O$) can be linked covalently. This strategy was shown to yield the stable immobilization of the enzyme and to favor its adequate orientation for direct electron transfer from the electrode [114]. Figure 8.14 illustrates the steps involved in this process.

ATR–SEIRAS can be used to characterized the sequential steps in the covalent immobilization of an enzyme on a thin-film metal electrode, as shown below for the cases of laccase immobilized on Au electrodes modified either by a mixed

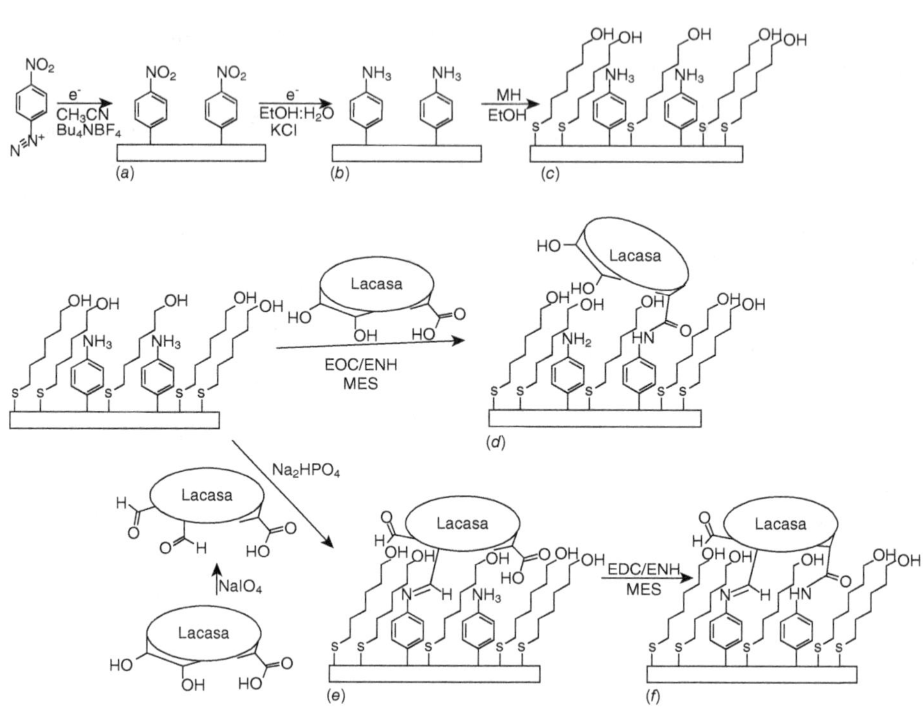


**Figure 8.14** Sequence of the steps followed for the modification of a Au surface with *p*-nitrophenyldiazonium salt and subsequent immobilization of laccase: (*a*) formation of nitrophenyl adlayer; (*b*) aminophenyl adlayer formed by electroreduction of the nitro groups of the nitrophenyl adlayer; (*c*) formation of a mixed aminophenyl/mercaptohexanol monolayer; (*d*) covalent attachment of laccase to the mixed monolayer by formation of amide bonds; (*e*) covalent attachment of laccase to the mixed monolayer by formation of imine bonds; and (*f*) covalent attachment of laccase to the mixed monolayer by formation of imine and amide bonds. Adapted from Ref. [115].

mercaptohexanol (MHO) and aminophenyl monolayer or by 4-aminotiophenol (4-ATP) [115]. The bands observed in the corresponding spectra are summarized in Tables 8.2–8.4.

## 8.5.1 Laccase Immobilization onto Mixed Aminophenyl–MHO-Modified Au Surface

The first step in the modification process consisted in the electroreduction of *p*-nitrophenyldiazonium cations, aimed at forming a Au–carbon bond and at creating a *p*-nitrophenyl-modified Au surface. The success of this first step is demonstrated by the observation of the characteristic bands of the nitrophenyl group in the ATR–SEIRA spectrum of the resulting surface, particularly the asymmetric $NO_2$ stretching ($\nu_s$-$NO_2$) at 1520 cm$^{-1}$ and the symmetric $NO_2$ stretching ($\nu_s$-$NO_2$) at 1350 cm$^{-1}$

**Table 8.2** Assignement of Bands Observed in Spectra of Chemically Modified Au Surfaces, prior to Immobilization of Laccase

| | Nitrophenyl-Modified Au Surfaces (cm$^{-1}$) | Aminophenyl-Modified Au Surfaces (cm$^{-1}$) | Mixed Aminophenyl–MHO-Modified Au Surfaces (cm$^{-1}$) | 4-ATP-Modified Au Surfaces (cm$^{-1}$) |
|---|---|---|---|---|
| $\nu_{C\text{-}N}$ | | | | 1290 |
| $\nu_{C\text{-}S}$ | | | | 1178 |
| $\nu_s$-$NO_2$ | 1350 | | | |
| $\delta_{OH(NHOH)}$ | | 1425 | 1425 | |
| $\nu_{C\text{-}C(aromatic)}$ | | 1507 | 1507 | 1488 |
| $\nu_{N=O}$ | | | | |
| $\nu_a$-$NO_2$ | 1520 | | | |
| $\nu_{C\text{-}C(aromatic)}$ | 1600 | 1558–1737 | 1558–1737 | 1584 |
| $\delta_{NH2}$ | | | | 1630 |
| $\delta_{NH(NHOH)}$ | | | | |
| $\nu_{C\text{-}H(aromatic)}$ | | | | 3020 |
| $\nu_{NH(NH2)}$ | | 3250 | 3250 | 3330 |
| $\nu_{NH(NHOH)}$ | | | | |
| $\nu_{OH(MHO)}$ | | | 3500 | |
| $\nu_{OH(NHOH)}$ | | 3733 | 3733 | |

**Table 8.3** Assignment of Bands Observed in Spectra of Aminophenyl–MHO-Modified Au Surfaces after Covalent Immobilization of Laccase

| | Laccase Immobilized on Aminophenyl–MHO-Modified Au Surfaces via a Schiff Base Reaction (cm$^{-1}$) | Laccase Immobilized on Aminophenyl–MHO-Modified Au Surfaces via Amide Bond Formation (cm$^{-1}$) | Laccase Immobilized on Aminophenyl–MHO-Modified Au Surfaces via a Schiff Base Reaction Followed by Amide Bond Formation (cm$^{-1}$) |
|---|---|---|---|
| $\delta_{OH(NHOH)}$ | 1425 | | 1425 |
| **Amide II** | 1526 | 1547 | 1526 |
| **Amide I** | 1663 | 1660 | 1663 |
| $\nu_{C=O}$ | | 1736 | 1737 |
| $\nu_{NH(NH2)}$ | 3292 | 3000–3680 | 3292 |
| $\nu_{NH(NHOH)}$ | | | |
| $\nu_{OH(MHO)}$ | | | |
| $\nu_{OH(NHOH)}$ | 3733 | | 3733 |

**Table 8.4** Assignment of Bands Observed in Spectra of 4-ATP-Modified Au Surfaces after Covalent Immobilization of Laccase

| | Laccase Immobilized on 4-ATP-Modified Au Surfaces via a Schiff Base Reaction ($cm^{-1}$) | Laccase Immobilized on 4-ATP-Modified Au Surfaces via Amide Bond Formation ($cm^{-1}$) | Laccase Immobilized on 4-ATP-Modified Au Surfaces via a Schiff Base Reaction Followed by Amide Bond Formation ($cm^{-1}$) |
|---|---|---|---|
| $\nu_{C\text{-}N}$ | 1285 | 1285 | 1285 |
| $\nu_{C\text{-}C(aromatic)}$ | 1488 and 1584 | 1488 and 1584 | 1488 and 1584 |
| **Amide III** | 1400 and 1320 | 1400 and 1320 | 1400 and 1320 |
| **Amide II** | 1545 | 1545 | 1545 |
| **Amide I** | 1650 | 1650 | 1650 |
| $\nu_{C=N}$ | 1720 | | 1720 |
| $\nu_{C=O}$ | | 1736 | 1737 |

(Fig. 8.15, line *a*). The weaker band at 1600 $cm^{-1}$ corresponds to the aromatic ring stretching. It should be accompanied by another band around 1500 $cm^{-1}$, which is most likely obscured by the $\nu_a$-$NO_2$ band. Our results are in good agreement with previous infrared studies of nitrophenyl-funtionalized Au surfaces [116, 117]. The band at 1245 $cm^{-1}$, also observed by Ricci et al. [116], is due to silicon oxides that appear on the surface of the Si substrate during the electrochemical modification. The absence of a band at 2300 $cm^{-1}$ indicates the absence of physically adsorbed diazonium cations [117].

The second step consisted of the electroreduction of the nitro group, yielding an aminophenyl–functionalized surface (Fig. 8.14) to which the enzyme can be chemically bonded via (i) a Schiff base reaction with aldehyde groups of the laccase obtained by oxidation of the sugar residues and (ii) amide bonds with the carboxylic acid groups of the laccase. The bands at 1520 and 1350 $cm^{-1}$ due to the asymmetric and symmetric $NO_2$ stretching disappear after the second step of the surface modification (Fig. 8.15, line *b*), clear evidence of the successful electrochemical reduction of the nitro groups. Electrochemical reduction of the surface-anchored nitrophenyl groups to yield an aminophenyl-functionalized Au electrode (Au–Ar–$NH_2$) should result in an infrared spectrum very similar to that obtained with a 4-ATP-modified Au surface (see below). However, a small portion of hydroxylamine (Ar–NHOH) and nitroso groups (Ar–NO) are also generated during the reduction process [118]. As a consequence, instead of two well-defined bands around 1600 $cm^{-1}$ corresponding to the aromatic ring and to the amino group, as in the case of 4-ATP, a broad feature extending from 1558 to 1737 $cm^{-1}$ that includes the ring stretching, the aminophenyl $NH_2$ scissoring, and the hydroxylamine NH bending modes was observed. A small band appears at 1507 $cm^{-1}$, which can be attributed to the ring band accompanying that at 1600 $cm^{-1}$ and/or to the N=O stretching of nitroso groups. The weak, broad symmetric band centered at 3250 $cm^{-1}$ includes the N–H stretching mode of the $NH_2$

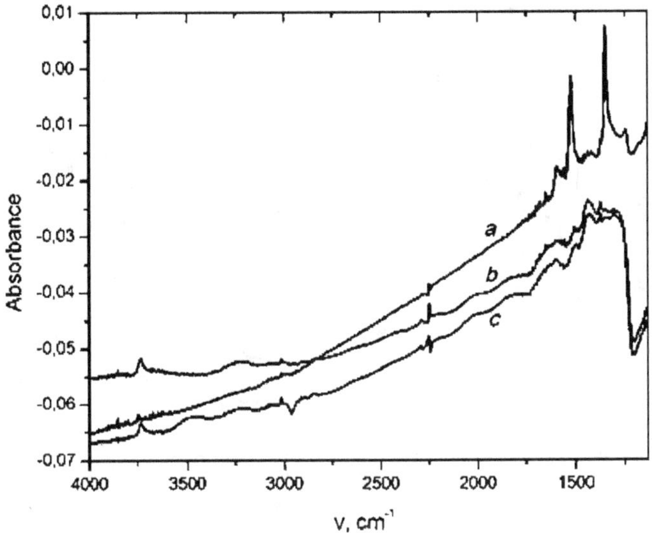


**Figure 8.15** ATR–SEIRA spectra of the nitrophenyl-modified (*a*), the aminophenyl-modified (*b*), and the mixed monolayer (aminophenyl–MHO)-modified (*c*) Au surface in $CH_3CN$, obtained using the bare Au surface as reference. From Ref. [115].

and the NHOH groups. The sharp band at 3733 and the band at 1425 cm$^{-1}$ correspond to the stretching and bending modes, respectively, of hydrogen-bond-free OH groups of hydroxylaminophenyl moieties.

The third step consisted of the formation of a mixed aminophenyl–MHO adlayer by immersion of the aminophenyl-modified Au surface in a 1-mM solution of MHO in absolute EtOH (Fig. 8.14). The addition of the thiol resulted in the formation of a SAM, which prevents laccase denaturation by blocking the bare Au surface. After this step, a new broad band appears around 3500 cm$^{-1}$, which was assigned to the O–H stretching mode of the alcohol group of MHO (Fig. 8.15, line *c*). The rest of the spectrum remains identical to that observed after reduction of the nitrophenyl groups.

Bonding of laccase by a Schiff base reaction between the amino groups in the monolayer and the aldehyde groups of the laccase obtained by chemical oxidation of the sugar residues originally present on the surface of the enzyme (Fig. 8.14) is evidenced by the appearance of two bands at 1663 and 1526 cm$^{-1}$ (Fig. 8.16), corresponding to the protein backbone vibrations amide I and amide II, respectively [119]. Upon attachment of laccase to the electrode surface, the broad bands at 3500 and 3250 cm$^{-1}$, which we had attributed to the O–H stretching of MHO and to the N–H stretching of the aminophenyl groups, respectively, are substituted by an asymmetric, more intense broad band with the absorption maximum at 3292 cm$^{-1}$ (Fig. 8.16, line *b*). No decrease in the 3733 and 1425 cm$^{-1}$ bands is detected, as expected if they correspond to the OH vibrations of hydroxylaminophenyl groups, which are not involved in the covalent immobilization via formation of a Schiff base.

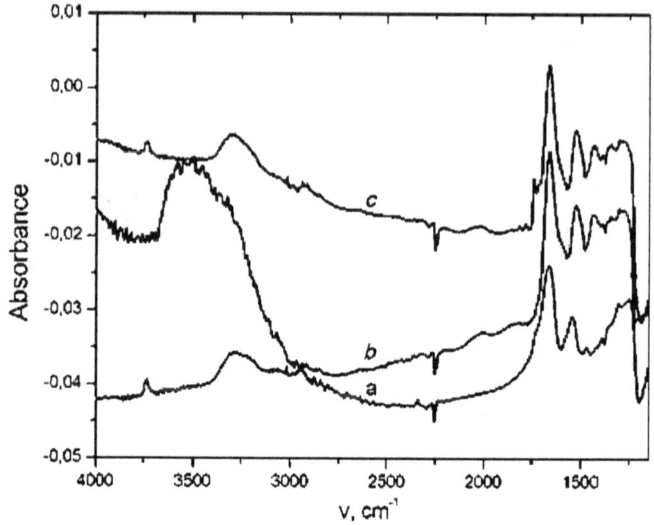


**Figure 8.16** ATR–SEIRA spectra in $CH_3CN$, calculated using the bare Au surface as reference, of the laccase immobilized onto the mixed monolayer (aminophenyl–MHO)-modified Au surface via amide bond formation (*a*), binding by Schiff base reaction (*b*), and binding by Schiff base reaction followed by amide bond formation (*c*). From Ref. [115].

The last step, that is, the treatment with N-hydroxysuccinimide (NHS) and 1-(3-dimethylaminopropyl)-3-ethylcarbodiimide hydrochloride (EDC) in 2-(N-morpholino) ethanesulfonic acid (MES, pH 6.0), aimed at catalyzing the formation of amide bonds between the amino groups of the aminophenyl moieties on the surface of the chemically modified Au electrode and the carboxylic acid groups of the laccase (Fig. 8.14) results in the emergence of a new band at 1737 $cm^{-1}$, typical of a carbonyl group, thus confirming the formation of covalent amide bonds between the chemically modified surface and the laccase (Fig. 8.16, line *c*). Some differences can be observed in the ATR–SEIRA spectrum when the Schiff base reaction does not precede the formation of an amide bond (Fig. 8.16, line *a*): Attachment of the enzyme to the electrode surface is again evidenced by peaks at 1660 and 1547 $cm^{-1}$, corresponding to the protein backbone vibrations amide I and amide II, and a shoulder at 1736 $cm^{-1}$ confirms the formation of a covalent amide bond between the chemically modified surface and laccase. However, in this case the bands at 3733 and 1425 $cm^{-1}$, which we have attributed to the O–H vibrations of hydroxylaminophenyl groups, disappear after the reaction, and a broad band between 3680 and 3000 $cm^{-1}$, with an absorption maximum around 3500 $cm^{-1}$, substitutes the bands at 3500 and 3250 $cm^{-1}$, that we had attributed, respectively, to the O–H stretching of MHO and to the N–H stretching of the aminophenyl groups. These differences indicate that, although the two procedures used to immobilize the enzyme are successful, the resulting surfaces are different. This is probably due to the different laccase residues involved in both immobilization methods. Schiff bases formation involves sugar residues, which are mostly placed next to the Cu T1 site, whereas carboxylic residues are more abundant in the opposite region of the laccase [120]. Therefore, a different orientation of the enzyme with respect to the Au surface is expected when laccase is immobilized only via amide bonds instead of the two-step method. As a consequence, the interaction between the OH groups of the MHO adlayer and the laccase residues is different in these two cases, generating the observed differences in the spectra. These ATR–SEIRA results are in agreement with previous electrochemical results, which suggested that the formation of imine bonds via a Schiff base reaction prior to the amide bonds formation promotes a specific laccase orientation that facilitates a better communication between the T1 center and the metal surface and, consequently, a higher activity at low overpotentials for the direct bioelectrocatalytic reduction of $O_2$ to $H_2O$ [114].

### 8.5.2 Laccase Immobilization onto 4-ATP-Modified Au Surface

The ATR–SEIRA spectrum of the 4-ATP-modified Au surface (Fig. 8.17, line *a*) is similar to that reported for a 4-ATP-modified platinum electrode [121]. The absence of a band around 2550 $cm^{-1}$ indicates the loss of the S–H bond and the formation of a S–Au bond. Bands corresponding to the stretching and scissoring modes of the amino group can be clearly observed at 3370 and 1630 $cm^{-1}$, respectively. The broad band at 3020 $cm^{-1}$ is typical of the C–H stretching modes of the aromatic ring, and

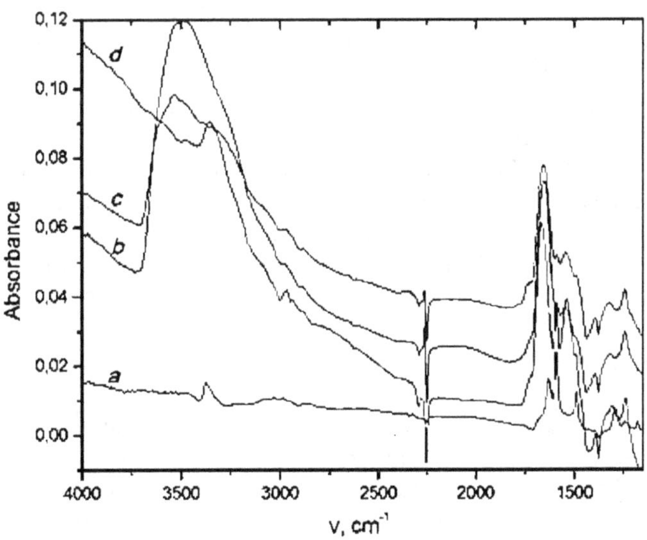


**Figure 8.17** ATR-SEIRA spectra of a Au surface in $CH_3CN$ after modification with 4-ATP (*a*), after immobilization of laccase via Schiff base reaction (*b*), after immobilization of laccase via Schiff base reaction followed by amide bond formation (*c*), and after immobilization of laccase via amide bond formation (*d*), calculated using the bare Au surface as reference. From Ref. [115].

the bands at 1585 and 1488 $cm^{-1}$ correspond to the C–C aromatic ring stretch. The band at 1290 $cm^{-1}$ is due to the C–N stretching and the band at 1178 $cm^{-1}$ could correspond to the C–S stretching.

Immobilization of laccase by formation of a Schiff base onto a 4-ATP-modified Au surface (Fig. 8.17, lines *b* and *c*) is evidenced by the appearance of two bands due to the protein backbone vibrations amide I (1650 $cm^{-1}$) and amide II (1545 $cm^{-1}$), as described above for the immobilization onto a mixed aminophenyl–MHO-modified Au electrode. Other new bands appear at 1400, 1320, and 1240 $cm^{-1}$. The 1400 and 1320 $cm^{-1}$ bands can be attributed to the amide III mode of the protein backbone (1200–1400 $cm^{-1}$) [119]. The band at 1320 $cm^{-1}$ is quite broad because it includes that at 1285 $cm^{-1}$, attributed to the C–N single bond of the amino group. The band at 1240 $cm^{-1}$ is possibly due to silicon oxides, which are formed on the surface of the Si substrate during the modification. A small shoulder appears at about 1720 $cm^{-1}$, which can be assigned to the C=N double bond formed after the Schiff base reaction. In the high-frequency region, a band centered at 3350 $cm^{-1}$ emerges, probably due to a shift of the stretching mode of the amino group after laccase immobilization. Formation of the amide bond is evidenced by the shoulder at 1736 $cm^{-1}$ (Fig. 8.17, lines *c* and *d*).

Amide III bands could be observed in the ATR–SEIRA spectra when laccase is immobilized onto the 4-ATP-modified surface, but not if the immobilization is made on a Au surface modified with a mixed aminophenyl–MHO adlayer. The similar intensities of the amide I and amide II bands in the spectra in Figures 8.16 and 8.17 suggest that this is not due to a higher coverage by immobilized laccase onto the 4-ATP-modified Au surface, and that, on the contrary, the spectral differences might be due to a different enzyme orientation. Since the functional groups to which the enzyme was anchored and the immobilization procedure (Schiff base reaction followed by amide bond formation) was the same on the two chemically modified Au surfaces, the different orientation must be the result of the presence of a large amount of MHO on the surface of the aminophenyl–MHO-modified Au electrodes. The important role of MHO in adequately orienting the laccase units is supported by the fact that significantly lower activity for direct electrocatalytic reduction of $O_2$ was obtained when thiols with different terminal functional groups, particularly cystamine and mercaptopropionic acid, were used instead of MHO [114].

## 8.6 CONCLUSIONS

During the approximately 30 years that have elapsed since Bewick's seminal work, in situ infrared spectroscopy has significantly impacted interfacial electrochemistry and our understanding of the structure and composition of the electrochemical double layer. This is due not only to the technique's ability to provide information regarding the chemical nature of the species present in the electrochemical double layer, their bonding to the electrode surface, and their orientation or conformation but also to its relative simplicity and inexpensiveness, which make it affordable to a large number of laboratories, and allow one to rapidly achieve the minimum necessary skills. In spite of this simplicity, it must not be forgotten that the spectral shape is always affected by the optical properties of the electrode surface, and that adsorbates on electrode materials with nonmetallic optical properties may often appear with unexpected band shapes, that can though be perfectly explained within the frame of classical electromagnetism and Fresnel's reflection laws.

By itself or in combination with other in situ nonelectrochemical techniques, in situ IR spectroscopy has provided very valuable information regarding the structure and potential-dependent structural changes of electrode surfaces and of adsorbate layers, as I have tried to illustrate with results regarding a classical system, CO adsorbed on Pt, from my own group. I have also tried to show how the developments introduced in the last decade of the last century, aimed at improving the technique's sensitivity and time resolution, have made IR spectroscopy an invaluable tool for studying the adsorption of biological molecules on chemically modified electrodes, and to solve the most intricate details of the mechanisms of electrocatalytic reactions. Due to this latter ability, and together with the introduction of new, more sophisticated advances, in situ IR spectroscopy will certainly play in the future an important role in the study of dynamics at the electrode–electrolyte interface, this being out of any doubt one of the new frontiers of interfacial electrochemistry.

## ACKNOWLEDGMENTS

I wish to thank my co-workers, Dr. Ana López-Cudero, Dr. Gema Cabello, and Dr. Cristina Vaz-Domínguez, for performing most of the experimental work reviewed here; Dr. Carlos Pecharromán for the brief but fruitful collaboration; and Prof. Claudio Gutiérrez for his continuous support.

## REFERENCES

1. Stedman, M. *Symp. Faraday Soc.* **1970**, *4*, 64–71.
2. Plieth, W. J. *Symp. Faraday Soc.* **1970**, *4*, 137–144.
3. Memming, R.; Mollers, F. *Symp. Faraday Soc.* **1970**, *4*, 145–156.
4. McIntyre, J. D. E.; Kolb, D. M. *Symp. Faraday Soc.* **1970**, *4*, 99–113.
5. Mark, H. B.; Randall, E. N. *Symp. Faraday Soc.* **1970**, *4*, 157–172.
6. Hansen, W. N. *Symp. Faraday Soc.* **1970**, *4*, 27–35.
7. Cahan, B. D.; Horkans, J.; Yeager, E. *Symp. Faraday Soc.* **1970**, *4*, 36–44.
8. Bewick, A.; Tuxford, A. M. *Symp. Faraday Soc.* **1970**, *4*, 114–125.
9. Barrett, M. A.; Parsons, R. *Symp. Faraday Soc.* **1970**, *4*, 72–84.
10. McIntyre, J. D. E.; Peck, W. F. *Faraday Discuss. Chem. Soc.* **1973**, *56*, 122–137.
11. Kolb, D. M. *Faraday Discuss. Chem. Soc.* **1973**, *56*, 138–151.
12. Conway, B. E.; Angerstein-Kozlowska, H.; Ho, F. C.; Klinger, J.; MacDougall, B.; Gottesfeld, S. *Faraday Discuss. Chem. Soc.* **1973**, *56*, 210–227.
13. Broman, R. F.; Heineman, W. R.; Kuwana, T. *Faraday Discuss. Chem. Soc.* **1973**, *56*, 16–27.
14. Aylmer-Kelly, A. W. B.; Bewick, A.; Cantrill, P. R.; Tuxford, A. M. *Faraday Discuss. Chem. Soc.* **1973**, *56*, 96–107.
15. Fleischmann, M.; Hendra, P. J.; McQuillan, A. J. *Chem. Phys. Lett.* **1974**, *26*, 163–166.
16. Albrecht, M. G.; Creighton, J. A. *J. Am. Chem. Soc.* **1977**, *99*, 5215–5217.
17. Jeanmaire, D. L.; Van Duyne, R. P. *J. Electroanal. Chem.* **1977**, *84*, 1–20.

18. Bewick, A.; Kunimatsu, K.; Pons, B. S. *Electrochim. Acta* **1980**, *25*, 465–468.
19. Reed, A. H.; Yeager, E. *Electrochim. Acta* **1970**, *15*, 1345–1354.
20. Clarke, J. S.; Kuhn, A. T.; Orville-Thomas, W. J.; Stedman, M. *J. Electroanal. Chem.* **1974**, *49*, 199–208.
21. Osawa, M.; Ikeda, M. *J. Phys. Chem.* **1991**, *95*, 9914–9919.
22. Matsuda, N.; Yoshii, K.; Ataka, K.-I.; Osawa, M.; Matsue, T.; Uchida, I. *Chem. Lett.* **1992**, *21*, 1385–1388.
23. Osawa, M.; Ataka, K.-I.; Yoshii, K.; Nishikawa, Y. *Appl. Spectrosc.* **1993**, *47*, 1497–1502.
24. Osawa, M.; Ataka, K.-I.; Yoshii, K.; Yotsuyanagi, T. *J. Electron Spectrosc. Relat. Phenom.* **1993**, *64–65*, 371–379.
25. Osawa, M. *Bull. Chem. Soc. Jpn.* **1997**, *70*, 2861–2880.
26. Osawa, M. In-situ surface-enhanced infrared spectroscopy of the electrode/solution interface, in *Advances in Electrochemical Science and Engineering, Vol. 9*, R. C. Alkire, D. M. Kolb, J. Lipkowski, and P. N. Ross, Eds., Wiley-VCH: Weinheim, **2006**, pp. 269–314.
27. Hartstein, A.; Kirtley, J. R.; Tsang, J. C. *Phys. Rev. Lett.* **1980**, *45*, 201–204.
28. Guyot-Sionnest, P.; Tadjeddine, A. *Chem. Phys. Lett.* **1990**, *172*, 341–345.
29. Tolstoy, V. P.; Chernyshova, I. V.; Skryshevsky, V. A. *Handbook of Infrared Spectroscopy of Ultrathin Films*, Wiley: Hoboken, NJ, **2003**.
30. Kolb, D. M. UV-visible reflectance spectroscopy, in *Spectroelectrochemistry: Theory and Practice*, R. J. Gale, Ed., Plenum Press: New York, **1988**, pp. 87–188.
31. Hummel, R. E. *Optische Eigenschaften von Metallen und Legierungen*, Springer: Berlin, **1971**.
32. Wooten, F. *Optical Properties of Solids*, Academic Press: New York, **1972**.
33. Hansen, W. N. *J. Opt. Soc. Am.* **1968**, *58*, 380–388.
34. Hansen, W. N. in *Advances in Electrochemistry and Electrochemical Engineering, Vol. 9*, P. Delahay and C. W. Tobias, Eds., Wiley: New York, **1973**, pp. 1–60.
35. McIntyre, J. D. E. in *Advances in Electrochemistry and Electrochemical Engineering, Vol. 9*, P. Delahay and C. W. Tobias, Eds., Wiley: New York, **1973**, pp. 61–166.
36. Sperline, R. P.; Jeon, J. S.; Raghavan, S. *Appl. Spectrosc.* **1995**, *49*, 1178–1182.
37. Porter, M. D.; Bright, T. B.; Allara, D. L.; Kuwana, T. *Anal. Chem.* **1986**, *58*, 2461–2465.
38. Ortiz, R.; Cuesta, A.; Márquez, O. P.; Márquez, J.; Méndez, J. A.; Gutiérrez, C. *J. Electroanal. Chem.* **1999**, *465*, 234–238.
39. Mielczarski, J. A. *J. Phys. Chem.* **1993**, *97*, 2649–2663.
40. Mielczarski, J. A. *Langmuir* **1997**, *13*, 878–880.
41. Mielczarski, J. A.; Mielczarski, E.; Zachwieja, J.; Cases, J. M. *Langmuir* **1995**, *11*, 2787–2799.
42. Mielczarski, J. A.; Yoon, R. H. *J. Phys. Chem.* **1989**, *93*, 2034–2038.
43. Christensen, P. A.; Hamnett, A.; Munk, J.; Troughton, G. L. *J. Electroanal. Chem.* **1994**, *370*, 251–258.
44. Lu, G.-Q.; Sun, S.-G.; Chen, S.-P.; Cai, L.-R. *J. Electroanal. Chem.* **1997**, *421*, 19–23.
45. Bjerke, A. E.; Griffiths, P. R.; Theiss, W. *Anal. Chem.* **1999**, *71*, 1967–1974.
46. Park, S.; Tong, Y.; Wieckowski, A.; Weaver, M. J.; *Electrochem. Commun.* **2001**, *3*, 509–513.
47. Gong, H.; Chen, S.; Zhou, Z.; Sun, S. *Chin. Sci. Bull.* **2001**, *46*, 1612–1614.
48. Pecharromán, C.; Cuesta, A.; Gutiérrez, C. *J. Electroanal. Chem.* **2002**, *529*, 145–154.
49. Pecharromán, C.; Cuesta, A.; Gutiérrez, C. *J. Electroanal. Chem.* **2004**, *563*, 91–109.
50. Pecharromán, C.; Iglesias, J. E. *Phys. Rev. B* **1994**, *49*, 7137–7147.
51. Bruggeman, D. A. G. *Ann. Phys.* **1935**, *24*, 636–664.
52. Edwards, D. F. *Handbook of Optical Constants of Solids*, E. D. Palik, Ed., Academic Press: Orlando, FL, **1985**.
53. Nishikawa, Y.; Fujiwara, K.; Ataka, K.; Osawa, M. *Anal. Chem.* **1993**, *65*, 556–562.
54. Kitamura, F.; Takeda, M.; Takahashi, M.; Ito, M. *Chem. Phys. Lett.* **1987**, *142*, 318–322.
55. Furuya, N.; Motto, S.; Kunimatsu, K. *J. Electroanal. Chem.* **1988**, *239*, 347–360.
56. Kitamura, F.; Takahashi, M.; Ito, M. *J. Phys. Chem.* **1988**, *92*, 3320–3323.
57. Leung, L. W. H.; Wieckowski, A.; Weaver, M. J. *J. Phys. Chem.* **1988**, *92*, 6985–6990.
58. Chang, S. C.; Leung, L.-W. H.; Weaver, M. J. *J. Phys. Chem.* **1989**, *93*, 5341–5345.

59. Chang, S.-C.; Weaver, M. J. *Surf. Sci.* **1990**, *238*, 142–162.
60. Chang, S. C.; Jiang, X.; Roth, J. D.; Weaver, M. J. *J. Phys. Chem.* **1991**, *95*, 5378–5382.
61. Chang, S. C.; Weaver, M. J. *J. Phys. Chem.* **1990**, *94*, 5095–5102.
62. Chang, S.-C.; Weaver, M. J. *J. Chem. Phys.* **1990**, *92*, 4582–4594.
63. Chang, S.-C.; Weaver, M. J. *Surf. Sci.* **1990**, *230*, 222–236.
64. Villegas, I.; Weaver, M. J. *J. Chem. Phys.* **1994**, *101*, 1648–1660.
65. López-Cudero, A.; Cuesta, A.; Gutiérrez, C. *J. Electroanal. Chem.* **2005**, *579*, 1–12.
66. Rodes, A.; Gómez, R.; Feliu, J. M.; Weaver, M. J. *Langmuir* **1999**, *16*, 811–816.
67. López-Cudero, A.; Cuesta, Á.; Gutiérrez, C. *J. Electroanal. Chem.* **2006**, *586*, 204–216.
68. Cuesta, A.; Couto, A.; Rincón, A.; Pérez, M. C.; López-Cudero, A.; Gutiérrez, C. *J. Electroanal. Chem.* **2006**, *586*, 184–195.
69. Cuesta, A.; del Carmen Pérez, M.; Rincón, A.; Gutiérrez, C. *ChemPhysChem* **2006**, *7*, 2346–2351.
70. López-Cudero, A.; Cuesta, A.; Gutiérrez, C. *J. Electroanal. Chem.* **2003**, *548*, 109–119.
71. Gómez, R.; Feliu, J. M.; Aldaz, A.; Weaver, M. J. *Surf. Sci.* **1998**, *410*, 48–61.
72. Climent, V.; Gómez, R.; Orts, J. M.; Rodes, A.; Aldaz, A.; Feliu, J. M. in *Interfacial Electrochemistry*, A. Wieckowski, Ed., Marcel Dekker: New York, **1999**, pp. 463–475.
73. Müller, E. *Z. Elektrochem.* **1923**, *29*, 264–274.
74. Breiter, M. *Electrochemical Processes in Fuel Cells*, Springer: Berlin, **1969**.
75. Capon, A.; Parson, R. *J. Electroanal. Chem.* **1973**, *44*, 1–7.
76. Capon, A.; Parsons, R. *J. Electroanal. Chem.* **1973**, *45*, 205–231.
77. Capon, A.; Parsons, R. *J. Electroanal. Chem.* **1973**, *44*, 239–254.
78. Beden, B.; Lamy, C.; Bewick, A.; Kunimatsu, K.; *J. Electroanal. Chem.* **1981**, *121*, 343–347.
79. Beden, B.; Bewick, A.; Lamy, C. *J. Electroanal. Chem.* **1983**, *148*, 147–160.
80. Solomun, T. *Surf. Sci.* **1986**, *176*, 593–602.
81. Chen, Y. X.; Miki, A.; Ye, S.; Sakai, H.; Osawa, M. *J. Am. Chem. Soc.* **2003**, *125*, 3680–3681.
82. Miki, A.; Ye, S.; Senzaki, T.; Osawa, M. *J. Electroanal. Chem.* **2004**, *563*, 23–31.
83. Miki, A.; Ye, S.; Osawa, M. *Chem. Commun.* **2002**, 1500–1501.
84. Chen, Y. X.; Heinen, M.; Jusys, Z.; Behm, R. J. *Langmuir* **2006**, *22*, 10399–10408.
85. Chen, Y. X.; Heinen, M.; Jusys, Z.; Behm, R. J. *Angew. Chem. Int. Ed.* **2006**, *45*, 981–985.
86. Osawa, M.; Komatsu, K.-I.; Samjeské, G.; Uchida, T.; Ikeshoji, T.; Cuesta, A.; Gutiérrez, C. *Angew. Chem. Int. Ed.* **2011**, *50*, 1159–1163.
87. Cuesta, A.; Cabello, G.; Osawa, M.; Gutiérrez, C. *ACS Catal.* **2012**, *2*, 728–738.
88. Cuesta, A.; Cabello, G.; Gutierrez, C.; Osawa, M.; *Phys. Chem. Chem. Phys.* **2011**, *13*, 20091–20095.
89. Osawa, M. in *Diffraction and Spectroscopic Methods in Electrochemistry*, R. C. Alkire, D. M. Kolb, J. Lipkowski, and P. H. Ross, Eds., Wiley-VCH: Weinheim, **2006**.
90. Samjeské,G.; Komatsu, K.-I.; Osawa, M. *J. Phys. Chem. C* **2009**, *113*, 10222–10228.
91. Heinen, M.; Chen, Y. X.; Jusys, Z.; Behm, R. J. *Electrochim. Acta* **2007**, *52*, 5634–5643.
92. Cuesta, A. *J. Am. Chem. Soc.* **2006**, *128*, 13332–13333.
93. Grozovski, V.; Climent, V.; Herrero, E.; Feliu, J. M. *Phys. Chem. Chem. Phys.* **2010**, *12*, 8822–8831.
94. Cuesta, A. *Surf. Sci.* **2004**, *572*, 11–22.
95. Cuesta, A.; Cabello, G.; Hartl, F. W.; Escudero-Escribano, M.; Vaz-Domínguez, C.; Kibler, L. A.; Osawa, M.; Gutiérrez, C. *Catal. Today* **2013**, *202*, 79–86.
96. Beltramo, G. L.; Shubina, T. E.; Koper, M. T. M. *ChemPhysChem* **2005**, *6*, 2597–2606.
97. Osawa, M.; Komatsu, K.-I.; Samjeské, G.; Uchida, T.; Ikeshoji, T.; Cuesta, A.; Gutiérrez, C. *Angew. Chem. Int. Ed.* **2011**, *50*, 1159–1163.
98. Wang, H.-F.; Liu, Z.-P. *J. Phys. Chem. C* **2009**, *113*, 17502–17508.
99. Neurock, M.; Janik, M.; Wieckowski, A. *Faraday Discuss.* **2008**, *140*, 363–378.
100. Gao, W.; Keith, J. A.; Anton, J.; Jacob, T. *J. Am. Chem. Soc.* **2010**, *132*, 18377–18385.
101. Samjeské, G.; Miki, A.; Ye, S.; Osawa, M. *J. Phys. Chem. B* **2006**, *110*, 16559–16566.
102. Samjeské, G.; Miki, A.; Ye, S.; Yamakata, A.; Mukouyama, Y.; Okamoto, H.; Osawa, M. *J. Phys. Chem. B* **2005**, *109*, 23509–23516.

103. Samjeské, G.; Osawa, M. *Angew. Chem. Int. Ed.* **2005**, *44*, 5694–5698.
104. Crépy, G.; Lamy, C.; Maximovitch, S. *J. Electroanal. Chem.* **1974**, *54*, 161–179.
105. Grozovski, V.; Vidal-Iglesias, F. J.; Herrero, E.; Feliu, J. M. *ChemPhysChem* **2011**, *12*, 1641–1644.
106. Willner, I.; Katz, E. *Angew. Chem. Int. Ed.* **2000**, *39*, 1180–1218.
107. Delamar, M.; Hitmi, R.; Pinson, J.; Saveant, J. M. *J. Am. Chem. Soc.* **1992**, *114*, 5883–5884.
108. Allongue, P.; Delamar, M.; Desbat, B.; Fagebaume, O.; Hitmi, R.; Pinson, J.; Savéant, J.-M. *J. Am. Chem. Soc.* **1997**, *119*, 201–207.
109. Delamar, M.; Désarmot, G.; Fagebaume, O.; Hitmi, R.; Pinsonc, J.; Savéant, J. M. *Carbon* **1997**, *35*, 801–807.
110. Rüdiger, O.; Abad, J. M.; Hatchikian, E. C.; Fernandez, V. M.; De Lacey, A. L. *J. Am. Chem. Soc.* **2005**, *127*, 16008–16009.
111. Alonso-Lomillo, M. A.; Rüdiger, O.; Maroto-Valiente, A.; Velez, M.; Rodríguez-Ramos, I.; Muñoz, F. J.; Fernández, V. M.; De Lacey, A. L. *Nano Lett.* **2007**, *7*, 1603–1608.
112. Polsky, R.; Harper, J. C.; Dirk, S. M.; Arango, D. C.; Wheeler, D. R.; Brozik, S. M. *Langmuir* **2006**, *23*, 364–366.
113. Radi, A.-E.; Muñoz-Berbel, X.; Cortina-Puig, M.; Marty, J.-L. *Electroanalysis* **2009**, *21*, 696–700.
114. Pita, M.; Gutierrez-Sanchez, C.; Olea, D.; Velez, M.; Garcia-Diego, C.; Shleev, S.; Fernandez, V. M.; De Lacey, A. L. *J. Phys. Chem. C* **2011**, *115*, 13420–13428.
115. Vaz-Domínguez, C.; Pita, M.; De Lacey, A. L.; Shleev, S.; Cuesta, A. *J. Phys. Chem. C* **2012**, *116*, 16532–16540.
116. Ricci, A.; Bonazzola, C; Calvo, E. J. *Phys. Chem. Chem. Phys.* **2006**, *8*, 4297–4299.
117. Shewchuk, D. M.; McDermott, M. T. *Langmuir* **2009**, *25*, 4556–4563.
118. Gui, A. L.; Liu, G.; Chockalingam, M.; Le Saux, G.; Luais, E.; Harper, J. B.; Gooding, J. J. *Electroanalysis* **2010**, *22*, 1824–1830.
119. Barth, A. *Biochim. Biophys. Acta* **2007**, *1767*, 1073–1101.
120. Polyakov, K. M.; Fedorova, T. V.; Stepanova, E. V.; Cherkashin, E. A.; Kurzeev, S. A.; Strokopytov, B. V.; Lamzin, V. S.; Koroleva, O. V. *Acta Cryst. D* **2009**, *65*, 611–617.
121. Rosario-Castro, B. I.; Fachini, E. R.; Hernández, J.; Pérez-Davis, M. E.; Cabrera, C. R. *Langmuir* **2006**, *22*, 6102–6108.

# Chapter 9

# Vibrational Stark Effect at Halide Precovered Cu(100) Electrodes

**Melanie Röefzaad,[1] Duc Thanh Pham,[1] and Klaus Wandelt[1,2]**

[1]Institute of Physical and Theoretical Chemistry, University of Bonn, Bonn, Germany

[2]Institute of Experimental Physics, University of Wroclaw, Wroclaw, Poland

## 9.1 INTRODUCTION

The self-assembling of molecular building blocks on well-defined single-crystal surfaces is a promising strategy to form complex supramolecular structures following the so-called bottom-up approach. Electrostatical and sterical interactions have directional character and can be used to manipulate the long-range structure among the molecular building blocks [1–4]. The complex interplay of competing adsorbate–adsorbate and adsorbate–substrate interactions controls the two-dimensional (2D)-phase behavior. As long as enhanced adsorbate–substrate interactions govern the 2D-phase formation, the surface can serve as a template in the ordering process of the molecular building blocks. Placed into an electrochemical environment, the binding properties between the adsorbate and the substrate, serving as electrode, can be manipulated by changing the potential applied to the electrode. Not only can the binding strength of the adsorbate to the electrode surface thereby be changed, but also the molecular vibration modes can be affected due to interactions of the molecules with the variable electrostatic field near the surface. The resultant shifts in vibration frequencies can be detected via spectroscopy methods such as Raman and infrared (IR) spectroscopy. The shift of absorption bands ($d\nu$) due to changes of the

*Vibrational Spectroscopy at Electrified Interfaces*, First Edition.
Edited by Andrzej Wieckowski, Carol Korzeniewski, and Björn Braunschweig.

applied potential (*dP*) is based on the vibrational Stark effect and can be quantified by the Stark tuning rate ($d\nu/dP$) [5, 6]. The extent of the variation of a vibration frequency with the electrode potential depends on the adsorption geometry, the surface coverage of the molecules, and the electrostatic properties of the electrode surface.

In this chapter we present an in situ infrared absorption study on the *vibrational Stark effect* on diphenylviologen (1,1′-diphenyl-4,4′-bipyridinium dichloride, abbreviated as DPV) monolayers adsorbed on anion precovered Cu(100) electrode surfaces. Halide anions ($Cl^-$, $Br^-$, $I^-$) are known to form a condensed and highly ordered monolayer on single-crystal electrode surfaces such as Cu(100). Specifically adsorbed from the supporting electrolyte, these adlayers can alter both the geometric and electronic structure of the electrode surface and, as a consequence, the entire charge distribution perpendicular to the surface in the near-surface region [7]. Therefore, they can serve not only as a structural template but also as an electronic template for the adsorption of positively charged organic molecules [8–13]. Owing to their different electronegativity and, hence, different polarizability, the three halide species influence the structural and electrostatic properties of the surface differently [14–19]. In particular, the charge state of the respective specifically adsorbed anion is expected to be important for the coadsorption of cationic organic species, especially when electrostatic attraction is the main driving force for their surface bonding. Conversely, the adsorption of organic molecules can serve to probe different charge states of a surface by virtue of vibrational frequency shifts and their potential dependency. Therefore, in this work the *Stark tuning rate* of DPV molecules adsorbed on a Cu(100) electrode surface precovered with different halide interlayers has been studied via substractively normalized interfacial Fourier transform infrared spectroscopy (SNIFTIRS).

## 9.2 EXPERIMENTAL

All spectroscopic measurements were performed in a home-built transfer chamber [20] that allows the contamination-free transfer of the sample from the spectroelectrochemical cell located in an argon-filled glove box into an ultrahigh vacuum (UHV) chamber ($2 \times 10^{-10}$ mbar) and vice versa. The Cu(100) crystal was cleaned under UHV conditions as described elsewhere [21], and surface cleanliness was controlled by Auger electron spectroscopy (AES) and low-energy electron diffraction (LEED), followed by the transfer into the glove box.

The copper sample (working electrode) was placed in the spectroelectrochemical cell with a Pt wire as counterelectrode and a reversible hydrogen electrode in such a way that only the (100) surface was in contact with the electrolyte ("hanging miniscus configuration"). Cyclic voltammograms (CVs) were recorded with a scan rate of 10 mV/s.

For the FT-IR measurements a Nicolet Magna560 spectrometer (Thermo Fisher Scientific, Waltham, MA) was used with a nitrogen-cooled mercury cadmium telluride (MCT) detector. The infrared (IR) measurements were performed under external reflection conditions. The spectroelectrochemical cell is designed so as to establish a thin-layer configuration. Therefore, the cell window is a $CaF_2$ prism beveled at 60° from the surface normal [22]. For each measurement 128 scans per spectrum with a resolution of 4 $cm^{-1}$ were averaged. The polarization of the incident light can be varied between *s*- and *p*-polarization. For background correction a reference spectrum was recorded at +30 mV so that the background-corrected absorption is then given by the formula

$$\Delta \text{Abs} = -\log \frac{R_{\text{sample}}}{R_{\text{reference}}} \tag{9.1}$$

The obtained SNIFTIR spectra represent absorption differences between sample and reference spectrum.

The surface structures were determined by in situ electrochemical scanning tunneling microscopy measurements using our home-designed and -built Bonn-ECSTM described in detail in Ref. [23].

For all solutions, high-purity water (Milli-Q purification system; conductivity <18 MΩ·cm; total organic carbon (TOC) <5 ppb) and reagent-grade chemicals were used. All electrolyte solutions were routinely deoxygenated with argon gas (5.0) for an hour before use. In situ measurements were carried out in 10 mmol/L HCl (Merck, Germany) solution as basic electrolyte. For bromide and iodide precovered electrodes KBr and KI (1 mmol/L) (Merck, Germany) are added, respectively. All potentials in this chapter refer to a reversible hydrogen electrode (RHE) and a Pt wire serves as the counterelectrode. All experiments routinely started with an electrochemical characterization of the Cu(100) surface in one of the electrolytes. The probe molecule is introduced in the form of $DPVCl_2$ salt (Sigma Aldrich, Germany) into the above electrolyte systems under control of the working potential $E_{\text{work}} = 0$ mV with a concentration of 0.1 mmol/L unless stated otherwise.

## 9.3 RESULTS

### 9.3.1 Electrochemical Properties

#### *Halide Precovered Cu(100) Electrode*

The interaction of chloride, bromide, and iodide with Cu(100) is well documented in the literature. Halide anions adsorb on the Cu(100) surface at high potentials with the formation of an ordered adlayer. The binding energy of the halide species to the Cu surface decreases with increasing electronegativity of the adsorbate. Consequently, a chloride adlayer will be replaced completely by bromide or iodide if present in the electrolyte. Therefore, hydrochloric acid (10 mmol/L) was used as base electrolyte while a KBr or KI solution (1 mmol/L each) was added for the formation of a closed bromide or iodide adlayer, respectively.

Choride and bromide form a highly ordered adlayer structure described by a $c(2 \times 2)$-X or $(\sqrt{2} \times \sqrt{2})$R45°-X unit cell (X = Cl, Br) with a resulting saturation coverage of $\theta = 0.5$ ML (normalized to the topmost substrate layer) [17, 24, 25]. This structure (Fig. 9.1*a*,*b*) is stable in the potential regime (+250 to −400 mV)

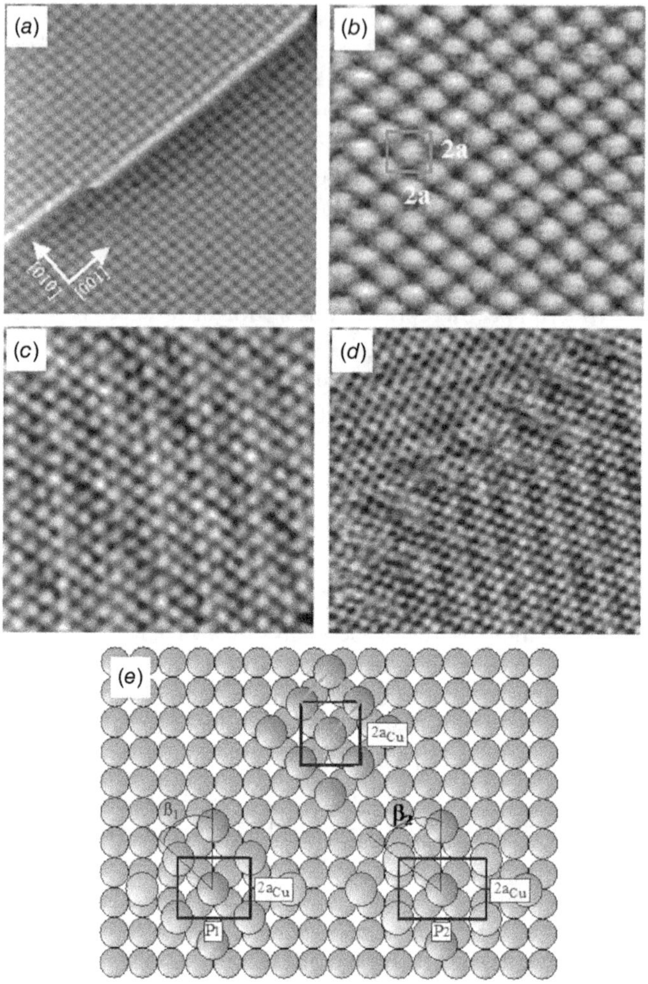


**Figure 9.1** In situ STM images showing the atomically resolved structures of (*a,b*) bromide and (*c,d*) iodide on a Cu(100) electrode surface. While bromide (like chloride) anions form a $c(2 \times 2)$ structure (*e*, top), iodide anions form a uniaxially expanded, incommensurate structure (*c*), which for symmetry reasons occurs in two orthogonal orientations (*d* and *e*, bottom).

limited by the Cu dissolution reaction (CDR) at high potentials and the hydrogen evolution reaction (HER) at low potentials (see dashed trace in Fig. 9.3). At the onset of the HER order–disorder transitions within the adlayer structure have been observed in electrochemical scanning tunneling microscopy (EC–STM) studies and have been assigned to the starting chloride (bromide) desorption from the electrode surface [7]. In contrast to chloride and bromide, iodide adlayers exhibit the phenomena of electrocompression, structure transitions, and the formation of 2D and three-dimensional (3D) CuI phases depending on the applied electrode

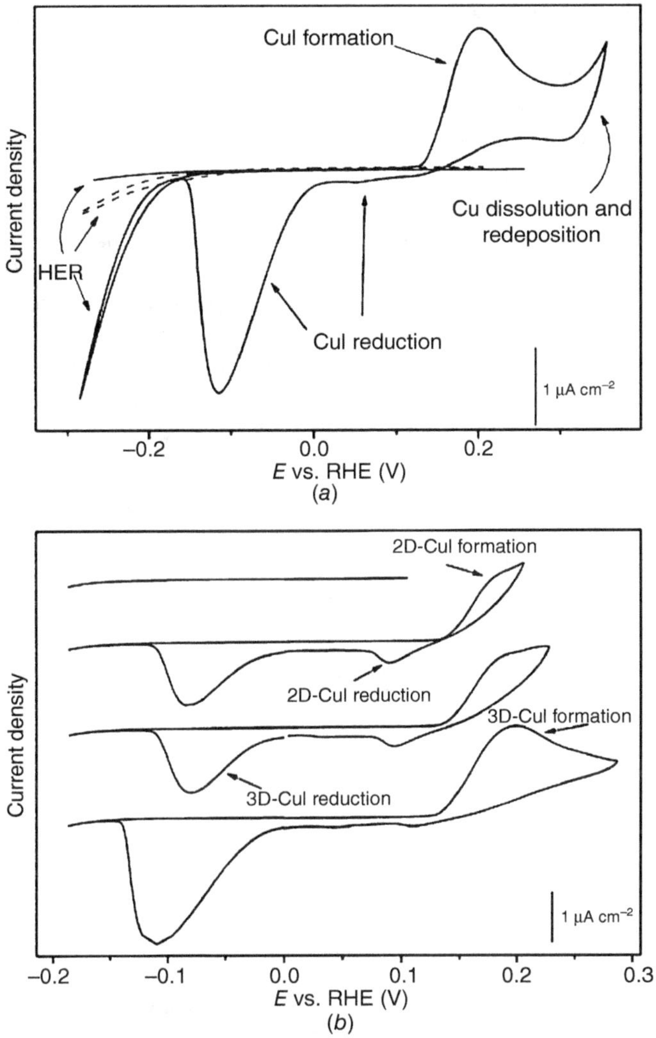


**Figure 9.2** (*a*) Cyclic voltammogram of Cu(100) in 10 mM HCl (—) and additional 1 mM KBr (---) or 1 mM KI (—). In (*b*) a sequence of CVs of Cu(100) in 10 mM HCl + 1 mM KI is shown with different limiting anodic potentials.

potential [18]. The solid trace in Figure 9.2*a* shows a steady-state CV of Cu(100) in iodide containing electrolyte in the potential regime from +300 to −300 mV. At potentials around +200 mV, just before the copper dissolution process, the broad anodic current wave reflects the formation of 3D CuI on the electrode surface. After reversing the scan direction toward lower potential, two separate signals at +45 and −115 mV emerge. While the first signal has been identified as the reduction of a 2D CuI phase, the large signal at −115 mV arises from the reduction of the formerly grown 3D CuI phase followed by the hydrogen evolution reaction [14, 16, 26]. Both the formation of CuI and significant structure changes of the iodide adlayer can be avoided if the range of the applied potential is limited to +70 to −200 mV as shown in Figure 9.2*b*, upper trace). Here a sequence of CVs displays the appearance of the CuI signals depending on the anodic limit. As long as the applied potential is kept below +70 mV CuI formation can be excluded. In this potential regime, +70 to −200 mV, iodide forms just an ordered adlayer. However, the larger *van der Waals* radius of iodide compared to chloride and bromide results in a slight lattice mismatch with respect to the lattice of the topmost Cu layer. As a consequence the formed iodide adlayer structure can be described by an uniaxially expanded $c(p \times 2)$-I unit cell and, of course, exists on the Cu(100) surface in equivalent orthogonal domains (Figs. 9.1*c*,*d*). p is larger than two times the nearest-neighbor Cu–Cu distance and dependent on the applied potential. With decreasing potential the iodide structure expands in the [100] direction (Fig. 9.1*e*). Thus, the surface concentration of iodide is potential dependent and ranges from 0.4 to 0.36 ML with respect to the Cu lattice [18].

### *Electrochemistry of Probe Molecule, Diphenylviologen*

Figure 9.3*a* depicts the structure and dimensions of the DPV molecule used for the adsorption experiments reported here. These molecules are redox active and may, depending on the electrode potential, undergo two one-electron transfer reactions:

$$\mathrm{DBV}^{0} \underset{-e^{-}}{\overset{+e^{-}}{\underset{2}{\rightleftharpoons}}} \mathrm{DBV}^{-\cdot} \underset{-e^{-}}{\overset{+e^{-}}{\underset{1}{\rightleftharpoons}}} \mathrm{DBV}^{2-} \tag{9.2}$$

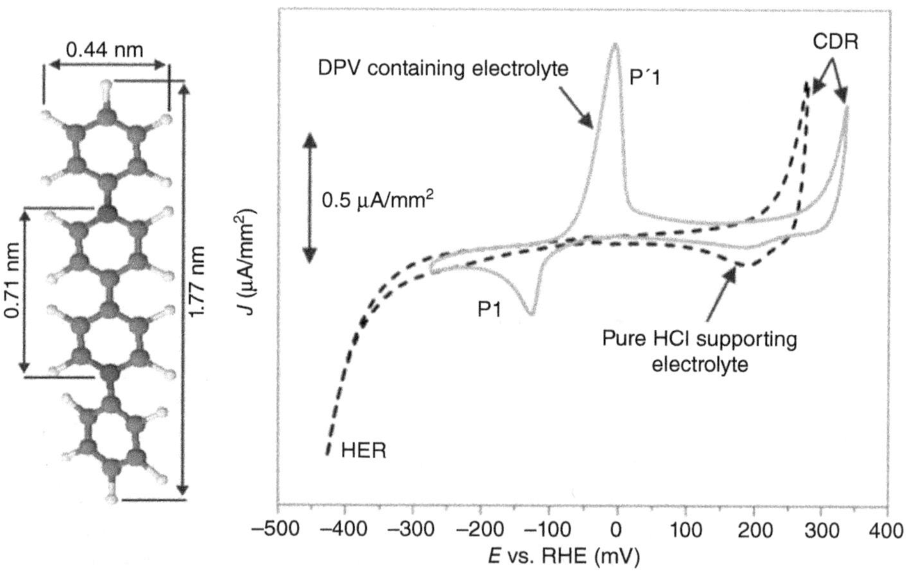


(*a*) (*b*)

**Figure 9.3** (*a*) Structure of the diphenylviologen molecule and (*b*) CV of a Cu(100) electrode in pure HCl solution (dashed line) and in $DPV^{2+}$ containing electrolyte (full line).

where the second reaction is much less reversible. Upon electron transfer the bipyridinium moiety of the Viologen undergoes an intramolecular conformation change [27, 28] as demonstrated by quantum chemical density functional theory (DFT) calculations for the DBV molecule [11]. The dihedral angle between the two pyridinium ring planes in the dicationic form is substantially different from zero [$\varphi(DBV^{2+}) = 40.1°$], at least in the absence of external forces. This angle decreases to a value close to zero in the isolated radical monocation [$\varphi(DBV^{+\bullet}) = -5.4°$], making the two pyridinium rings almost co-planar, and it is this co-planarity that enables the formation of stacked assemblies as the characteristic structural motif for viologens in their radical monocationic state at the electrode surface (see Fig. 9.5*c*) [29].

The electrochemical potential–current response of the system due to the presence of $DPV^{2+}$ in HCl solution is presented in Figure 9.3*b* and is very similar for the chloride-, bromide-, or iodide-precovered Cu(100) electrode, as shown in Figure 9.4, trace 1–3, respectively. The peak system P1/P'1 corresponds to the first reduction step $DPV^{2+} \rightarrow DPV^{+\bullet}$ of the species in the respective solution, that is, $DPV^{2+}$ dications diffusing to the surface, being reduced at the surface and leaving the surface as $DPV^{+\bullet}$ monocation radicals. This is supported by the fact that the intensity

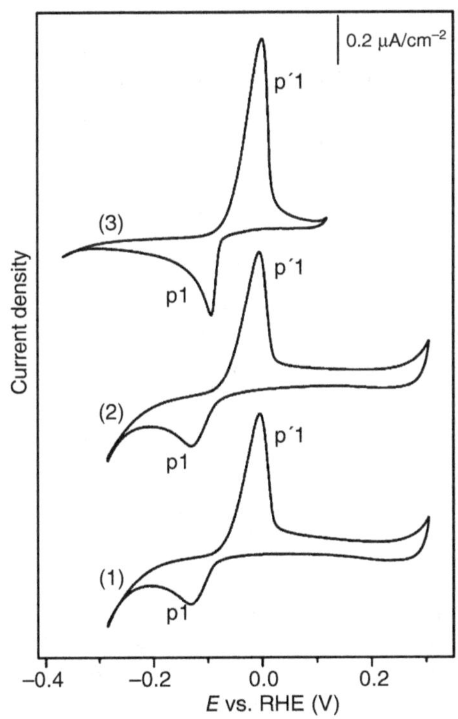


**Figure 9.4** CV of Cu(100) in halide and DPV containing electrolyte, trace (1): 10 mM HCl + 0.1 mM $DPVCl_2$, (2) 10 mM HCl + 1 mM KBr + 0.1 mM $DPVCl_2$, and (3) 10 mM HCl + 1 mM KI + 0.1 mM DPV.

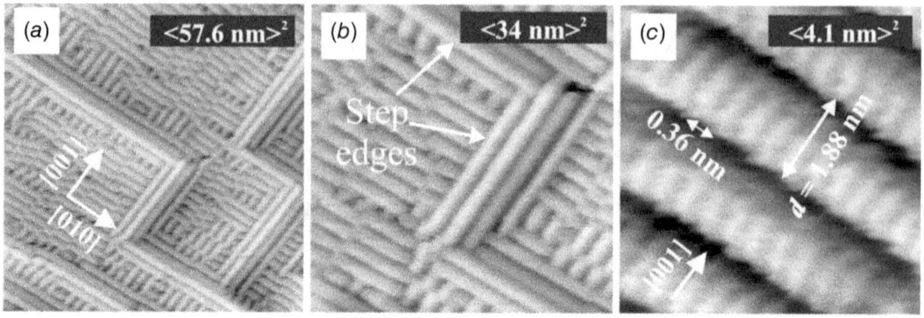


**Figure 9.5** In situ STM images of an ordered $DPV^{+\bullet}$ layer on a chloride-precovered Cu(100) electrode surface; increasing lateral resolution from (*a*) to (*c*). The stripes are composed of π–π stacked $DPV^{+\bullet}$ monocation radicals (*c*).

of both peaks depends linearly on the $DPV^{2+}$ concentration in the solution. In contrast to this solution species ex situ X-ray photoelectron spectroscopy (XPS) experiments demonstrated that the adsorbed DPV species is in the radical monocationic state even after preparation at potentials where the $DPV^{2+}$ in solution is stable [12]. Additionally, in the whole potential regime STM studies of the 2D surface structure revealed exclusively the characteristic structure motif of the reduced species, that is, π–π stacked $DPV^{+\bullet}$ assemblies visible as long stripes on the electrode surface [7, 12]. This structure motif as shown in Figures 9.5*a–c*, which does *not* change when the applied potential is varied, has equally been found on bromide- and iodide-precovered Cu(100) electrode surfaces [30]. Thus, upon adsorption the $DPV^{2+}$ dications are instantaneously reduced to the corresponding monocationic $DPV^{+\bullet}$ species, even at potentials above the redox potential of the respective solution species. This and the fact that no structural transitions are observed with EC–STM in the potential range in question (+70 to −200 mV) make this system a perfect probe to measure changes of the adsorbate properties, which are *exclusively* due to variations of the applied potential because changes of the optical signal due to redox processes and reorientation in the organic adlayer can be excluded. Furthermore, it is possible to study the influence of different halide interlayers since the structure motive of the DPV adlayer and its potential independence is the same on the chloride-, bromide- or iodide-precovered Cu(100) surface.

### 9.3.2 FTIR Studies

The SNIFTIRS technique provides the opportunity to directly compare the absorption characteristics of the DPV adsorbate layer at different potentials. The reference spectrum was registered right after the adsorption of DPV at +30 mV. Afterward the potential was gradually decreased and a sample spectrum was collected after every potential step. With the help of Eq. (9.1) absorption difference spectra were calculated, representing only the changes provoked by the variation of the electrode potential. The considered spectral region ranges from 1500 to 1150 $cm^{-1}$. In this fre-

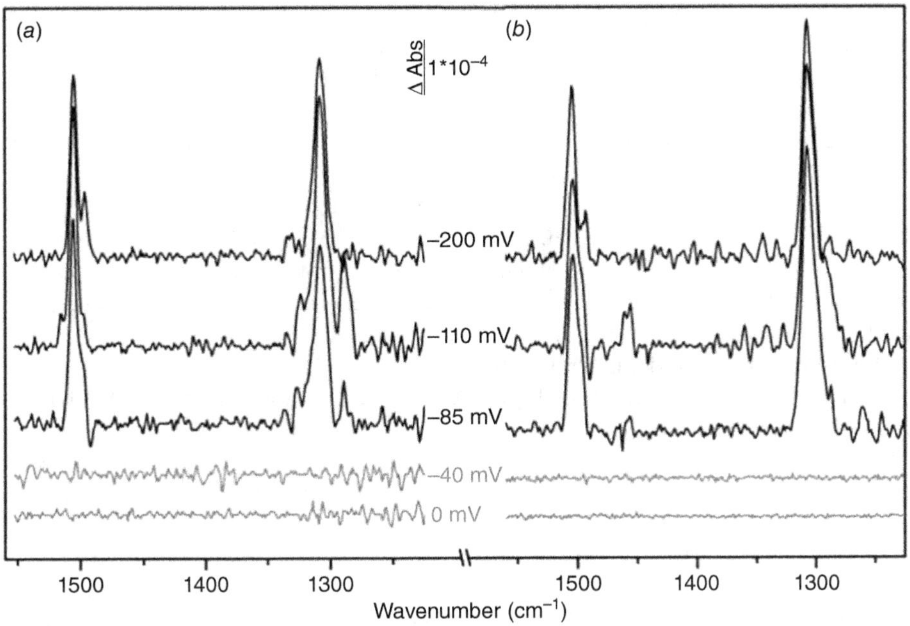


**Figure 9.6** SNIFTIR spectra of a DPV monolayer adsorbed on (*a*) chloride- and (*b*) bromide-precovered Cu(100) electrode surfaces after stepwise decrease of the electrode potential; reference spectra were collected at +30 mV after adsorption process.

quency regime mainly the ring vibration modes of heteroatomic aromatic ring systems containing nitrogen atoms such as pyridine and bipyridine can be detected [31–33].

### *Interlayer-Dependent Stark Effect on DPV Monolayer*

Figures 9.6*a*,*b* show the absorption differences obtained between +30 mV and 0, −40, −85, −110, and −200 mV in the case of the chloride- (*a*) and bromide- (*b*) precovered Cu(100) electrode, detected with *p*-polarized light. Both systems show nearly the same development of the absorption properties with decreasing potential. While at potentials ≥−40 mV no detectable differences compared to the system at +30 mV can be observed, switching the applied potential to −85 mV causes the rise of two distinct absorption bands at 1505 and 1309/1308 $cm^{-1}$. A further decrease of the potential causes only some slight changes in intensity but not in frequency. Likewise neither bipolar bands originating from frequency shifts are observable nor are additional absorption bands emerging. Since these difference signals do not show up in spectra collected with *s*-polarized light (not shown here), but only with *p*-polarized light, they arise from changes directly next to the electrode surface ("surface selection rule"). As described above in situ EC–STM and ex situ XPS data provide clear evidence for the existence of a DPV monolayer composed of *radical monocations* adsorbed on the electrode surface *in the whole potential window*. A reorientation and restructuring within the formed DPV+• monolayer can also be excluded by SNFITIR spectroscopy. The observed absorption bands are related to the reduction processes of the $DPV^{2+}$ dications diffusing from the electrolyte bulk into the electrochemical double layer, which will be explained in more detail later in this chapter.

The same experiments have also been done with *iodide*-precovered Cu(100) electrodes and result in a different picture. Figure 9.7 shows the differences of the absorption signals in the ring vibration region monitored with *p*-polarized light in the potential regime between +30 and −200 mV. The applied potential was never increased above the starting potential (+30 mV) in order to avoid CuI compound formation (Fig. 9.2). Note that in this case the concentration of DPV was reduced to 0.02 mmol/L instead of 0.1 mmol/L, as it was used in the case of the chloride and bromide precovered Cu(100) surface.

The difference spectrum in Figure 9.7 shows predominantly bipolar bands, indicating a potential induced shift of absorption bands to higher wave numbers. The emerging bipolar bands can be assigned to two different kinds of vibration modes: Bands centered at 1551, 1447, 1227, and 1151 $cm^{-1}$ are caused by vibrations within the bipyridine moieties. More intense, however, are the bands around 1595, 1498, and 1300 $cm^{-1}$, which are known to be induced by vibronic coupling in the so-called charge-transfer (CT) complexes consisting of at least two radical monocations ($DPV^{+\bullet}$). The formation of these complexes is driven by a pairing of the lonely radical electrons when coupling the radicalic bipyridine moieties face to face [34–38]. These high-intensity absorption bands have been explained in terms of the *vibronic coupling mechanism* [39], which was first proposed for solid-state CT complexes [40, 41]. The totally symmetric ($A_g$) vibrational modes of the bipyridine ring are inherently infrared inactive. However, the $A_g$ vibrations of adjacent molecules within the CT complexes can couple in phase or out of phase. The in-phase

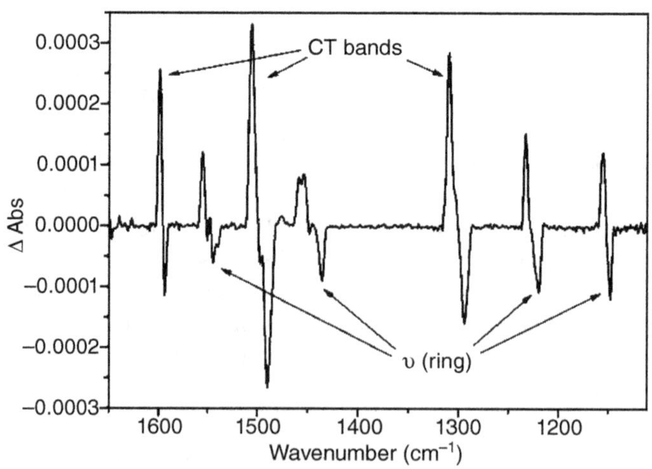


**Figure 9.7** SNIFTIR spectra with *p*-polarized light of DPV on an iodide-precovered Cu electrode at −200 mV with a decreased concentration of 0.02 mmol/L in the electrolyte, containing 1 mM KI and 10 mM HCl. Reference spectrum was collected at +30 mV.

vibration is completely decoupled from the radicalic electron system, whereas the out-of-phase vibration is accompanied by an intermolecular CT transition between the bipyridine units. As a consequence a large charge polarization normal to the ring planes is thereby induced, which generates a change of the transition dipole moment in the same direction [39]. Absorption signals caused by this coupling mechanism will be denoted as *charge-transfer*-bands (CT bands) in the following.

As can be seen in Figure 9.7, all vibration frequencies concerning the bipyridine part of the molecule are affected in the same way when the potential is decreased. Moreover, the lower the applied potential, the higher the frequencies of the mentioned vibration modes are as shown in Figure 9.8. This points to a strengthening

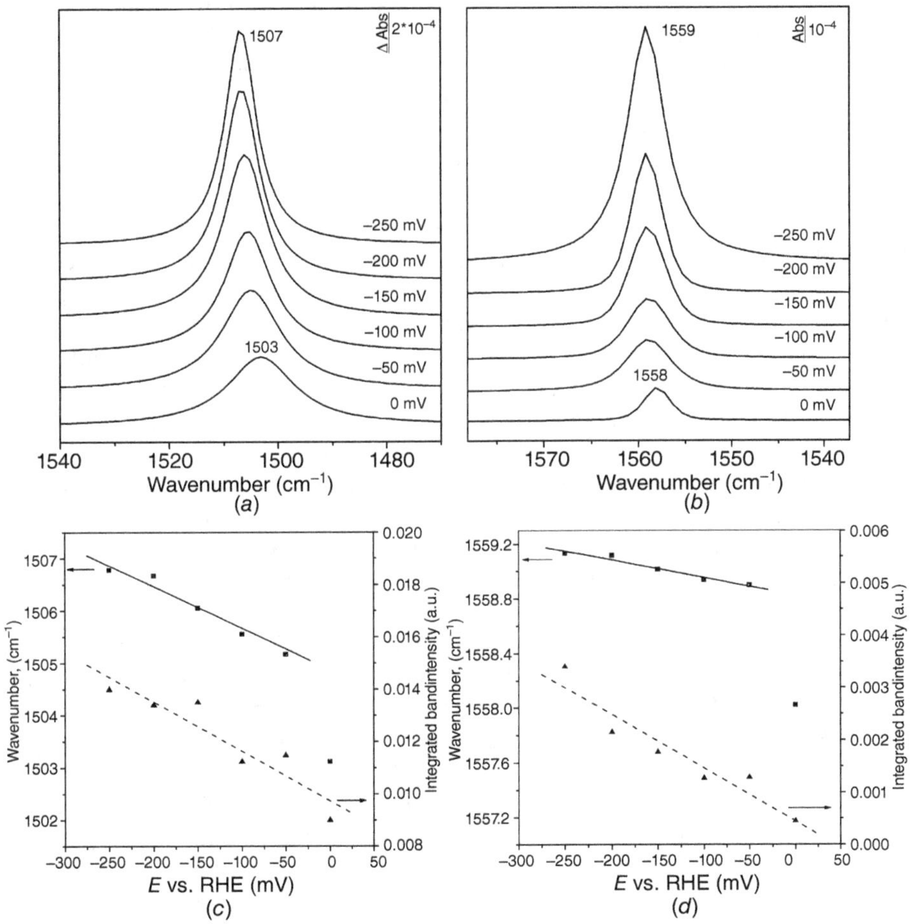


**Figure 9.8** Calculated absorption bands after subtraction of the reference signals at +30 mV instancing the two vibration types: (*a*) CT band at 1507 $cm^{-1}$, (*b*) ring vibration at 1559 $cm^{-1}$; (*c*) and (*d*) depict the dependency of the band center frequency and integrated band intensity on the applied potential corresponding to the CT band and the ring vibration, respectively.

of inter- and intramolecular bonds, which is, in particular, in the case of the CT bands associated with an increase in electron charge density transferred into the DPV layer from the electrode surface.

In order to obtain undisturbed information about the potential dependence of the band center position, it is necessary that the absorption bands of the sample spectra under investigation are not distorted by the signal in the corresponding reference spectrum. Therefore, in order to extract the absolute band positions of the ring vibration modes and the signals of vibronic coupling, bipolar features in the difference spectra have been deconvoluted assuming Lorentz-type line shapes. The maximum experimental separation between both peaks is obtained at –250 mV (spectrum not shown here). Hence, the deconvolution of both, negative and positive, bands can be performed at this potential with lowest uncertainty. The negative bands correspond to absorption signals at the reference potential +30 mV, and their positions are assumed to remain constant during the collection of all the subsequent spectra in a series since only one reference spectrum was collected for the complete set of spectra. The Lorentzian curves obtained for the reference potential were subtracted from the difference spectra at each potential. The results are displayed in Figures 9.8*a*,*b*, showing the absolute absorption signals of the CT band at 1506 $cm^{-1}$ and the ring vibration band at 1559 $cm^{-1}$, respectively, representing the two types of vibration. The main source of inaccuracy appears in the determination of the bandwith, while the uncertainty in the band center position and the integrated band intensity is low. Figures 9.8*c*,*d* show both the peak frequency and the integrated band intensity as a function of the applied potential for the calculated positive-going bands plotted in Figures 9.8*a*,*b*, respectively. Both signals clearly constantly gain intensity when the potential is decreased. Due to the large CT complexes in the present case, namely the polymeric stripes of π–π stacked monocationic radicals, an extended delocalized π-electron system within the organic adlayer is formed and causes a high polarizability of the adlayer. With decreasing potential the static dipole moment between the substrate and the iodide layer grows stronger, enhancing the polarization of the viologen layer in the direction of the surface normal, which increases the interaction strength with the electric vector of the incident light, and thereby the intensity (Fig. 9.8*d*).

A similar tendency can be observed for the center frequency (Fig. 9.8*c*). At negative potentials between –50 and –250 mV the CT band shifts with a negative tuning rate of $-7.9\ cm^{-1}\ V^{-1}$, while the tuning rate for the ring vibration only adds up to $-1.3\ cm^{-1}\ V^{-1}$. The values of the tuning rates are quiet small but well within the range of large, aromatic molecules such as naphthalene [42]. Indeed the most significant shift occurs around 0 mV versus RHE for both the CT and ring vibration bands, breaking the linear correlation between peak position and applied potential at more negative potentials.

Putting together the results obtained with different halides acting as interlayer between the electrode surface and the organic, polycationic $DPV^{+\bullet}$ adlayer, a Stark effect can only be observed with an iodide interlayer. It is well known that halides form pursuant to their electronegativity a more or less polarized bond to metallic substrates such as copper [15]. Chloride retains most of its partial charge upon adsorption, whereby the Cu–Cl bond obtains a strong ionic character. In contrast to that the Cu–I bond is found to have a significant covalent component due to lower electronegativity and good polarizability of iodide. The interfacial structure including surface relaxation phenomena (interlayer spacings parallel to the surface normal) has been determined by in situ surface X-ray diffraction (SXRD) techniques. The Cu–Cl bond is found to be even 4% longer than the Cu–I bond while the first Cu layer is relaxed outward by 2.2% with respect to the ideal bulk layer spacing due to depletion of electrons from the copper surface and therefore weakening of the interaction between the first two Cu layers [14]. Moreover, in the case of chloride precoverage the interfacial spacings turned out to be potential independent, pointing to an already highly polarized bond that is not further affected by potential changes [19]. By contrast, the interfacial structure in the presence of bromide and iodide is influenced by the electrode potential. While bromide induces an outward relaxation of the first Cu layer such as chloride, with iodide the interlayer spacing between the first two Cu layers is shortened by about 3% with respect to the bulk layer spacing, indicating a not negligible charge transfer from the iodide layer into the substrate [14, 19]. The same applies for the Cu–Br and Cu–I distance. With decreasing potential both bromide- and iodide-precovered electrodes show an increase in the interlayer spacings. Both the relaxation phenomena and the potential dependence are a consequence of the low electronegativity and high polarizability of iodide. In the present context it is important to note that not only the bond length and the binding strength but also the vibration frequency of the halide against the substrate can be affected in the same manner. While the static dipole moment in a Cu–Cl or Cu–Br bond keeps the direction throughout the whole potential window, the static dipole moment in the Cu–I bond is found to be reversed when the point of zero charge ($E_{pzc}$) in the system is exceeded [43]. At potentials larger than $E_{pzc}$ iodide acts as an electron donor and transfers charge density into the substrate. However, at potentials lower than $E_{pzc}$, iodide gains charge density, acting as an electron acceptor. This potential-dependent electron donor and/or acceptor characteristic of the iodide layer, of course, heavily affects the binding nature of the DPV overlayer.

Both the electrode underneath and the polycationic $DPV^{+\bullet}$ layer on top compete for the electron density of the iodide in between. At high (positive) potentials the copper substrate gains the charge from the interlayer, which strengthens the Cu–I bond and weakens the interactions with the DPV layer. Low (negative) potentials, however, cause a charge transfer into the DPV layer. Strong evidence for this reasoning is the blue shift of the ring vibration and vibronic coupling frequencies observed in this study (Fig. 9.8). Higher frequencies indicate a stronger inter- and intramolecular interaction in the organic layer. Due to the large polymeric CT complexes formed by π–π interaction and radical coupling among the $DPV^{+\bullet}$ radical monocations charge transfer into the layer affects especially the *frequency* of the vibronic coupling, indicated by the largest tuning rate. But also the ring vibration modes are observed to be sensitive, which, indeed, is not surprising because of the general involvement of the π-electron system in the related ring vibration modes. Around 0 mV versus RHE the tendency to a higher tuning rate indicates the adjacency of $E_{pzc}$. In this potential regime the largest effect of the switching of the electron donor–acceptor characteristics of the iodide interlayer on the absorption properties of the DPV monolayer should be observable (Fig. 9.8*c*,*d*). Considering the large π-electron system within the $DPV^{+\bullet}$ stripe phase also explains the general increase in signal *intensity* because with decreasing potential the static dipole moment on the surface grows stronger. This enhances the polarization of the viologen layer in the direction of the surface normal, which, in turn, strengthens the interaction with the vector of the incident light.

### *Electrochemical Double-Layer Region: Formation of Multilayer*

The difference spectrum of DPV on an iodide-precovered surface in Figure 9.7 has been obtained with a reduced amount of DPV in the electrolyte (0.02 mmol/L) because with a higher concentration of DPV (0.1 mmol/L) even the slightest decrease in electrode potential caused significant changes in the absorption spectra, which completely cover up the bipolar signals due to the Stark effect described in the previous section. Figure 9.9*a* plots the absorption differences detected with *p*-polarized light after adsorption of DPV from the 0.1-mmol/L solution at +30 mV and stepwise potential decrease. As can be seen, strong negative difference bands show up when the potential falls below –40 mV. Comparison with literature leads to an assignment of the main negative bands at 1636, 1586, 1490, 1260, 1210, and 1182 $cm^{-1}$ to the radical monocation $DPV^{+\bullet}$ [34]. Furthermore, some bipolar signals are observed and at least two distinct absorption bands at 1279 and 1174 $cm^{-1}$ gain considerable intensity with decreasing potential. These frequencies can clearly be assigned to the noncharged species $DPV^0$ after the second reduction step [Eq. (9.2)] [44]. Accordingly, a not negligible part of the radical monocations, existent near the electrode surface due to reduction of solution species, is reduced further to the noncharged species when the potential reaches negative values. The concomitant loss of $DPV^{+\bullet}$ species in the surface near electrochemical double layer becomes obvious by comparing spectra from *s*- and *p*-polarized light, plotted in Figure 9.9*b*. Significant intensity losses are only apparent in the solid trace collected with *p*-polarized light. Since these signals are absent in the *s*-spectrum (dashed line), only the DPV species near the electrode surface is affected even at potentials where in solution radical monocations are also present. On the other hand, the $DPV^{+\bullet}$ monolayer on the iodide-precovered Cu electrode remains unchanged in the potential window between +30 and –200 mV as evidenced by EC-STM [30]. The intensity gaining signals in the *s*-spectrum point to an increasing amount of the noncharged species diffusing into the electrolyte bulk after reduction near the surface. A closer inspection of the ratio between intensity loss due to reduction of the radical monocation and intensity increase due to the formation of $DPV^0$ reveals a discrepancy, which points to a higher loss of $DPV^{+\bullet}$ than $DPV^0$ formation. But note that, first of all, the signal intensity of the species next to the electrode is enhanced since the vector of the *p*-polarized light is amplified normal to the electrode surface in reflection mode, and, second, the effective concentration of DPV on the electrode is higher than in the "bulk solution" after diffusion.

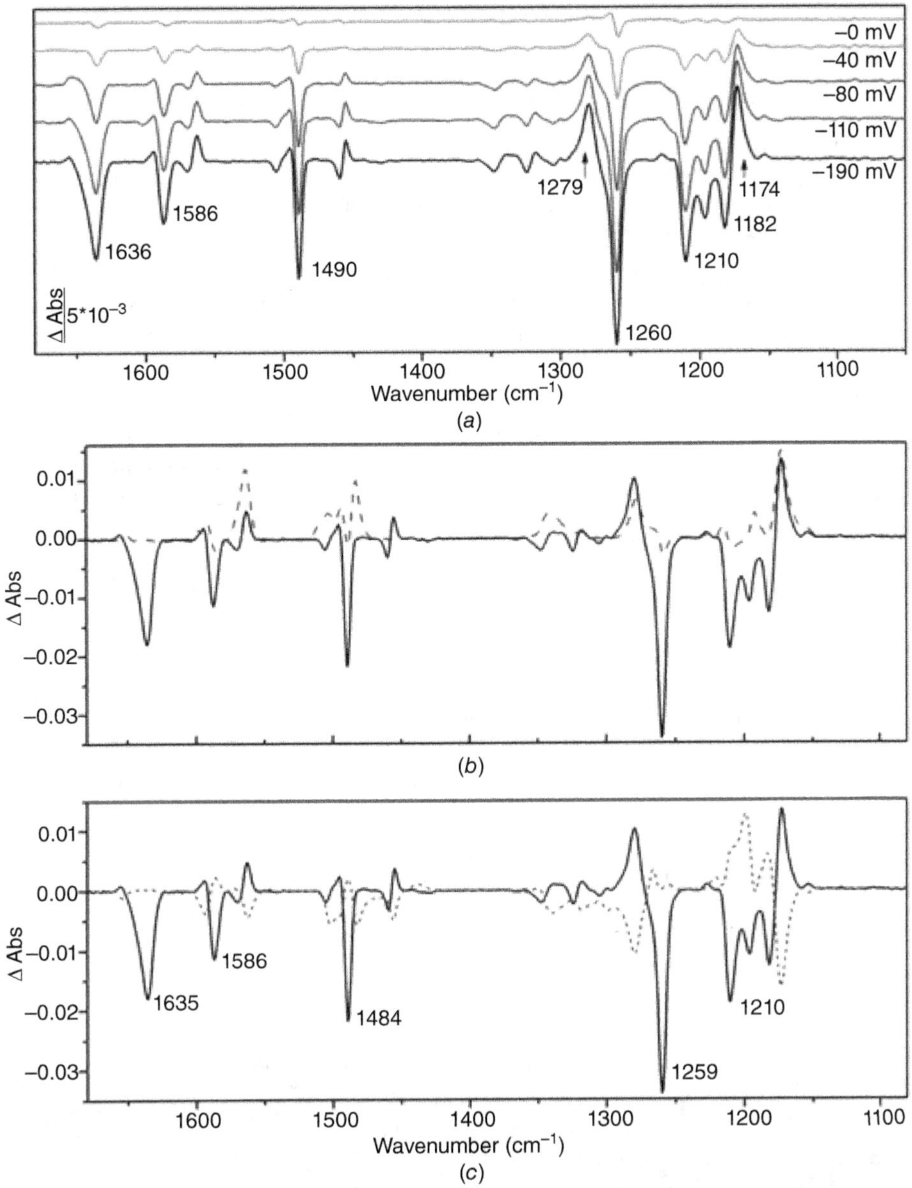


**Figure 9.9** SNIFTIR spectra of DPV (0.1 mM) adsorbed on a iodide-precovered Cu(100) electrode surface: (*a*) *p*-spectra after stepwise decrease of the potential, reference spectrum was collected at +30 mV after adsorption process; (*b*) comparison of the *p*- (—) and *s*- (---) spectra at –190 mV with reference at +30 mV; (*c*) difference spectra with *p*-polarized light after the cathodic sweep from +30 to –190 mV (reference at +30 mV) (—) and after the following anodic sweep back to the starting potential at +30 mV (reference at –190 mV) (---).

The noncharged species is known to form hardly soluble salts with halides after the comproportionation reaction with the dicationic species [27].

$$DPV^{0}_{(aq)} + DPV^{2+}_{(aq)} + 2X^{-}_{(aq)} \rightarrow 2\left[DPV^{+}X^{-}\right]_{s} \tag{9.3}$$

Since all these components are present in the electrolyte bulk, the formation of the salt cannot be excluded. To gain an idea of the reversibility of the observed processes on the iodide-precovered Cu electrode, in Figure 9.9*c* difference spectra after the cathodic sweep and the following anodic sweep are compared. The solid trace depicts the changes in the system after the potential was decreased from +30 to −190 mV, the dashed line, in turn, plots the changes when the system was set back to the start potential. As can be easily seen, especially the intense negative bands at 1635, 1586, 1484, 1259, and 1210 $cm^{-1}$ are not reproduced in the reverse potential sweep. This makes clear that the loss of $DPV^{+\bullet}$ is highly irreversible.

Also in the case of the *chloride*- and *bromide*-precovered Cu(100) surface changes in the SNIFTIR spectra could be observed, which are not assignable to the merely $DPV^{+\bullet}$ containing monolayer. As mentioned before, once the potential drops at least below −80 mV, two distinct absorption bands at 1505 and 1309/1308 $cm^{-1}$ show up in the difference spectra (Fig. 9.6*a*,*b*). Under the same conditions STM studies revealed a disordered *second* layer consisting of small "dots" at steps and defects in the underlying compact monolayer (Fig. 9.10). At these locations the electrode is under reactive conditions and enables the reduction process of solution species, which can also be observed in the CV indicated by the onset of the current signal in the cathodic sweep. The reduction of $DPV^{2+}$ diffusing from the electrolyte bulk toward the electrode surface creates $DPV^{+\bullet}$ species that again tend to form dimers for a matter of stabilization. Those dimers formed and accumulated on the monolayer are held responsible for the bright dots in Figure 9.10 and for the absorption bands in the spectra (Fig. 9.6) since the wavelength fits well to the frequencies of the CT complexes due to vibronic coupling:

$$2DPV^{2+}_{(aq)} + 2e^{-} \rightarrow 2DPV^{+\bullet}_{(aq)} \rightarrow \left[DPV_{2}\right]^{2+}_{(aq)} \tag{9.4}$$

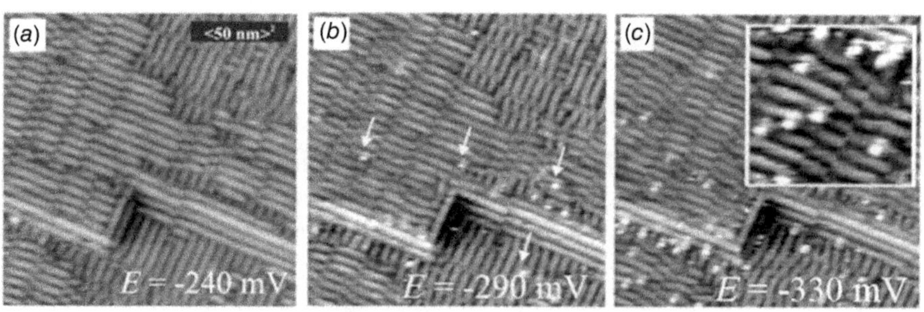


**Figure 9.10** In situ STM images of a *c*(2 × 2)Cl-precovered Cu(100) electrode surface at different electrode potentials. The growing number of bright dots with decreasing potential corresponds to the precipitation of $(DPV)_2^{+2}$ dimers on the $DPV^{+\bullet}$ monolayer.

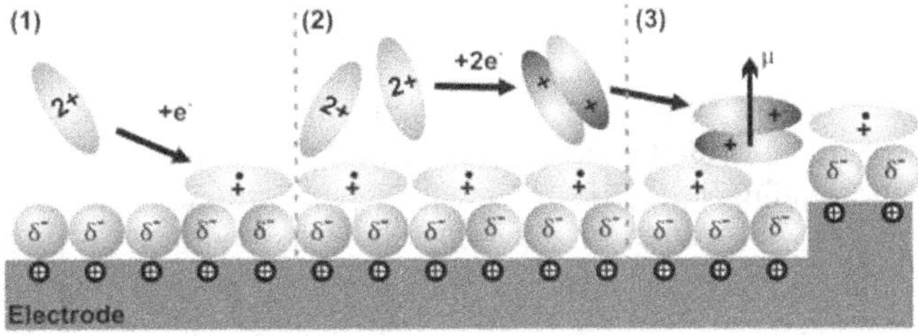


**Figure 9.11** Out-of-plane structure model of the interface illustrating the different electron transfers steps for (1) reduction during the adsorption process, (2) reduction of $DPV^{2+}$ present in the electrochemical double layer, and (3) formation of the dimeric CT complexes.

The one-electron transfer step occurring during the adsorption process at positive potentials is illustrated schematically in Figure 9.11 in step (1). A reduction process within the adsorbed DPV monolayer after its formation can be excluded. But since the CT bands gain intensity at potentials where the reduction process of the solution species starts, these bands are assigned to CT-(dimer) complexes formed in the electrochemical double layer *near* the electrode surface as depicted in Figure 9.11 in steps (2) and (3).

These twofold positively charged dimers accumulate on top of the first formed $DPV^{+\bullet}$ monolayer (Fig. 9.10) because of the charge profile within the electrochemical double layer.

$$[DPV_2]^{2+}_{(aq)} \rightarrow [DPV_2]^{2+}_{(ads)} \tag{9.5}$$

The actual surface concentration of dimers as seen in the STM images (Fig. 9.10) is lower than suggested by the intensity of the absorption bands and no other absorption bands due to increasing concentration of DPV near the electrode surface are observable. This leads to the conclusion that the dimeric complexes are arranged with the ring planes parallel to the electrode surface (face-on), because this enhances the interactions of the vector of the incident light with the transition dipole moment of the vibronic coupling perpendicular to the molecular ring plane and minimizes the contribution from the ring vibrations, which mostly lie in plane of the bipyridine rings. Additionally, absorption bands due to vibronic coupling are known to exhibit an enhanced transition dipole moment so that CT bands are expected to be more intense and better detectable than normal asymmetric vibration modes, particularly when the dipole moment is directed along the vector of the exiting light [36].

This leads us to assign the observed absorption bands in the difference spectra of Figure 9.9 to dimeric CT complexes (dots in Fig. 9.10), which arrange in a face-on configuration on top of the $DPV^{+\bullet}$ monolayer when the applied potential allows the solution species to be reduced as depicted in Figure 9.11 step (3). The transition dipole moment μ of the dimers for vibration coupling modes is indicated with an arrow directed along the surface normal.

## 9.4 CONCLUSION

In this study we employed cyclic voltammetry, in situ EC-STM, and in situ SNIFTIR spectroscopy to gain information about the effect of potential and nature of the halide interlayer on the binding properties of organic, polycationic viologen monolayers. Changes in the binding strength of molecules and their interaction with the electrostatic field on electrode surfaces generally have an effect on the absorption properties, resulting in shifts of vibration frequencies. This effect is denoted as electrochemical Stark effect. This has been unambiguously shown here for diphenyviologen monolayers on Cu(100) electrode surfaces modified by different halides, namely chloride, bromide, and iodide monolayers, because the DPV forms a stable monolayer whose redox state and structure is potential independent, as confirmed by ex situ XPS and in situ EC-STM studies. The observed differences in the absorbate properties in these systems can, thus, clearly be assigned to variations in the applied potential only.

Cyclic voltammetry reveals only one set of current waves, attributed to the first electron-transfer step ($DPV^{2+}/DPV^{+\bullet}$) of the *bulk solution species*. The shape of the CV was found independent from the nature of the preadsorbed halide layer, that is, chloride, bromide, or iodide, as long as the potential did not exceed +70 mV in the presence of iodide in order to prevent CuI formation. With the help of the SNIFTIRS technique we had the possibility to observe potential induced changes in absorption frequencies only in the case of the iodide-precovered Cu(100) electrode. With a chloride or bromide interlayer no charge transfer in the viologen layer could be observed. Only the precipitation of dimeric DPV CT complexes on top of the DPV monolayer was observed when the potential falls below the reduction potential of the solution DPV species.

The high polarizability and low electronegativity of the iodide interlayer allow the electrode surface underneath and the DPV layer on top to compete for its charge density depending on the applied potential. In contrast to chloride and bromide the potential-dependent electron donor–acceptor characteristic of iodide enables a charge transfer between substrate and DPV layer, explaining the higher influence of the iodide-precovered Cu(100) electrode. Only with iodide we observed a blue shift of absorption bands with more negative potentials that are assigned to ring vibration modes of individual pyridine rings as well as vibrational coupling within the CT complexes composed of polymeric chains of radical monocations. Clearly the strongest influence of the potential was observed for those CT bands revealing a Stark tuning rate of $-7.9\ cm^{-1}\ V^{-1}$. This enhanced influence on CT bands points to a charge transfer from the substrate into the DPV layer causing an enrichment of the charge density in the large delocalized $\pi$-electron system, when negative potential is applied. A further strong evidence for this reasoning is the massive loss of radical monocations from the electrochemical double layer when present with high DPV concentrations in the electrolyte. This reduction to the noncharged species and precipitation of DPV salt was not observed for chloride- and bromide-precovered Cu(100) electrodes in the same potential regime.

## ACKNOWLEDGMENTS

This work has been financially supported by the Stiftung Stipendien Fonds of the alliance of the chemical industry and the collaborative research center 624 of the German Science Foundation.

## REFERENCES


1. VÖGTLE, F. *Supramolecular Chemistry*, Wiley: Chichester, **1991**.
2. LEHN, J. M. *Angew. Chem.* **1988**, *100*, 91.
3. KAIFER, A. E.; GOMEZ-KAIFER, M. *Supramolecular Electrochemistry*, Wiley-VCH: Weinheim, **1999**.
4. BOULAS, P. L.; GOMEZ-KAIFER, M.; ECHEGOYEN, L. *Angew. Chem.* **1998**, *110*, 226.
5. LAMBERT, D. K. *Electrochim. Acta* **1996**, *41*(5), 623.
6. ANDERSON, A. B. *J. Electroanal. Chem.* **1990**, *280*, 37.
7. PHAM, D. T.; et al. *Chimia* **2009**, *63*(3), 115.
8. WANG, D.; WAN, L. J. *J. Phys. Chem. C* **2007**, *111*, 16109.
9. BAE, S.-E.; GEWIRTH, A. A. *Langmuir* **2006**, *22*, 10315.
10. SAFAROWSKY, C.; WANDELT, K.; BROEKMANN, P. *Langmuir* **2004**, *20*, 8261.
11. PHAM, D. T.; et al. *New J. Chem.* **2006**, *30*, 1439.
12. BREUER, S.; et al. *New J. Phys.* **2008**, *10*, 125033.
13. HAI, N. T. M.; et al. *Phys. Chem. Chem. Phys.* **2009**, *11*, 5422.
14. HUEMANN, S.; et al. *J. Phys. Chem. B* **2006**, *110*, 24955.
15. MAGNUSSEN, O. M. *Chem. Rev.* **2002**, *102*, 679.
16. HOMMES, A.; et al. *Surf. Sci.* **2003**, *547*(1–2), 239.
17. BROEKMANN, P.; et al. *J. Electroanal. Chem.* **2001**, *500*, 241.
18. BROEKMANN, P.; et al. *Surf. Sci.* **2002**, *517*(1–3), 123.
19. SARACINO, M.; et al. *Phys. Rev. B* **2009**, *79*, 115448.
20. LENNARTZ, M. *Vergleichende in-situ und ex-situ spektroskopische Untersuchungen elektrochemischer Grenzflächen, Dissertationsschrift, Rheinische Friedrich-Wilhelms-Universität, Bonn*, **2000**.
21. BRANDT, K.; et al. *J. Electroanal. Chem.* **2004**, *570*, 47.
22. LENNARTZ, M.; et al. *Surf. Sci.* **2000**, *461*, 98.
23. WILMS, M.; KRUFT, M.; BERMES, G.; WANDELT, K. *Rev. Sci. Instrum.* **1999**, *70*, 3641.
24. VOGT, M. R.; LACHENWITZER, A.; MAGNUSSEN, O. M. *Surf. Sci.* **1998**, *399*, 49.
25. SUGGS, D. W.; BARD, J. *J. Phys. Chem.* **1995**, *99*, 8349.
26. BROEKMANN, P.; HAI, N. T. M.; WANDELT, K. *Surf. Sci.* **2006**, *600*, 3971.
27. BIRD, C. L.; KUHN, A. T. *Chem. Soc. Rev.* **1981**, *10*, 49.
28. MONK, P. M. S. *The Viologens: Physicochemical Properties, Synthesis and Applocations of the Salts of 4,4′-Bipyridines*, Wiley: Chichester, **1998**.
29. PHAM, D. T.; et al. *J. Phys. Chem.* **2007**, *111*(44), 16428.
30. GENTZ, K.; BROEKMANN, P.; WANDELT, K. in preparation.
31. TOPACLI, A.; AKYÜZ, S. *Spectrochim. Acta* **1995**, *51A*(4), 633.
32. WIBERG, K. B.; et al. *J. Phys. Chem.* **1984**, *88*(24), 6067.
33. MAYER, D.; et al. *J. Electroanal. Chem.* **2002**, *524–525*, 20.
34. BRIENNE, S. H. R.; COONEY, R. P.; BOWMAKER, G. A. *J. Chem. Soc. Faraday Trans.* **1991**, *87*(9), 1355.
35. KITAMURA, F.; OHSAKA, T.; TOKUDA, K. *J. Electroanal. Chem.* **1993**, *353*, 323.
36. CHRISTENSEN, P. A.; HAMNETT, A. *J. Electroanal. Chem.* **1989**, *263*, 49.
37. OSAWA, M.; SUETAKA, W. *J. Electroanal. Chem.* **1989**, *270*, 261.
38. ITO, M.; SASAKI, H.; TAKAHASHI, M. *J. Phys. Chem.* **1987**, *91*, 3932.
39. OSAWA, M. *Bull. Chem. Soc. Jpn.* **1997**, *70*, 2861.

40. Rice, M. J. *Solid State Commun.* **1979**, *31*, 93.
41. Girlando, A.; Marzola, F.; Pecile, C. *J. Chem. Phys.* **1983**, *79*(2), 1075.
42. Korzeniewski, C.; Shirts, R. B.; Pons, S. *J. Phys. Chem.* **1985**, *89*, 2297.
43. Wasileski, S. A.; Weaver, M. J. *J. Electroanal. Chem.* **2002**, *524–525*, 219.
44. Ghoshal, S.; et al. *Spectrochim. Acta* **1988**, *44A*(7), 651.

# Chapter 10

# Vibrational Spectroscopy of the Ionomer–Catalyst Interface

**Ian Kendrick, Jonathan Doan, and Eugene S. Smotkin**
Department of Chemistry and Chemical Biology, Northeastern University, Boston, Massachusetts

## 10.1 INTRODUCTION

Ionomer membranes have a wide and rapidly growing range of applications [1–10]. Ionomers have a backbone structure with side chains terminated by an anion or cation exchange group. Nafion, a sulfonated tetrafluoroethylene copolymer, and its short side-chain derivative ionomer revolutionized low-temperature fuel cell development. Although there have been over 7500 publications on Nafion since 1975 [11], relatively few focus on the Nafion–metal interface, which is the single most important factor in fuel cell electrode kinetics. Electrocatalytic surfaces undergo dynamic structural transformations upon changes in potential. The active state of an ionomer–catalyst interface exists only during the catalysis [12]. Thus interfacial characterization within an operating device under relevant reaction conditions (operando) is advantageous. A fuel cell based on Nafion 117 is shown in Figure 10.1. The "11" designates an equivalent weight of 1100 g and the "7" the thickness in millimeters. Fuel cell catalysts (e.g., metal blacks, Pt/C, PtRu/C, etc.) are dispersed within alcoholic solutions of solubilized ionomer (ink) [13]. The inks can be cast onto Teflon and decal transferred to the Nafion membrane, painted directly onto the membrane [i.e., catalyst-coated membrane (CCM)] or deposited directly onto carbon gas diffusion layers (GDLs) to yield gas diffusion electrodes (GDEs). The GDE is hot pressed onto

*Vibrational Spectroscopy at Electrified Interfaces*, First Edition.
Edited by Andrzej Wieckowski, Carol Korzeniewski, and Björn Braunschweig.

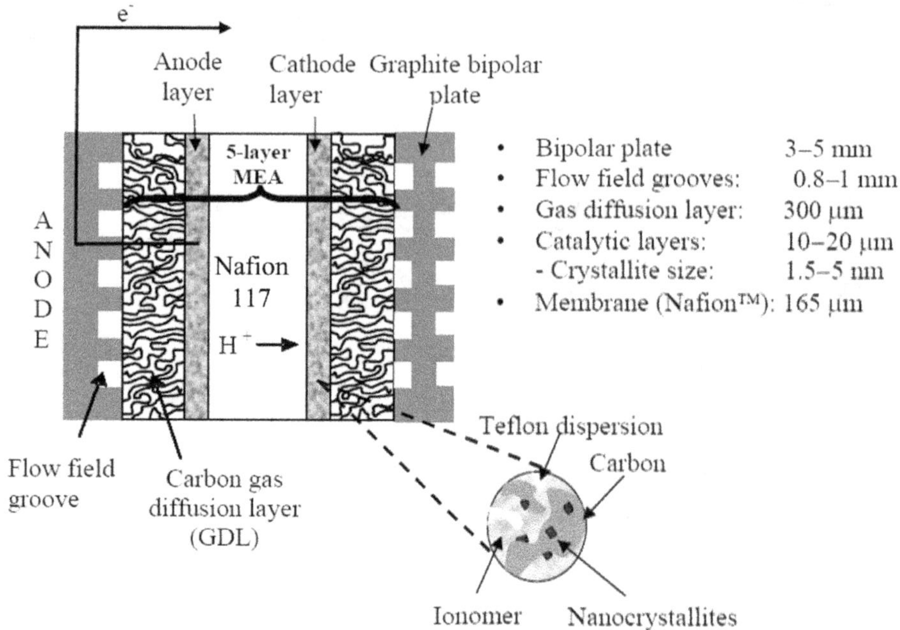


**Figure 10.1** Fuel cell with Nafion catalyzed on both sides, between two gas diffusion layers and flow fields for reactant distribution. Interior flow fields (bipolar plate on right side) enable stack assembly.

the Nafion membrane to yield a five-layer membrane electrode assembly (MEA). All MEA fabrication processes yield catalyst particles with an ionomer coating that serves as an electrolyte and also enhances the surface for catalysis [14]. The final step of MEA preparation occurs within the fuel cell during a "conditioning" process. The large variety of potentiostatic or galvanostatic conditioning processes likely address the same issue: The humidified gaseous flow streams hydrate the MEA and affect wetting of the catalyst surface by the ionomer (i.e., optimization of the ionomer–metal interface) [15]. Fabrication methods have been thoroughly reviewed by Gottesfeld [13, 16]. The ionomer–catalyst interface is beneath a carbon gas diffusion layer and the flow field, precluding the use of scanning probe techniques such as atomic force and scanning electron microscopy. Infrared (IR) spectroscopy is an ideal probe of interfacial structure because the nondestructive beam and detector are provided by "hardware at a distance." A slit in the gas diffusion layer over a CCM and a $CaF_2$ window in the flow field provide a path for the IR beam to the ionomer–metal interface of the catalytic layer. Cell designs for fuel cell operando IR (and X-ray) spectroscopy have been described and are commercially available [17–19]. This chapter highlights the benefits of augmenting operando spectroscopy with density functional theory (DFT) calculations and polarization-modulated IR spectroscopy of model surfaces.

## 10.2 VIBRATIONAL SPECTROSCOPY OF BULK NAFION

Nonlinear molecules of N atoms have 3N-6 normal modes of vibrations. Although a large molecular weight ionomer would be expected to have a complex spectrum, the attenuated total reflectance (ATR) spectrum of Nafion is simple (Fig. 10.2).

Bower and Maddams rationalize the simplicity by considering polymer spectra as those of ensembles of chemical repeat units interacting with the IR radiation [20]. A large number of chemical repeat units can fit within the length scales of the absorbed IR (~$2 \times 10^{-6}$ m). For a noncrystalline polymer with minimal geometrical regularity, a first approximation is to regard the chemical repeat units as separate molecules in their own environment: The number of normal modes can be approximated as $3n$ or less where $n$ is the number of atoms in the chemical repeat unit. The Nafion 55-atom chemical repeat unit (Fig. 10.2, inset) was used for DFT calculation of 159 Nafion normal modes [21]. DFT lines with intensity are superimposed upon the ATR spectrum to confirm the reasonableness of the selected repeat unit. Ideally, ionomer normal-mode analyses should be based on physical repeat units that are consistent with an overall model for ionomer structure. The cluster network model

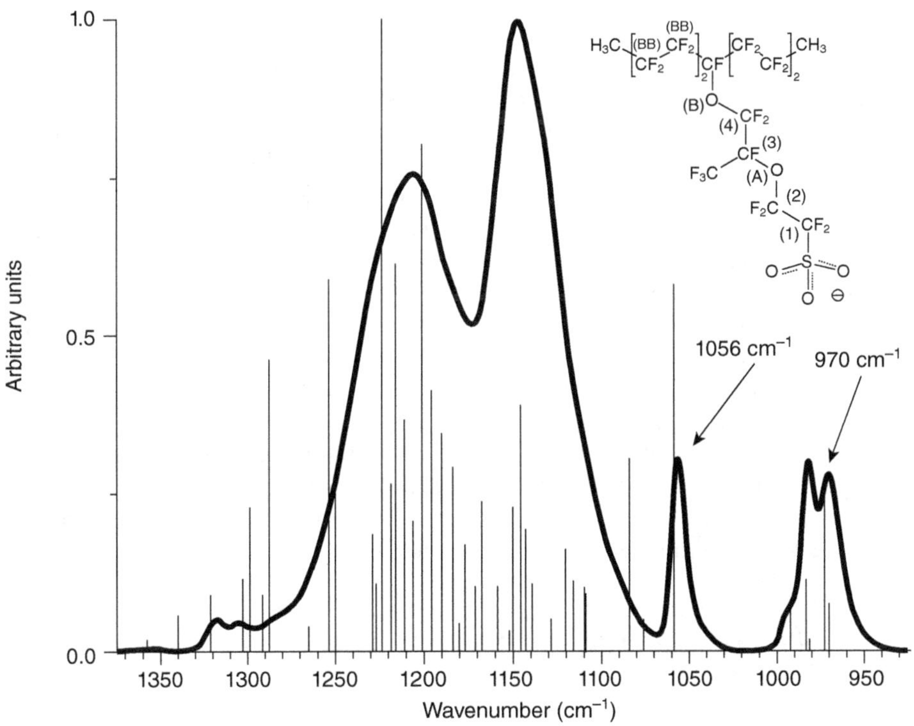


**Figure 10.2** Nafion ATR spectrum (Vertex 80V spectrometer, Bruker Optics Inc, Billerica, MA, and MIRacle ZnSe ATR accessory) and DFT-calculated normal modes (black lines).

of Hsu and Gierke [22], based on a very limited amount of experimental data, served as a starting point for more recent models. The Yeager and Steck three-phase model includes clusters with no defined geometrical definition and transitional interphases with lower degrees of order than spheres and cylinders [23]. Other models include a core–shell structure [24], a lamellar model [25], a sandwich-like model [26], and local order models [27, 28]. Mauritz and Moore [5] believe that the most plausible model is three phases (i.e., hydrophilic and hydrophobic phases separated by "fuzzy" boundary phases). Even though all models incorporate extended aqueous phases within the confines of an ionomeric framework, there is no consensus on the model. Proposed physical repeat unit(s) corresponding to a model, including effects of $\lambda$ [29] (water/sulfonate ratio) variations, are absent from the literature. An alternative is to develop more reasonable physical repeat units based on coordinates obtained from molecular dynamic calculations (vide infra). A realistic physical repeat unit should include interactions between exchange groups (beyond that of adjacent groups) and should vary with $\lambda$.

There is a normal-mode animation associated with each DFT line in Figure 10.2. Maestro (Schrodinger Inc., Portland, OR) converts Jaguar output files to vibrational mode animations. The full animations of selected normal modes of this report are in the supporting information of Kendrick et al. [30]. The analyses of the DFT-calculated spectra are facilitated by considering each normal mode as composed of mechanically coupled "pure modes." The normal modes of $CF_3SO_3^-$ and $CF_3OCF_3$ are examples of "pure modes" that can be mechanically coupled to model group modes involving the Nafion sulfonate and COC(A) groups, respectively. Snapshots of the $CF_3SO_3^-$ symmetric stretch [$\nu_s(A_1)$] and the $CF_3OCF_3$ asymmetric [$\nu_{as}(B_2)$] and rocking modes [$\rho_r(B_2)$] are shown in Figure 10.3, top row.

The $\nu_s(A_1)$, $\nu_{as}(B_2)$, and $\rho_r(B_2)$ are the "pure modes" discussed above. The center row shows that $983^*$ cm$^{-1}$ mode results from the coupling of the $\nu_s(A_1)$, $\nu_{as}(B_2)$, and $\rho_r(B_2)$ with the dominant mode being the $\nu_s(A_1)$. Webber et al. [21] and Kumari et al. [31] established that peaks at $969^{TR}$ cm$^{-1}$ and $970^{ATR}$ cm$^{-1}$ (referred to as $\overline{\nu}_{lf}$), respectively, are sensitive to ion exchange and to state of hydration. This enables the association of the $\overline{\nu}_{lf}$ peak with the $983^*$ cm$^{-1}$. The dominant pure mode of the $983^*$ cm$^{-1}$ is the symmetric stretch of the sulfonic acid group [i.e., $\nu_s(A_1)$]. The bottom row snapshots show that the dominant pure mode of the $1059^*$ cm$^{-1}$ peak is actually the $CF_3OCF_3$ $\nu_{as}(B_2)$ mode. The dominant pure modes of the ~1060 cm$^{-1}$ and $\overline{\nu}_{lf}$ are the $\nu_{as}(B_2)$ ether mode, and the $\nu_s(A_1)$ sulfonic acid mode, respectively, in contrast to decades of literature [21, 30, 32–34]. More importantly the ~1060 cm$^{-1}$ and $\overline{\nu}_{lf}$ peaks both include internal coordinates from both pure modes and cannot be described by single functional group assignments.

## 10.3 STATE OF HYDRATION OF NAFION: FTIR AND DFT ANALYSIS

The transmission spectra of Nafion 112 were obtained in the fully hydrated state and under vacuum. The detector saturates between 1100 and 1350 cm$^{-1}$ due to the strong

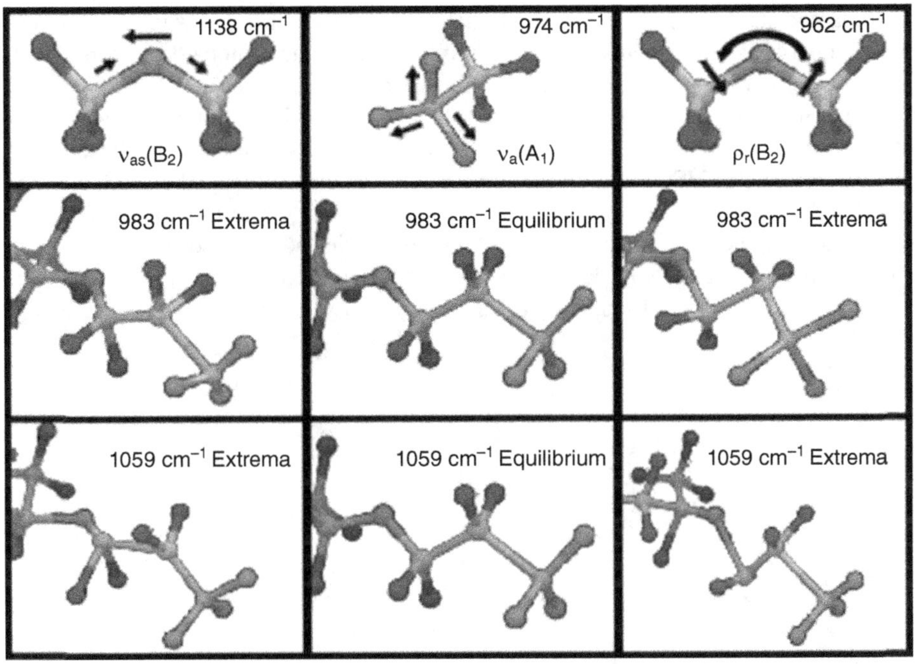


**Figure 10.3** Maestro normal-mode animation snapshots: (*Top row*) Small molecule "pure" modes; (*middle row*) 55-atom repeat unit, 983 $cm^{-1}$; (*bottom row*) 55-atom repeat unit 1059 $cm^{-1}$, AVIs in the supporting information of Kendrick et al. [30].

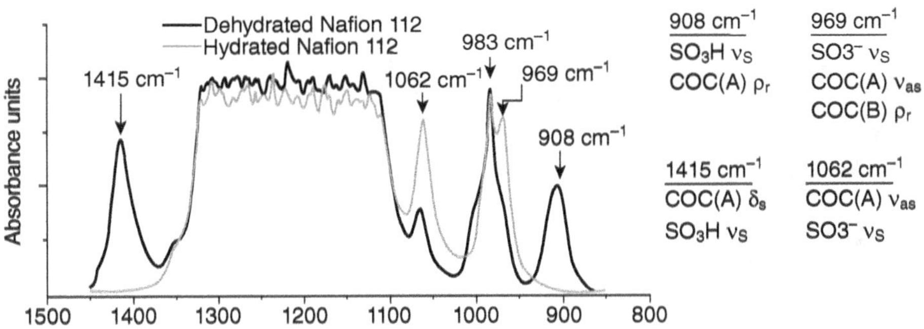


**Figure 10.4** Transmission IR spectra of Nafion 112 showing the evolution of 1415 and 908 $cm^{-1}$ bands upon dehydration.

absorption of the $-(CF_2)_n$ backbone and side-chain functional groups. Figure 10.4 emphasizes four peaks that vary with the state of hydration. The $1062^{TR}$ and $969^{TR}$ $cm^{-1}$ peaks of hydrated Nafion vanish concurrently with evolution of peaks at $1415^{TR}$ and $908^{TR}$ $cm^{-1}$ during dehydration. As mentioned above, the $1062^{TR}$ and $969^{TR}$ $cm^{-1}$ peaks are group modes derived from the COC(A) and $SO_3^-$ pure modes. As Nafion is dehydrated, the local symmetry of the sulfonic acid group changes from

$C_{3v}$ (unassociated) to $C_1$ (protonated): The $1062^{TR}$ and $969^{TR}$ $cm^{-1}$ peaks give way to the $908^{TR}$ and $1415^{TR}$ $cm^{-1}$ peaks. The latter modes involve the protonated sulfonic acid.

The dehydration of Nafion has been modeled by unrestricted DFT [35, 36]. The hybrid X3LYP [37] functional was used for geometry optimization and calculations of the normal-mode frequencies and corresponding IR spectra for triflic acid, the Nafion side chain (NSC), and the entire 55-atom repeat unit (Fig. 10.5) [21].

The X3LYP extension of the B3LYP [32] functional yields more accurate heats of formation. The all-electron 6-311G**++ Pople triple-ζ basis set was used in all calculations ("**" and "++" denote polarization [33] and diffuse [34] basis set functions, respectively). The calculations were done at λ [29] values from 0 to 10. Figure 10.6

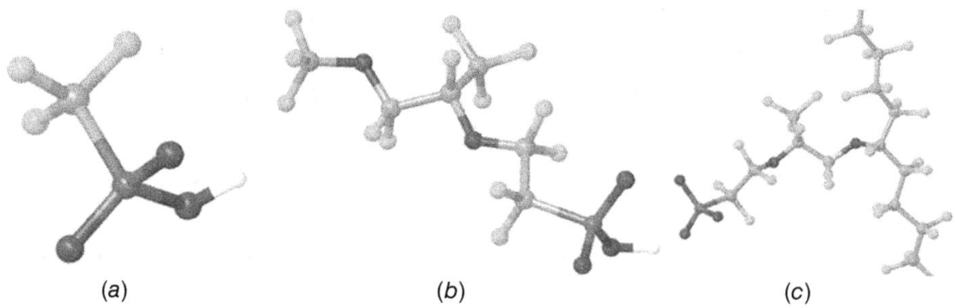


**Figure 10.5** Maestro image of (*a*) triflic acid, (*b*) Nafion side chain, and (*c*) 55-atom Nafion repeat unit.

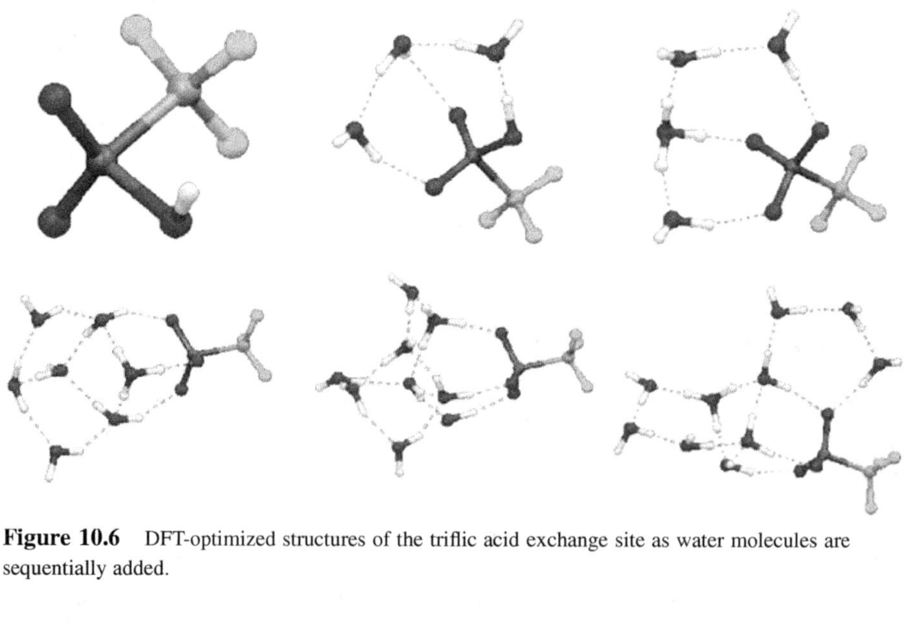

**Figure 10.6** DFT-optimized structures of the triflic acid exchange site as water molecules are sequentially added.

**Table 10.1** Comparison of Acidic Proton O–H distances, and the O–O Distances about the Central Hydrogen Ion, for Triflic Acid and the Nafion Side Chain (NSC)

| | SO–H (Å) | | SO–H–OH (Å) | |
|---|---|---|---|---|
| $\lambda$ | Triflic Acid | NSC | Triflic Acid | NSC |
| **0** | 0.970 | | | |
| **1** | 1.004 | | 2.628 | |
| **2** | 1.037 | | 2.531 | |
| **3** | 1.074 | 1.077 | 2.465 | 2.464 |
| **4** | **1.522** | **1.531** | **2.526** | **2.541** |
| **5** | 1.481 | | 2.503 | |
| **6** | 1.49 | | 2.519 | |
| **7** | 1.511 | 1.526 | 2.523 | 2.533 |
| **8** | 1.523 | | 2.539 | |
| **9** | 1.523 | 1.532 | 2.539 | 2.545 |
| **10** | **3.677** | | **4.372** | |

shows DFT-optimized structures of the triflic acid exchange site as water molecules are sequentially included in the solvation sphere. A dielectric was not used because the effect would be small compared to that of the sequential addition of water molecules. The triflic acid calculations reveal a threshold ($\lambda_d$) where the SO–H bond dissociates (Fig. 10.6, top right), and a $\lambda_{i\text{-}o}$, where the $H_3O^+$ loses a direct hydrogen bond to the exchange site (i.e., moves from the inner to outer solvation sphere). Paddison used B3LYP/6-31G** to calculate $\lambda_d$ and $\lambda_{i\text{-}o}$ of 3 and 6 (Table 10.1, bold values), respectively [38]. Our higher values of 4 and 10, respectively, result from the use of diffuse basis set functions, which are more realistic when hydrogen bonding is important (Table 10.1). Spitznagel et al. [39] observed significant changes to optimized geometries involving anions and proton affinity when using diffuse functions.

Extension of the $\lambda$-dependent calculations to the Nafion side chain and the 55-atom repeat unit confirmed that triflic acid is an adequate model for determination of $\lambda_d$ and $\lambda_{i\text{-}o}$. The similarities between the O–H and O–O distances of solvated triflic acid and the Nafion side chain, versus $\lambda$, suggests that at the levels of hydration calculated, the sulfonic acid group behavior is adequately modeled by triflic acid. The SO–H and SO–H–OH distances versus $\lambda$ are given in Table 10.1.

The reduction of $\lambda$ can affect proton transport in two ways. Ionic conductivity is governed by carrier density and mobility. As $\lambda$ drops below $\lambda_{i\text{-}o}$, the activity coefficient for protons would be expected to decrease as they enter the first solvation sphere. Below $\lambda_d$ the protons are largely associated, further lowering the proton activity. In addition, the lowering of $\lambda$ below threshold values diminishes the "bulk" water phase and thus the contribution of Grotthus hopping to the proton transport mechanism. Thus the carrier densities and mobilities diminish as $\lambda$ decreases.

## 10.4 OPERANDO SPECTROSCOPY OF IONOMER–Pt INTERFACE

Operando IR spectroscopy of the ionomer–metal interface is enabled by a small slit in the GDL, which enable beam access to the catalyst-coated membrane. Adsorption processes at the interface can be studied as a function of fuel cell operating temperature and potential. Adsorbates can cause surface restructuring where surface atoms can be entirely displaced in order to make stronger surface-adsorbate bonds [40]. Ionomer functional groups are also adsorbates. It addition to serving as an electrolyte, the ionomer has also been shown to enhance electrocatalysis [41]. The relationship between the nature of adsorbing ionomer functional groups and electrode kinetics is not well understood and warrants investigation.

The current cell design (Fig. 10.7) combines features of the Viswananthan cell [42]. with the specular reflectance infrared cell reported by Fan et al. [18]. The infrared/X-ray absorption spectroscopy (IR-XAS) cell enables operando XAS (in transmission and fluorescence) and specular reflectance Fourier transform infrared (FTIR) spectroscopy. The top flow field accommodates a $CaF_2$ window for IR reflectance studies [19, 43]. A pin-style upper flow field optimizes flow distribution around the $CaF_2$ window inset. The cell slides into a Pike Technologies (Watertown, WI) diffuse reflectance accessory.

Figure 10.8 shows a cyclic voltammogram background (solid) and the CO stripping wave (dashed) obtained with the cell at 50°C. The CO stripping wave extends from 600 to 900 mV. The potential-dependent spectra of CO adsorbed at 100 mV versus normal hydrogen electrode (NHE) are shown in Figure 10.9. The corresponding Stark tuning plots for CO adsorbed at 100, 200, 300, and 400 mV (Fig. 10.9) show the effects of co-adsorption on the CO stretching frequencies. The Stark tuning rates of this work, $6.8 \pm 0.6$ $cm^{-1}/mV$, are within the range of 2.5 $cm^{-1}/V$ to 18.3 $cm^{-1}/V$ of previous operando Stark tuning studies of direct methanol fuel cells at negative potentials of 0.5 V versus NHE [19].

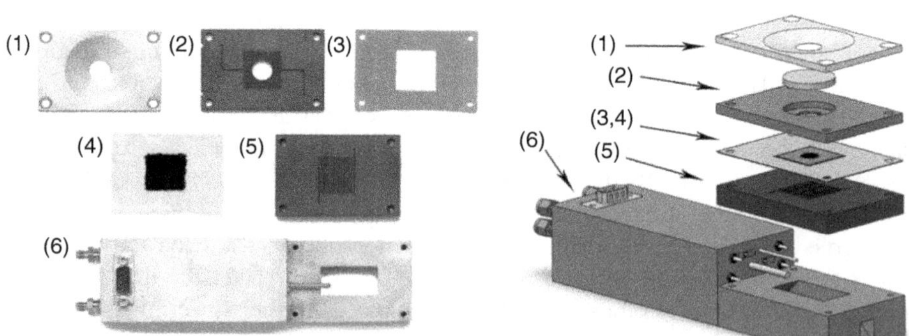


**Figure 10.7** Photograph (*left*) and exploded diagram (*right*) of the IR-XAS cell and components: (1) top plate, (2) top flow field, (3) membrane electrode assembly gasket, (4) membrane electrode assembly, (5) lower flow field, and (6) slider housing.

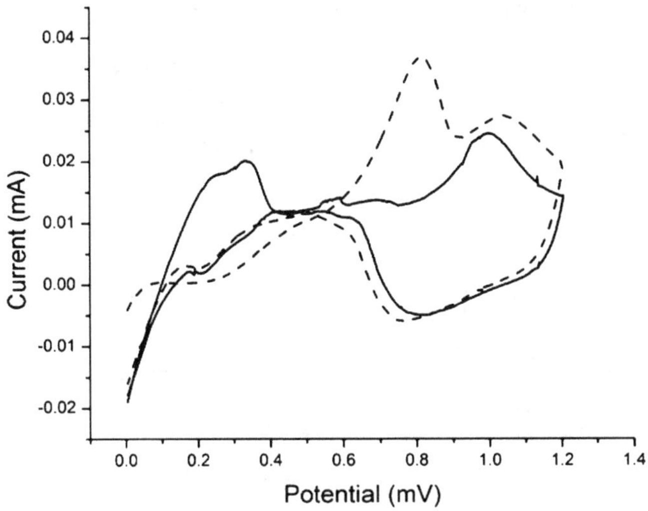


**Figure 10.8** Cyclic voltammetry (10 mV/s) of fuel cell working electrode (5 $cm^2$) under humidified $N_2$ at 50°C (solid). Humidified $H_2$ (50 sccm) at the counterreference. CO stripping wave (dashed).

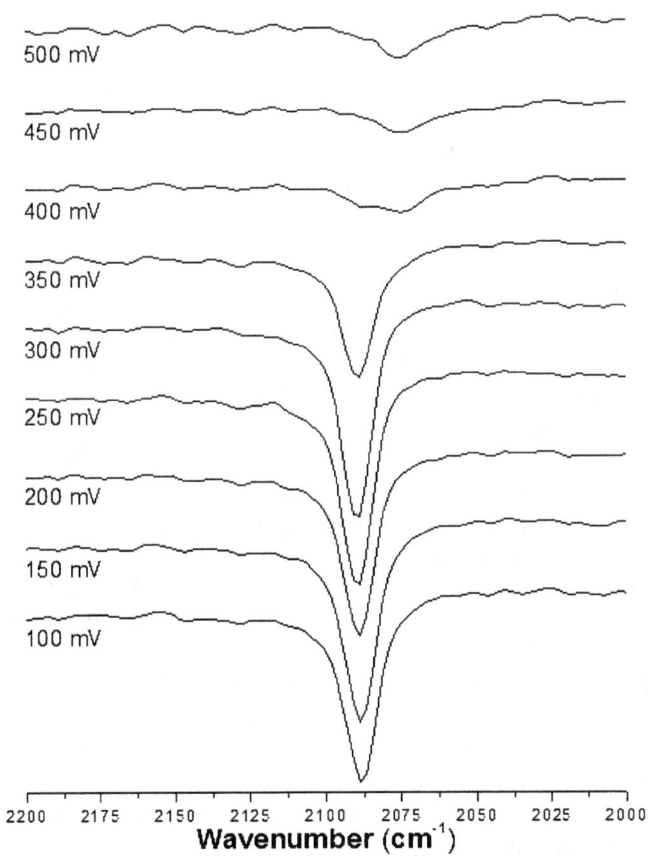


**Figure 10.9** Potential-dependent spectra of CO on Pt; adsorption potential of 100 mV operating at 50°C.

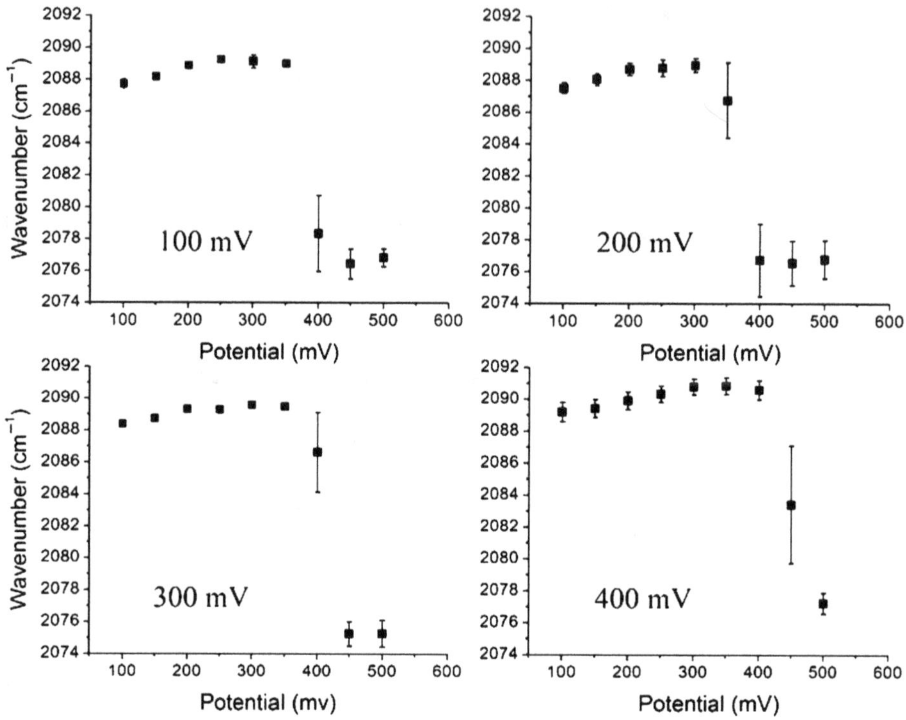


**Figure 10.10** Stark tuning plots of linearly adsorbed CO on Pt at 50°C.

Stamenkovic et al. reported similar Stark tuning behavior of CO on a Pt(111) electrode in $H_2SO_4$ [44]. The complex Stark tuning of $CO_{ads}$ stretching frequencies ($\nu_{CO}$) in Figure 10.10 are attributed to the compression/dissipation of $CO_{ads}$ islands: A linear Stark tuning region has a subtle $\nu_{CO}$ blue shift from the extrapolated linear region. We believe this shift is due to adsorption of ionomer functional groups such as the sulfonic acid and possibly the $-CF_3$ group. The subtle upturn is followed by a precipitous drop in $\nu_{CO}$, followed by an upturn. The potential where $\nu_{CO}$ precipitously drops ($E_{onset}$) correlates with $CO_{ads}$ oxidation. $CO_{ads}$ oxidation, induced by $OH^-$ adsorption, diminishes dipole–dipole coupling and thus precipitously decreases $\nu_{CO}$. The upturn is ascribed to increased adsorption of Nafion functional group species relative to $OH^-$. This reestablishes repulsive dipole interactions that compress $CO_{ads}$ islands and increases $\nu_{CO}$.

Figure 10.11 shows the dependency of $E_{onset}$ on the adsorption potential ($E_{ads}$). At 400 mV, $E_{onset}$ and $E_{ads}$ coincide because at higher potentials $CO_{ads}$ favors adsorption sites with higher enthalpies of adsorption (e.g., steps and kinks) [45, 46], and there is no thermodynamic force driving migration to sites of lower $\Delta H_{ads}$. At an $E_{ads}$ of 300 mV, the $E_{onset}$ exceeds $E_{ads}$ by 60 mV because the distribution of adsorption sites are more heterogeneous and there is a driving force for $CO_{ads}$ on terraces to migrate to steps, kinks, and ad atoms. Thus the difference between $E_{onset}$ on $E_{ads}$

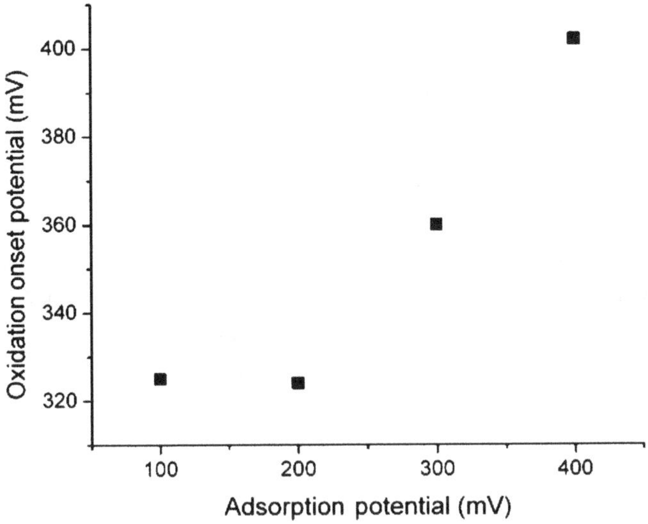


**Figure 10.11** Co-electrooxidation onset potentials versus adsorption potentials.

would be expected to increase as $E_{ads}$ is decreased. At $E_{ads}$ of 200 mV and lower, the $E_{onset}$ is leveled to 320 mV.

## 10.5 POLARIZATION-MODULATED IR REFLECTION ABSORPTION SPECTROSCOPY OF NAFION–Pt INTERFACES

Polarization modulation–infrared reflection adsorption spectroscopy (PM–IRRAS) enhances vibrational modes of functional groups ordered by the metal substrate. Figure 10.12 shows the ATR spectrum (dotted) and the PM–IRRAS spectra of the Nafion–H/Pt (gray line) and Nafion–Li/Pt (black line) interfaces. The $1056^{ATR}$ $cm^{-1}$, the Nafion–H/Pt at $1061^{PM\text{-}I}$ $cm^{-1}$, and the Nafion–Li/Pt at $1077^{PM\text{-}I}$ $cm^{-1}$ (Fig. 10.12) confirm that cations (e.g., $H^+$, $Li^+$) at the Pt surface induce frequency *shifts*, as do extent of hydration [21] and ion exchange of bulk Nafion [47]. Calculated IR normal-mode lines are superimposed upon the spectra. The lines most enhanced by the PM–IRRAS, highlighted in bold, point to the functional groups responsible for the self-assembly of Nafion onto the Pt surface.

A robust convention for correlating PM–IRRAS enhanced peaks to the calculated normal modes would enable identification of functional groups ordered by the Pt surface: The association of observed PM–IRRAS peaks with DFT peaks, assigned by visualization of mechanically coupled internal coordinates [21, 48], provides the basis for such a convention. Normal-mode coordinate animations explicitly show how functional groups are mechanically coupled to yield group modes. The calculated internal coordinates are viewed in the context of the previously discussed pure

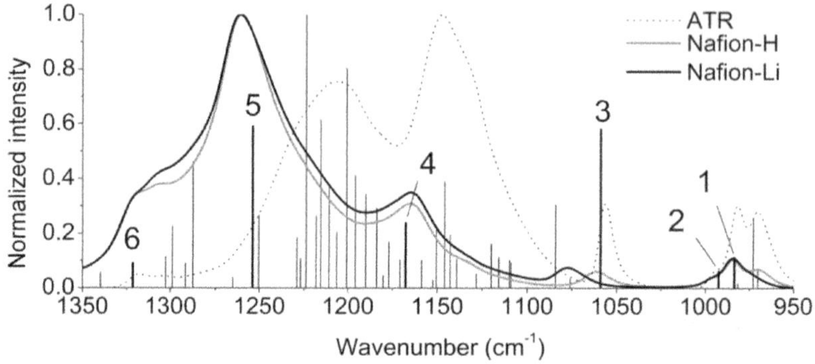


**Figure 10.12** Nafion ATR spectrum (dashed). PM–IRRAS spectra of Nafion–H/Pt interface (solid gray) and Nafion–Li/Pt interface (solid black).

modes. The pure modes serve as the basis elements for assigning DFT-calculated normal modes associated with observed peaks. Figure 10.13 shows the Maestro animation snapshots and ensuing assignments of the six selected normal modes. The atoms contributing to the dominating motion (black circles) and the next most significant atom motions (dotted circles) comprise pure modes that form the basis for the assignments. The correlation of the DFT to PM–IRRAS peaks (Fig. 10.12) and the resulting assignments in terms of the mechanically coupled pure modes are tabulated in Table 10.2.

The pure-mode assignments elucidate functional groups ordered by the Pt surface. The pure-mode assignments are supported by the excellent overlap of the PM–IRRAS peaks with the DFT peak positions. Consider the DFT and PM–IRRAS peaks of Table 10.2 in the context of the bulk-Nafion ATR (Fig. 10.2): The $1056^{ATR}$ and $970^{ATR}$ $cm^{-1}$ peaks shift with alterations of the sulfonate group environment. Less explicit than the $970^{ATR}$ and $1056^{ATR}$ $cm^{-1}$ peaks are bulk-Nafion modes, which are convoluted within the Nafion ATR envelop region (1100–1300 $cm^{-1}$). The animation of the $1254^{*}$ $cm^{-1}$ (Fig. 10.12, line-5), associated with PM–IRRAS peaks at $1260^{PM\text{-}I}$ $cm^{-1}$, suggests that the $-CF_3$ internal coordinates dominate that normal mode. The insensitivity of the $1254^{*}$ $cm^{-1}$ peak, to ion exchange, suggests that the internal coordinates are essentially uncoupled to the sulfonate group. The $1260^{PM\text{-}I}$ $cm^{-1}$ intensity is over 10-fold greater than that of the cluster of peaks (i.e., associated with theoretical lines 1 and 2) that are mechanically coupled to the sulfonate pure mode: The $-CF_3$ is a co-adsorbate of comparable importance to the sulfonate group in the self-assembly of Nafion onto Pt. This is supported by the repeat-unit-atom atomic charges obtained by Mulliken population [49] analysis. Table 10.3 shows the average charges of the backbone, side chain, $-CF_3$ fluorine atoms, and the sulfonate oxygen atoms. The charges for chemically equivalent atoms (e.g., $-CF_3$ fluorine and sulfonate oxygen atoms) differ because the calculations correspond to the lowest energy Newman projections where the atomic environments are different for chemically equivalent atoms because of the absence of symmetry in the full molecule. The

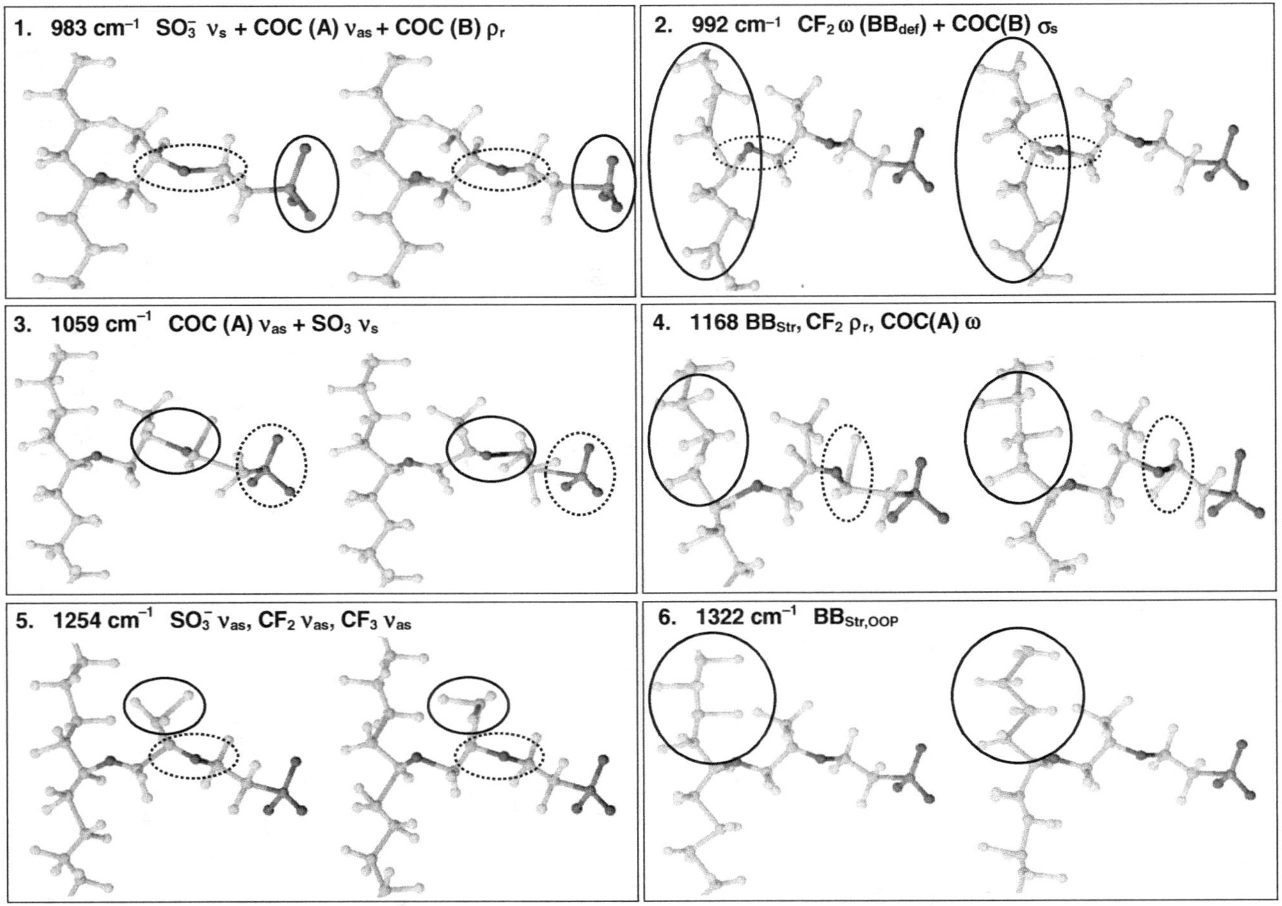


**Figure 10.13** Normal-mode animation snapshots of the Nafion side-chain anion and backbone fragment. Left and right views are extrema positions of the vibrational mode. Functional groups associated with the dominant internal coordinates and next most significant motions are designated by solid and dotted boundary lines, respectively.

**Table 10.2** PM-IRRAS and DFT IR Adsorption Peaks and Assignments

| | Wavenumber ($cm^{-1}$) | | |
|---|---|---|---|
| | PM-IRRAS | DFT | Pure Mode Components |
| 1 | 971 | 983 | $SO_3^-$ $\nu_s$, COC(A) $\nu_{as}$, COC(B) $\rho_r$ |
| 2 | 984 | 992 | $BB_{Str}$, $BB_{Str,OOP}$, COC(A) $\omega$ |
| 3 | 1061 | 1059 | COC(A) $\nu_{as}$, $SO_3^-$ $\nu_s$ |
| 4 | 1164 | 1168 | $BB_{Str}$, $CF_2$ $\rho_r$, COC(A) $\omega$ |
| 5 | 1260 | 1254 | $CF_3$ $\nu_{as}$, COC(A) $\delta_s$, COC(B) $\delta_s$ |
| 6 | 1322 | 1322 | $BB_{Str,OOP}$ |

$\nu_s$, symmetric stretching; $\nu_{as}$, asymmetric stretching; $\omega$, wagging; $\delta_s$, scissoring; $\tau$, twisting; $\rho_r$, rocking; Str, unspecified stretching; Str, OOP: stretching out-of-plane.

**Table 10.3** Average Partial Charges of Selected Nafion Segments

| Parameter | Backbone (F) 13 atoms | Side Chain (F) 8 atoms | $CF_3$ (F) 3 atoms | Sulfonate (O) 3 atoms |
|---|---|---|---|---|
| Avg. partial charge | −0.0665 | −0.0816 | −0.0876 | −0.4879 |
| Standard deviation | 0.027 | 0.055 | 0.013 | 0.012 |

chemically equivalent atoms have smaller charge standard deviations as would be expected. The average charge of the $–CF_3$ fluorine atoms are about 18% that of the sulfonate oxygens. The aggregate of operando spectroscopy, ATR spectroscopy of Nafion 117, PM–IRRAS of Nafion spin coated onto Pt and DFT-calculated normal modes suggest a model for the Pt-Nafion interface the includes the Nafion–$CF_3$ group as an important co-adsorbate at the ionomer–Pt interface.

Gaussian 03 Viewer (Gaussian, Wallingford, CT) was used to construct a flexible (i.e., enables dihedral angle rotation while maintaining the native functional group bond angles) 2-repeat unit model. The $–CF_3$ and $–SO_3^-$ groups, oriented with planes defined by their fluorine and oxygen atoms parallel to a Pt surface, effect ordering of the $–CF_2–$ backbone segments with respect to the Pt surface (Fig. 10.14). Thus, peaks due to the backbone $–CF_2–$ would be expected in the PM–IRRAS: The peak at $1164^{PM\text{-}I}$ $cm^{-1}$ is associated with the $1168^*$ $cm^{-1}$ peak. The $1168^*$ $cm^{-1}$ animation shows that the $–CF_2–$ backbone internal coordinates dominate the mode, supporting the suggestion of ordered $–CF_2–$ groups. The numbers associate DFT-calculated IR peaks (line 1–6, Fig. 10.12) and associated PM–IRRAS peaks with regions of order induced by the $–CF_3$ and $–SO_3^-$ functional group adsorbates.

The Stark tuning data (Fig. 10.10), PM–IRRAS and DFT calculations support the model for the self-assembly of Nafion onto Pt. The details of exactly how adsorbed $CF_3$ functional groups influence the operando Stark tuning curves (Fig. 10.10) are

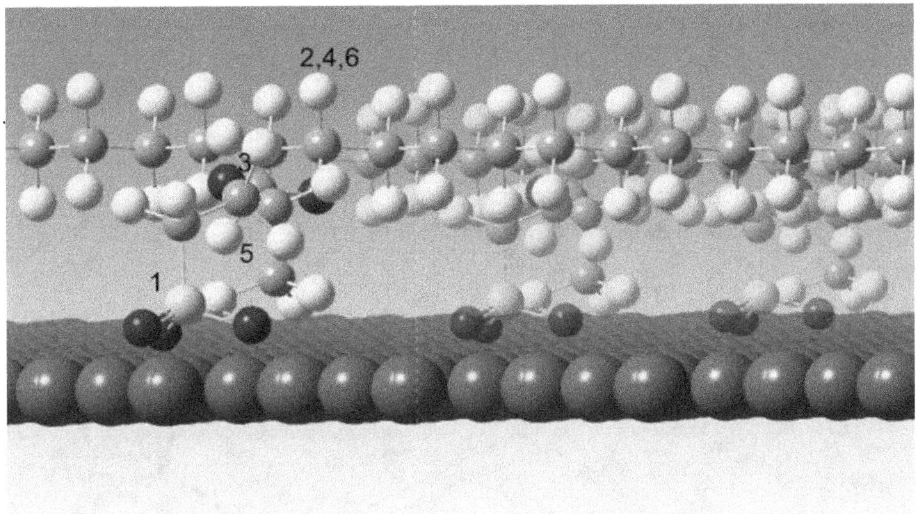


**Figure 10.14** Gaussian 03 viewer Nafion–Pt interface model.

not yet established. The low density of functional group adsorption sites, relative to the number of backbone $-CF_2-$ groups suggests an explanation as to why Nafion is observed to enhance electrode processes [8, 14]. The anchoring functional groups represent about 25% of the footprint of the adsorption model. Neutron reflectivity experiments may provide information concerning water within free volume between the metal surface and the Nafion backbone [50].

Figure 10.15 summarizes how the methodology described in this chapter can be used to develop models for ionomer–metal interfaces. The ATR of the bulk ionomer is obtained (Fig. 10.15*a*). A chemical or physical ionomer repeat unit is selected (middle of figure) and 3*n*-6 normal-mode lines are calculated and overlaid upon the ATR spectrum (Fig. 10.15*b*) to ensure reasonableness of the selected repeat unit. Important DFT lines are selected and assigned based on visualization of the normal mode animations, ion exchange, and state-of-hydration spectroscopy (Fig. 10.15*c*). PM–IRRAS is used to determine what DFT lines are relevant to the adsorption process (Fig. 10.15*d*). Animations are viewed to assign the relevant lines and identify functional groups ordered by the surface. A Gaussian viewer is used to build the model based on identified adsorbing functional groups (Fig. 10.15*f*).

## 10.6 FUTURE WORK

Ideally, a realistic physical repeat unit should be used instead of a chemical repeat unit. One strategy is to use coordinates obtained from molecular dynamic calculations. This should provide a repeat unit that varies with the value of $\lambda$ and considers interactions between nonadjacent exchange group sites. The second improvement requires removal of the subjectivity that comes with visualization of animations.

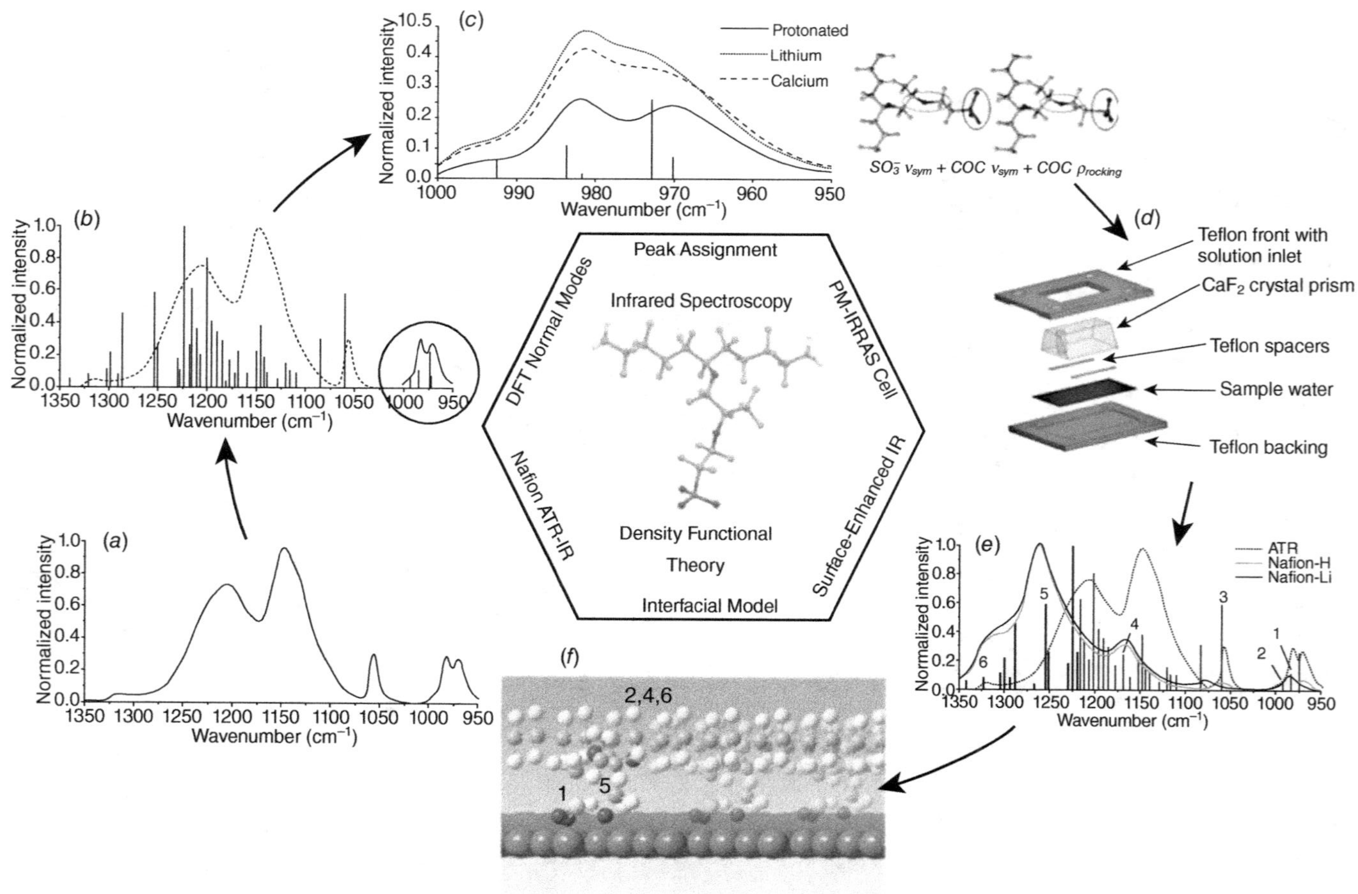


**Figure 10.15** Depiction of the process of correlating DFT calculations and infrared spectra to form an elegant interfacial model of Nafion adsorption on a Pt surface.

This is now being addressed by defining functional groups as subsets of the normal-mode generalized coordinates. The contribution of a functional group to a normal mode can then be quatified using the subset coordinates.

## REFERENCES


1. Hickner, M. A. *Mat. Today* **2010**, *13*, 34.
2. Hamrock, S. J.; Yandrasits, M. A. *Polym. Rev.* **2006**, *46*, 219.
3. Li, S. K.; Zhu, H.; Higuchi, W. I. *Pharma. Res.* **2006**, *23*, 1857.
4. Allegrezza, A. E., Jr.; Parekh, B. S.; Parise, P. L.; Swiniarski, E. J.; White, J. L. *Desalination* **1987**, *64*, 285.
5. Mauritz, K. A.; Moore, R. B. *Chem. Rev.* **2004**, *104*, 4535.
6. Elabd, Y. A.; Napadensky, E.; Walker, C. W.; Winey, K. I. *Macromolecules* **2006**, *39*, 399.
7. Elabd, Y. A.; Hickner, M. A. *Macromolecules* **2011**, *44*, 1.
8. Ploense, L.; Salazar, M.; Gurau, B.; Smotkin, E. S. *J. Am. Chem. Soc.* **1997**, *119*, 11550.
9. Liu, Z. J.; Arnold, R. G.; Betterton, E. A.; Smotkin, E. *Environ. Sci. Technol.* **2001**, *35*, 4320.
10. Hernandez-Pagan, E. A.; Vargas-Barbosa, N. M.; Wang, T. H.; Zhao, Y. X.; Smotkin, E. S.; Mallouk, T. E. *Energy Environ. Sci.* **2012**, *5*, 7582.
11. ISI Web of Knowledge Home Page, 2009.
12. Topsoe, H. *J. Catal.* **2003**, *216*, 155.
13. Wilson, M. S.; Gottesfeld, S. *J. Appl. Electrochem.* **1992**, *22*, 1.
14. Liu, L.; Viswanathan, R.; Liu, R. X.; Smotkin, E. S. *Electrochem. Solid State Lett.* **1998**, *1*, 123.
15. Viswanathan, R.; Hou, G.; Liu, R.; Bare, S. R.; Modica, F.; Mickelson, G.; Segre, C. U.; Leyarovska, N.; Smotkin, E. S. *J. Phys. Chem. B* **2002**, *106*, 3458.
16. Gottesfeld, S.; Zawodzinski, T. A. *Advances in Electrochemical Science and Engineering*, Wiley-VCH: **1997**, Vol. 5.
17. Fan, Q.; Pu, C.; Lay, K. L.; Smotkin, E. S. *J. Electrochem. Soc.* **1996**, *143*, L21.
18. Fan, Q. B.; Pu, C.; Smotkin, E. S. *J. Electrochem. Soc.* **1996**, *143*, 3053.
19. Bo, A. L.; Sanicharane, S.; Sompalli, B.; Fan, Q. B.; Gurau, B.; Liu, R. X.; Smotkin, E. S. *J. Phys. Chem. B* **2000**, *104*, 7377.
20. Bower, D. I.; Maddams, W. F. *The Vibrational Spectroscopy of Polymers*, Cambridge University Press: Cambridge, **1989**.
21. Webber, M.; Dimakis, N.; Kumari, D.; Fucillo, M.; Smotkin, E. S. *Macromolecules* **2010**, *43*, 5500.
22. Hsu, W. Y.; Gierke, T. D. *J. Memb. Sci.* **1983**, *13*, 307.
23. Yeager, H. L.; Steck, A. *J. Electrochem. Soc.* **1981**, *128*, 1880.
24. Fujimura, M.; Hashimoto, T.; Kawai, H. *Macromolecules* **1982**, *15*, 136.
25. Litt, M. H. *Polym. Prepr. (Am. Chem. Soc., DiV. Polym. Chem.)* **1997**, *38*, 2.
26. Haubold, H. G.; Vad, T.; Jungbluth, H.; Hiller, P. *Electrochim. Acta* **2001**, *46*, 1559.
27. Dreyfus, B.; Gebel, G.; Aldebert, P.; Pineri, M.; Escoubes, M.; Thomas, M. *J. Phys. France* **1990**, *51*, 1341.
28. Gebel, G.; Moore, R. B. *Macromolecules* **2000**, *33*, 4850.
29. Zawodzinski, T. A.; Springer, T. E.; Davey, J.; Jestel, R.; Lopez, C.; Valerio, J.; Gottesfeld, S. *Journal of the Electrochemical Society* **1993**, *140*, 1981.
30. Kendrick, I.; Kumari, D.; Yakaboski, A.; Dimakis, N.; Smotkin, E. S. *J. Am. Chem. Soc.* **2010**, *132*, 17611.
31. Kumari, D.; Kendrick, I.; Rivera, A.; Webber, M.; Dimakis, N.; Mion, T.; Smotkin, E. S. In preparation
32. Becke, A. D. *J. Chem. Phys.* **1993**, *98*, 5648.
33. Frisch, M. J.; Pople, J. A.; Binkley, J. S. *J. Chem. Phys.* **1984**, *80*, 3265.
34. Clark, T.; Chandrasekhar, J.; Spitznagel, G. W.; Schleyer, P. *J. Computat. Chem.* **1983**, *4*, 294.

35. Hohenberg, P.; Kohn, W. *Physical Review B* **1964**, *136*, B864.
36. Kohn, W.; Sham, L. J. *Physical Review* **1965**, *140*, 1133.
37. Xu, X.; Zhang, Q.; Muller, R. P.; Goddard, W. A. *Journal of Chemical Physics* **2005**, *122*, 14.
38. Paddison, S. J. *J. New Mat. Electrochem. Syst.* **2001**, *4*, 197.
39. Spitznagel, G. W.; Clark, T.; von Ragué Schleyer, P.; Hehre, W. J. *J. Comput. Chem.* **1987**, *8*, 1109.
40. Somorjai, G. A. *Introduction to Surface Chemistry and Catalysis*, Wiley: New York, **1994**.
41. Liu, L.; Viswanathan, R.; Liu, R.; Smotkin, E. S. *Electrochem. Solid-State Lett.* **1998**, *1*, 123.
42. Viswanathan, R.; Liu, R.; Smotkin, E. S. *Review of Scientific Instruments* **2002**, *73*, 2124.
43. Gasteiger, H. A.; Kocha, S. S.; Sompalli, B.; Wagner, F. T. *Appl. Cataly. B: Environ.* **2005**, *56*, 9.
44. Stamenkovic, V.; Chou, K. C.; Somorjai, G. A.; Ross, P. N.; Markovic, N. M. *J. Phys. Chem. B* **2005**, *109*, 678.
45. Kim, C. S.; Korzeniewski, C. *Analyt. Chem.* **1997**, *69*, 2349.
46. Yoshinobu, J.; Tsukahara, N.; Yasui, F.; Mukai, K.; Yamashita, Y. *Phys. Rev. Lett.* **2003**, *90*, 248301.
47. Cable, K. M.; Mauritz, K. A.; Moore, R. B. *J. Polym. Scie., Part B: Polym. Phys.* **1995**, *33*, 1065.
48. Warren, D. S.; McQuillan, A. J. *J. Phys. Chem. B* **2008**, *112*, 10535.
49. Mulliken, R. S. *J. Chem. Phys.* **1955**, *23*, 1833.
50. Dura, J. National Institute of Standards and Technology, personal communication, 2011.

# Chapter 11

# In Situ PM–IRRAS Studies of Biomimetic Membranes Supported at Gold Electrode Surfaces

**Annia H. Kycia,[1] ZhangFei Su,[1] Christa L. Brosseau,[2*] and Jacek Lipkowski[1]**
[1]Department of Chemistry, University of Guelph, Guelph, Ontario, Canada
[2]Department of Chemistry, Saint Mary's University, Halifax, Nova Scotia, Canada

## 11.1 INTRODUCTION

This chapter describes studies in which polarization modulation–infrared reflection absorption spectroscopy (PM–IRRAS) was combined with electrochemical methods to monitor potential-dependent behavior of biomimetic membranes at a single-crystal gold surface. Biomimetic membranes, composed of phospholipid, cholesterol, and glycolipid, were deposited at the Au(111) electrode surface by vesicle fusion and using the Langmuir–Blodgett and Langmuir–Schaefer transfer techniques. Electrochemical methods were employed to study the stability of the membrane, and PM–IRRAS was used to obtain information regarding the conformation, orientation, and hydration of the membrane. The objective of this chapter is to provide a detailed review of the PM–IRRAS studies of biomimetic membranes initiated by our group, and thus this chapter describes results obtained chiefly in our laboratory.

*Present address: Department of Chemistry, Saint Mary's University, Halifax, Nova Scotia, Canada

*Vibrational Spectroscopy at Electrified Interfaces*, First Edition.
Edited by Andrzej Wieckowski, Carol Korzeniewski, and Björn Braunschweig.

## 11.2 WHAT IS PM–IRRAS?

PM–IRRAS is a spectroscopic method used for the analysis of thin films adsorbed at smooth metal surfaces [1–4]. The high sensitivity demonstrated in PM–IRRAS with regard to the detection of surface species on metals is due to the fast modulation of the incident beam between two orthogonal directions: (a) parallel (*p* polarization) and (b) perpendicular (*s* polarization) to the plane of incidence of the infrared beam. According to the surface selection rules, when a smooth metal surface is illuminated with *p*-polarized light at a grazing angle of incidence, the electric field component perpendicular to the substrate is enhanced. Conversely, upon illumination with *s*-polarized light, the electric field component parallel to the substrate nearly vanishes at the surface (see Fig. 11.1) [5]. Therefore, the reflectivity of *s*-polarized light ($R_s$) is insensitive to the presence of a film of adsorbed molecules at a metal surface and can be used to obtain the background spectrum, while the reflectivity of *p*-polarized light ($R_p$) is sensitive to molecular adsorption at the surface (having a component of the transition dipole perpendicular to the substrate) and can be used to obtain the spectrum of the film. Therefore, in PM–IRRAS the difference between $|R_p - R_s|$ is measured and normalized by the average of the two reflectivities to give the reflection absorption spectrum $\Delta R/\langle R\rangle$ defined as [6]

$$\frac{\Delta R}{\langle R\rangle} = \frac{|R_p - R_s|}{(R_s + R_p)/2} \tag{11.1}$$

By determining the difference reflectivity surface signal $|R_p - R_s|$, any common-mode signal, such as that due to absorption by atmospheric $CO_2$ and $H_2O$, is removed such that only the surface species is detected.

PM–IRRAS can be applied for studying thin organic films at both metal–air and metal–solution interfaces. However, in the case of studying films at the metal–solution interface, the strong absorbance of IR radiation by aqueous electrolyte needs

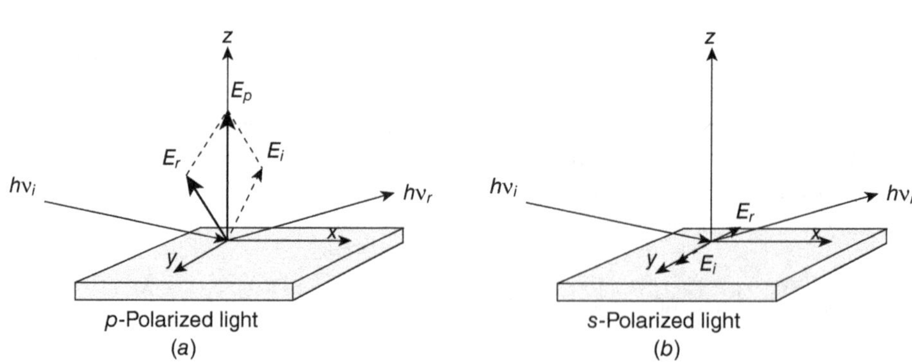


**Figure 11.1** Schematic of the reflection of *p*- and *s*-polarized light at the metal surface. Due to the constructive interference between incident and reflected beams, the electric field vector of *p*-polarized light is enhanced; and due to the destructive interference, the vector of *s*-polarized light nearly vanishes.

to be minimized. This is accomplished by employing a thin-layer cell configuration. In this configuration the single-crystal electrode [Au(111)] is pressed against an IR transparent window until it is separated from the window by a layer of electrolyte only several micrometers thick. The beam is transmitted through the window–solution interface, travels through the thin layer of the electrolyte, and is reflected from the electrode surface. The spectrum of the analyzed film is superimposed on the spectrum of the electrolyte. The magnitude of IR absorption due the electrolyte in the experimental spectrum is minimized by a proper choice of a solvent ($D_2O$ or $H_2O$) and by optimizing the alignment of the optical setup to achieve maximum cancelation of the absorbance for *p*- and *s*-polarized photons by solvent molecules within the thin-layer cavity. However, it is difficult to achieve a complete cancellation of the signal from molecules within the thin-layer cavity. Therefore, in the case of studying metal–solution interfaces, the background due to the aqueous electrolyte can never be completely removed from the $\Delta R/\langle R\rangle$ spectrum [6]. For this reason PM–IRRAS is an excellent technique to study insoluble films, biological membranes, or irreversibly adsorbed monolayers [7–11] as well as self-assembled thiols [12–16]. It is not a good technique to study reversible adsorption of soluble organic molecules that are present both at the surface and in the bulk of the thin-layer cavity. These problems were recently discussed by Blatt et al. [17].

In short, PM–IRRAS is based upon the metal surface selection rules where only *p*-polarized radiation is absorbed by molecules at the metal surface. Consequently, the active vibrations that can be detected in PM–IRRAS must have a component of the transition dipole moment in the direction normal to the surface. In order to obtain IR spectra of molecules adsorbed at a gold surface with a good signal-to-noise (S/N) ratio, the experimental conditions must be optimized. It is necessary to perform a correction for the nonideal performance of the photoelastic modulator (PEM) and the background due to IR absorption by the electrolyte. Once the spectra are corrected, the unique information concerning the orientation and conformation of the molecules adsorbed at the electrode surface can be determined as a function of applied electrode potential. The details of the PM–IRRAS experimental method and analysis of the spectra are discussed in the following sections.

## 11.3 EXPERIMENTAL

### 11.3.1 Spectroelectrochemical Cell

Figure 11.2 shows a block diagram of the layout of the PM–IRRAS setup modeled after that described by Barner et al. [18] and Green et al. [19]. Figure 11.3 shows a schematic of the PM–IRRAS setup assembled in the tabletop optical mount (TOM) box. The collimated infrared beam from the spectrometer [Fourier transform infrared (FTIR) spectrometer using a Michelson interferometer] is redirected using a flat mirror (Nicolet) and is refocused using a parabolic mirror ($f$ = 6 inches, Thermo Nicolet, Madison, WI). The beam then passes through a static polarizer (1-inch diameter, Specac) and then enters a photoelastic modulator (PEM) [Hinds Instruments (Portland, OR),

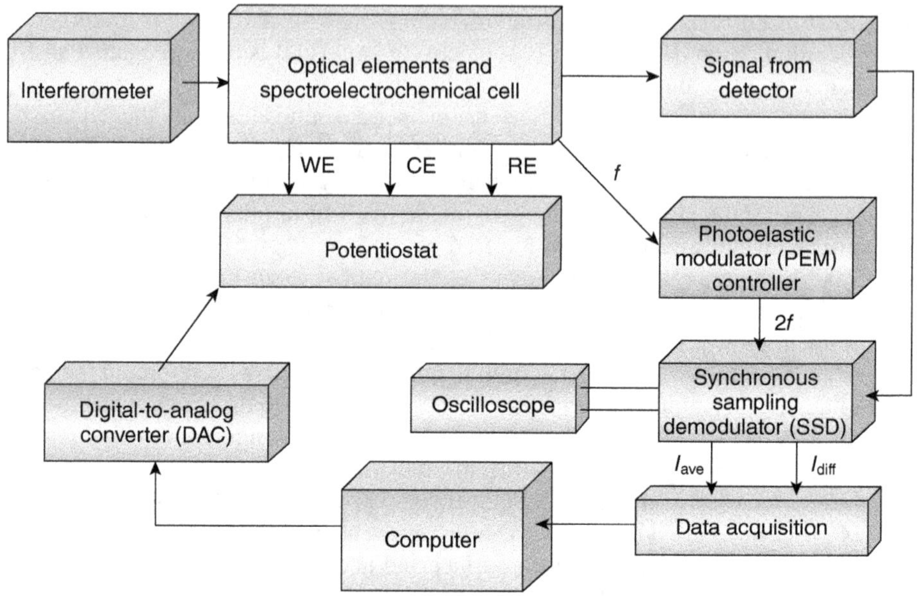


**Figure 11.2** Block diagram of the instrumentation used in the PM–IRRAS experiment (RE: reference electrode, WE: working electrode, CE: counter electrode). Adapted from Ref. [20].

PM-90 with an II/ZS50 ZnSe (50 kHz) optical head]. The beam is then reflected by the working electrode surface and is refocused with a ZnSe lens ($f = 1$ and 1.5-inch diameter with an antireflective coating, Janos Technology, Keene, NH) onto an MCT-A detector [time-resolved step-scan (TRS) 50-MHz bandwidth, Thermo Nicolet]. The signal from the detector was input to a synchronous sampling demodulator (SSD) (GWC Instruments, Madison, WI), while simultaneously a signal from the PEM controller was used as a reference signal and input to the SSD. An oscilloscope (Tektronix 2225, Beaverton, OR) was used to phase the SSD with the PEM. The signal was demodulated by the SSD, the outputs being the average and difference signals of the *p*- and *s*-polarized radiations, which were input into the computer.

The spectroelectrochemical cell made of glass was connected to the IR transparent window ($BaF_2$ or $CaF_2$ 1-inch equilateral prism, Harrick Scientific Corporation, Pleasantville, NY) via a Teflon flange. The working electrode was a disk-shaped Au(111) single crystal with a diameter of 15 mm connected to a gold wire, which was threaded through a Teflon sealer inserted into a glass tube. The glass tube was fed through a plastic screw, and, using a micrometer screw, this tube could be pushed very close to the IR window. A platinum foil served as the counter electrode and was sealed into the glass cell. The reference electrode (Ag/AgCl) was connected to the cell via a thin glass capillary. 0.1 M NaF in $H_2O$ or $D_2O$ was used as the supporting electrolyte. Prior to measurements, the cell was filled with electrolyte and purged with argon (BOC gases, Linde Canada, Guelph, Ontario) for approximately 1 h to remove oxygen from the solution. The potential was controlled by a potentiostat (EG&G, PAR Model 362, Princeton, NJ) using in-house software, an Omnic macro

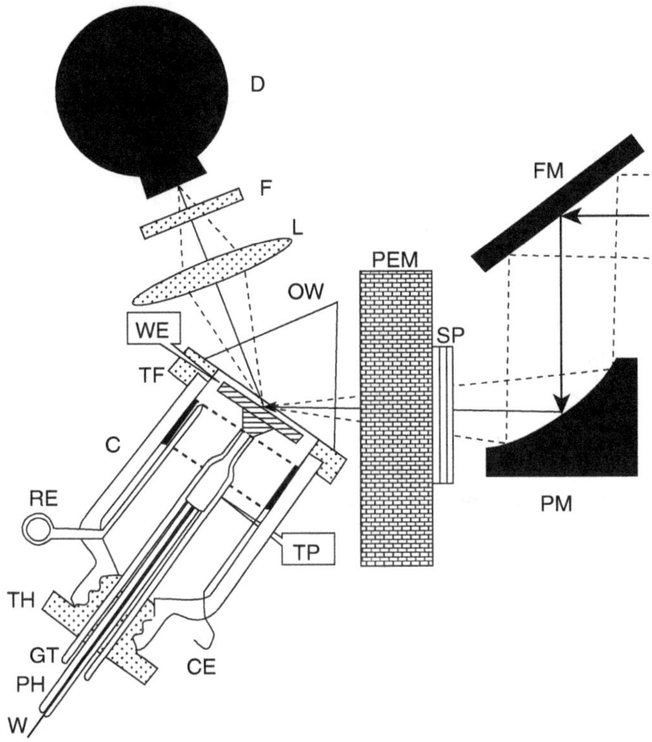


**Figure 11.3** Schematic of the experimental PM–IRRAS setup showing the details of the glass spectroelectrochemical cell (FM, flat mirror; PM, parabolic mirror; SP, static polarizer; PEM, photoelastic modulator; OW, optical window; L, lens; F, optical filter; D, detector; WE, working electrode; TF, Teflon flange; C, cell; RE, reference electrode; TH, Teflon holder, GT, glass tube; PH, piston handle; W, gold wire; CE, counter electrode; TP, Teflon piston.) Adapted from Ref. [6].

and a digital-to-analog convertor (Omega, Stamford, CT). A Whatman FTIR pure gas generator was used to purge the FTIR and the TOM box with dry and $CO_2$-free air.

### 11.3.2 Optimization of Angle of Incidence and Thin-Layer Cavity

The intensity of the IR band of adsorbed molecules is proportional to the mean-squared electric field strength (MSEFS) of *p*-polarized light at the metal surface [5]. Therefore it is necessary to calculate the MSEFS to determine the experimental conditions for the best S/N for the IR spectra of adsorbed species. In the spectro-electrochemical experiment the electrode is separated from the IR window by a thin layer of electrolyte, such that reflection takes place in a three-phase system. The first medium is an infrared transparent window ($BaF_2$ or $CaF_2$), while the other two phases are electrolyte and the gold electrode. The optical properties of these phases

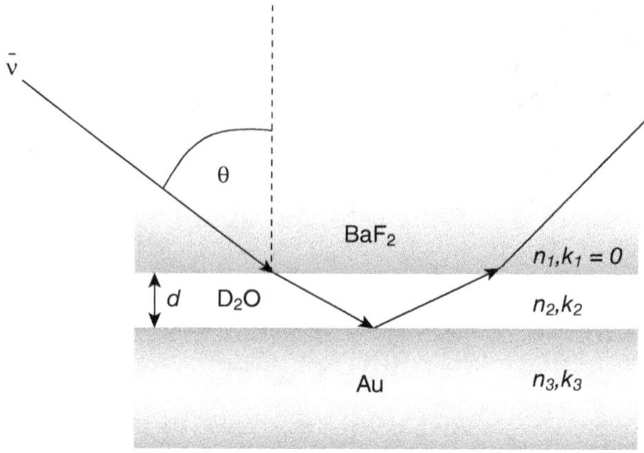


**Figure 11.4** Stratified medium used to model the thin-layer cell used in PM–IRRAS. The symbols $\bar{\nu}$ and θ represent the wavenumber and angle of incidence of the infrared beam, respectively; *n* and *k* represent the refractive index and attenuation coefficient of the medium, respectively; and *d* is the thickness of the electrolyte layer. Adapted from Ref. [20].

are defined by their optical constants, whose values are available in the literature [6]. The MSEFS of the IR photons at the metal surface can be calculated by solving Fresnel equations for a stack of homogeneous planar layers with parallel interfaces (stratified medium) by the transfer matrix method [21, 22]. The stratified medium for a gold (Au) electrode, $D_2O$ electrolyte, and a $BaF_2$ optical window is shown in Figure 11.4. A software program to solve Fresnel equations for this complex interface was written by Zamlynny [6, 23]. This software allows one to calculate the MSEFS using input parameters such as the wavenumber of IR photons, the angle of incidence of the incoming radiation, the thickness of the thin layer of electrolyte, and the optical constants of the different phases. Figure 11.5 shows a three-dimensional plot of the MSEFS as a function of the gap thickness and the angle of incidence calculated using this software. The position of the global maximum of the MSEFS in the graph yields the optimal angle of incidence and gap thickness for a specific experiment for the wavenumber region of interest. For example, for the wavenumber region centered at 1600 $cm^{-1}$, the MSEFS plot in Figure 11.5 displays a global maximum at an incident angle of 66° and a thin-cavity thickness of 3.4 μm. Since the position of the global maximum on the MSEFS surface is dependent upon the frequency of radiation, it is important that optimization of the angle of incidence and thin-cavity layer be performed for each wavenumber region studied. A more complete description of the details regarding optimization of experimental conditions can be found in Ref [6].

The calculations in Figure 11.5 give the maximum value of the MSEFS at the gold surface. However, to eliminate the background due to absorption by the solvent in the bulk of the thin-layer cavity, it is sometimes convenient to run experiments at the optimum cancelation of the average intensities of the *p*- and *s*-polarized photons within the thin-layer cavity. This point is illustrated by Figure 11.6 in which the solid

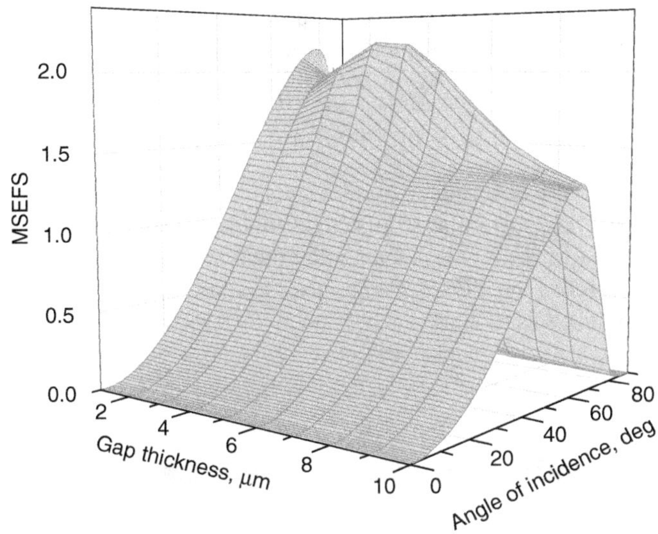


**Figure 11.5** Mean-squared electric field strength (MSEFS) at the gold surface for a *p*-polarized beam as a function of the angle of incidence and the thin cavity (gap) thickness. Calculated for the convergent (±6°) radiation in 1600 cm$^{-1}$. For stratified medium: $CaF_2/D_2O/Au$. Adapted from Ref [20].

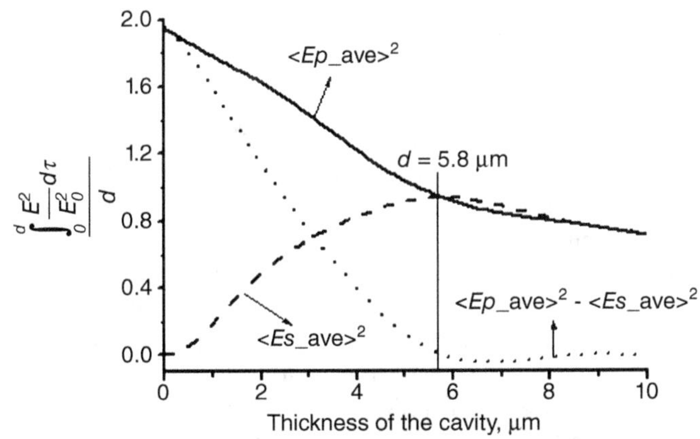


**Figure 11.6** The MSEFS of *p*- (solid) and *s*- (dash) polarized photons averaged over the thickness of the thin-layer cavity plotted as a function of the cavity thickness. The dotted line represents the difference of the average MSEFS of the *p*- and *s*-polarized photons. The best cancelation of the averaged MSEFS of *p*- and *s*-polarized photon is observed for the cavity thickness of 5.8 μm. Calculated for the convergent (±6°) radiation in 1600 cm$^{-1}$. For stratified medium: $CaF_2/D_2O/Au$.

and dashed lines plot the MSEFS of the *p*- and *s*-polarized photons averaged over the thickness of the thin-layer cavity ($\left[\int_o^d (E^2/E_0^2)d\tau\right]/d$) as a function of the cavity thickness. Figure 11.6 shows that at a gap thickness of 5.8 μm the average intensities of *p*- and *s*-polarized photons inside the gap are equal. The dotted line shows that at this value of the gap thickness the background due to the absorption by the solvent inside the thin-layer cavity is canceled and the measured signal

$$\frac{\Delta R}{\langle R \rangle} = \frac{|R_p - R_s|}{(R_s + R_p)/2}$$

corresponds only to the absorbance by molecules adsorbed at the electrode surface. However, Figure 11.5 shows that at this cavity thickness the MSEFS of *p*-polarized photons at the metal surface is less than at the thickness corresponding to the global maximum. Therefore, the signal form adsorbed molecules will be attenuated. However, in some cases it may be beneficial to remove the background due to absorption by the solvent even at the cost of the loss of the signal intensity for molecules adsorbed at the surface.

### 11.3.3 Determination of Thin-Layer Cavity Thickness

While the angle of incidence can be accurately set without much difficulty, obtaining the thin cavity thickness with a precision on the order of fractions of a micrometer can be challenging. To accomplish this, a procedure has been developed that allows for the determination of the actual angle of incidence and thin cavity thickness using the measured reflectivity spectra [24]. An overview of this method is presented in this section for an experiment in which $BaF_2$ was used as the window, $D_2O$ as the solvent, and gold as the electrode.

To calculate the true angle of incidence and thin cavity thickness, the experimental reflectivity curve is compared to a theoretical curve calculated using Fresnel equations for a system consisting of three homogeneous phases. The experimental reflectivity curve is taken as the ratio of the intensity of the beam reflected at the $BaF_2$–electrolyte–Au interface ($R$) to the intensity of the beam reflected at the $BaF_2$ surface in the absence of electrolyte ($R_0$). In the absence of electrolyte, the incident beam is totally reflected because the angle of incidence is above the critical angle for the $BaF_2$–air interface. Therefore, $R_0$ is a good measure of the intensity of the incident beam. An experimental reflectivity curve is shown in Figure 11.7 as the solid line. In the calculations of the reflectivity spectrum, the thickness and the angle of incidence are adjusted until a best fit to the experimental curve is obtained. Figure 11.7 shows the comparison of the experimental and calculated reflectivity curve. When good agreement between the two curves is achieved the incident angle and thin cavity thickness are determined as the best-fit parameters.

### 11.3.4 Determination of PEM Response Functions

The PEM is the heart of the PM–IRRAS setup. It generates the alternating linear light polarization states (*p*- and *s*-polarized components). As mentioned above, the PM–IRRAS experimental setup consists of a static polarizer and a PEM that is

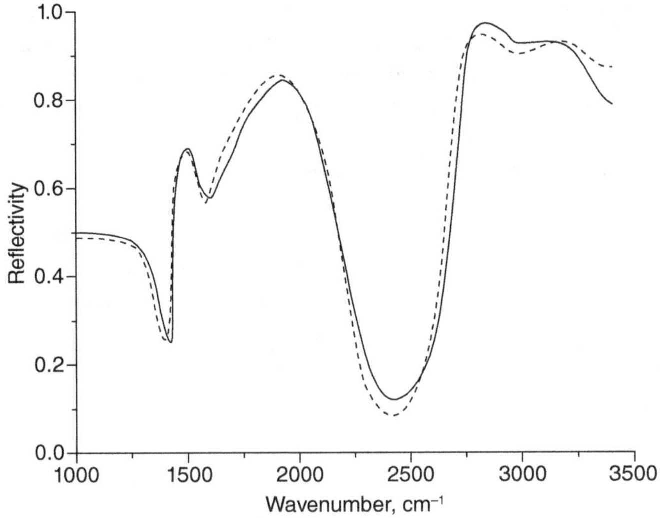


**Figure 11.7** Reflectivity spectra for the thin-layer cavity spectroelectrochemical cell ($BaF_2/D_2O$/Au). Experimental spectrum is represented by the solid line, calculated spectrum is represented by the dashed line ($\theta = 62°$, $d = 2.6$ μm). Adapted from Ref. [20].

mounted 45° with respect to the plane of the optical bench. Once the incident beam crosses the static polarizer, the radiation becomes linearly polarized in the direction parallel to the optical bench and at 45° with respect to the optical axis of the PEM. Therefore, the beam of linearly polarized radiation can be regarded as the sum of two components, one parallel and one perpendicular to the optical axis of the PEM.

The PEM is made of a piezoelectric transducer that is attached to a ZnSe crystal. The piezoelement converts a periodic voltage into a periodic mechanical (acoustic) wave that compresses or expands the crystal. Based on the photoelastic effect, the mechanically stressed sample exhibits birefringence proportional to the resulting strain. The birefringence introduced in the crystal retards one of the linearly polarized components more than the other such that the PEM acts as a tunable wave plate. The PEM transducer is usually set to the resonant frequency of the optical element (50 kHz). For PM–IRRAS experiments the amplitude of the applied voltage is set such that the PEM imposes the maximum relative phase shift of 90° (half-wave retardation), causing the incident beam to change its polarization every 90°. During one cycle of the applied voltage, the half-wave retardation occurs twice: once when the optical element becomes compressed (the beam is linearly polarized at an angle of +45° with respect to the incident beam) and once when the optical element becomes stretched (the beam is then linearly polarized at an angle of –45° with respect to the incident beam) [6]. Therefore, the PEM can be employed to change the polarization of the incident linearly polarized beam at a frequency that is twice the frequency of the voltage applied to the piezoelement.

It is important to note that the PEM is optimized for a selected wavelength, $\lambda_0$. At this wavelength the beam is periodically changed between $p$ and $s$. At wavelengths other than $\lambda_0$, the polarization of the incident beam becomes circular (light with both $s$ and $p$ components). As a result, experimental raw spectra contain artifacts introduced by the PEM at wavelengths other than $\lambda_0$. Therefore, to perform quantitative analysis for an entire spectral region, one must remove the artifacts introduced by the PEM. This correction method is described below.

In a PM–IRRAS experiment, the synchronous sampling demodulator (SSD) yields two signals, the intensity average $I_A(\omega)$ and the intensity difference $I_D(\omega)$ signals. These two signals are described by the following equations [18]:

$$I_A(\omega) = \frac{I_s(\omega)+I_p(\omega)}{2} + \frac{I_p(\omega)-I_s(\omega)}{2} J_0(\varphi_0) = \langle I \rangle + \frac{1}{2}\Delta I J_0(\varphi_0) \qquad (11.2)$$

and

$$I_D(\omega) = |I_p(\omega) - I_s(\omega)| J_2(\varphi_0) = \Delta I J_2(\varphi_0) \qquad (11.3)$$

where $I_p$ and $I_s$ are the intensities of $p$- and $s$-polarized light, respectively, $\varphi_0$ is the maximum retardation of the incident radiation imposed by the PEM, and $J_0(\omega)$ and $J_2(\omega)$ are the PEM response functions (zero-order and second-order Bessel functions), which account for the slow variation of the retardation imposed by the PEM. To determine the ratio $\Delta I/\langle I \rangle$, the PEM response functions must be calculated and removed from $I_A(\omega)$ and $I_D(\omega)$.

To calculate the $J$ functions a similar procedure to the one developed by Buffeteau et al. [1] was employed. A $BaF_2$ equilateral prism was placed in a holder and mounted in the cell holder in the TOM box. This prism serves as a dielectric total external reflection mirror. A second static polarizer (Harrick Scientific Corporation, Pleasantville, NY) was then positioned just after the PEM and set to transmit only $p$-polarized light. The first reference spectrum was recorded with the PEM switched off to obtain the intensity of the $p$-polarized light $I_p^{\mathrm{cal}}(\omega)$, which passes through the entire optical bench. The PEM was then turned on and after demodulation two calibration signals $I_A^{\mathrm{cal}}(\omega)$ and $I_D^{\mathrm{cal}}(\omega)$ were recorded. With the use of the second polarizer allowing only $p$-polarized light through, $I_s^{\mathrm{cal}}(\omega)$ is equal to zero. Therefore, Eqs. (11.2) and (11.3) can be simplified to

$$I_A(\omega)^{\mathrm{cal}} = \frac{I_p(\omega)^{\mathrm{cal}}}{2} + \frac{I_p(\omega)^{\mathrm{cal}}}{2} J_0(\varphi_0) \qquad (11.4)$$

and

$$I_D(\omega)^{\mathrm{cal}} = I_p(\omega)^{\mathrm{cal}} J_2(\varphi_0) \qquad (11.5)$$

The PEM response, represented by $J_2(\varphi_0)$ and $J_0(\varphi_0)$, can then be calculated from the experimentally measured calibration signals using the following equations [25]:

$$J_0(\varphi_0) = 2\frac{I_A(\omega)^{\mathrm{cal}}}{I_p(\omega)^{\mathrm{cal}}} - 1 \qquad (11.6)$$

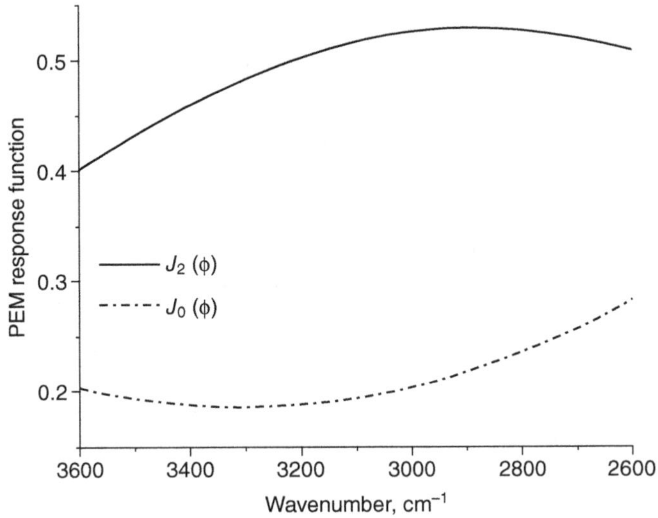


**Figure 11.8** PEM response as a function of wavelength for $J_0$ and $J_2$. PEM was optimized for 2900 cm$^{-1}$. Adapted from Ref. [20].

$$J_2(\varphi_0) = \frac{I_D(\omega)^{\text{cal}}}{I_p(\omega)^{\text{cal}}} \tag{11.7}$$

Plots of $J_0(\omega)$ and $J_2(\omega)$ determined using this calibration procedure are shown in Figure 11.8, where the PEM was optimized for 2900 cm$^{-1}$. Knowing the PEM response functions, the measured signals $I_A(\omega)$ and $I_D(\omega)$ can be corrected with the average $\Delta I$ and difference $<I>$ signals calculated from rearrangement of Eqs. (11.2) and (11.3).

### 11.3.5 Background Correction

Although the optical setup is optimized for the cancelation of the average *p*- and *s*-polarized light within the thin-layer cavity, such optimization is never perfect and the spectra of adsorbed molecules are superimposed on a background due to residual absorption by the solvent. Figure 11.9*a* shows a PM–IRRAS spectrum with a solvent background. Therefore, the PM–IRRAS spectra need to be corrected for this broadband absorption. Since it is impossible to record a background spectrum under conditions identical to those employed to measure the analyte film, Zamlynny et al. [25] developed a method similar to that proposed by Barner et al. [18] to remove the background. In this method, a baseline is created for the experimental spectra using spline interpolation as shown in Figure 11.9*a*. Successful interpolation of the spectra requires knowledge of the exact location of absorption bands and some experience. Once the background has been subtracted from the spectra, as shown in

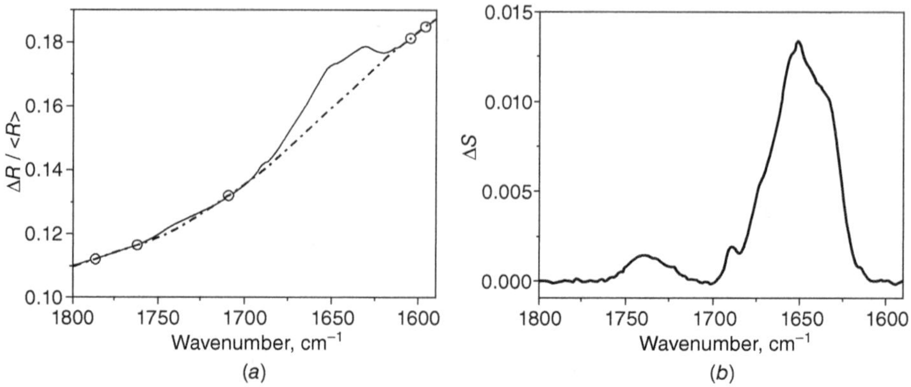


**Figure 11.9** Baseline correction for an experimental PM–IRRAS spectrum: (*a*) spline interpolation to create baseline (circles represent template points for building the spline) and (*b*) resulting spectrum obtained after subtraction of the background. Adapted from Ref. [20].

Figure 11.9*b*, the processing of the raw spectra is complete, and the bands can be deconvoluted and integrated to obtain peak areas, which are required to determine the orientation of species adsorbed on the Au(111) surface.

### 11.3.6 Band Deconvolution

Quantitative analysis of organic films by infrared spectroscopy requires very precise knowledge of the band positions of all vibrations associated with the film. For biological samples, this is difficult due to the overlap of bands arising from different vibrations. Two methods that have been utilized to aid in resolving overlapping peaks are Fourier self-deconvolution [26, 27] and two-dimensional correlation spectroscopy (2D-COS) [28].

#### *Fourier Self-Deconvolution*

Fourier self-deconvolution (FSD) is a spectral processing technique that mathematically sharpens a band profile to resolve the structure hidden by the overlap of neighboring peaks. It does not actually increase the signal resolution but instead improves the differentiation of spectral features. A very good description of this procedure is given by Surewicz and Mantsch [29]. This procedure is explained in Figure 11.10. When an IR band $A(\nu)$ has Lorentzian shape described by

$$A(\nu) = \frac{\gamma^2 + (\nu - \nu_0)^2}{A_0 \gamma^2} \tag{11.8}$$

where $A_0$ is the band maximum, $\nu$ is the wavenumber, $\nu_0$ is the wavenumber at band maximum, and $\gamma$ is the half-width at half-maximum (HWHM), the Fourier transform of this band [of Eq. (11.8)] gives

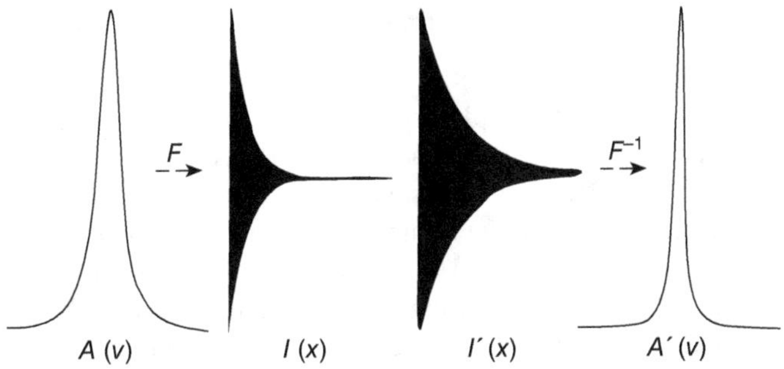


**Figure 11.10** Principle of band narrowing by Fourier self-deconvolution; $A(\nu)$ is the IR band of Laurentzian shape, $I(x)$ is its cosine Fourier transform; $I'(x)$ is the result of multiplication with $\exp(x\gamma/2.5)$; $A'(\nu)$ is a new band whose HWHM is $\gamma/2.5$ where $\gamma$ is the HWHM of the band $A(\nu)$. Taken with permission from Ref. [29].

$$I(x) = 0.5A_0\gamma\cos(2\pi\nu_0 x)e^{-2\pi\gamma x} \tag{11.9}$$

The exponential decay of function $I(x)$ is determined by the term $\exp(-2\pi\gamma x)$. It shows that the wider the IR band (larger $\gamma$) the more rapidly the function $I(x)$ decays. In the FSD procedure one multiplies $I(x)$ by an exponential function $\exp(2\pi\gamma' x)$ to give a new function $I'(x)$ with a reduced decay rate:

$$I'(x) = 0.5A_0\gamma\cos(2\pi\nu_0 x)e^{2\pi(\gamma'-\gamma)x} \tag{11.10}$$

The new function $I'(x)$ is often smoothed by multiplication with an apodization function. The reverse transformation of $I'(x)$ is then computed [30]. The reverse FT of $I'(x)$ gives the spectrum $A'(\nu)$ whose HWHM is reduced to the difference $(\gamma - \gamma')$. If such a procedure is applied to a broad band that contains multiple peaks, the FSD spectrum consists of narrower peaks than the original [30]. The ratio $K = \gamma/\gamma'$ is defined as the resolution enhancement.

When properly applied, FSD is an excellent tool to provide the number and positions of the peaks under a spectral band to a first approximation. These parameters then become initial input for iterative peak fitting procedures for the further extraction of quantitative information from the spectral region. The FSD tool is available in Omnic v.6.0a or higher edition (Thermo software). FSD was employed assuming a typical bandwidth between 10 and 12 $cm^{-1}$ and a resolution enhancement usually set at 1.5 or less. Bandwidth here refers to an estimation of the HWHM of the overlapped peaks in the experimental spectrum, whereas enhancement refers to the degree to which the spectral features are resolved, which can be set between 1 and 5. The value of these two parameters is determined as the minimum value that would not give negative side lobes in the FSD spectrum. Although this is a very important technique in spectral processing, one must be aware that this technique is susceptible to artifacts. Griffiths and Pariente have discussed how these pitfalls can be avoided [30].

### *Generalized Two-Dimensional Correlation Spectroscopy*

Generalized two-dimensional correlation spectroscopy (2D-COS) is a technique that simplifies the visualization of complex spectra consisting of overlapping bands and enhances the spectral resolution by spreading spectral peaks over the second dimension [28]. In generalized 2D-COS, 2D spectra are generated from a set of sequentially obtained spectral data for a given sample influenced by an external perturbation that induces variations in the spectral features (intensity). For the studies described in this chapter, the external perturbation is the potential applied to the gold electrode ($E$). To generate the 2D correlation spectra, dynamic spectra must be calculated first. A dynamic spectrum [($\tilde{y}(\upsilon, E)$)] is defined as [28]

$$\tilde{y}(\upsilon, E) = \begin{cases} y(\upsilon, t) - \bar{y}(\upsilon) & \text{for } E_{\min} \le E \le E_{\max} \\ 0 & \text{otherwise} \end{cases} \tag{11.11}$$

where $y(\upsilon, E)$ is the spectral intensity variation observed as a function of a spectral variable $\upsilon$ (wavenumber) during an interval of the external variable $E$ between $E_{\min}$ and $E_{\max}$ and $\bar{y}(\upsilon, E)$ is the reference spectrum of the system. The dynamic spectra are transformed into two-dimensional spectra by cross-correlation analysis. This analysis involves a quantitative comparison of the patterns of spectral intensity variations along the external variable $E$ observed at two independently chosen wavenumbers, $\upsilon_1$ and $\upsilon_2$, over the interval $E_{\min}$ to $E_{\max}$. The resulting 2D correlation spectrum can be expressed as [28]

$$X(\upsilon_1, \upsilon_2) = \langle \tilde{y}(\upsilon_1, E) \cdot \tilde{y}(\upsilon_2, E') \rangle \tag{11.12}$$

where the symbol $\langle\rangle$ is a cross-correlation function defined as

$$C(E, \tau) = \langle \Phi(\tau) | \Psi(E) \rangle \tag{11.13}$$

Two types of spectra, synchronous and asynchronous correlation spectra, are generated from the dynamic spectra. These two spectra are obtained by treating the 2D correlation spectrum as a complex number function:

$$X(\upsilon_1, \upsilon_2) = \Phi(\upsilon_1, \upsilon_2) + i\Psi(\upsilon_1, \upsilon_2) \tag{11.14}$$

where $\Phi(\upsilon_1, \upsilon_2)$ corresponds to the synchronous component, while $\Psi(\upsilon_1, \upsilon_2)$ is defined as the asynchronous component. The synchronous 2D correlation intensity $\Phi(\upsilon_1, \upsilon_2)$ represents the simultaneous changes of a measured spectral series, while the asynchronous 2D correlation intensity $\Psi(\upsilon_1, \upsilon_2)$ represents the sequential changes in the spectral intensity variations. Synchronous spectra are always symmetrical along the diagonal and have peaks along the diagonal ($\upsilon_1 = \upsilon_2$), defined as autocorrelation peaks or autopeaks (Fig. 11.11*a*). The intensity of these peaks represents the strength of the band. The peaks off the diagonal are defined as cross peaks and represent a degree of correlation. Generally, a correlation square is made by joining the pair of cross peaks located at opposite sides of the diagonal line drawn through the corresponding autopeaks. The correlation square illustrates the existence of coherent variation of spectral intensities at these spectral variables (Fig. 11.11*a*). When the cross peak is positive, then both peaks are changing in the same direction; peaks are both increasing or decreasing at the same time. Negative cross peaks

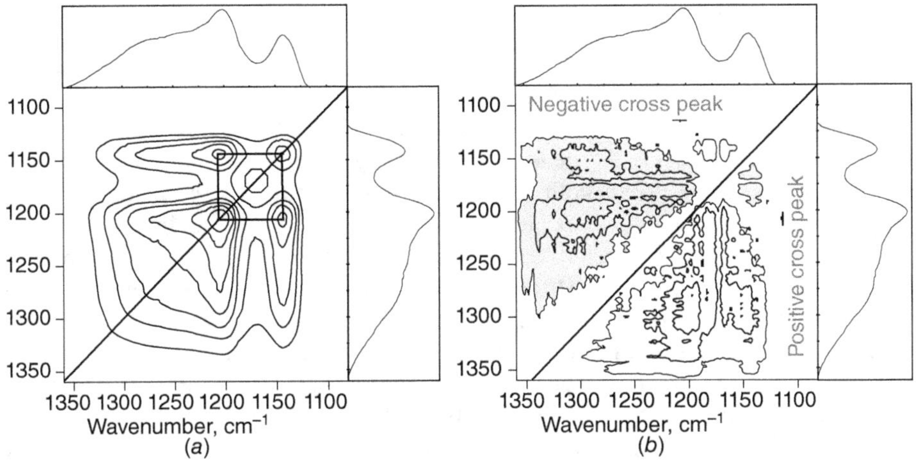


**Figure 11.11** (*a*) Synchronous and (*b*) asynchronous spectra with marked correlation squares and peaks of a floating bilayer lipid membrane (fBLM) composed of DMPC/DMPE-PEG350/chol at a β-Tg-modified Au(111) surface (see Section 11.12 for details).

indicate that the peaks are changing in opposite directions, with one decreasing while the other is increasing. Asynchronous spectra are always antisymmetric along the diagonal and do not have peaks on the diagonal. An asynchronous correlation square is drawn by extending the lines from the spectral coordinates of the cross peaks to the corresponding diagonal positions. When the cross peak is positive, then a band from the first spectra is growing earlier or more intensivly then a band from second spectra and vice versa. Based on the interpretation of the 2D-COS spectra, poorly resolved bands (specifically the amide region) can be successfully deconvoluted [31].

Figure 11.11 gives an example of the 2D-COS procedure applied to deconvolute the broad band corresponding to the overlapping vibrations of the phosphate and the polyethylene glycol groups in a floating bilayer composed of 1,2-dimyristoyl-*sn*-glycero-3-phosphocholine (DMPC), 1,2-dimyristoyl-*sn*-glycero-3-phosphoethanolamine-N-[methoxy(polyethylene glycol)-350] (DMPE-PEG350), and cholesterol. The correlation square for two strong bands at 1142 and 1202 $cm^{-1}$ can be observed. The correlation square consists of two autocorrelation peaks (1142,1142) and (1202,1202), which are located on the diagonal, and two cross-correlation peaks (1202,1142) and (1142,1202), which are located off diagonal. By increasing the resolution, other correlation squares can be found in the 2D-COS spectrum of this spectral region, allowing one to identify additional peak positions. Figure 11.11*b* shows the asynchronous component of the correlation. This is much more complex and difficult to analyze due to complicated patterns that arise from peaks and background shifting as a function of potential. This issue has been discussed by others, in particular by Czarnecki [32]. Therefore, the asynchronous component of the correlation was not analyzed in our studies. We will show later how 2D-COS is helpful in identifying the number of bands involved in the broad band associated with various amide vibrational modes in the gramicidin–DMPC bilayer deposited on gold [33].

***Band Fitting (Gaussian–Lorentzian)***

Once the number of peaks and their positions agree for the two independent determinations, FSD and 2D-COS, the peaks can be assigned to the corresponding modes of vibration based on the published literature. Lastly, using PeakFit software, depending on the type of peaks, the peaks were fit using purely Gaussian or Lorentzian band shapes or a mixture of both to a deconvolution of $R^2 \geq 0.99$. Gaussian band shapes are used for peaks associated with vibrational modes in inhomogeneous samples. An example of such a case is when a distribution of hydrated states involving hydrogen bonding may exist (i.e. carbonyl stretching modes). Lorentzian band shapes are used for vibrational modes in homogeneous samples where all molecules experience the same state. For example, the CH stretching region of acyl chains of phospholipid molecules in the gel state of a bilayer. Mantsch et al. [34] described how to use a mixture of the two types of band shape. They explained that although the true profile of an IR band approaches a Lorentzian curve, several physical factors that perturb this true band profile have a Gaussian distribution. The result is that the spectroscopic bands of liquids often approximate to a profile that is at least 90% Lorentzian and 10% Gaussian. Additionally, the presence of slightly different conformations of the same molecule broadens the corresponding IR band. To account for this effect the Gaussian component of the band shape profile is increased [34–36].

## 11.3.7 Determination of Tilt Angles and Order Parameter

In PM–IRRAS the integrated intensity of the absorption band ($A$) is proportional to the square of the absolute value of the dot product of the transition dipole vector ($\vec{\mu}$) and the electric field vector ($\vec{E}$) of the incident radiation, which is expressed as [37, 38]

$$\int A\, d\nu \propto \left|\vec{\mu} \cdot \vec{E}\right|^2 = |\mu|^2 \langle E^2 \rangle \cos^2 \Theta \tag{11.15}$$

where $\Theta$ is the angle between the two vectors, $|\mu^2|$ is the square of the absolute value of the transition dipole, and $\langle E^2 \rangle$ is the mean square of the electric field of the photon. Since the direction of the electric field vector of the photon is normal to the metal surface, the integrated band intensity is proportional to $\cos^2 \Theta$, where $\Theta$ is the angle between the surface normal and the average direction of the transition dipole moment. To calculate the absolute value of $\Theta$ for a particular vibrational mode a method developed by Allara et al. can be employed [37–39]. A simulated reflection absorption spectrum of a film of randomly oriented molecules on the gold surface is calculated using the same parameters as employed during the collection of experimental spectra (thin-cavity thickness, angle of incidence, optical window material, etc.). In the case of absorption of a linearly polarized beam by randomly oriented molecules, $\cos^2 \Theta = \frac{1}{3}$. Therefore, the ratio of the integrated intensity of the absorption band obtained experimentally ($A_{exp}$) to that obtained from the simulation ($A_{calc}$) yields the following expression:

$$\cos^2 \Theta = \frac{1}{3} \frac{\int A_{\text{exp}} dv}{\int A_{\text{calc}} dv} \tag{11.16}$$

The above expression allows calculation of the absolute value of Θ directly.

To simulate the spectrum of a lipid bilayer with randomly oriented molecules at a gold surface, the optical constants (refractive index, $n$, and attenuation coefficient, $k$) are required for each component used in the PM–IRRAS experiment (window, electrolyte, metal, and lipid bilayer). The optical constants for the window, electrolyte, and metal can be obtained from the literature [40–43]. To model a film of randomly oriented molecules on the surface the optical constants must be obtained for an isotropic sample. These isotropic optical constants are determined by collecting the transmission spectrum of vesicles in aqueous solution as explained in Ref. [6]. The simulated reflectance spectrum for a particular PM–IRRAS experiment is then calculated from the various optical constants as described in Ref. [6]. In order to apply this method, one also has to determine the surface concentration of the adsorbed species or the thickness of the film, using an independent method.

Once Θ is calculated, the orientation of the molecule at the metal surface can be determined by taking into account the direction of the transition dipole of the vibrational mode with respect to the coordinates of the adsorbed molecule. The directions of the transition dipoles are either available in the literature or can be determined from ab initio normal coordinate calculations. This method has been extremely useful to determine the average tilt angle of lipid acyl chains from PM–IRRAS spectra. To calculate the hydrocarbon tilt angle, the peaks corresponding to the vibrational modes of the symmetric and asymmetric methylene stretches are used [e.g., $\nu_s(CH_2)$ and $\nu_{as}(CH_2)$, respectively]. Using Eq. (11.16), the angle between the surface normal and the average direction of the transition dipole moment is calculated for these two modes ($\theta_s$ and $\theta_{as}$). Figure 11.12 shows the directions of the

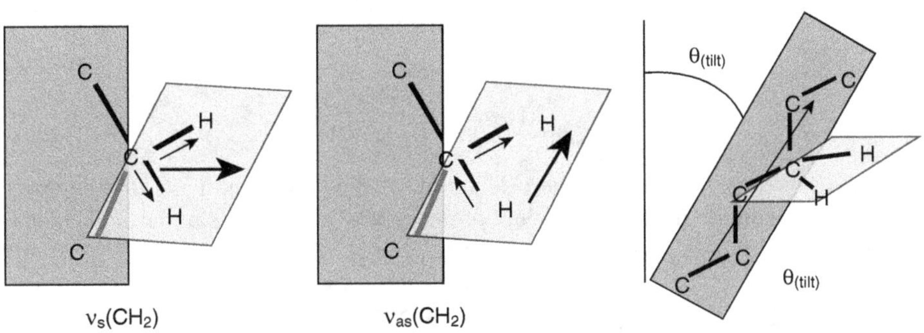


**Figure 11.12** Schematic of the direction of the transition dipoles for (*a*) symmetric and (*b*) asymmetric methylene stretches for a lipid acyl chain. (*c*) Schematic of the tilt angle ($\theta_{\text{tilt}}$) of acyl chains.

transition dipoles for $\nu_s(CH_2)$ and $\nu_{as}(CH_2)$. The direction of the transition dipole of the symmetric methylene stretch lies along the bisector of the $CH_2$ plane, while the direction of the transition dipole of the asymmetric stretch lies along the line joining the two hydrogen atoms of the $CH_2$ group [44]. The vectors of the transition dipoles of the symmetric and asymmetric methylene stretches are perpendicular to each other and are perpendicular to the line of a fully extended all-trans fragment of the acyl chain. Therefore, the tilt angle ($\theta_{tilt}$) of the trans fragments of acyl chains can be calculated using the following equation [45]:

$$\cos^2\theta_{as} + \cos^2\theta_s + \cos^2\theta_{tilt} = 1 \tag{11.17}$$

It is important to note that the above equation gives the average orientation of trans segments of a hydrocarbon chain. It can therefore only be used as an approximation of the chain tilt angle in most cases, since chain melting may occur.

In nuclear magnetic resonance (NMR) or attenuated total reflection(ATR) literature, the orientation of acyl chains is frequently described in terms of the order parameter ($S$). Therefore, for comparison sake it is useful to calculate the order parameter from the PM–IRRAS spectra. For the direction of the dipole moment of an IR band, the order parameter is defined as [46]

$$S_{CH_2} = \frac{1}{2}\left(3\langle\cos^2\theta\rangle - 1\right) \tag{11.18}$$

where $\theta$ is determined from Eq. (11.16) and the angular brackets denote ensemble averaging over all absorbing dipoles in the sample. Equation (11.18) implies that the order parameter can have values that range between unity and –0.5 when the transition dipoles are oriented parallel or perpendicular with respect to the normal to the surface, respectively. When $S = 0$, the system is completely disordered with randomly orientated molecules, corresponding to an average tilt angle equal to the magic angle (54.7°). Since the angle between transition dipoles of $CH_2$ stretches and a trans fragment of the acyl chain is 90°, the order parameter of the transition dipoles can be used to calculate the order parameter of the trans segments of the acyl chains using the transformation [47]:

$$S_{chain} = S_{CH_2}\left[\frac{1}{2}\left(3\langle\cos^2 90\rangle - 1\right)\right]^{-1} = -2S_{CH_2} \tag{11.19}$$

The order parameter, $S_{chain}$, is defined as

$$S_{chain} = \frac{1}{2}\left(3\langle\cos^2\theta_{chain}\rangle - 1\right) \tag{11.20}$$

$S_{chain}$ is an average measure of the orientation of the chain segments that depends both on the chain tilt and on the amount of the gauche conformers in the acyl chains. Using the above expression the value of $\theta_{chain}$ can be obtained as well.

## 11.4 ELECTROCHEMICAL STUDIES OF GOLD-ELECTRODE-SUPPORTED BILAYERS

Model biological membranes supported on a Au(111) electrode need to be characterized using electrochemical techniques before PM–IRRAS experiments can be performed. In order to understand the relevance of these measurements, it is important to review the electric potential profile across a natural cell membrane and its electrical properties. This is discussed in the following sections.

### 11.4.1 Electric Potential Profile across Cell Membrane

The surface of the cell membrane is composed of various charged or dipolar groups such as negatively charged or zwitterionic lipid head groups, water dipoles, and dipolar lipid residues. These charged groups generate a Volta or outer potential ($\Psi$), and these surface charges are compensated by ions in solutions on the two sides of the membrane. This results in a potential drop that has a long-range character. In contrast, oriented surface dipoles generate a surface potential $\chi$ that has a local character. As a result, the electrical potential does not change linearly across a cell membrane but instead follows a complex profile as shown in Figure 11.13 [48, 49]. The value of $\Psi$ at any point is a result of contributions from three electrical potential differences: the transmembrane potential ($\Delta\Psi$), the potential due to surface charges ($\Psi_s$), and the dipole potential ($\chi_M$).

The transmembrane potential ($\Delta\Psi$) is the overall difference in potential due to ionic concentration differences between the two bulk phases separated by the membrane and selective ionic permeability of the membrane to one or several ions (Donnan potential). The transmembrane potential across a membrane 5–7 nm thick is typically in the range of −10 to −250 mV and imposes an electric field strength of approximately $-2 \times 10^6$ to $-5 \times 10^7$ V m$^{-1}$ on molecules within the cell membrane

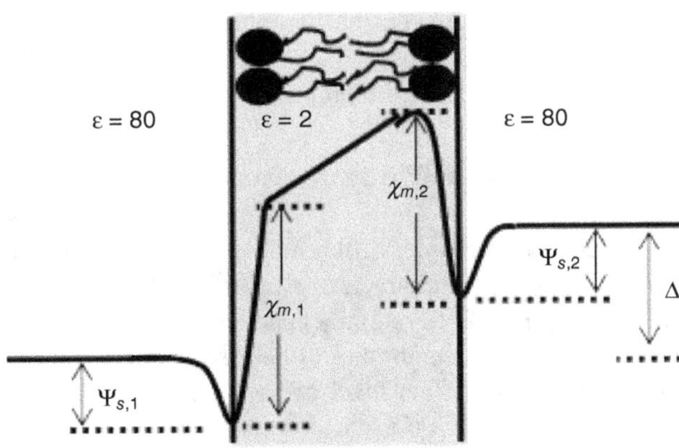


**Figure 11.13** Electrical potential ($\Psi$) profile across a cell membrane. The three contributions to the electrical potential are shown: the transmembrane potential ($\Delta\Psi$), the Volta potential ($\Psi_s$), and the dipole potential ($\chi_m$). Subscripts 1 and 2 denote the potentials associated with the two leaflets of the lipid membrane. The dielectric constants ($\varepsilon$) for the lipid core ($\varepsilon = 2$) and the exterior sides of the membrane ($\varepsilon = 80$) are given.

[50]. As mentioned above, the Volta potential or double-layer potential ($\Psi_s$) arises from charged lipid head groups (generally negatively charged) at the membrane surface. These fixed charges at the membrane surface control the concentration of counterions at the membrane–solution interface, resulting in a diffuse double layer. This electrical double layer gives rise to a potential of approximately 60 mV, corresponding to an electric field strength of up to $2 \times 10^7$ V m$^{-1}$, within 1 nm from the surface [50]. The third contribution, the membrane dipole potential ($\chi_m$), arises from the alignment of dipolar residues of the lipids and/or water dipoles in the region between the aqueous phases and the hydrocarbon-like interior of the membrane [48, 49]. Unlike the transmembrane and Volta potentials, which can be determined directly from electrochemical experiments, the dipole potential can only be theoretically calculated or inferred indirectly from experiment [48, 49]. Such theoretical calculations for a fully saturated phosphatidylcholine bilayer yield a dipole potential in the range 220–280 mV, positive in the interior of the membrane [48, 49]. Since the dipole potential drop across the head group region of the membrane is a relatively small distance, the electric field strength produced is very large ($10^8$–$10^9$ V m$^{-1}$) in comparison to the other two contributions [48, 49].

### 11.4.2 Capacitance of Cell Membranes

A bilayer membrane can be modeled electrically as a thin slab of nonconducting material separating two aqueous electrolyte solutions. Therefore, the membrane can then be viewed as a simple parallel-plate capacitor where charge can accumulate at either interface. The capacitance ($C_d$) is defined by

$$C_d = \frac{\varepsilon\varepsilon_0}{d} \tag{11.21}$$

which depends on the distance between the two surfaces ($d$) and on the dielectric constant of the insulating hydrocarbons ($\varepsilon = 2$). The measured specific capacitance (capacitance per unit area) of biological membranes is around 1 μF cm$^{-2}$ [51].

### 11.4.3 Electrochemical Measurements

In the research described in this chapter differential capacitance and chronocoulometry techniques were employed to characterize model biological membranes supported on Au(111) electrodes. Differential capacitance measurements allow the quality of supported bilayer lipid membranes to be assessed, providing useful information on the compactness and defect level of the membrane. Chronocoulometry measurements, on the other hand, are used to determine the charge density at the electrode surface. From charge density plots, the potential drop across the membrane ($\Delta\phi_{M-2}$) can be estimated. These methods are described in more detail in the following sections. The electrochemical measurements described in this chapter were carried out in an all-glass three-electrode cell using the hanging meniscus configuration [52].

### *Differential Capacitance*

In electrochemical measurements, the capacitance of an electrode changes with potential, and these changes reflect the variation of coverage by specifically adsorbed ions or molecules or reorientation of dipolar molecules. The charge of a metal electrode with adsorbed species is a function of Gibbs surface excess and the applied potential [$\sigma_M = f(\Gamma, E)$]. Therefore, the exact differential of the charge yields

$$d\sigma_M = \left(\frac{\partial \sigma_M}{\partial E}\right)_\Gamma dE + \left(\frac{\partial \sigma_M}{\partial \Gamma}\right)_E d\Gamma \tag{11.22}$$

The expression for the differential capacitance can be determined by taking the derivate of the Eq. (11.22) with respect to the applied potential:

$$C = \frac{d\sigma_M}{dE} = \left(\frac{\partial \sigma_M}{\partial E}\right)_\Gamma + \left(\frac{\partial \sigma_M}{\partial \Gamma}\right)_E \frac{d\Gamma}{dE} \tag{11.23}$$

The capacitance of an electrode in the presence of specifically adsorbed ions or molecules has two contributions, the true or infinite frequency capacitance ($C_{\text{true}/\infty}$) and the pseudocapacitance ($C_{\text{pseudo}}$):

$$C_{\text{true}/\infty} = \left(\frac{\partial \sigma_M}{\partial E}\right)_\Gamma \qquad C_{\text{pseudo}} = \left(\frac{\partial \sigma_M}{\partial \Gamma}\right)_E \frac{d\Gamma}{dE} \tag{11.24}$$

The measured differential capacitance curve contains features that can be related to these two terms. In the range of potentials with constant surface coverage $d\Gamma/dE = 0$ and $C = C_{\text{true}/\infty}$. The capacitance is potential independent and is generally small in magnitude (1–2 μF $cm^{-2}$). However, at certain potentials large peaks are observed that are related to the $C_{\text{pseudo}}$ term. These peaks are called pseudocapacitive peaks and are attributable to a change in the Gibbs surface excess of the adsorbed species (equivalent to surface coverage) or a change in the charge on the metal per adsorbed molecule. Therefore, the pseudocapacitive peaks correspond to a phase transition in the adsorbed film or to adsorption–desorption of species at the metal electrode surface.

This is illustrated by Figure 11.14 which shows differential capacitance curves for the Au(111) electrode without a membrane present and with a mixed 70% DMPC and 30% cholesterol bilayer deposited onto the metal surface. For the membrane-covered electrode, the capacitance is nearly potential independent within the range –0.1 to 0.2 V. Its value is low (~2 μF $cm^{-2}$) and represents chiefly the true capacitance. At more negative potentials the capacity increases and a peak could be seen at ~–0.9 V. These changes at more negative potentials are due to pseudocapacitance caused by progressive electrodewetting of the metal surface from the bilayer. At potentials more negative than −1.0 V, the capacitance of the electrode initially covered by the membrane becomes equal to the capacitance of the electrode that was membrane free. At this point the membrane is totally detached from the metal. However, neutron reflectivity experiments have shown that it remains in close vicinity to the metal surface separated from it by an ~1-nm-thick layer of the electrolyte [53, 54].

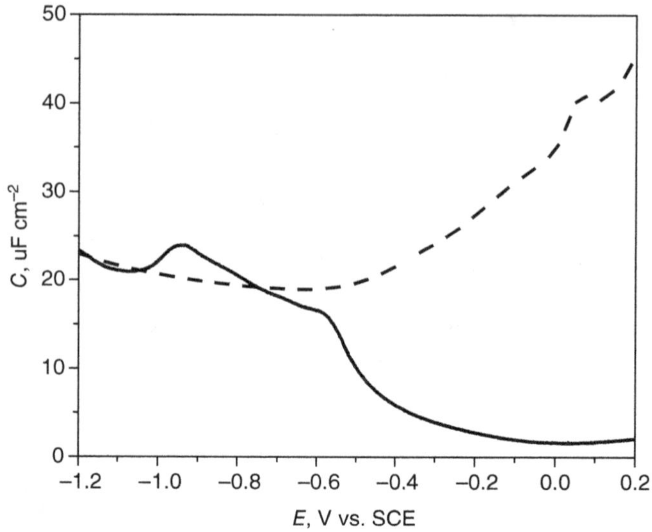


**Figure 11.14** Differential capacitance curves determined for a Au(111) electrode in 0.1M NaF solution: (dashed line) without membrane; (solid line) with a mixed 70% DMPC and 30% cholesterol bilayer deposited using the combination of Langmuir–Blodgett and Langmuir–Schaefer techniques.

It should be stressed that in regard to bilayer lipid membranes supported on a Au(111) electrode, the differential capacitance measurement is a qualitative tool to assess the stability of the lipid bilayer as a function of the applied potential. In general, three important properties of the lipid bilayer can be determined from the differential capacitance curves; (i) the potential region in which the lipid bilayer is stable on the gold surface (this is the region in which the capacitance remains relatively constant); (ii) within this stable region, the minimum capacitance provides information on the quality of the bilayer, where a value greater than that of biological membranes (1 μF cm$^{-2}$) indicates the presence of defects in the bilayer; and (iii) the potential region where the membrane is detached from the gold surface.

### *Charge Density Measurements*

Chronocoulometry measures the difference between the charge density at a potential $E_i$, where the bilayer lipid membrane is adsorbed onto the Au(111) surface, and at a potential $E_{des}$, where total desorption (detachment) of the film takes place, defined as follows [55]:

$$\Delta\sigma_M(E_i) = \sigma_M(E_{des}) - \sigma_M(E_i) \tag{11.25}$$

where $E_i$ and $E_{des}$ are determined from the differential capacitance curve of the bilayer as shown in Figure 11.14. A schematic outline of the chronocoulometry experiment is shown in Figure 11.15. First, a sequence of potential steps is applied to the gold working electrode as illustrated in Figure 11.15*a*. The Au(111) electrode

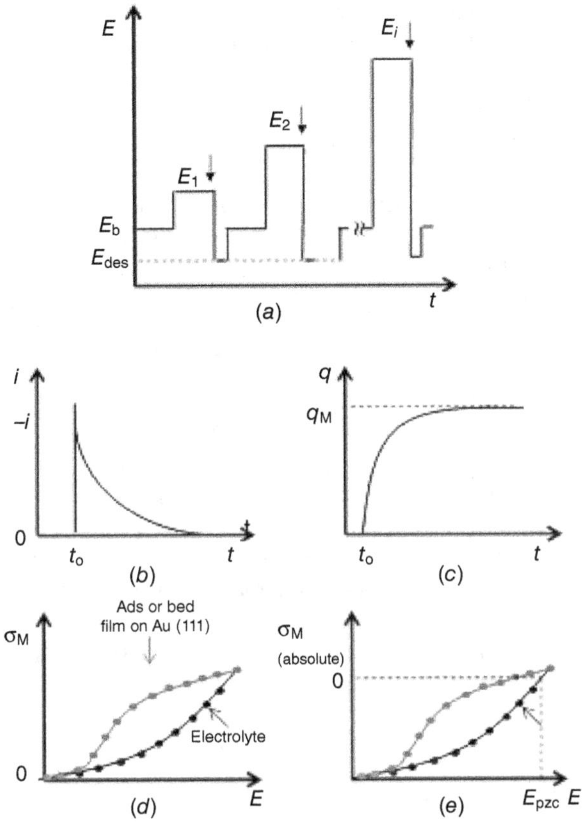


**Figure 11.15** Schematic diagram of the chronocoulmetry experiment. (*a*) The sequence of potential steps applied to the working electrode as a function of time, where current transients are measured at the time indicated by the arrows. (*b*) A measured current transient associated with the step from a variable potential ($E_i$) to the desorption potential ($E_{des}$). (*c*) Charge transient calculated from the integration of current transient curve. (*d*) The charge density ($\sigma_M$), and (*e*) absolute ($\sigma_{M(\text{absolute})}$) charge density plotted against the variable potentials. In (*d*) and (*e*) charge density curves are shown for bare Au(111) in supporting electrolyte (black circles) and for a Au(111) electrode with an adsorbed film (gray circles).

is initially held at a base potential ($E_b$) for a specific time ($t_b$) while the film is adsorbed and allowed to equilibrate. After this time the potential is stepped to a variable potential ($E_i$). It is held at this variable potential until equilibrium is achieved between the film and the gold solution interface ($t_i$). The potential is then stepped to the desorption potential ($E_{des}$). For each step, $E_i$ to $E_{des}$, a current transient is recorded for a brief time period ($t_{des}$) corresponding to the desorption of the film and recharging of the interface (Fig. 11.15*b*). The potential $E_i$ is varied sequentially using various potential step widths covering the double-layer region. Integration of the current transients yields the difference between charge densities at potentials $E_i$ and

$E_{des}$ (Figs. 11.15*c*,*d*). To determine the charge for the background solution, the same experiment is performed for the Au(111) in the pure supporting electrolyte (see Figs. 11.15*c*, *d*).

Equation (11.25) shows that in order to calculate the absolute charge densities $[\sigma_M(E_i)]$, the absolute charge density at the desorption potential $[\sigma_M(E_{des})]$ must be known. For the membrane-free electrode the value of $\sigma_M(E_{des})$ can be calculated from the measured $\Delta\sigma_M(E_i)$ and independently determined potential of zero charge ($E_{pzc}$). The potential of zero charge ($E_{pzc}$) is determined from the position of the diffuse layer minimum on a differential capacitance curve for a dilute solution of the pure supporting electrolyte [56]. Using Eq. (11.25), $\Delta\sigma_M(E_{pzc})$ can be expressed as [55]

$$\Delta\sigma_M(E_{pzc}) = \sigma_M(E_{des}) - \sigma_M(E_{pzc}) = \sigma_M(E_{des}) \tag{11.26}$$

Since $\sigma_M(E_{pzc}) = 0$, the absolute charge density at the desorption potential $[\sigma_M(E_{des})]$ is equivalent to $\Delta\sigma_M(E_{pzc})$. In addition, the charge, $\sigma_M$, at $E_{des}$ is independent of the presence of molecules in solution since there is no adsorption at $E_{des}$. Therefore, the absolute charge densities for the entire potential range $[\sigma_M(E_i)]$ can be calculated from $\Delta\sigma_M(E_i)$ and $\sigma_M(E_{des})$ using Eq. (11.25).

The charge density data can be used to estimate the potential drop across the lipid membrane. Figure 11.16 shows a schematic of the electrical potential profile across a lipid bilayer attached to the gold surface at the metal–electrolyte interface. The potential drop across the metal and the bulk electrolyte solution ($\Delta\phi_{M\text{-}S}$) is described by the formula [57]

$$\Delta\phi_{M\text{-}S} = \frac{\sigma_M}{C} + \chi_M \tag{11.27}$$

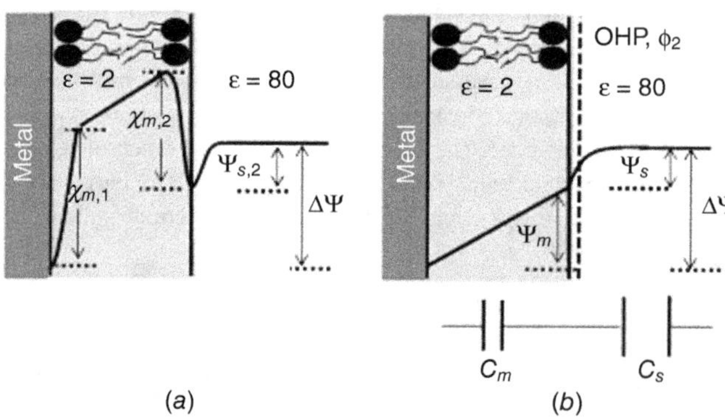


**Figure 11.16** Electrical potential ($\psi$) profiles across a bilayer lipid membrane attached to a metal electrode [(*a*) and (*b*)]. The three contributions to the electrical potential are shown: the transmembrane potential ($\Delta\Psi$), the surface potential ($\Psi_s$), and the dipole potential ($\chi_m$). In the schematic shown in (*b*) the dipole potentials have been omitted from the electrical profile. Below (b) an equivalent electric circuit of the membrane is shown where $C_m$ and $C_s$ represent the capacitance of the membrane and diffuse layer, respectively. The subscripts 1 and 2 denote the potentials associated with the two leaflets of the lipid membrane. The dielectric constants ($\varepsilon$) for the lipid core ($\varepsilon = 2$) and the exterior sides of the membrane ($\varepsilon = 80$) are given.

where $\chi_M$ is the dipole potential of the membrane and $C$ is the total capacitance of the interface. The capacitance $C$ is calculated by numerical differentiation of the charge density curve. The lipid membrane in contact with electrolyte solution is equivalent to two capacitors connected in series: a capacitor with capacitance $C_s$ associated with the diffuse layer and a capacitor with capacitance $C_m$ representing the membrane (see Fig. 11.16*b*). Therefore, Eq. (11.27) can be rewritten as

$$\Delta\phi_{M\text{-}S} = \sigma_M\left(\frac{1}{C_m} + \frac{1}{C_s}\right) + \chi_M = \Delta\psi_{M\text{-}2} + \Delta\psi_{2\text{-}S} + \chi_M \tag{11.28}$$

where

$$\Delta\psi_{M\text{-}2} = \frac{\sigma_M}{C_m} \tag{11.29}$$

is the potential drop across the membrane due to the presence of charge on the metal and $\Psi_{2\text{-}S} = \Psi_2$ is the outer Helmholtz plane potential (note that $\phi = \Psi_2 + \chi_M$ is the inner or Galvani potential). The potential $\psi_2$ can be calculated using the diffuse layer theory and the expression [56]

$$\psi_2 = \left(\frac{2RT}{zF}\right)\sinh^{-1}\left[\frac{\sigma_M}{(8RT\varepsilon c^b)^{1/2}}\right] \tag{11.30}$$

where $R$ is the gas constant, $T$ is temperature, $F$ is Faraday's constant, $z$ is the ionic charge, $\sigma_M$ is the charge density at the metal surface covered with the lipid layer, and $c$ is the ionic concentration in the bulk of solution.

Independently, the value of $\Delta\phi_{M-2}$ can also be calculated using the following expression [58]:

$$\Delta\phi_{M-2} = E - E_{\text{pzc}} - \psi_2 + \chi_M \tag{11.31}$$

The first three terms in Eq. (11.31) are equal to $\Delta\psi_{M-2} = E - E_{\text{pzc}} - \psi_2$. These terms are either experimental quantities or a quantity that could be calculated from charge density data. To calculate $\Delta\phi_{M-2}$ one needs to know the dipole potential $\chi_M$. The dipole potential can only be theoretically calculated or estimated indirectly from experiment [57]. Therefore, in the estimation of the potential drop across the membrane, the contribution from $\chi_M$ has been frequently omitted and $\Delta\Psi_{M\text{-}s}$ is used as an approximate measure of $\Delta\phi_{M-2}$. A schematic of the simplified electrical profile is shown in Figure 11.16*b*. According to Eqs. (11.29) and (11.31), $\sigma_M/C_m = E - E_{\text{pzc}} - \psi_2$. To a good approximation $C_m \approx C$, and therefore the potential drop across the membrane can be either estimated using $\sigma_M/C$ or $(E - E_{\text{pzc}} - \psi_2)$. However, if the dipole potential $\chi_M$ is neglected, $\phi_2 = \psi_2$.

This point is illustrated in Figure 11.17, which shows the $\sigma_M/C$ ratio using $C$ calculated from the charge density data for a DMPC bilayer at the Au(111) electrode surface [58]. In addition, the open squares plot the values of $(E - E_{\text{pzc}} - \phi_2)$ where $E_{\text{pzc}}$ is the potential of zero charge for the bilayer-covered electrode and $\psi_2$ is the outer Helmholtz plane potential calculated according to Eq. (11.30). Figure 11.17 shows that the agreement between these two sets of data is good. The DMPC bilayer is adsorbed at the Au(111) surface in the potential range from –0.4 to 0.45 V (versus

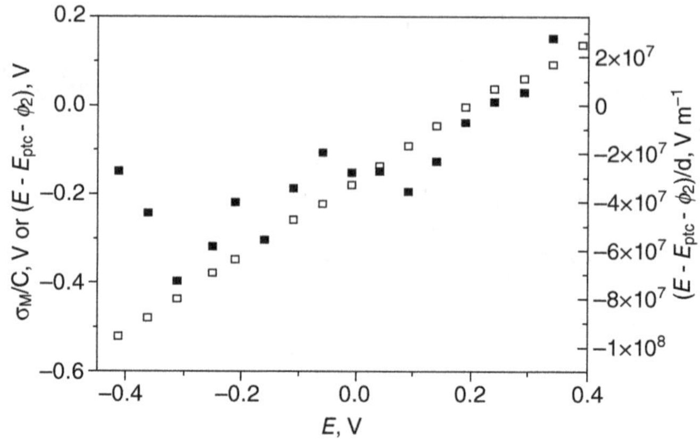


**Figure 11.17** Left-hand side vertical axis: (filled square) $\sigma_M/C$ calculated from the charge density data for a DMPC bilayer at the Au(111) electrode surface; (open square) plot of $(E - E_{pzc} - \phi_2)$. Right-hand side vertical axis: the electric field across the bilayer corresponding to $\sigma_M/C_m$ or $(E - E_{pzc} - \phi_2)/d$. Potentials were measured versus the Ag/AgCl reference electrode. Adapted from Ref. [58].

Ag/AgCl), and the corresponding potential drop across the membrane is ~−0.4 to ~0.2 V. Considering the thickness of the bilayer is ~5 nm, the electric field across the bilayer changes between ~−1.0 × $10^8$ V $m^{-1}$ and ~2 × $10^7$ V $m^{-1}$ in the potential region where the bilayer is adsorbed at the metal surface.

## 11.5 IR STUDIES OF ELECTRODE-SUPPORTED MEMBRANES

Fourier transform infrared spectroscopy (FTIR) is a powerful technique for biomembrane characterization. IR spectra can be obtained for a wide range of environments, including aqueous solutions and varying temperatures. In addition, there is no need for the use of special spin labels or fluorescent probes, and there is no size limitation for the molecules being studied. In addition, the availability of modern computers has enabled rapid and powerful FTIR data processing and conversion. Natural components of bilayer membranes have many vibrational modes that are IR active, and hence FTIR is a useful tool for studying structure and dynamics in membrane systems. The application of FTIR for studying two significant components of biomembranes, namely lipids and proteins, is discussed in the following sections.

### 11.5.1 Membranes Lipids

Figure 11.18 shows the isotropic optical constants of a typical membrane lipid, DMPC [59]. In order to concisely convey the most important features of lipid IR

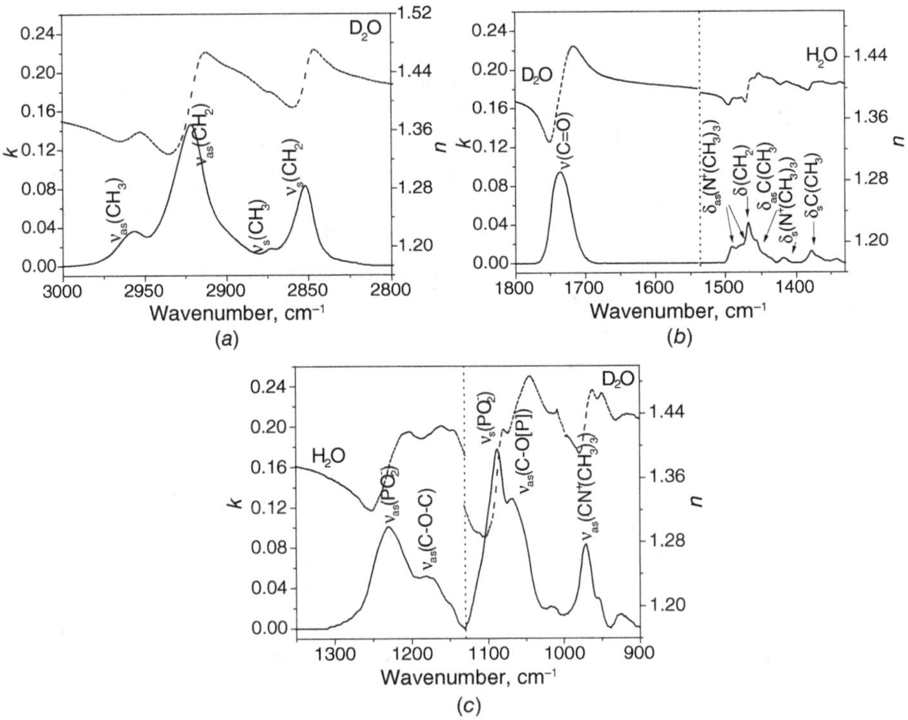


**Figure 11.18** Isotropic optical constants of DMPC determined from the transmittance IR experiment in $D_2O$ or $H_2O$ solution. (*a*) C–H stretching region, (*b*) C=O stretching region, and (c) P–O stretching region, *n* (dash line) refractive index, *k* (solid line) attenuation coefficient. Adapted from Ref. [59].

spectroscopy, the absorption bands of DMPC can be discussed in three parts according to their position within the lipid structure.

### *Acyl Chain Vibrational Modes (Fig. 11.18a)*

Infrared absorption bands due to hydrocarbon chains are easily identified on the basis of well-studied internal vibrations of polymethylene chain compounds [60, 61]. The C–H stretching vibrations of the acyl chain in the lipid molecules ($CH_3$, $CH_2$, etc.) generally fall in the wavenumber region between 2800 and 3100 $cm^{-1}$. Methyl aliphatic symmetric and asymmetric stretching vibrations occur at 2865–2885 and 2950–2975 $cm^{-1}$, respectively. Methylene aliphatic symmetric and asymmetric stretching vibrations occur at 2840–2870 and 2915–2940 $cm^{-1}$, respectively. The acyl chains give rise to C–H asymmetric and symmetric deformation modes which arise at 1440–1460 and 1370–1390 $cm^{-1}$, respectively. The asymmetric deformation band often overlaps with the $CH_2$ scissor vibration, which occurs at 1460–1480 $cm^{-1}$. The $CH_2$ rocking and wagging vibrations located in the region from

720–740 to 1340–1370 $cm^{-1}$ tend to be weak and not particularly useful from a structural analysis point of view.

The $CH_2$ asymmetric [$\nu_{as}(CH_2)$] and symmetric [$\nu_s(CH_2)$] stretching bands are the most intense bands in the lipid IR spectra and are commonly used to detect and monitor lipid-phase transitions. Lipids are known to exist in one or more polymorphic forms, and, depending on the conditions, conversions between these forms (known as phase transitions) can readily occur. The gel ($L_\alpha$) to liquid-crystalline ($L_c$) phase transition of phospholipids has been characterized by monitoring the changes in the band position and bandwidths of the $CH_2$ stretching bands when the phospholipid membrane crosses its chain melting phase transition temperature ($T_m$) [62–64]. Regardless of how the phase transition is induced, such transitions are accompanied by changes in both the peak position and the bandwidth of the $CH_2$ stretching bands. These changes accompany increases in conformational hydrocarbon chain disorder and mobility that occur during hydrocarbon chain-melting phase transitions, which are diagnostic of the incorporation of more gauche conformers into the acyl chains. While both the $\nu_{as}(CH_2)$ and $\nu_s(CH_2)$ bands can be used to monitor phase transitions, the $\nu_{as}(CH_2)$ band is less frequently used because it overlaps with contributions from the $CH_3$ groups, as well as Fermi resonance overtone vibrations from the methylene scissoring vibration. The $\nu_s(CH_2)$ band, in contrast, arises purely from the $CH_2$ symmetric stretching vibration and continues to be the main marker for lipid-phase transitions. In general, the melting of the hydrocarbon chains is accompanied by a 1.5–2.5 $cm^{-1}$ increase in the position of the $\nu_s(CH_2)$ band, and the magnitude of this shift is dependent on the type of acyl chains, the nature of the polar head group, and the presence of chemical additives [62–64]. The incorporation of a larger number of gauche conformers also increases the bandwidth of the peak, a result of increased lipid mobility [65, 66].

The $CH_2$ scissoring vibration, $\gamma(CH_2)$, located at 1470 $cm^{-1}$, can also be used to detect and characterize lipid chain-melting phase transitions. When the acyl chains are in an all-trans configuration, this mode gives rise to sharp absorption bands that are moderate in intensity. Upon chain melting, the band width increases. In addition, these bands are particularly useful because in the gel state, the band profile is very sensitive to the lateral packing interactions of the all-trans lipid chains. As a result, an examination of the contours of these bands can provide information regarding the mode of hydrocarbon packing prior to chain melting [67]. Methylene chains that are packed in a triclinic or hexagonal subcell arrangement give rise to scissoring vibrations at 1473 and 1469 $cm^{-1}$, respectively [68]. Chains that are packed in an orthorhombic perpendicular arrangement give rise to a split peak, with maxima at 1465 and 1473 $cm^{-1}$.

### Lipid Polar–Apolar Interface Vibrational Modes (Fig. 11.18b)

The polar–apolar interfacial region of lipids can contain several infrared-active vibrations, which can be used to obtain information regarding the structure of the lipids. In particular, many of the vibrational modes of the polar head groups are sensitive to hydrogen bonding and polarity effects and can be employed to observe direct or indirect hydration or dehydration effects, intra- or extramolecular binding of hydroxyl or amino groups, as well as ion-mediated changes at the bilayer–solution interface. Of these groups, the overwhelming majority of studies have been directed at a characterization of the IR absorption bands exhibited by the ester carbonyl stretching vibration of 1,2-diacylglycerolipids [69, 70]. From these studies, which have encompassed the liquid crystalline phases of a wide variety of diacylglycerolipids, the ester carbonyl stretching vibration has been found to be very complex and varies significantly depending on the nature of the polar head group, the environment, and the type and length of the acyl chains. For most of the common 1,2-diacylglycerolipids, the ν(C=O) band is actually a composite of two bands, centered at 1738–1742 and 1724–1729 $cm^{-1}$. Originally, these two bands were assigned to the *sn*-1 and *sn*-2 ester groups, respectively. However, later studies using $^{13}C{=}O$-labeled phospholipids suggested that the maxima of the *sn*-1 and *sn*-2 ν(C=O) should differ by no more than 4 $cm^{-1}$, and that the low- and high-frequency components of the ester band, at least for the case of hydrated phospholipids, arises from the differences in the degrees of hydration of, and/or hydrogen bonding to, the ester carbonyl groups [69–71]. The ester carbonyl stretching vibration is also sensitive to changes in the phase state of the membrane, with the low-frequency component of this band increasing in intensity as the membrane undergoes a gel to liquid crystalline phase transition [72].

Another useful vibration in the polar–apolar interfacial region is the C–C(O)–O–C group vibration [$\nu_{as}$(C–O–C)], which comes from both the γ and β chains [44]. This vibration, which is found between 1160 and 1180 $cm^{-1}$ depends significantly on the conformation of the ester group. If this moiety is planar, this band appears at 1180 $cm^{-1}$, and deviation from this conformation shifts the band to lower wavenumbers. The ester group in the γ chain exhibits a planar configuration of the C–C(O)–O–C frame. The corresponding absorption band is near 1180 $cm^{-1}$ and the dipole moment directs along the hydrocarbon chain. The C–C(O)–O–C fragment in the β chain has no planar arrangement and leads to a band shift from 1180 to 1165 $cm^{-1}$. Structural information concerning the polar–apolar interface region of phospholipid membranes can be extracted from the band position of this vibrational mode.

### Polar Head Group Vibrational Modes (Fig. 11.18c)

A number of characteristic infrared bands originate from the head group region of lipid molecules. Phospholipids contain several characteristic phosphate group vibrations, particularly the strong infrared bands between 1220 and 1260 $cm^{-1}$, which arise from the asymmetric $PO_2^-$ stretching mode [$\nu_{as}(PO_2^-)$], and the medium infrared band between 1085 and 1110 $cm^{-1}$, which is due to the symmetric $PO_2^-$ stretching mode [$\nu_s(PO_2^-)$] [73]. The frequencies of both vibrations depend strongly on the degree of hydration and intermolecular interactions. The C–O single bond of the phosphate ester [ν(C–OP)] gives rise to an absorption band located in the spectral region of 1040–1090 $cm^{-1}$, the exact frequency of this band depends significantly on the conformation of the C–O–P–O–C moiety [74]. The vibrations that arise from the phosphate region provide information on the degree of hydration of the head group. Studies on oriented phospholipid bilayers formed by drop deposition onto an ATR reflection plate revealed that the $\nu_{as}(PO_2^-)$ is extremely sensitive to hydration, shifting in frequency from 1250 $cm^{-1}$ for a dry lipid bilayer to 1230 $cm^{-1}$ for a hydrated lipid bilayer [73]. Similar behavior was observed in this spectral region for various phosphodiester-containing lipids, including phosphatidylcholine, phosphatidylethanolamine, and cardiolipin [75, 76]. These studies indicated that the phosphate head groups of hydrated lamellar phospholipids tend to retain comparable levels of hydration irrespective of the chemical structure of the head group and the phase state of the lipid.

The vibrational modes arising from the choline group in lipid molecules include the asymmetric C–N stretching of the methyl groups of the choline head group [$\nu_{as}(CN^+(CH_3)_3]$ at 950–970 $cm^{-1}$, the symmetric C–N stretching [$\nu_s(CN^+(CH_3)_3)$] at 860–930 $cm^{-1}$, and the asymmetric bending mode of the choline group [$\delta_{as}(CN^+(CH_3)_3]$ at 1470–1490 $cm^{-1}$ [74]. The frequencies of the vibrational modes of the choline group depend significantly on the conformation of the O–C–C–N frame. It has been reported that the $\nu_s(N(CH_3)_3)$ vibrational mode shows two bands, one at ~860 $cm^{-1}$ and the other at ~900 $cm^{-1}$ if the O–C–C–N fragment of choline has a gauche conformation, and only one band around 930 $cm^{-1}$ if the O–C–C–N fragment of choline has a trans conformation. These bands have been used to estimate the gauche/trans ratio of the O–C–C–N fragment in DPPC [74]. The interaction between water and the choline moiety in 1,2-dipalmitoyl-*sn*-glycero-3-phosphocholine (DPPC) has been investigated by Wong and Mantsch [77]. Direct hydrogen bond formation between water and the $CN^+(CH_3)_3$ does not occur. However, water molecules are expected to associate with the choline groups in fully hydrated DPPC through polar interactions between the positive charge of the choline group and the dipole and multipoles of water molecules [77]. These polar interactions are similar to those between water molecules and cations present in the electrolyte. This kind of "solvent effect" causes a decrease in the overall volume of the choline groups.

### 11.5.2 Proteins

Most proteins fold into unique three-dimensional structures, referred to as the native structure of the protein. There are four levels of protein structure [78]. At the simplest level, the primary structure of a protein is the sequence of amino acids that form the protein backbone. The secondary structure is the regularly repeating local structures, which are stabilized mainly by hydrogen bonding. Some of the common secondary structural elements include the α-helix, the β-sheet, the random coil, and the turn structure. Proteins can include more than one secondary structural element. Tertiary structure refers to the overall shape of the protein or the spatial relationship of different secondary structural elements. At the most complex level, the quaternary structure of proteins is the shape or structure that results from the interaction of one protein with another. Proteins are biopolymers of amino acids linked together via amide bonds. As a result, the major infrared feature of proteins comes from nine amide vibrational modes, which are listed in Table 11.1. Not all of the amide vibrational modes are suitable for IR studies of protein structure. The most useful modes are the amide I and amide II vibrations, discussed in detail below.

**Table 11.1** Characteristic IR Bands of Amide Vibrational Modes [79, 80]

| Vibrational Mode Designation | Frequency Region ($cm^{-1}$) | Nature of Vibration |
|---|---|---|
| Amide A | 3300 | N–H stretching in resonance with amide II overtone |
| Amide B | 3100 | |
| Amide I | 1600–1690 | 80% C=O stretching, 10% C–N stretching, 10% N–H bending |
| Amide II | 1480–1580 | 60% N–H bending, 40% C–N stretching |
| Amide III | 1230–1300 | 30% C–N stretching, 30% N–H bending, 10% C=O stretching, 10% O–C–N bending, 20% other |
| Amide IV | 625–770 | 40% O–C–N bending, 60% other |
| Amide V | 640–800 | Out-of-plane N–H bending |
| Amide VI | 530–610 | Out-of-plane C=O bending |
| Amide VII | 200 | Skeletal torsion |

### *Amide I Vibration*

The amide I vibration, which is composed of ~80% C=O stretching coupled to the in-plane N–H bending and C–N stretching modes, is by far the most useful amide band for the structural analysis of proteins. The exact frequency of this band is dependent on the nature of hydrogen bonding experienced by the carbonyl group and the coupling between transition dipoles, which in turn is dictated by the secondary structural element adopted by the protein. Proteins often contain more than one secondary structural element, as mentioned above, and as a result, the amide I band is actually a composite band for this reason, containing many bands resulting from different structural conformations. These bands often overlap, causing the amide I band to appear intense and very broad. In order to determine the component band positions, generally Fourier self-deconvolution (FSD), second-order derivitization, and 2D-COS are employed for the band deconvolution [26–28, 79, 80]. Table 11.2 lists the characteristic amide I band frequencies and their designated protein secondary structures based on normal-mode analysis of model peptides and proteins. While the amide I band can be a very useful probe of protein secondary structure, it also has several limitations. First, the water bending vibration centered at 1640 $cm^{-1}$ is a major interference and necessitates the use of $D_2O$ for in situ studies. Using $D_2O$ as the solvent causes only a minimal shift (<3 $cm^{-1}$) in the amide I composite bands. An additional complication of using the amide I band for analysis is that an accurate band deconvolution can be difficult to obtain, especially for a protein of unknown secondary structure.

### *Amide II Vibration*

The amide II vibration, which originates mainly from N–H bending, can provide useful information on the tertiary structure of the protein. The position of this band is very sensitive to deuteration and shifts from ~1550 to ~1460 $cm^{-1}$ upon isotopic

**Table 11.2** Amide I Frequencies of Protein Secondary Structure in $H_2O$ [81, 82]

| Frequency ($cm^{-1}$) | Assignment |
|---|---|
| 1621–1627 | β-sheet |
| 1628–1634 | β-sheet |
| 1635–1640 | β-sheet |
| 1646–1650 | Random coil |
| 1651–1657 | α-Helix |
| 1660–1666 | $3_{10}$ helix |
| 1668–1671 | β-turn |
| 1671–1676 | β-turn |
| 1678–1682 | β-turn |
| 1683–1687 | β-turn |
| 1689–1698 | β-sheet |

exchange of the N–H hydrogen for deuterium (N–D, named amide II′ mode) [81]. Unfortunately, for in situ studies, the shifted N–D band overlaps with the H–O–D bending vibration at ~1438 $cm^{-1}$. However, the residual component of the 1550 $cm^{-1}$ band (N–H, which does not undergo isotopic exchange) can provide information regarding solvent inaccessibility to the peptide backbone. For example, hydrophobic environments or tightly ordered structures such as sheets and helices will impede the exchange of the amide N–H proton. Therefore, protein folding–unfolding events can be monitored using this band.

## 11.6 PREPARATION OF BIOMIMETIC MEMBRANES

There are several experimental methodologies for the fabrication of model biomimetic membranes. For example: (i) unilamellar vesicles, which are hollow spheres (25 nm to 100 μm in diameter) enveloped by a bilayer of lipid molecules; (ii) lipid monolayers at the air–water interface; (iii) black lipid membranes suspended over an aperture between two aqueous phases; (iv) Langmuir–Blodgett and Langmuir–Schaefer transfer methods, which allow the transfer of lipid mono- and multilayers from the air–water interface to a solid support; (v) self-assembled monolayers (SAMs, thiolipids on gold); and (iv) polymer coatings with tethers followed by the spontaneous spreading of liposomes. The most common procedures used to build biomimetic membranes at gold electrode surfaces are the vesicle fusion (VF) and Langmuir–Blodgett (LB) and Langmuir–Schaefer (LS) transfer methods, both of which are discussed below.

### 11.6.1 Vesicle Fusion

Vesicles are spherical lipid bilayers that enclose an aqueous solution. The procedure for vesicle fusion consists of the adhesion and fusion of small unilamellar vesicles

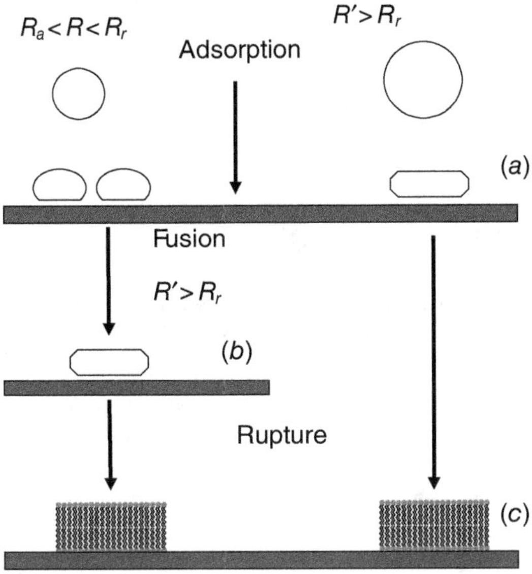


**Figure 11.19** Model of the early stages of supported lipid bilayer formation via vesicle fusion. (*a*) Vesicle adsorption to a solid surface (black). At large *R*, an adsorbed vesicle resembles a pancake. (*b*) Fusion of surface-bound vesicles leads to an increase in the vesicle radius. Consequently, the adsorbed vesicle becomes more pancake-like and ruptures. (*c*) Rupture of surface-bound vesicles with $R > R_r$ leads to the formation of single-bilayer disks. Adapted from Ref. [83].

(SUVs, 20–50 nm in diameter) at a solid substrate from an aqueous vesicle dispersion. At the Au(111) surface, vesicle fusion forms a lipid bilayer by rupturing of the vesicles and unrolling and spreading onto the substrate.

A theory depicting the adhesion, fusion and rupture of vesicles at solid surfaces has been developed by Lipowsky and Seifert [84] and Lipowsky [85]. The validity of this theory has been experimentally confirmed by Reviakine and Brisson [83]. The theory assumes that the adsorption, fusion, and rupture of SUVs depend on the vesicle size. Figure 11.19 demonstrates the model of the early stages of supported lipid bilayer formation. Vesicles with a radius $R < R_a$ (critical adsorption radius) do not adsorb on the substrate, and on the other hand, vesicles can rupture and form a single bilayer disk directly (Fig. 11.19*c*) if $R > R_r$ (the minimum rupture radius). Adsorbed vesicles with radius between $R_a$ and $R_r$ fuse first to form a larger disk (Fig. 11.19*b*), which later ruptures. The $R_a$ and $R_r$ can be determined by the bending rigidity and line tension of the bilayer.

In studies presented in this chapter, the procedure described by Barenholtz et al. [86] was used to prepare the SUVs. A 10-mg/mL chloroform solution of lipid (e.g., DMPC) was used as the stock solution. The stock solution was dried by vortexing in a test tube under a flow of argon. To remove any remaining residue of the solvent, the dried lipid film was placed in a vacuum desiccator over night. Next, 0.1 M NaF electrolyte was added to the dry lipid and the mixture was sonicated at

35°C for at least 1 h. The solutions of vesicles were added to a PM–IRRAS cell using a syringe such that the final concentration of lipid was about $1 \times 10^{-4}$ M in the IR cell. Dynamic light scattering (DLS) measurement [87] demonstrated that, in DMPC vesicle solution, vesicles with a diameter range between 16 and 30 nm are dominant.

### 11.6.2 Langmuir–Blodgett and Langmuir–Schaefer Deposition

The Langmuir–Blodgett (LB) technique involves the vertical movement of a solid substrate through a compressed monolayer at the air–solution interface. The surface pressure and temperature of the monolayer at the surface must be controlled such that the organic film is in a condensed and stable state. Generally, this is for surface pressures ranging from 10 to 40 mN m$^{-1}$ for the compressed monolayer film. Figure 11.20*a* shows the most common case of LB deposition, where the substrate is hydrophilic and the first monolayer is transferred like a carpet onto the substrate surface as it is raised through the compressed monolayer. When the substrate is placed into the subphase before the monolayer is spread, the resulting multilayers that are deposited are referred to as Y-type layers. While this is the most common case of deposition, there are many cases of amphiphiles that will only transfer on a downstroke (downward movement of the substrate through the interface) or an upstroke (upward movement of the substrate through the interface). DMPC, for example, will transfer only on an upstroke, making direct LB deposition of a Y-type bilayer structure impossible.

Film transfer is characterized by the measurement of the transfer ratio (or deposition ratio), τ. The transfer ratio is defined as the decrease in the area occupied by the monolayer (held at constant pressure) on the subphase surface, divided by the coated area of the solid substrate as [88]

$$\tau = \frac{A_L}{A_S} \tag{11.32}$$

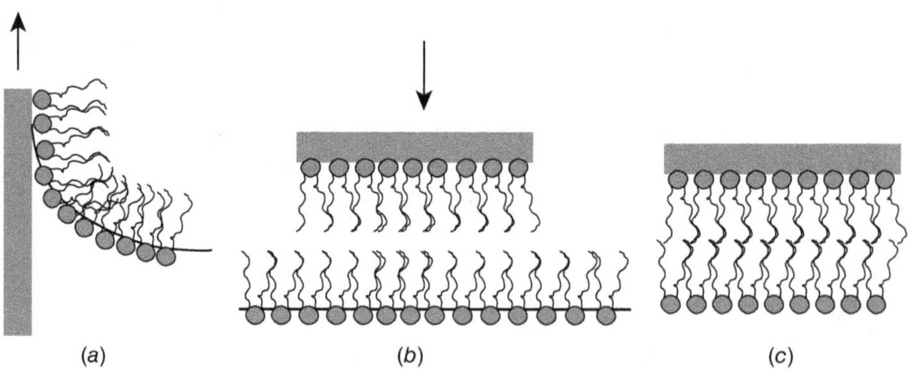


**Figure 11.20** Schematic diagrams of the (*a*) LB and (*b*) LS depositions. (*c*) Schematic structure of a Y-type bilayer. Adapted from Ref. [58].

where $A_L$ is the decrease in the area occupied by the monolayer on the subphase surface and $A_S$ is the coated area of the substrate. The transfer ratio for a perfectly transferred monolayer should be 1.0, and transfer ratios that are outside of the range of 0.9–1.1 suggest poor film homogeneity. In order to achieve a good transfer ratio, the deposition speed (speed at which the substrate is brought through the compressed monolayer) should be chosen appropriately. To improve the deposition process, it is often suggested that the dipping process be halted after an upstroke to allow the transferred monolayer to dry completely before continuing the deposition cycle.

The schematic of the Langmuir–Schaefer (LS), or horizontal touch, deposition is shown in Figure 11.20*b*. The first step in the LS deposition involves compressing a monolayer at the air–solution interface to a desired pressure in a Langmuir–Blodgett trough. The substrate, which is oriented parallel to the surface, is then brought into contact with the compressed monolayer and then immediately withdrawn at a very slow speed. This procedure allows for the transfer of a single monolayer onto the surface, which is oriented such that the hydrocarbon chains face the substrate and the polar head groups extend into the air. A combination of the LB and LS deposition methods, therefore, result in a Y-type bilayer on the Au(111) surface (Fig. 11.20*c*).

An advantage of the LB–LS deposition over vesicle fusion is that it allows for the construction of an asymmetric bilayer. In the LB–LS deposition, different components can be used for the inner and outer leaflet, for example, construction of a hydrophilic spacer molecule in only the inner leaflet bilayer while maintaining a more natural phospholipid environment in the outer leaflet [89]. On the other hand, an advantage of vesicle fusion over the LB–LS depositions is that vesicles may easily incorporate membrane proteins, forming proteoliposomes.

## 11.7 PM–IRRAS STUDIES OF ONE-COMPONENT BIOMIMETIC MEMBRANES

The works by Horswell et al. [90] and Zamlynny et al. [25] were the first quantitative in situ PM–IRRAS studies of thin organic films at a gold electrode surface. However, important improvements of both the technique and the data analysis have been made in the past 10 years. The first generation of biomimetic membranes supported at the electrode surfaces consisted of pure DMPC [58, 59, 90, 91]. Figure 11.21 shows the schematic diagram of the DMPC molecule and the directions of the transition dipole moments of the major IR bands. DMPC is a zwitterionic molecule, consisting of two 14-carbon saturated acyl chains, a glycerol backbone, a negatively charged phosphate group, and a positively charged choline group. It is relatively stable in air and its liquid crystalline–gel phase transition temperature ($T_m$) is about 24C [96]. Therefore, it is easy to assemble DMPC bilayers at a solid surface that are either in the liquid crystalline phase or gel phase. The DMPC bilayer can be deposited at the Au(111) surface by the vesicle fusion (VF) [59, 90] and LB–LS methods [58, 91]. However, the LB–LS method allows one to form bilayers in which the structure is better controlled than bilayers formed by the VF method. Therefore,

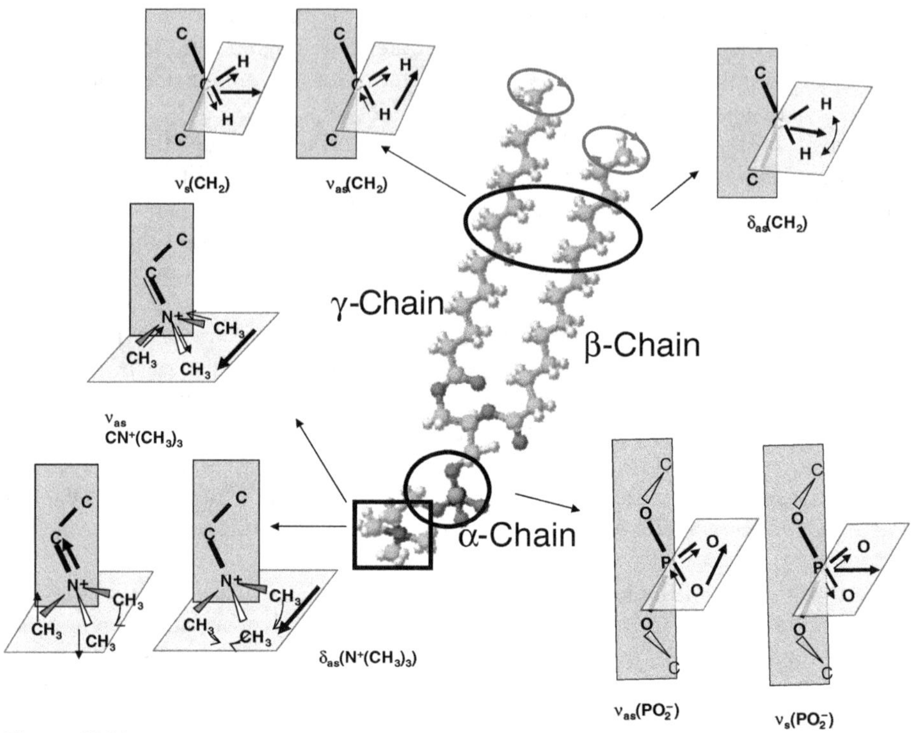


**Figure 11.21** Schematic diagram of the DMPC molecule and the directions of the transition dipole moments of the major IR bands. Adapted from Ref. [59].

the DMPC bilayer prepared by the LB–LS method will be used as a model to illustrate what information about a biomimetic membrane one can obtain from combined electrochemical and PM–IRRAS studies.

Figure 11.22*a* plots the charge density curves for the bare Au(111) electrode in a 0.1-M NaF solution and for the electrode with DMPC deposited at a Au(111) surface by the LB–LS method at the surface pressure of 42 mN $m^{-1}$ [58]. The bottom horizontal axis plots the potential applied to the gold electrode versus the reference electrode. The top horizontal axis plots $(E - E_{pzc} - \phi_2)$, which is a measure of potential drop across the membrane, equivalent to the transmembrane potential. The data illustrate that the DMPC bilayer is adsorbed on the metal surface when the absolute value of the charge on the metal is less than 10 μC $cm^{-2}$. The plateau section on the curve corresponding to the bilayer-covered electrode has a small step at $E \approx -0.1$ V. This behavior suggests that the adsorbed bilayer exists in different states in the range of potentials between $-0.4 < E < -0.1$ V and $-0.1 < E < 0.4$ V versus the saturated calomel electrode (SCE) (corresponding to transmebrane potentials between –0.5 and –0.2 V and –0.2 and 0.1 V, respectively). Independent neutron reflectivity studies demonstrated that the membrane is stable in the –0.2 to 0.1 V region of the transmembrane potential and that electroporation of the adsorbed membrane takes

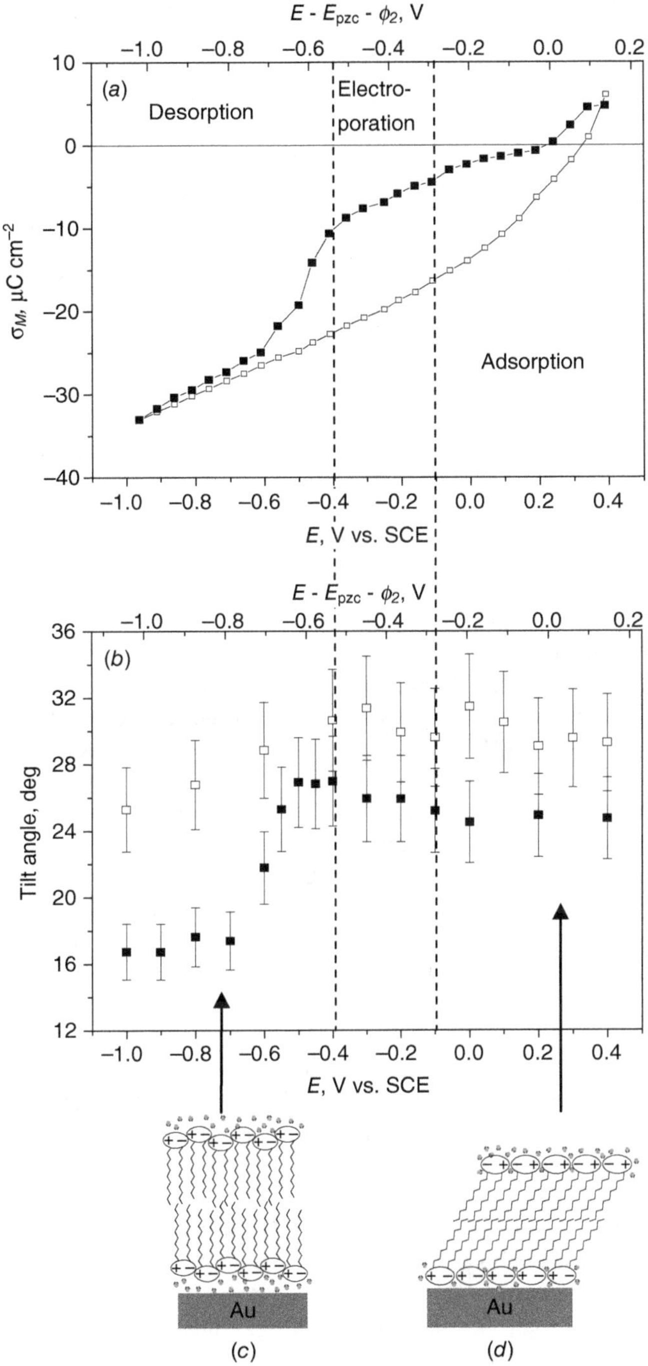


**Figure 11.22** (a) Charge density curves for a DMPC bilayer covered (filled squares) and bare (open squares) Au(111) electrode in 0.1 M NaF solution. (*b*) The tilt angle of the acyl chains in the DMPC bilayer transferred at the surface pressure of 42 mN m$^{-1}$ (filled squares) and 40 mN m$^{-1}$ (open squares). The dashed line represents the potential at which the DMPC bilayer begins to desorb from Au(111) surface. The bottom two illustrations show the packing of the DMPC molecules in the desorption (*c*) and adsorption (*d*) states. Adapted from Ref. [58].

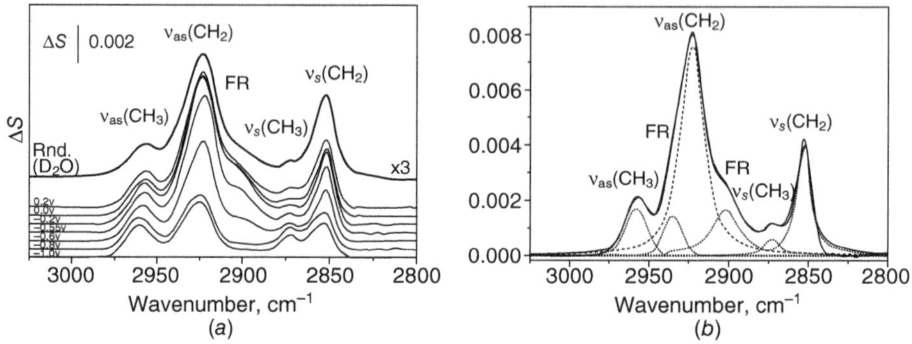


**Figure 11.23** (*a*) PM–IRRAS spectra in the C–H stretching region for a DMPC bilayer at Au(111) electrode in 0.1 M NaF/$D_2O$ solution. The top trace plots the spectrum for a 5.4-nm-thick film of randomly oriented DMPC molecules calculated using optical constants for DMPC vesicles in $D_2O$ solution. (*b*) Band deconvolution result of the PM–IRRAS spectrum at $E = -0.2$ V. Adapted from Ref. [58].

place at transmembrane potentials between −0.2 and −0.5 V. When the charge density is more negative than −10 μC $cm^{-2}$ (transmembrane potential is more negative than −0.5 V), the DMPC bilayer begins to desorb from the Au(111) surface. The charge density curve of the bilayer-covered electrode merges with that of the bare electrode when $E \leq -1.0$ V, indicating that DMPC is completely detached from the gold surface at these very negative potentials to float on an ~1-nm-thick cushion of electrolyte that separates the bilayer from the gold surface [53, 54].

We will now illustrate the wealth of molecular level information one can obtain regarding potential-induced changes to the bilayer by combining electrochemical and PM–IRRAS measurements. Figure 11.23*a* plots the PM–IRRAS spectra in the C–H stretching region for the DMPC bilayer at the Au(111) surface, at select electrode potentials. Figure 11.23*b* illustrates the band deconvolution of the PM–IRRAS spectrum of the DMPC bilayer C–H stretching vibrations at $E = -0.2$ V. The spectrum consists of four distinct bands, which correspond to the $CH_3$ asymmetric stretching ($\nu_{as}(CH_3)$ at ~ 2960 $cm^{-1}$), $CH_2$ asymmetric stretching [$\nu_{as}(CH_2)$ at ~2924 $cm^{-1}$], $CH_3$ symmetric stretching [$\nu_s(CH_3)$ at ~2872 $cm^{-1}$] and $CH_2$ symmetric stretching [$\nu_s(CH_2)$ at ~2853 $cm^{-1}$] vibrational modes. The two shoulder peaks on the high- and low-frequency sides of the $\nu_{as}(CH_2)$ band result from the Fermi resonance of the bending modes of $CH_3$ and $CH_2$ groups.

The frequency of the $CH_2$ vibration provides useful information about the conformation of the acyl chains in the DMPC molecule. If the bilayer is in the gel state, in which the acyl chains of DMPC molecules are fully stretched and assume an all *trans* conformation, the positions of the $\nu_{as}(CH_2)$ and $\nu_s(CH_2)$ should be lower than 2920 and 2850 $cm^{-1}$, respectively [60, 92, 93]. In the spectra shown in Figure 11.23*a*, the positions of the $\nu_{as}(CH_2)$ and $\nu_s(CH_2)$ bands have higher wavenumbers than these reported values, and this blue shift indicates that the acyl chains of DMPC molecules in the bilayer are not in the all-trans state but also contain a small proportion of gauche conformers. The shape and position of the $\nu_{as}(CH_2)$ and $\nu_s(CH_2)$ bands depends somewhat on the electrode potential. The bands are shifted to lower frequencies and are somewhat narrower at $E > -0.4$ V, indicating that the bilayer is more ordered when it is adsorbed onto the gold surface.

Figure 11.23 shows that the band intensity is lower at negative potentials where the bilayer is detached from the gold surface and increases at more positive potentials where the bilayer is adsorbed directly onto the gold surface. This behavior indicates that adsorption/detachment of the film has an influence on the orientation of DMPC molecules and, consequently, on the angle between the direction of the transition dipole of a given vibrational mode and the metal surface normal. Figure 11.22*b* plots the average values of the tilt angle, $\theta_{tilt}$, of the trans fragment of the acyl chains in the DMPC bilayer calculated from the integrated intensities of the $\nu_{as}(CH_2)$ and the $\nu_s(CH_2)$ bands with the help of Eq. (11.17). There are two sets of these values coming from two independent measurements. The first corresponds to a film transferred at 42 mN $m^{-1}$ at which the area per DMPC molecule was 0.45 $nm^2$ [58]. The second was transferred at 40 mN $m^{-1}$ and the area per DMPC molecule was 0.48 $nm^2$ [91]. The two sets of data display similar changes as a function of the applied potential. However, smaller values of the tilt angle are observed for the bilayer transferred at 42 mN $m^{-1}$. The changes of the $\theta_{tilt}$ with potential correlate very well with the changes in the surface charge density. The IR data show that for transmembrane potentials between −0.2 and 0.1 V the average chain tilt is ~25°. The tilt increases to ~28° for transmembrane potentials between −0.2 and −0.5 V. The increase of the tilt angle in this region is a measure of partial disorder in the bilayer due to electroporation. The tilt angle drops down significantly at more negative potentials to attain a value ~18° in the detached state (at $E < -0.8$ V versus SCE), indicating that the structure of the bilayer is different in the adsorbed and desorbed states. In the adsorbed state, the polar heads of DMPC molecules are located in one plane (Fig. 11.22*c*); while in the detached state the bilayer floats on an ~1-nm-thick layer of the electrolyte [53, 54], and the polar heads are packed in a zigzag fashion (Fig. 11.22*d*), which allows for a smaller tilt of the acyl chains. The change of the acyl chain tilt angle is driven by the difference in the packing of the polar heads for the two states.

The LB–LS method allows for a layer-by-layer characterization of the supported bilayer. Garcia et al. [91] assembled DMPC bilayers with one leaflet consisting of deuterium and another leaflet by hydrogen-substituted acyl chains of the lipids. PM–IRRAS studies of such bilayers provide information concerning the orientation of the acyl chains in the individual leaflets of the bilayer. At the gold electrode surface, the bilayer is exposed to an asymmetric environment. The bottom leaflet is in contact with a hydrophobic metal surface, while the top leaflet is interacting with the aqueous solution. Figure 11.24 shows the plot of the average tilt of trans fragments of acyl chains in the top and bottom leaflets of this bilayer. The results show that the asymmetry of the environment has an effect on the orientation of the DMPC molecules in each leaflet. The tilt angle of the acyl chains of the DMPC molecules in the leaflet in contact with gold is ~10° smaller than in the top leaflet that is exposed to the solution. The bilayer deposited at the electrode surface is adsorbed onto the metal surface up to an applied potential of −0.4 V versus SCE or transmembrane potentials more positive than −0.5 V. As mentioned above, at more negative potentials,

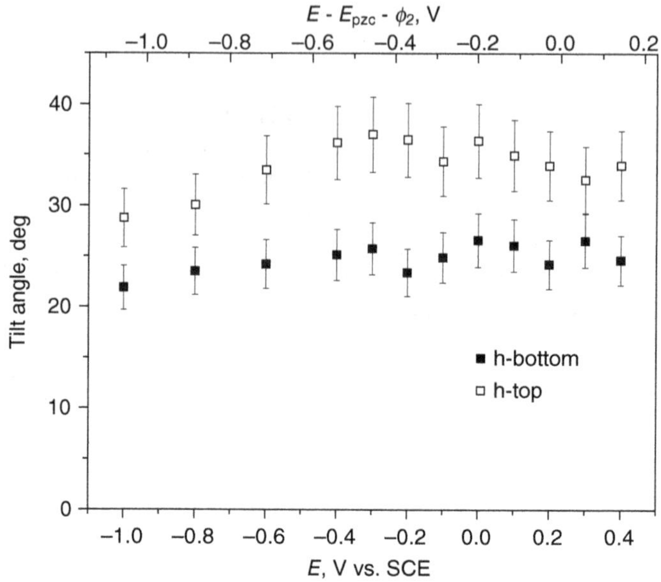


**Figure 11.24** Plots of the angle between the acyl chain of h-DMPC molecule and the surface normal as a function of the electrode potential. (Filled squares) for a bilayer with h-DMPC in the bottom layer, and (open squares) for a bilayer with h-DMPC in the top layer. Taken with permission from Ref. [91].

the bilayer is detached from the metal surface but remains in close proximity to the surface, separated from the metal by a thin cushion of electrolyte. In the detached state, the two leaflets are in contact with water and the bilayer is exposed to a much more symmetric environment. In response to this symmetric environment, the difference in the tilt angles for the two leaflets becomes smaller (~5°). This result provided direct evidence that the structure of a phospholipid bilayer supported at an electrode surface is affected by the interaction with the metal. The tilt angle data for the bilayer transferred at 40 mN m$^{-1}$ in Figure 11.22 is the average of the tilt angle in the individual leaflets shown in Figure 11.24.

The IR bands originating from the polar head group region provide information not only about the orientation and conformation of this fragment of the molecule but also about its hydration state. This is illustrated by the IR bands of the glycerol fragment of the molecule. Two bands correspond to the vibrations of the glycerol ester group: the C=O stretching [ν(C=O)] at ~1740 cm$^{-1}$ and the asymmetric C–O–C stretching [$\nu_{as}$(C-O-C)] at ~1180 cm$^{-1}$. The ν(C=O) bands can provide useful information concerning the potential-induced changes in the local environment of the carbonyl ester group and particularly changes in the hydration of the ester group. Figure 11.25*a* shows the PM–IRRAS spectra of the DMPC bilayer in the C=O stretching region. In the dispersion of DMPC vesicles in $D_2O$, the ν(C=O) band is broad because it is composed of two overlapping bands with maxima at 1744 and 1734 cm$^{-1}$, which correspond to the presence of non-hydrogen-bonded and

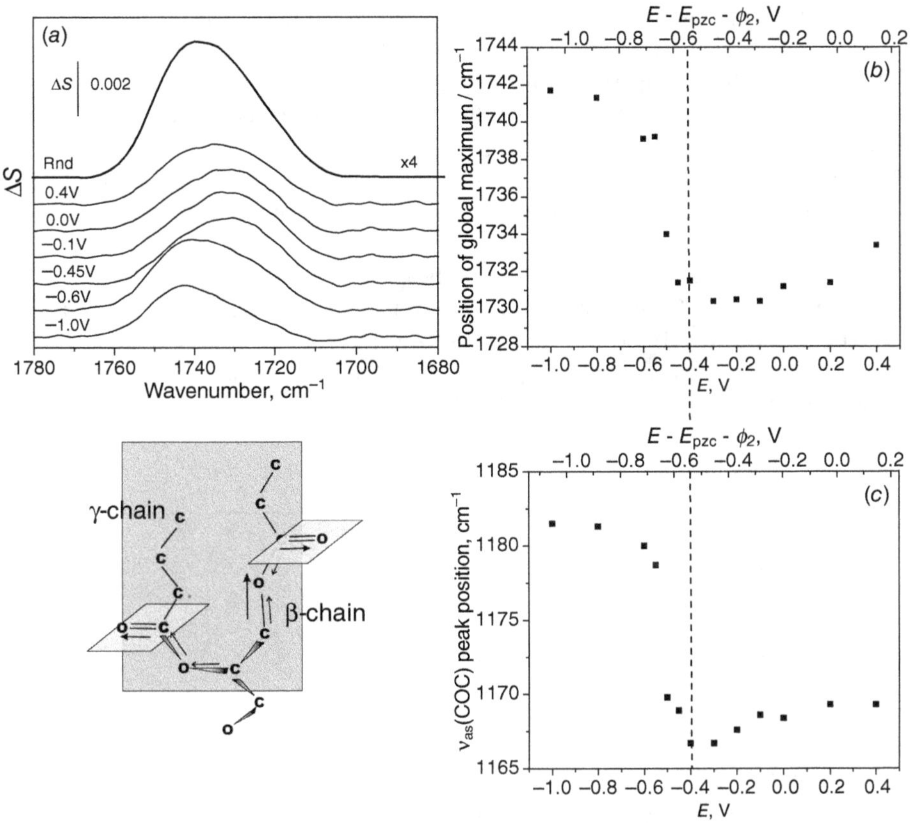


**Figure 11.25** (*a*) PM–IRRAS spectra of a DMPC bilayer at a Au(111) surface in the C=O stretching region in 0.1 M NaF/$D_2O$ solution. The top trace plots the spectrum (thick line) for a 5.4-nm-thick film of randomly oriented DMPC molecules calculated using optical constants of DMPC vesicles in $D_2O$ solution. (*b*) Position of the global maximum of the ν(C=O) band in the DMPC bilayer. The scheme of the glycerol moiety in phosphatidylcholine with marked directions of the transition dipole moment vectors of the ν(C=O) and the $\nu_{as}$(COC) modes. (*c*) Position of the $\nu_{as}$(COC) band in the DMPC bilayer. Dashed line indicates the potential at which the onset of desorption is observed to begin. Adapted from Ref. [58]. Potentials measured versus Ag/AgCl/$Cl^-$ electrode.

hydrogen-bonded ester groups, respectively [70]. Figure 11.25*b* plots the variation of the ν(C=O) band position as a function of electrode potential for the DMPC bilayer. In the detached bilayer, the maximum of the ν(C=O) band is located at 1742 $cm^{-1}$, indicating that the non-hydrogen-bonded state dominates. In the potential region where adsorption of the bilayer is indicated ($E > -0.4$ V versus SCE and transmembrane potentials more positive than −0.5 V), the ν(C=O) band position undergoes a significant red shift to 1731$cm^{-1}$, illustrating that the ester group is well hydrated in the adsorbed state.

The IR data demonstrate that the adsorption–desorption of the bilayer leads to significant changes in the hydration state of the ester group. The ester carbonyl group is less hydrated in the desorbed state and more hydrated in the adsorbed state. When the bilayer is in direct contact with the metal surface, more water penetrates into the polar head group region. Interestingly, the positions of the band maximum have somewhat higher values at transmembrane potentials between −0.2 and 0.1 V than between potentials −0.5 to −0.2 V. This behavior correlates well with the onset of electroporation observed in the latter region in neutron reflectivity experiments [53, 54]. Electroporation apparently increases the hydration of the interfacial region of the bilayer.

The asymmetric C–O–C stretching of the ester groups in the β and γ acyl chains is observed at ~1190 cm$^{-1}$ and overlaps with the asymmetric $PO_2^-$ stretching at ~1240 cm$^{-1}$. Figure 11.26*a* shows PM–IRRAS spectra of a DMPC bilayer in this spectral region and the calculated spectra for the film of randomly oriented DMPC molecules using optical constants of DMPC vesicles in $D_2O$. The band deconvolu-

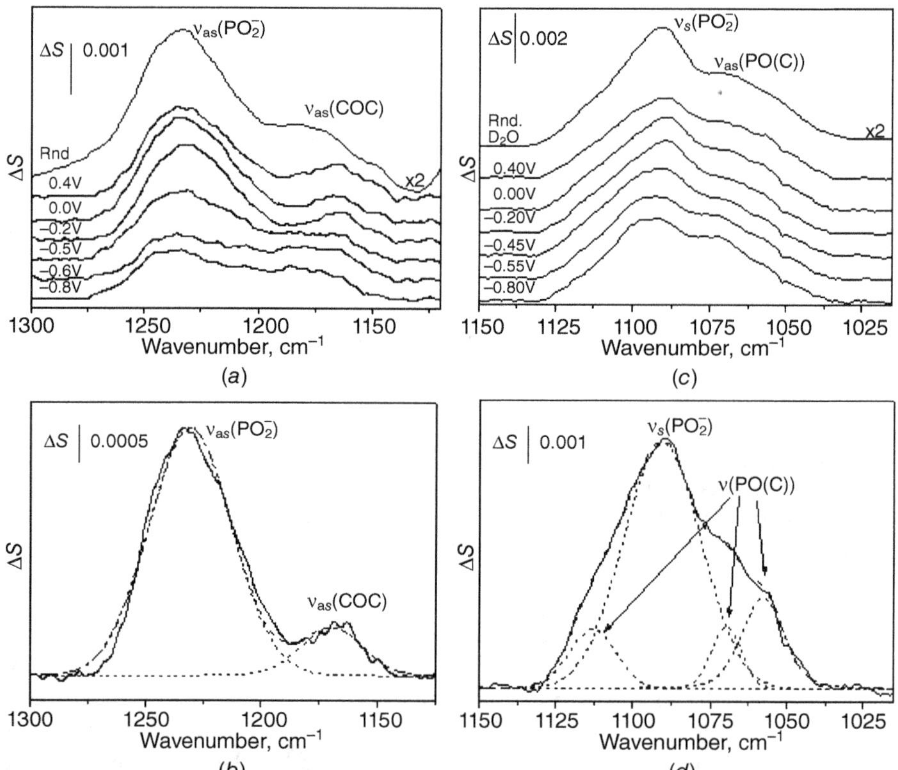


**Figure 11.26** PM–IRRAS spectra in the phosphate stretching region of (*a*, *b*) $\nu_{as}(PO_2^-)$ and (*c*, *d*) $\nu_s(PO_2^-)$ of a DMPC bilayer at Au(111) surface in 0.1 M NaF/$H_2O$ solution. The top trace plots spectrum (thick line) for 5.4-nm-thick film of randomly oriented DMPC molecules calculated using optical constants of DMPC vesicles in $D_2O$ solution. (*b*, *d*) Deconvoluted spectra at $E = -0.2$ V with band assignment. Adapted from Ref. [58].

tion is shown in Figure 11.26*b*. The ab initio calculation shows that the $\nu_{as}$(C–O–C) of the ester group is strongly coupled with the wagging motions of the tails [94]. The calculations also show that the coupling of the vibrations of the –C(O)–O–C– groups in the β and γ chains is rather weak. The ester group in the β chain is coplanar with the chain, while the ester group of the γ chain is out of the plane of the chain (see cartoon in Fig. 11.25*c*). The $\nu_{as}$(C–O–C) band is observed at ~1180 $cm^{-1}$ when the C–C(O)–O–C frame of the ester group has a planar conformation, and a deviation from the planar conformation leads to a red shift of this band to ~1160 $cm^{-1}$ [44]. Figure 11.25*c* plots the peak of the $\nu_{as}$(C–O–C) band maximum as a function of the electrode potential. The position of this band changes from ~1180 $cm^{-1}$ in the desorbed state to ~1167 $cm^{-1}$ in the adsorbed state. This behavior suggests that a significant deviation from planarity takes place upon moving from the desorbed to the adsorbed state. The chains are coplanar in the desorbed state but are stressed and twisted in the adsorbed state. This is an indication that the DMPC bilayer adsorbed at the Au(111) surface is stressed.

The bands of the phosphate group include the asymmetric $\nu_{as}(PO_2^-)$ and symmetric $\nu_s(PO_2^-)$ stretching of the $PO_2^-$ group and two bands corresponding to the complex asymmetric vibrations of the phosphate ester group [$\nu_{as}$(C–O[P])]. The $\nu_{as}(PO_2^-)$ overlaps with the C–O–C stretch of the ester group, and these bands are shown in Figure 11.26*b*. The $\nu_{as}(PO_2^-)$ band has a maximum at 1255 $cm^{-1}$ in the $CCl_4$ solution, and in the vesicle solution in $D_2O$, the band is broader and its maximum is red shifted to 1237 $cm^{-1}$ [59]. In fact, this is a composite band that has two components centered at 1228 and 1246 $cm^{-1}$. The lower-frequency component is assigned to the hydrated phosphate group, while the higher-frequency band is assigned to the non-hydrogen-bonded phosphate group [95]. In the PM–IRRAS spectra, the band center shifts from ~1238 $cm^{-1}$ in the desorption state to ~1231 $cm^{-1}$ in the adsorption state. This behavior indicates the phosphate group is more hydrated in the adsorption state than the desorption state, consistent with the behavior of the C=O group.

Figure 11.26*c* shows the PM–IRRAS spectra of the DMPC bilayer in the 1125–1000 $cm^{-1}$ region, which contains the $\nu_s(PO_2^-)$ at ~1090 $cm^{-1}$ overlapping with two asymmetric $\nu_{as}$(C–O[P]) bands at ~1070 and ~1055 $cm^{-1}$. The band deconvolution is shown in Figure 11.26*d*. The $\nu_s(PO_2^-)$ band has a maximum at 1091 $cm^{-1}$ in the vesicle solution in $D_2O$. In the DMPC bilayer at a Au(111) surface, the maximum of the $\nu_s(PO_2^-)$ band shifts from ~1090 to ~1094 $cm^{-1}$ when the electrode potential changes from positive to negative. Consistent with the results for the $\nu_{as}(PO_2^-)$, these changes indicate that the phosphate group is more hydrated when the bilayer is adsorbed at the electrode surface.

The transition dipoles of the $\nu_s(PO_2^-)$ and $\nu_{as}(PO_2^-)$ are located in the plane of the nonesterified group and are perpendicular to each other. They are also perpendicular to the line connecting the two esterified oxygen atoms of the phosphate group. Hence, Eq. (11.16) can be used to calculate the angle between the C–O–P–O–C line (line joining the two oxygen atoms in the phosphate group) and the surface normal. In the adsorbed state, this angle is ~22°; while in the desorbed state, the angle decreases to ~20°. These differences are very small and within experimental

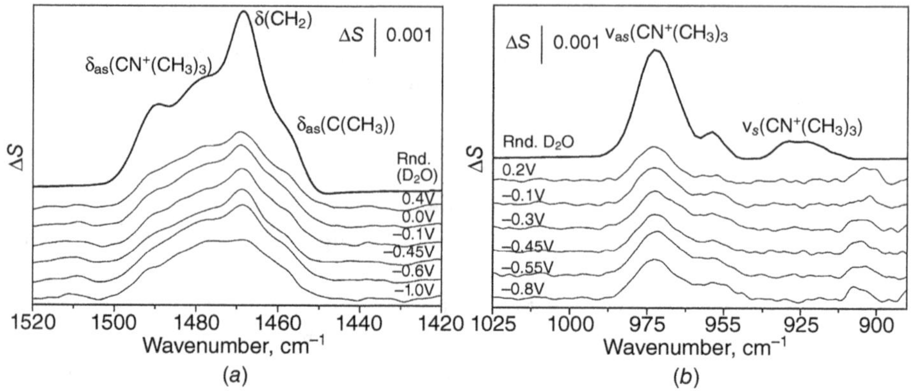


**Figure 11.27** PM–IRRAS spectra region for a DMPC bilayer at a Au(111) electrode in the (*a*) C–H bending and the (*b*) C–N stretching region. The top trace plots spectrum for 5.4-nm-thick film of randomly oriented DMPC molecules calculated using optical constants of DMPC vesicles in $D_2O$ solution. Adapted from Ref. [58].

uncertainty. They suggest that the orientation of the phosphate group changes little with the electrode potential.

Information concerning the choline group can be extracted from the 1500–1350 $cm^{-1}$ region where C–H bending modes of the $CH_3$ groups in the $CN^+(CH_3)_3$ moiety are located and in the 850 and 1000 $cm^{-1}$ regions where the C–N stretching vibrations can be found. Figure 11.27 shows the PM–IRRAS in the two regions. In the 1500–1350 $cm^{-1}$ region, the C–H bending bands of the choline group overlap with the bending modes of the terminal $CH_3$ groups of the acyl chains. The asymmetric bending modes of the $CH_3$ groups $\delta_{as}[CN^+(CH_3)_3]$ attached to the nitrogen atom in the choline group appear at ~1490 and ~1480 $cm^{-1}$ and the symmetric modes are at ~1420 and ~1380 $cm^{-1}$. The symmetric bands are much weaker and more difficult to separate from the background in the PM–IRRAS spectra. In the spectrum of the DMPC bilayer supported at the Au(111) surface, the center of the high-frequency and low-frequency band is observed at 1490 and 1478 $cm^{-1}$, respectively. The band shape and position for the choline group are nearly independent of the electrode potential. Due to the cut-off frequency of the $BaF_2$ prism (~800 $cm^{-1}$) and low performance of the PEM in the low wavenumber region (<1000 $cm^{-1}$) only those C–N stretching bands with frequencies higher than 900 $cm^{-1}$ can be measured using PM–IRRAS. The position of the symmetric $CN^+(CH_3)_3$ stretches provides information on the conformation of the O–C–C–N frame of the choline moiety. In the trans O–C–C–N conformation, the $\nu_s[CN^+(CH_3)_3]$ peaks are observed at 925 and 875 $cm^{-1}$, and in the gauche O–C–C–N conformation, these two bands appear at 900 and 860 $cm^{-1}$ [44]. Since in PM–IRRAS we are not able to see bands below 900 $cm^{-1}$, only the presence of a band at either 925 or ~900 $cm^{-1}$ is used as the evidence of the trans or gauche conformations, respectively.

Figure 11.27*b* shows the PM–IRRAS spectra in the C–N stretch region for a DMPC bilayer at a Au(111) electrode in 0.1 M NaF/$D_2O$ solution. The $\nu_{as}[CN^+(CH_3)_3]$

bands are located at ~970 and ~948 $cm^{-1}$, which are slightly red shifted with respect to the bands in the spectrum of DMPC vesicles. In the spectrum recorded for a solution of vesicles, the C–N stretch vibration is observed at ~925 $cm^{-1}$, indicating that the O–C–C–N has a trans conformation. However, for the LB-LS bilayer, the $\nu_s[CN^+(CH_3)_3]$ band is significantly red-shifted to ~905 $cm^{-1}$. This large red shift indicates that the O–C–C–N frame of the choline moiety in the DMPC bilayer prepared by the LB–LS method assumes a gauche conformation. The position of the $\nu_s[CN^+(CH_3)_3]$ and $\nu_{as}[CN^+(CH_3)_3]$ bands is independent of the electrode potential, consistent with the behavior of the C–H bending modes of the choline group.

The directions of the transition dipoles of the choline group vibrations were determined by Fringeli [44] and are shown in Figure 11.21. The angles between the direction of these transition dipoles of the $CN^+(CH_3)_3$ group and the surface normal can be calculated from the integrated intensity of these bands and using Eq. (11.17). These angles are reported in Figure 11.28 using models that display angles between different fragments of DMPC molecules and the surface normal for the bilayer formed by the LB–LS method, in both the adsorbed and desorbed states. For comparison the orientation of DMPC molecules in the bilayer formed by VF is also shown in Figure 11.28. There are significant differences between the tilt angles of the acyl chains, the –O–P–O– line of the phosphate group, and the C–N bond of the choline moiety for the two bilayers. All these angles are significantly smaller in the LB–LS bilayer than those in the VF bilayer. For the LB–LS bilayer, the tilt angles of the acyl chains are ~30° and ~17° in the adsorbed and desorbed state, respectively, which is about 25° smaller than the respective angles in the VF bilayer. The frequencies for the $CH_2$ stretching bands in the LB–LS bilayer and VF bilayer are quite comparable, indicating the same physical state of the acyl chains for the two bilayers. The angles between the –O–P–O– line and the surface normal are ~25° in the LB–LS bilayer, which is about 20° smaller than in the VF bilayer. The tilt angles of the choline group with respect to the surface normal are ~62 to ~55° for the LB–LS bilayer and ~80 to ~75° for the VF bilayer. The angle between the average direction of the transition dipoles of the C=O group and the surface normal is ~80° in the LB–LS bilayer, while in the VF bilayer, this value is ~70 to ~60°. Since the direction of the C=O group is nearly perpendicular to the acyl chain of the DMPC molecules, a large angle of the C=O group corresponds to the small tilt angle of the acyl chain.

In addition, the peak maximum of the ν(C=O) band in the LB–LS bilayer is red shifted compared to that in the VF bilayer, indicting that the ester group is more hydrated in the LB–LS bilayer. The peak frequency of the $\nu_{as}$(COC) band in the LB–LS bilayer is also ~30 $cm^{-1}$ smaller than in the VF bilayer, demonstrating that the conformation of the C–C(O)–O–C frame is significantly different in the two bilayers.

The difference in the orientation and hydration of the two bilayers reflect significant differences in packing densities of phospholipid molecules in these two systems [58, 59]. The packing densities are lower in the VF bilayer, where the area per molecule changes from 0.47 to 0.66 $nm^2$ on going from the desorbed to the adsorbed state. In contrast, these numbers vary from 0.41 to 0.45 $nm^2$ for the LB–LS bilayer. Thus, the LB–LS method allows one to form bilayers with a higher packing density, resulting in a smaller tilt angle of the acyl chains. Atomic force microscopy

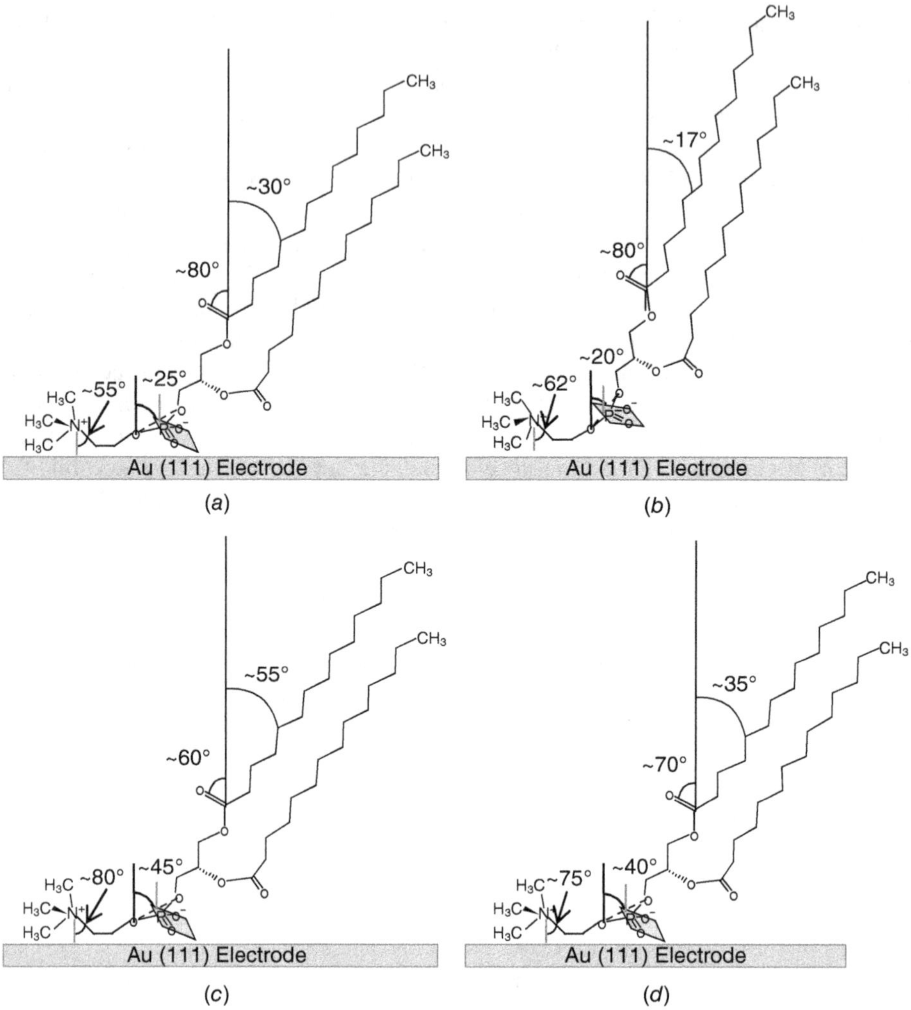


**Figure 11.28** Orientation of DMPC molecules in the bilayer formed by the LB–LS method: (*a*) $E = 0.3$ V (adsorbed film) and (*b*) $E = -0.8$ V (detached film); and in the bilayer formed by the vesicle fusion: (*c*) $E = 0.3$ V (adsorbed film) and (*d*) $E = -0.8$ V (detached film). Adapted from Ref. [58].

(AFM) studies of the bilayers at the Au(111) surface [96] provided additional evidence that the tilt angle of the acyl chains in the LB–LS bilayer is significantly smaller than in the VF bilayer. All these investigations demonstrated that the LB–LS method offers better control over film structure and allows for the formation of more ordered films.

PM–IRRAS has been employed to study several other phospholipid bilayer systems. Zawisza et al. [97, 98] investigated a 1,2-dioyl-*sn*-glycero-3-phosphatidylcholine (DOPC) bilayer at a Au(111) surface prepared by vesicle fusion. More recently

Zawisza et al. applied PM–IRRAS to study lipid bilayers deposited on gold covered by a thin layer of silica [13] or a titanium oxide layer [99]. Hillman et al. employed this technique to study a bilayer of anionic lipids supported at a gold electrode [100]. In addition, Madrid and Horswell [101] employed PM–IRRAS to compare potential-dependent behavior of gold-supported bilayers assembled from dimyristoylphosphatidylserine (DMPS), dimyristoylphosphatidylethanolamine (DMPE), and DMPC. These studies illustrate that PM–IRRAS has become a popular and versatile technique for characterizing the structure of model membranes at electrode surfaces.

## 11.8 TWO-COMPONENT BILAYERS

Cholesterol (chol) is a ubiquitous constituent of biological membranes. It belongs to the family of neutral lipids having a steroid ring system. Incorporation of cholesterol into phospholipid bilayers is known to increase the fluidity of the gel state and reduce the fluidity of the liquid crystalline state. In this way cholesterol releases stress within the membrane, creating a more uniform and homogeneous film [102, 103]. This effect of cholesterol on lipid bilayers is attributed to its molecular structure. Cholesterol inserts into a membrane bilayer normal to the bilayer plane, with its hydroxyl group in close vicinity to the ester carbonyl of phospholipids, and its alkyl side chain extending toward the bilayer center. The hydroxyl group of cholesterol plays an important role in the phospholipid–cholesterol interactions. It has been suggested that cholesterol may form a hydrogen bond with either of the two C=O groups or the phosphate group of phospholipid molecules [104–106]. As mentioned previously, bilayers formed from pure DMPC are stressed, disordered, and contain numerous defects, with a minimum capacity value in the range of ~5 to ~8 $\mu F\ cm^{-2}$. To obtain a high-quality biomimetic membrane with fewer defects, the DMPC–chol (70 : 30 mol% ratio) bilayers were deposited at the Au(111) surface using the LB–LS method [107] and vesicle fusion [108, 109]. These bilayers were then characterized by electrochemistry and PM–IRRAS.

The shape of aggregates formed by lipids is governed by the packing parameter $p$ defined as [110]

$$p = \frac{V}{\pi r_0^2 h} \tag{11.33}$$

where $V$ is the volume of the cylinder circumcised by the motion of the tails to the motion of the volume of a cylinder with radius $r_0$ corresponding to the radius of the head group. Figure 11.29 shows expected geometries of the aggregates formed by amphiphilic molecules depending on the value of their packing parameter. A bilayer is formed when $p = 1$. Figure 11.29 shows that the DMPC molecule is conical and its packing parameter is ~0.75. In contrast, cholesterol has the shape of an inverted cone, and its packing parameter is 1.21 [111]. The packing parameter of the 70 : 30 mol% mixture is equal to 0.89 and is close to the value expected for a bilayer. The improved value of the packing parameter explains why insertion of cholesterol releases the stress in the bilayer and improves its mechanical properties.

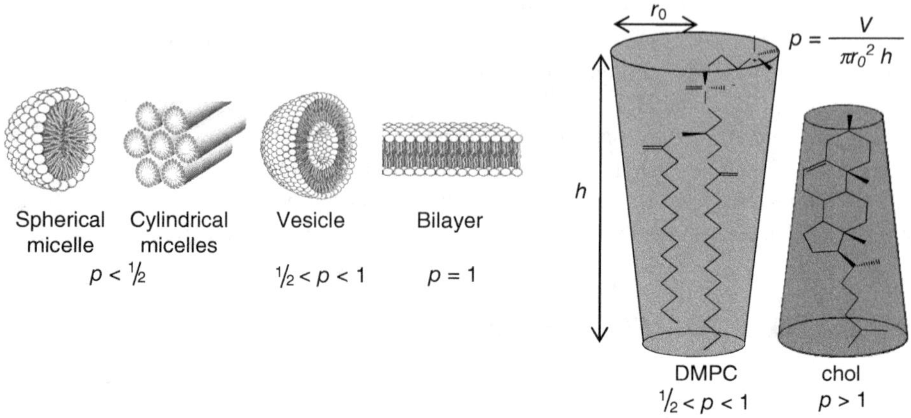


**Figure 11.29** Shape of aggregates formed by phospholipid molecules for different values of the packing parameter and shapes and estimation of packing parameter for DMPC and cholesterol molecules. Courtesy of Ryan Seenath.

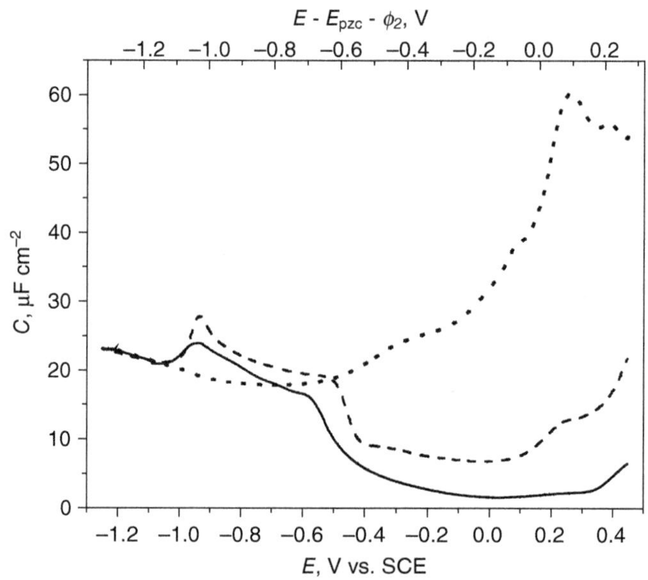


**Figure 11.30** Differential capacity curves for the bare Au(111) electrode (dotted line), DMPC bilayer (dash line) and DMPC–chol bilayer (solid line) covered Au(111) electrode in 0.1 M NaF. Adapted from Ref. [107].

Figure 11.30 plots the differential capacity curves for a bare Au(111) surface (dotted line), a pure DMPC bilayer (dashed line), and the DMPC–chol bilayer containing cholesterol in both leaflets (thick solid line). The bilayers were deposited at a Au(111) surface by the LB–LS method at a transfer pressure of 40 mN m$^{-1}$. When cholesterol is present in both leaflets, there is a significant lowering of the capacitance to ~1.7 μF $cm^{-2}$. This result indicates that cholesterol improves the quality of the membrane considerably when it is present in both leaflets of the bilayer. Figure 11.31 plots the surface charge density as a function of applied potential for the DMPC–chol bilayer. For comparison, the charge density curve for the gold electrode covered by the pure DMPC bilayer is also included in Figure 11.31. In both cases, the bilayers are adsorbed at the electrode surface when the absolute value of the charge density is less than 10 μC $cm^{-2}$. The charge density curves display a similar shape; however, lower absolute values of the surface charge density are observed for the bilayer containing cholesterol in both leaflets.

These electrochemical studies indicate that the presence of cholesterol improves the properties of the bilayer lipid membrane (BLM). To further investigate this, the PM–IRRAS technique was used to characterize changes in the molecular structure of the phospholipids that leads to these improved properties of the bilayer. The IR bands in the CH stretch region provide information about the orientation and conformation of acyl chains. The average tilt of the trans fragments of the acyl chains of DMPC molecules in the two-component bilayer are compared to the tilt in the pure DMPC bilayer in Figure 11.31. In the desorbed state the tilt angles have comparable magnitude for the two bilayers. However, in the adsorbed state the tilt of acyl chains is about 5° smaller in the presence of cholesterol, indicating that the lipids are more tightly packed in the mixed bilayer. Further, in the two-component bilayer the difference between the tilt angle in the adsorbed and desorbed states is small, indicating that cholesterol provides additional free space for the polar heads of DMPC molecules.

Figure 11.32 plots the position of the $\nu_s(CH_2)$ and $\nu_{as}(CH_2)$ bands as a function of electrode potential. The average values of the band frequency are 2922 and 2852 $cm^{-1}$ for the $\nu_{as}(CH_2)$ and $\nu_s(CH_2)$ modes, respectively. Within experimental error these numbers are equal to the frequencies of $\nu_{as}(CH_2)$ at ~2924 $cm^{-1}$ and $\nu_s(CH_2)$ at ~2853 $cm^{-1}$ for the pure DMPC bilayer. These values indicate that the DMPC–chol bilayer prepared by the LB–LS method consists of a mixture of lipids in the all-trans configuration and gauche configuration. However, there is no apparent difference in the number of gauche conformations upon incorporation of cholesterol into this supported bilayer.

The addition of cholesterol has some effect on the hydration of the polar head group. Figure 11.33 compares the position of the global maximum of the carbonyl stretching vibration in the mixed DMPC–Cholesterol and pure DMPC bilayers. In the adsorbed state the global minimum of the $\nu(C{=}O)$ band in the DMPC–chol bilayer is shifted to higher wavenumbers with respect to that of the pure DMPC bilayer. This shift is particularly large for the DMPC–chol bilayer formed by vesicle fusion. In the desorbed state, the $\nu(C{=}O)$ band maximum shifts to lower wavenumbers for the DMPC–chol bilayer. In this region the shift is larger for the bilayer formed by the LB–LS method. This behavior demonstrates that the carbonyl group of DMPC molecules in the DMPC–chol bilayer is more hydrated than in the pure DMPC bilayer in the desorbed state and less hydrated in the adsorbed state. In the adsorbed state the hydration is somewhat lower for transmembrane potentials between −0.2 and 0.1 V and higher in the electroporation region (transmembrane potentials between −0.5 and −0.2 V).

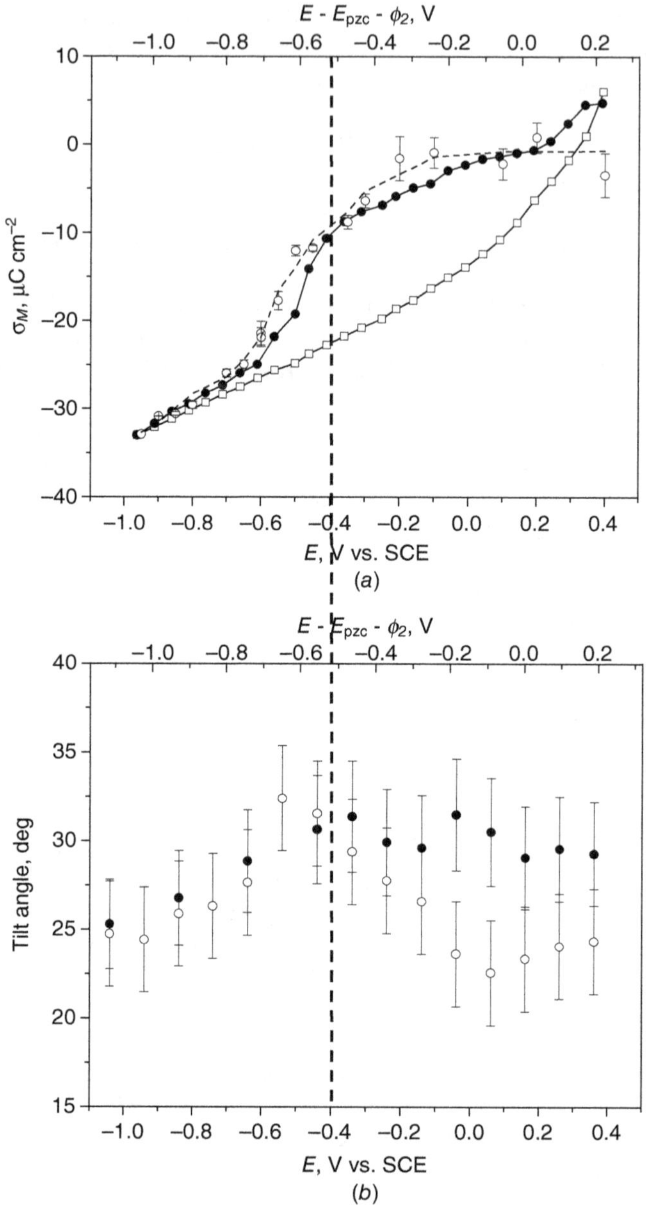


**Figure 11.31** (*a*) Charge density curves for DMPC bilayer (filled circles), DMPC–chol bilayer (open circles) covered and bare (open squares) Au(111) electrode in 0.1 M NaF solution. (*b*) Comparison of the tilt angle of the acyl chains in DMPC bilayer (filled circles) and DMPC–chol (open circles) bilayers. Dashed line indicates the potential at which the onset of desorption is observed to begin. Adapted from Ref. [107]. The charge density data for the DMPC–chol bilayer are unpublished results by Christa Brosseau.

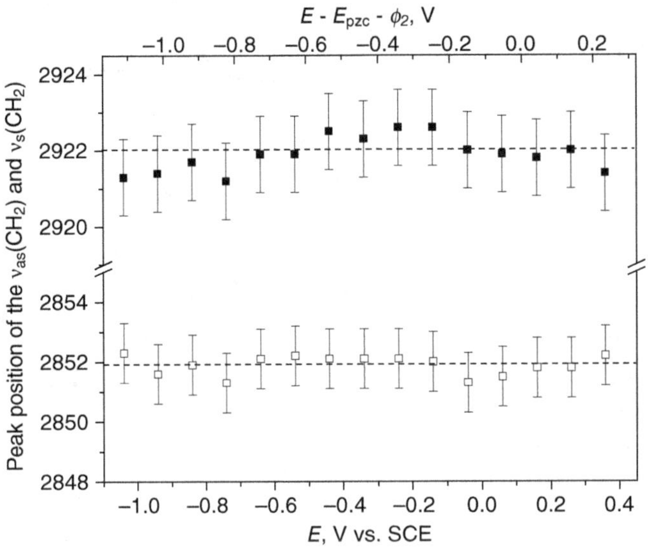


**Figure 11.32** Peak position as a function of potential for the $\nu_s(CH_2)$ (open squares) and $\nu_{as}(CH_2)$ (filled squares) bands for the LB–LS DMPC–chol bilayer assembled by the LB–LS method at the Au(111) surface in a 0.1 M NaF/$D_2O$. Dashed lines indicate the calculated average values. Adapted from Ref. [107].

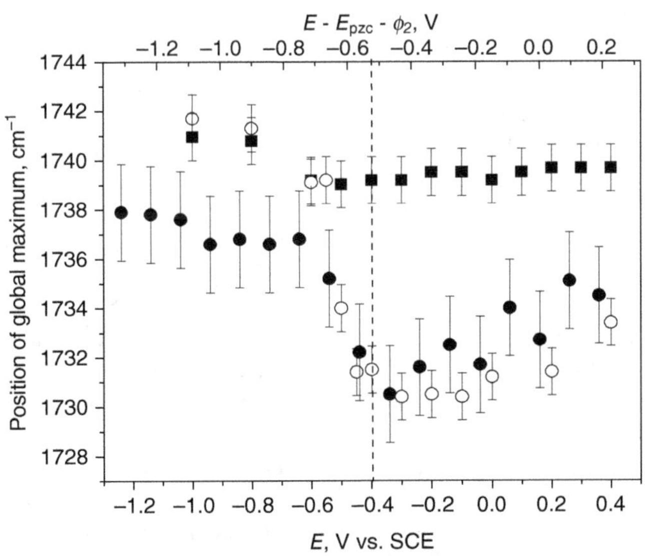


**Figure 11.33** Position of the global maximum of the ν(C=O) band in the DMPC–chol bilayer prepared by the LB–LS method (filled circles; unpublished data determined by Zhangfei Su), pure DMPC bilayer prepared by the LB–LS method (open circles) [58], and DMPC–chol bilayer prepared by vesicle fusion (filled squares) [109]. Dashed line indicates the potential at which the onset of desorption is observed to begin.

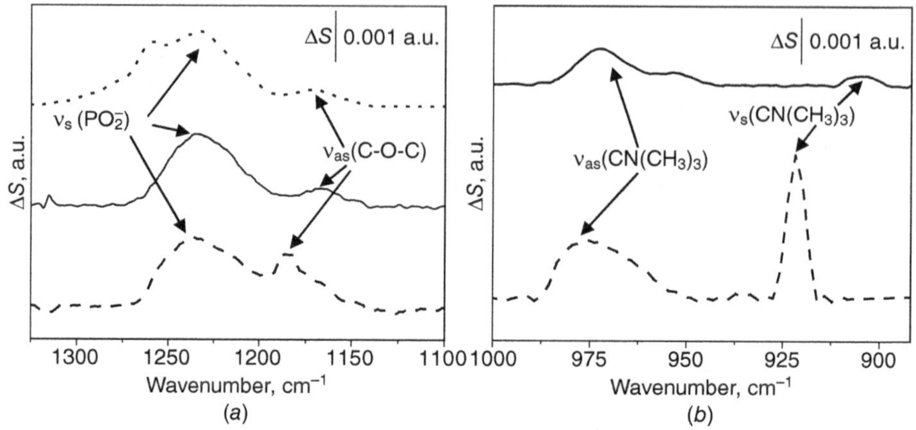


**Figure 11.34** (*a*) Comparison of the PM–IRRAS spectra for the pure DMPC (solid line), the DMPC–chol (dash line), the DMPC–chol–GM1 (dotted line) bilayer at potential $E = 0.36$ V in the $\nu_s(PO_2^-)$ region. All the bilayers are prepared by the LB–LS method. (*b*) Comparison of the PM–IRRAS spectra for the pure DMPC (solid line) and the DMPC–chol (dash line) bilayer at potential $E = 0.36$ V in the C–N stretching region.

The presence of cholesterol has also a notable impact on the conformation and orientation of other fragments of the polar head region. Figure 11.34*a* compares the spectra corresponding to the $\nu_{as}(PO_2^-)$ and the ester $\nu_{as}$(C–O–C) vibration. The bands in the DMPC–chol bilayer are visibly blue shifted with respect to the bands in the pure DMPC bilayer. The position of the $\nu_s(PO_2^-)$ band maximum shifts from ~1231 cm$^{-1}$ in pure DMPC to ~1233 cm$^{-1}$ in the DMPC–chol bilayer, indicating that the phosphate group is less hydrated in the presence of cholesterol. In the mixed DMPC/chol bilayer, the frequency of the $\nu_{as}$(C–O–C) is essentially potential independent and its average value amounts to ~1090 cm$^{-1}$, indicating that the two acyl chains are co-planar. Figure 11.34*b* shows that the C–N stretching band shifts from ~905 cm$^{-1}$ in pure DMPC to ~920 cm$^{-1}$ in the two-component bilayer. This shift indicates that the conformation of the –O–C–C–N moiety changes from gauche to trans in the presence of cholesterol. The presence of cholesterol increases the space between the head groups of DMPC molecules and hence affects the packing of acyl chains, as well as the conformation and hydration of the polar head groups of the phospholipid molecules.

In another example of a two-component bilayer, Matyszewska et al. [112] employed a hybrid DMPC bilayer with hydrogen-substituted acyl chains in the top leaflet (solution side) and deuterium-substituted acyl chains in the bottom leaflet (gold side) to study the influence of perfluorinated compounds, specifically perfluorooctanoic acid (PFOA) and perfluorooctanesulfonic acid (PFOS), on the structure of the biomimetic membrane. These compounds were incorporated into the top leaflet of the bilayer. The PM–IRRAS spectra showed that the incorporation of perfluorinated compounds causes a broadening of the $CH_2$ bands and a blue shift of the band positions. The average tilt angle of trans fragments of the acyl chains decreased upon incorporation of the perfluorinated compounds (thickness of the membrane increased). The average tilt angle is smaller upon insertion of PFOS (~24°) than in the presence of PFOA (~30°). In summary, the incorporation of the perfluorinated compounds increased the membrane fluidity and thickness. Their effect on the membrane structure was similar to cholesterol.

## 11.9 THREE-COMPONENT BILAYERS

In addition to phospholipids and cholesterol, specific types of glycolipids, known as gangliosides, are present in biological membranes. These glycolipids play an important role in many cellular processes, such as cell growth and cell–cell interactions. Glycolipids are involved in the regulation of cell signaling by binding to important proteins, for example, interferon and insulin [113, 114]. Ganglioside monosialic acid 1 (GM1) is one of the most biologically relevant gangliosides because it has been implicated in important cellular processes such as lipid raft formation whereby localized ordered domains are formed within a fluid lipid matrix [115–118]. In addition, GM1 binds specifically to the cholera toxin [119], so this three-component sBLM can be used to study the interaction of the binding unit of cholera toxin with a model membrane [120, 121]. The structure of GM1 is shown in Figure 11.35*a*.

To obtain a closer model to a real biological membrane, a three-component DMPC–chol–GM1 bilayer was assembled at the Au(111) surface by the combination of the LB and LS techniques [122]. The inner leaflet, composed of DMPC and cholesterol in a 70 : 30 mol% ratio was transferred to the Au(111) surface using the LB method. The outer leaflet, containing DMPC, cholesterol and GM1 (in a 60 : 30 : 10 mol% ratio) was deposited using the LS method. The transfer pressure for both inner and outer leaflets was 40 mN $m^{-1}$. Figure 11.35*b* plots the differential capacitance curves for a bare Au(111) surface (dotted line) and the Au(111) surface covered with a DMPC–chol bilayer (dashed line) and a DMPC–chol–GM1 bilayer (solid line). It can be observed that all the membranes are adsorbed on the gold surface in the potential region from –0.2 to +0.4 V. The minimum values of the membrane capacitance are ~1.7 μF $cm^{-2}$ for both the DMPC–chol and DMPC–chol–GM1 bilayers [123]. For the three-component membrane, the low-capacity region is observed in somewhat the extended potentials range, indicating that the bilayer is stable in a broader range of potentials. This indicates that the three-component BLM containing phospholipid, cholesterol, and glycolipid is functioning as a good model of a cell membrane.

Figure 11.35*c* shows the surface charge density plotted as a function of applied potential for the two- (open squares) and three-component (solid squares) BLMs. Each point on these two curves corresponds to a freshly prepared bilayer. The three-component membrane shows small changes in the charge density as a function of potential from –0.2 to +0.3 V, and then a phase transition is observed to occur at about –0.2 V versus SCE, which results in another phase that is present from –0.3 to –0.5 V. When $E < -0.5$ V, the charge density decreases rapidly, indicating the

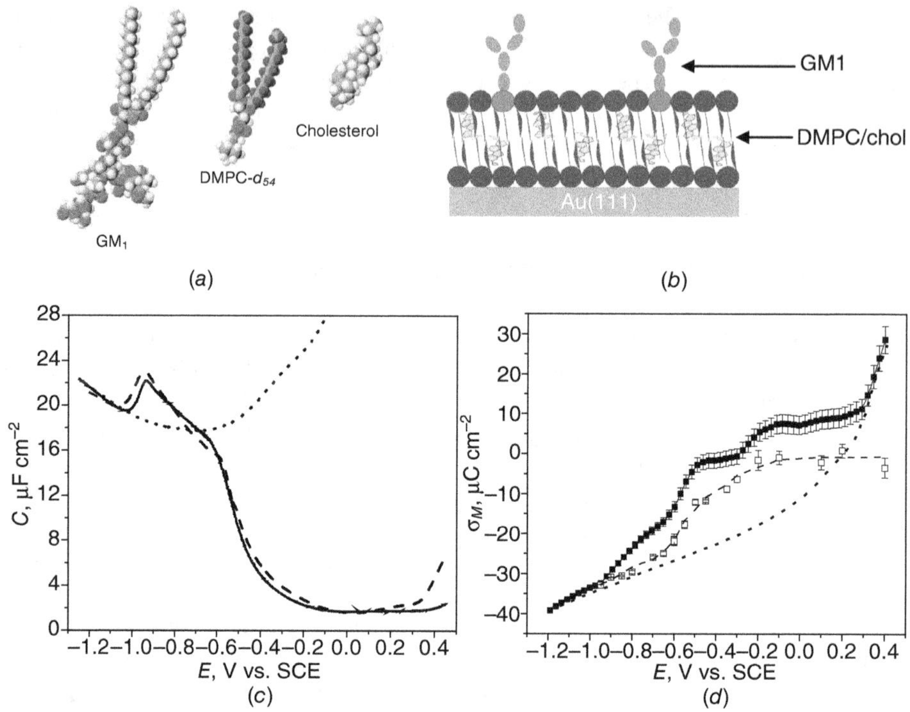


**Figure 11.35** (*a*) Space-filling models of GM1, DMPC-*d*54, and cholesterol used for building three-component biomimetic membranes. (*b*) Schematic structure of the DMPC–chol–GM1 bilayer. (*c*) Differential capacity plots for a bare Au(111) electrode (dotted line) and for a DMPC–chol bilayer (dash line) and a DMPC–chol–GM1 bilayer (solid line) in 0.1 M NaF solution. (*d*) Surface charge density for a bare Au(111) electrode (dotted line) and a DMPC–chol bilayer (open squares) and a DMPC–chol–GM1 bilayer (filled squares) supported at a Au(111) electrode in 0.1 M NaF solution. Adapted from Ref. [122].

onset of desorption where the bilayer begins to lift from the electrode surface. At very negative potentials ($E < -1.1$ V), the bilayer curve merges with the bare electrode curve, which indicates complete desorption of the bilayer from the electrode surface. The charge density of the DMPC–chol bilayer is observed to be significantly lower than that of the DMPC–chol–GM1 bilayer. In addition, the charge density for the three-component membrane is positive at potentials higher than −0.2 V, which is most likely due to the presence of the negatively charged glycolipid. The results from the electrochemical data show that a supported BLM composed of DMPC, cholesterol, and GM1 is well formed and has a low density of defects.

Since all three components (DMPC, cholesterol, and GM1) absorb IR radiation in the C–H stretching region, it is difficult to obtain quantitative data about the orientation and conformation of the bilayer in this region due to spectral overlap. Therefore, deuterium-substituted DMPC-*d*54 molecules were employed to determine information about the conformation and orientation of the acyl chains of the phospholipid molecules in the supported BLM (sBLM). The C–D stretching bands are shifted to lower frequencies and located in the region between 2000 and 2300 $cm^{-1}$, which is free of spectral overlap from other vibrational modes. DMPC-*d*54 has a main phase transition at 20.5°C; therefore, the PM–IRRAS measurements of the DMPC-*d*54–chol–GM1 bilayer were performed at 18 ± 1C, so that the deuterated acyl chains would predominantly adopt a trans conformation. Figure 11.36*a* shows the C–D stretching spectra of the DMPC-*d*54–chol–GM1 bilayer for selected potentials. Figure 11.36*b* shows the band deconvolution (Lorentzian fit) of the C–D stretching region including the spectral band assignment. The spectrum consists of four distinct bands, which correspond to the $CD_3$ asymmetric stretching [$\nu_{as}(CD_3)$ at

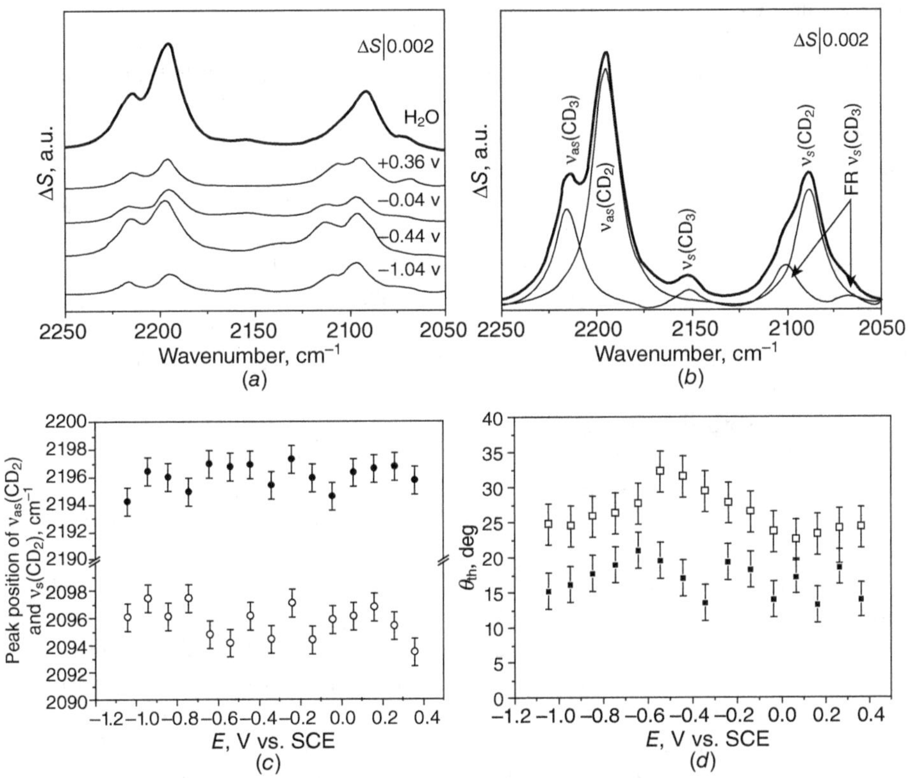


**Figure 11.36** (*a*) PM–IRRAS spectra for a DMPC-*d*54–chol–GM1 bilayer supported at the Au(111) surface in the C–D stretching region in a 0.1 M NaF/$H_2O$ solution at selected potentials. The top trace plots the spectra calculated for a 5.5-nm-thick bilayer using optical constants determined from the transmission IR. (*b*) Deconvolution of the C–D stretching bands for the calculated spectrum shown in the top trace of Figure 9.2*a* (Lorentzian fit). (*c*) Peak position of the $\nu_{as}(CD_2)$ (filled circle) and $\nu_s(CD_2)$ (open circle) bands in the deconvoluted spectra as a function of electrode potential for the DMPC–chol–GM1 bilayer. (*d*) Tilt angle of the acyl chains in DMPC-*d*54–chol (open squares) and DMPC-*d*54–chol–GM1 (filled squares) bilayers. Adapted from Ref. [122].

~2216 $cm^{-1}$], $CD_2$ asymmetric stretching [$\nu_{as}(CD_2)$ at ~2196 $cm^{-1}$], $CD_3$ symmetric stretching [$\nu_s(CD_3)$ at ~2154 $cm^{-1}$], and $CD_2$ symmetric stretching [$\nu_s(CD_2)$ at ~2091 $cm^{-1}$] vibrational modes. The two shoulders on the higher and lower frequency sides of the $\nu_s(CD_2)$ band result from the Fermi resonances of the bending modes of $CD_3$ and $CD_2$ [124, 125].

Figure 11.36*c* shows that positions of the $\nu_{as}(CD_2)$ and $\nu_s(CD_2)$ bands are nearly potential independent and their average values are 2197 and 2096 $cm^{-1}$, respectively. For the DMPC-*d*54 bilayer in the gel state, in which the acyl chains assume an all-trans conformation, the positions of the $\nu_{as}(CD_2)$ and $\nu_s(CD_2)$ should be 2191 and 2088 $cm^{-1}$, respectively [125]. The maxima of $\nu_{as}(CD_2)$ and $\nu_s(CD_2)$ bands in the sBLM have higher frequencies, suggesting that the three-component bilayer has a significant number of gauche conformers. Apparently GM1, in addition to cholesterol, is very efficient at modulating membrane fluidity and introduces considerable disorder into the gel state of the lipid.

Figure 11.36*d* plots the tilt angle of the acyl chains for a DMPC-*d*54–chol bilayer (open squares) and a DMPC-*d*54–chol–GM1 bilayer (filled squares) as a function of the electrode potential. The average value of the tilt angle for the three-component BLM is ~8° lower than that of the two-component sBLM, indicating that the presence of GM1 in the bilayer results in a smaller tilt of the lipid acyl chains with respect to the surface normal. This result is consistent with studies by Majewski et al. [126] that suggest incorporation of GM1 into phosholipid monolayers causes a reduction in the tilt angle of the acyl chains of the lipids. The tilt angles for the three-component BLM display a similar dependence on potential to that observed previously for a two-component membrane. Comparable tilt angles are observed in the adsorbed state at the most positive potentials and in the desorbed state at the most negative potentials. This suggests that the head groups of DMPC have sufficient space in the adsorbed state that allows them to pack without a need to increase the tilt of the chains. The tilt angle increases at potentials where the bilayer is detached from the metal. Apparently, desorption introduces disorder into the membrane.

Figure 11.37*a* shows the PM–IRRAS spectra for the DMPC–chol–GM1 sBLM for the C=O stretching region. Figure 11.37*b* shows the position of the global maximum of the $\nu(C{=}O)$ band for both a DMPC–chol bilayer (1) and a DMPC–chol–GM1 bilayer (2) as a function of potential. The position of the global maximum of the $\nu(C{=}O)$ displays a similar dependence on the electrode potential for the two bilayers. The frequencies of the $\nu(C{=}O)$ band are significantly lower in the adsorbed than in the desorbed state, indicating that the bilayers are more hydrated in the adsorbed state. In the adsorbed state, the band maximum for the bilayer shifts progressively to higher frequencies as the electrode potential becomes more positive. After the bilayer lifts from the electrode surface, the maximum shifts to higher frequencies. For the three-component bilayer, the shift to higher frequencies is observed at somewhat more positive potentials, indicating that lifting of the three-component bilayer takes place earlier, when the potential is moved in the negative direction.

The overall intensity of the $\nu(C{=}O)$ band appears to be independent of the electrode potential within experimental error. Because of the complex nature of the $\nu(C{=}O)$ band and significant changes in the hydration of the $\nu(C{=}O)$ group, it is

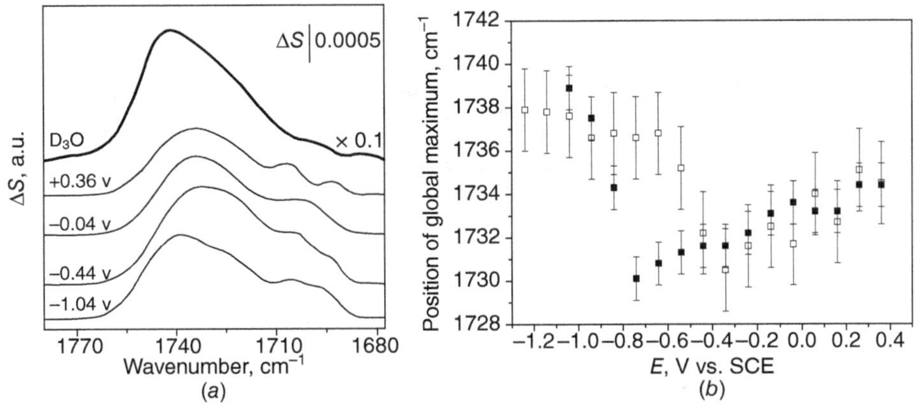


**Figure 11.37** (*a*) PM–IRRAS spectra for the ν(C=O) band for the DMPC–chol–GM1 bilayer in 0.1 M NaF/$D_2O$ at selected potentials. The top trace plots the spectrum calculated for a 5.5-nm thick-bilayer using optical constants determined from the transmission spectrum. (*b*) Position of the global maximum of the ν(C=O) band as a function of electrode potential for the DMPC–chol (open squares) and DMPC–chol–GM1 (filled squares) bilayers. Adapted from Ref. [122].

difficult to determine the angle between the transition dipole of this vibration and the surface normal. However, the band intensity in the PM–IRRAS spectra is about 10 times smaller than the intensity of this band in the spectrum for randomly oriented molecules calculated from optical constants. This behavior indicates that the angle between the transition dipole of this band and the surface normal is large. Taking the overall ν(C=O) band intensity, one can estimate that this angle is ~80°, which is consistent with the small tilt angle of the acyl chains.

For comparison with the one- and two-component bilayers, the spectra corresponding to the $\nu_{as}(PO_2^-)$ and the ester $\nu_{as}$(C–O–C) vibration are plotted in Figure 11.34*a* in the preceding section. The center of the $\nu_{as}$(C–O–C) band is observed at 1175 cm$^{-1}$, indicating that the two acyl chains are not perfectly parallel, although the deviation from parallel orientation is not large. Thus the acyl chains are parallel in the two-component bilayer, nearly parallel in the three-component bilayer, and are not parallel in the one-component bilayer. The band center of the phosphate group is observed at 1233 cm$^{-1}$, which is the same as in the two-component bilayer. In summary, the frequencies of the $\nu_{as}(PO_2^-)$ band show that the phosphate group is best hydrated in the one-component bilayer and somewhat less hydrated in the two- and three-component bilayers. Unfortunately, for the three-component bilayer the spectral range between 1100 and 900 cm$^{-1}$ contains additional bands of sugar groups from GM1. Therefore, it is impossible to determine the conformation of the –O–C–C–N fragment of the choline group in the DMPC molecules in this case.

Figure 11.38 compares the orientation of the acyl chains, glycerol ester, and choline head group with respect to the surface normal for a DMPC molecule in the DMPC (A), DMPC–chol (B) and DMPC–cholesterol–GM1 (C) sBLMs at the Au(111) electrode surface. The results show that addition of 10% GM1 to the top leaflet leads

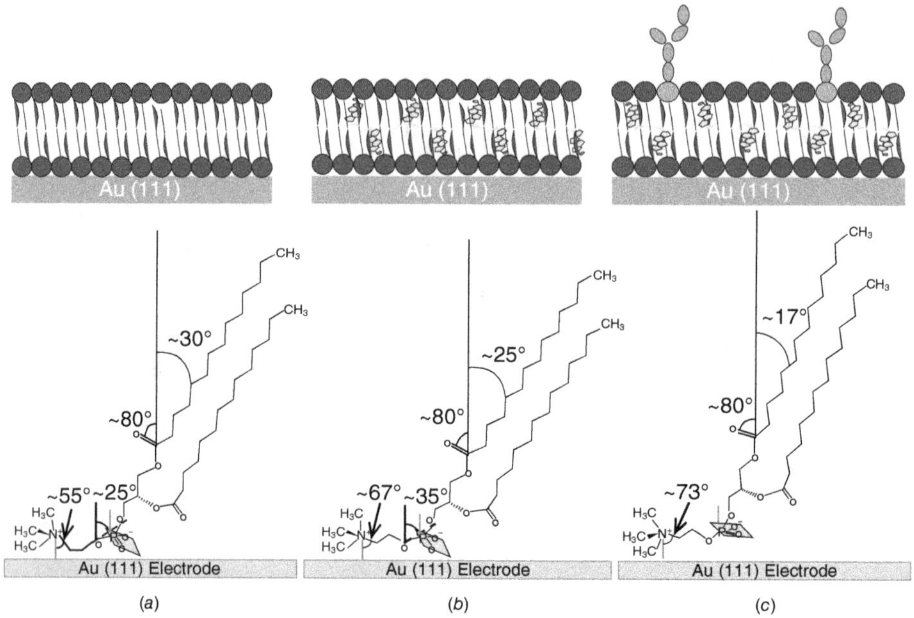


**Figure 11.38** Orientation of DMPC molecules for a (*a*) DMPC bilayer, a (*b*) DMPC–chol bilayer, and a (*c*) DMPC–chol–GM1 bilayer supported at Au(111) surface in the adsorption state. This is the average orientation of DMPC molecules in the two leaflets of the bilayer with respect to the surface normal. The illustrations show the schematic structure of the lipid bilayers.

to an ~8° decrease in the tilt angle of the acyl chains and about an ~5° increase in the angle between the $C_3$ axis of the choline group and the surface normal. The PM–IRRAS experiments also demonstrate that GM1 has a moderate effect on the conformation and hydration of the polar head group region of the bilayer. As a result, the bilayer that has only 10% GM1 molecules in the top leaflet is more condensed. The GM1 molecules present in the top leaflet provide binding sites for several proteins, which can be used for future membrane protein research.

## 11.10 BILAYERS WITH INCORPORATED PEPTIDES

The emergence of antibiotic resistance has necessitated the development of alternatives to traditional antibiotics [127, 128]. Many small peptides with 12–50 amino acid residues have been identified in recent years as having antimicrobial activity, and this biological activity is believed to be strongly related to their interactions within the cell membrane [129, 130]. A necessary step in understanding the mechanisms of bioactivity of antimicrobial peptides is the characterization of the structure of peptide–lipid assemblies. The linear gramicidins are the best understood peptide antibiotics [129, 130], and hence they are ideal model systems to investigate the properties of this class of peptides in the lipid environment. The natural mixture of gramicidins denoted as gramicidin D (GD) consists of ~85% gramicidin A, which is a pentadecapeptide: Formyl-L-Val1-Gly2-L-Ala3-D-Leu4-L-Ala5-D-Val6-L-Val7-D-Val8-L-Trp9-D-Leu10-L-Trp11-D-Leu12-L-Trp13-D-Leu14-L-Trp15-Ethanolamine. In lipid bilayers or monolayers, the alternating L-D amino acid sequence allows the molecule to fold as a right-handed $\beta^{6.3}$ helix with 6.3 residues per turn. The interior of the channel is formed by the polar peptide backbone with side chains projecting outward toward the lipids. The hydrogen bonds between the tryptophans and the lipid polar groups orient the helix with the C-terminus turned to the membrane surface and the N-terminus to its hydrophobic interior. When gramicidin molecules are present in the two leaflets of a bilayer, they may form a dimer stabilized by six hydrogen bonds between two N-end to N-end oriented molecules to create an ion conductive channel spanning the membrane. These features of GD lead to the formation of helices that have a β-sheet hydrogen bonding pattern with all the residues projecting from one side of the sheet. The peptide structure displays polymorphism and, depending on the environment, it forms either single-stranded or double-stranded helices denoted as β, where *m* is the total number of residues (L plus D) per turn. Here we will consider only the double-stranded $\beta^{5.6}$ and single-stranded $\beta^{6.3}$ structures. The single-stranded and double-stranded structures and models of the channel and the double-stranded conformation formed in a membrane are shown in Figure 11.39.

The insertion, conformation, and orientation of the peptide, as well as the orientation and configuration of the lipids in a phospholipid bilayer can be conveniently investigated with the help of IR spectroscopy. Laredo et al. [33] employed the PM–IRRAS technique to study the effect of the electrode potential on the orientation and conformation of GD and DMPC molecules in a mixed DMPC:GD bilayer (90:10 mol% ratio) assembled at a Au(111) surface using the LB–LS method.

The electrochemical data showed that the DMPC–GD bilayer was adsorbed and in direct contact with the metal surface when the potential across the membrane was more positive than –0.4 V and was lifted from the gold surface when the potential

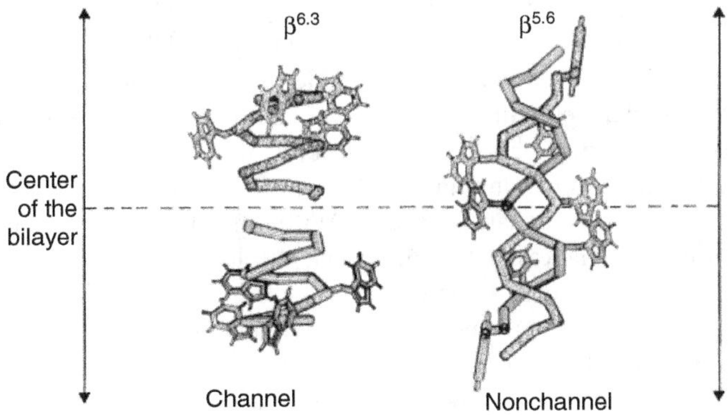


**Figure 11.39** Schematic representations of the channel and nonchannel conformations of gramicidin, indicating the location of tryptophan residues in a membrane bilayer. Adapted from Ref. [130].

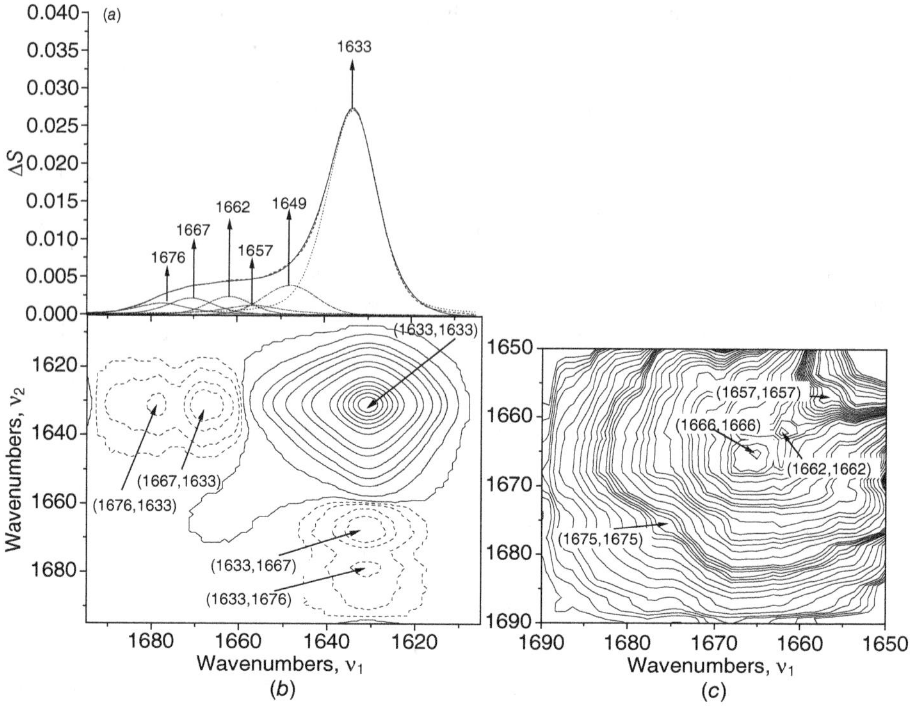


**Figure 11.40** (*a*) Deconvoluted spectrum of the amide I band of GD for a DMPC:GD bilayer at $E = -0.1$ V. (*b*) 2D synchronous spectrum of the amide I region for a large range of $z$ values. (*c*) A magnified portion of (*b*) with a reduced range of $z$ values. Four autopeaks can be distinguished at positions 1657, 1662, 1667, and 1676 $cm^{-1}$. Adapted from Ref. [33].

across the membrane was more negative than –0.4 V, similar to the behavior of the pure DMPC bilayer described in Section 11.7.

The information concerning the orientation and conformation of the peptide was obtained from the analysis of the amide I band. Figure 11.40 shows the PM–IRRAS spectrum of the amide I band for the DMPC : GD bilayer. The spectrum consists of a strong band at 1633 $cm^{-1}$ and a broad shoulder from 1640 to 1680 $cm^{-1}$. Because the band intensity changes with potential, one can apply 2D-COS spectroscopy to identify individual components in the envelope of the amide I band. The band deconvolution result is shown in Figures 11.40*b*,*c*, in which six sub-bands are identified under the envelope of the amide I band. With the help of the force field normal-mode calculations by Naik and Krimm [131], the low-frequency peak that is the main component of the amide I band at ~1633 $cm^{-1}$ can be assigned to the helical dimer (HD) $\beta^{6.3}$. The smaller bands under the higher-frequency shoulder at about 1649, 1656, 1662, 1668, and 1675 $cm^{-1}$ correspond to $\beta^{5.6}$ double helices.

The integrated intensities of the ~1633 $cm^{-1}$ band and Eq. (11.16) allow one to determine the angle ($\alpha$) between the direction of the transition dipole moment vector of the amide I band and the surface normal. Knowing the angle between the direction of the transition dipole moment of the amide I band and the helix axis, one can determine the orientation of the $\beta^{6.3}$ helix in the supported bilayer and hence one can investigate its orientational change as a function of the electrode potential. Laredo et al. [33] assumed that the angle of the effective transition dipole of the amide I band with respect to the helix axis in GD molecules is 32°, as suggested by Kota et al. [132]. This value was used to calculate the angle between the axis of the $\beta^{6.3}$ helix and the surface normal plotted as a function of the electrode potential in Figure 11.41. For comparison the average angle of the acyl chains of DMPC molecules in the DMPC–GD bilayer are also plotted in this figure.

The average tilt angle of the acyl chains in the DMPC–GD bilayer changes from ~35° in the adsorbed state to ~18° when it is detached from the gold surface. The

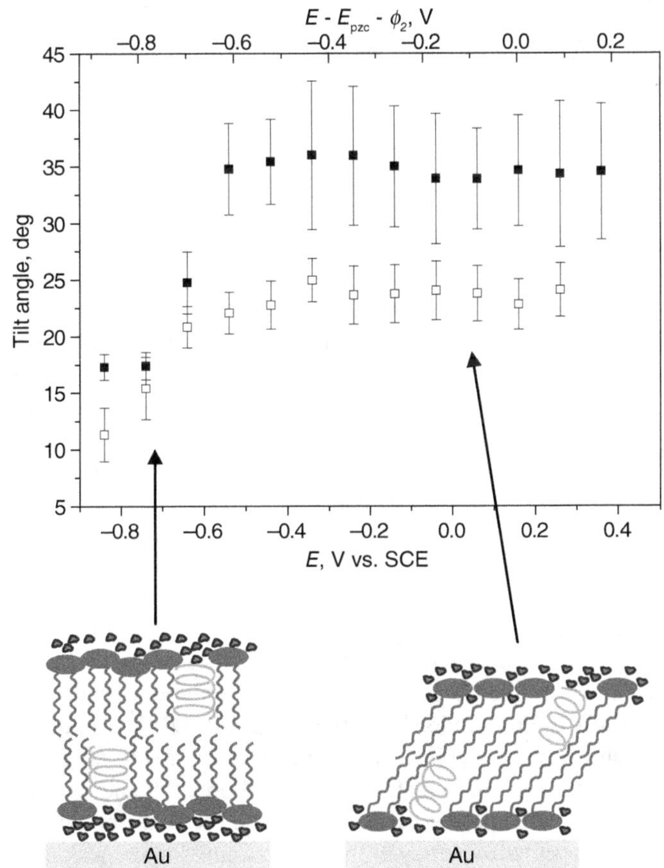


**Figure 11.41** Orientation of the acyl chains in the DMPC–GD bilayer (filled squares) versus potential as calculated from the order parameter $S_{CH_2}$. Open squares show the average tilt angle of the GD channel with respect to the surface normal as a function of electrode potential. The tilt angle of GD is calculated from the integrated band intensity and assuming an orientation of μ of 30° with respect to the helical axis. Adapted from Ref. [33].

data in Figure 11.41 indicate that the structure of the bilayer is different for the adsorbed and desorbed states. In the adsorbed state, all polar heads of lipids are in contact with the metal and are located in one plane, occupying a larger surface area occupied by the head groups, as shown by the cartoon in Figure 11.41. To compensate for this increase in the head, the chains tilt accordingly. The large change of the tilt angle between the adsorbed and the desorbed states is driven by the difference in the packing of the polar heads of DMPC molecules in the two states. The data in Figure 11.41 show that the stability of the bilayer is comparable to the stability of the membrane in the absence of the peptide.

Figure 11.41 shows that the change of the tilt of the $\beta^{6.3}$ helixes with potential parallels the tilt of the acyl chains of the DMPC molecules. The orientation of the peptide changes by about 10° in the potential range studied. In the adsorption state, the average tilt angle of the GD helix is 23 ± 2°. After the bilayer is detached from the gold surface, the peptide becomes more upright and the angle decreases to 12 ± 2°. The change in the tilt angle of the acyl chains is driven by the difference in the packing of the polar heads of DMPC molecules in the two states. The corresponding change in the tilt angle of the helix can be explained in terms of the hydrophobic matching of the thickness of the hydrophobic portion of the bilayer to the length of the peptide [133, 134]. The energetic cost of hydrophobic mismatch is proportional to the square of the difference between the thickness of the bilayer and the projected length of the peptide in the direction normal to the bilayer. The bilayer responds to the hydrophobic stress by changing the tilt of the peptide to better accommodate it within the bilayer core. The length of two all-trans stretched acyl chains of DMPC molecules is 3.4 nm. In the gel state, the maximum hydrophobic thickness of the bilayer is then given by the product of 3.4 nm and $\cos\langle\theta\rangle$, where $\theta$ is the angle between the axis of the peptide and the bilayer normal. Hence, the bilayer hydrophobic thickness changes from 2.7 ± 0.1 nm in the adsorbed state (corresponding to an angle of 35 ± 2°) to 3.2 ± 0.1 nm in the desorbed state (for an angle of 20 ± 3°). The length of the GD dimer is 2.6 nm (1.3 nm per monomer) [135]. The GD channel spans a thickness 2.6 $\cos\langle\theta\rangle$ equal to 2.35 ± 0.1 nm and 2.56 ± 0.1 nm in the adsorbed and desorbed states, respectively. Therefore, the hydrophobic mismatch in these two states corresponds to 0.35 ± 0.1 nm and 0.6 ± 0.1 nm, respectively. The hydrophobic matching appears to be smaller in the adsorbed state. However, the hydrophobic core thickness of the DMPC–GD bilayer is less than the above estimate since the acyl chains are not in an all-trans conformation but are partially melted. Indeed, the thickness of a DMPC bilayer decreases by ~0.5 nm by moving across the phase transition from the gel to the liquid crystalline state. This change is sufficient to match the hydrophobic thickness of the bilayer and the projected length of the peptide and to eliminate the hydrophobic stress.

Figure 11.42 plots the frequency of the deconvoluted main peak under the amide I band as a function of electrode potential. At positive potentials in the adsorbed state of the film, the peak is centered at ~1634 $cm^{-1}$. At negative potentials where the bilayer is desorbed, the maximum of adsorption for the amide I band is located at ~1632 $cm^{-1}$. The amide I band frequency depends strongly on the intramolecular hydrogen bonding pattern and on the transition dipole coupling (TDC) mechanism

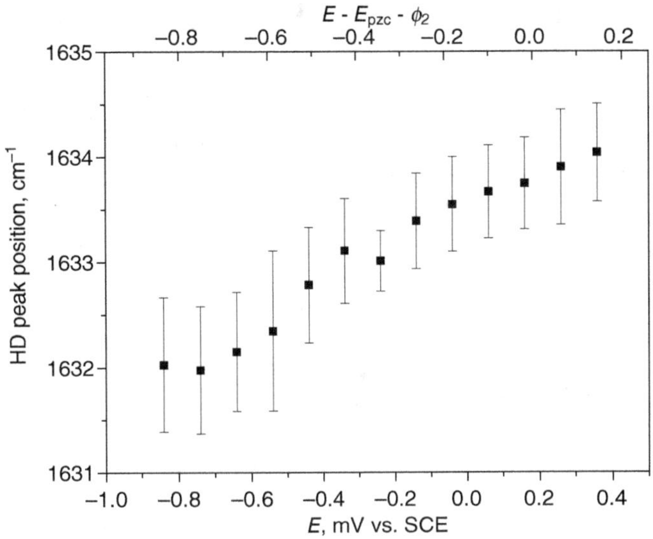


**Figure 11.42** Frequency shift of the deconvoluted main peak under the amide I band as a function of electrode potential. Adapted from Ref. [33].

[136]. The latter is the main mechanism that causes the amide I band to be sensitive to the secondary structure of the protein. The TDC involves interactions between the oscillating dipoles of neighboring amide groups that depend upon the relative orientation of and the distance between the dipoles. The blue shift of the amide I band indicates that the TDC and/or intramolecular hydrogen bonding are weakened in the adsorbed state. This suggests a change in the angle and/or in the distance between the oscillating dipoles of the amide groups when the bilayer is deposited at and in direct contact with the metal surface. This data suggests that the GD molecule is stressed in the adsorbed state of the bilayer. The stress is relaxed when the bilayer is lifted and separated from the metal by a thin layer of electrolyte at negative potentials.

In summary, the IR data indicated that the helical dimer ($\beta^{6.3}$) is the prevalent conformation of GD in the supported bilayer. The orientation of the helical axis changes with potential in parallel with the change of the tilt of the acyl chains of DMPC molecules. However, the average tilt of the helix is smaller than that of the DMPC chains by approximately 10°. The potential-controlled change in the helix orientation can be explained by hydrophobic matching. The orientation of the chains and the peptide are optimized to ensure that the thickness of the hydrophobic portion of the bilayer and the projected length of the peptide on the bilayer normal are matched. Consequently, the tilt angle of the helix is about 23° in the adsorbed state, which is about 10° more than in the desorbed state. At such a large tilt angle, the helix is stressed, and small changes in both the strength of its bonds and its bond angles occur that cause a blue shift of the amide I band of the helix. This stress is relaxed in the desorbed state.

## 11.11 PM–IRRAS STUDIES OF FLOATING BILAYER LIPID MEMBRANES

In BLMs supported at the Au(111) surface (sBLMs), the inner leaflet is in direct contact with the gold electrode surface. The membrane is in an asymmetric environment and hence the lipid molecules in the two leaflets assume different tilt angles and as a result the two leaflets of the bilayer are decoupled. Since there is no hydrophilic region between the gold substrate and the bilayer, peptides incorporated into the sBLM are stressed as shown by the example of gramicidin described in the previous section. Consequently, sBLMs cannot be used to study transmembrane proteins because such proteins may denature and lose their functionality when in contact with the metal surface [137]. To address this problem, tethered bilayer lipid membranes (tBLMs), in which the bilayer is separated from the substrate by a hydrophilic spacer molecule, have been designed for the investigation of the transmembrane proteins in lipid bilayers [138, 139]. The disadvantage of tBLMs is that the tethering lipids are often neither commercially available nor naturally occurring but rather are functionalized lipids typically synthesized via a complex procedure and hence difficult to obtain. As a solution to this problem, our group used novel floating bilayer lipid membranes (fBLMs) in which the lipid bilayer is separated from the substrate by a water-rich region using a physically adsorbed spacer molecule layer [140–142]. All the components of these fBLMs are commercially available molecules.

Two fBLM models were assembled and characterized. The first was composed of DMPC and cholesterol as the main bilayer components and the ganglioside GM1 [140, 141]. It had a similar composition to the three-component BLM discussed previously, but in this case GM1 was in the proximal leaflet facing the metal, and its content was higher (30%). The second model contained 1,2-dimyristoyl-*sn*-glycero-3-phosphoethanolamine-*N*-[methoxy(polyethylene glycol)-350] (DMPE-PEG350) [142] as a hydrophilic spacer in the inner leaflet. The two floating bilayers were deposited onto a gold surface modified by a self-assembled monolayer 1-thio-β-D-glucose (β-Tg). Figure 11.43 shows cartoon representations of these fBLM structures. 1-Thio-β-D-glucose (β-Tg) was self-assembled at the Au(111) surface to increase the overall hydrophilicity of the surface. The fBLM was deposited on the β-Tg self-assembled monolayer (SAM) using the LB–LS method. The inner leaflet, containing a mixture of DMPC, cholesterol, and $GM_1$ in a 40:30:30 mol% ratio or a mixture of DMPC, cholesterol, and DMPE-PEG350 in a 60:30:10 mol% ratio, was transferred to the β-Tg-modified Au(111) surface by the LB transfer method. The outer leaflet, composed of only DMPC and cholesterol in a 70:30 mol% ratio, was deposited using the LS touch. The carbohydrate head groups of the $GM_1$ molecules or the PEG group of the DMPE-PEG350 are physically adsorbed to the β-Tg SAM, forming a water-rich cushion between the fBLM and the substrate. The addition of this water region more closely mimics the natural environment of a biological membrane making the fBLM a desirable candidate for future in situ studies involving transmembrane proteins.

Figure 11.44 plots the differential capacitance curves of the bare Au(111) electrode (dash-dotted line), the β-Tg SAM (dotted line), the fBLM composed of

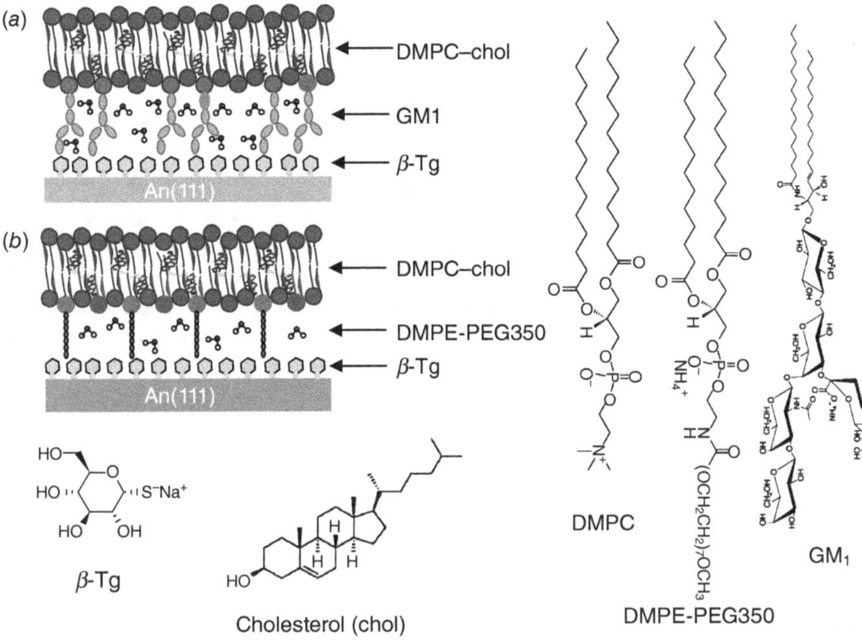


**Figure 11.43** Schematic of the fBLMs at the β-Tg-modified Au(111) surface containing (*a*) DMPC–chol–GM1 and (*b*) DMCP–chol–DMPE-EPG350. Chemical structures of the components used to build the fBLMs are shown.

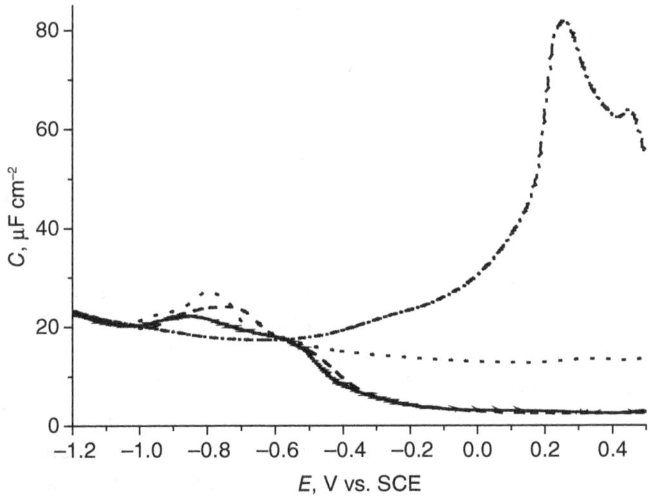


**Figure 11.44** Differential capacity curves of the bare Au(111) electrode (dash dotted line), β-Tg SAM (dotted line), the fBLM composed of DMPC–chol–GM1 (dash line) and the fBLM composed of DMPC–chol–DMPE-PEG350 (solid line).

DMPC–chol–GM1 (solid line), and the fBLM composed of DMPC–chol–DMPE-PEG350 (dashed line). The differential capacitance curves show that the fBLMs remain stable at the gold surface over a wide potential range from 0.5 to −0.10 V. The minimum capacitance value of the DMPC–chol–GM1 fBLM is ~2.2 μF $cm^{-2}$, while the minimum value of the DMPC–chol–DMPE-PEG350 fBLM is ~2.5 μF $cm^{-2}$, demonstrating the fBLM containing GM1 has somewhat less defects than the fBLM containing DMPE-PEG350. Although the capacitance values of the fBLMs are much lower than the minimum capacitance of the β-Tg SAM (~12.8 μF $cm^{-2}$), they are still higher than the capacitance of a biological membrane, which is ~1.0 μF $cm^{-2}$ [51]. These results imply that the fBLMs may contain some defects. The capacitance value increases slightly when the potential is scanned from −0.10 to −0.40 V and then increases rapidly at potentials that are more negative than −0.40 V, indicating that the fBLMs begin to desorb from the Au(111) surface. The capacitance curves of the fBLM-covered Au(111) electrodes merge with that of the bare electrode when $E$ is more negative than −1.0 V, demonstrating that the fBLMs are completely detached from the gold surface at these negative potentials. It should be stressed that once the fBLMs are detached from the gold surface, the small, water-soluble β-Tg molecules diffuse into the bulk solution, leading to irreversible changes in the membrane structure.

DMPC-*d*54 was used to probe the conformation and orientation of the acyl chains of DMPC molecules in the fBLMs since the IR absorption of the C–D stretching does not overlap with any other vibrations of the fBLMs components and the β-Tg SAM. The peak position of the $\nu_{as}(CD_2)$ and $\nu_s(CD_2)$ bands indicated that the acyl chains in the fBLMs were partially melted and had a certain number of gauche conformers. This result also demonstrated that incorporation of cholesterol, DMPE-PEG350, and $GM_1$ into the lipid matrix leads to an increase in the number of gauche conformers. The average tilt angle of the trans fragments of acyl chains in DMPC-*d*54 ($\theta_{tilt}$) was determined from the $\nu_{as}(CD_2)$ and $\nu_s(CD_2)$ band intensity using Eq. (11.17). The results are plotted versus potential in Figure 11.45. For comparison, the tilt angles of the three-component sBLM (DMPC–chol–GM1) are also plotted in this figure [122]. The data in Figure 11.45 show that the $\theta_{tilt}$ of the fBLMs are significantly smaller than that of the sBLM. In the adsorption state, the average tilt angle of the fBLM containing $GM_1$ is ~12°, while for the fBLM containing DMPE-PEG350, the average tilt angle decreases to ~6°. The tilt angle of the fBLMs is ~5°–15° smaller than that of the sBLM. In addition the tilt angles of fBLMs change much less with potential in comparison to sBLM. As was explained earlier, in the adsorbed state the polar heads of the lipids in the leaflet that is in contact with the metal are located in one plane, forcing the chains to tilt. The tilt is reduced in the desorbed state when the bilayer is separated from the metal by a cushion of electrolyte and the polar heads can be packed in a zigzag fashion. The small tilt of the acyl chains in fBLMs indicates that the bilayer is indeed separated from the metal by a water-rich region that allows the head group to pack in a zigzag fashion even in the adsorbed state. As a consequence the chains are more tightly packed in fBLMs than in sBLMs. Independent AFM studies of the DMPC–chol–$GM_1$ fBLM confirm that the acyl chains of DMPC molecules in the fBLM are indeed tightly packed [141].

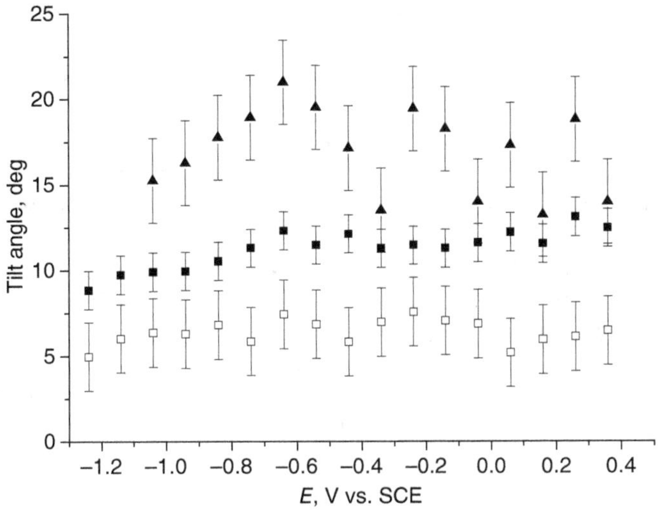


**Figure 11.45** Tilt angle of the acyl chains of DMPC molecules in the fBLM containing GM1 (filled squares) [141], the fBLM containing DMPE-PEG350 (open squares) [142], and DMPC–chol–GM1 sBLM (filled triangles) [122].

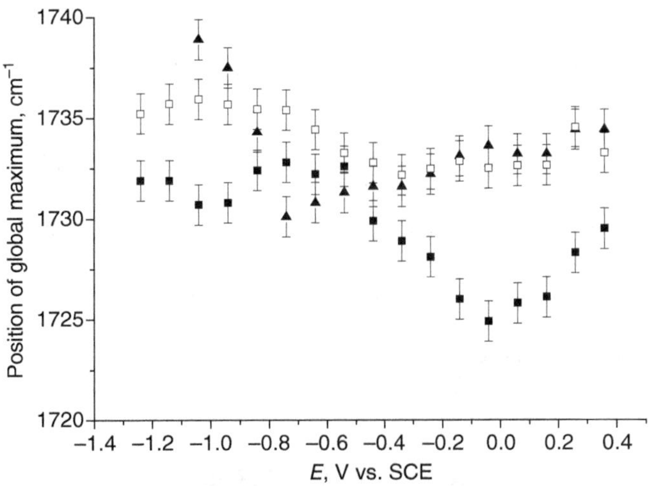


**Figure 11.46** Position of the global maximum of the C=O stretching band in the fBLM containing GM1 (filled squares) [141], the fBLM containing DMPE–PEG350 (open squares) [142], and the DMPC–chol–GM1 sBLM (filled triangles) [122] as a function of electrode potential.

Figure 11.46 plots the position of the global maximum of the ν(C=O) band in the fBLMs [141, 142] and the sBLM [122] as a function of electrode potential. It can be observed that the global maximum for the ν(C=O) band of the DMPC–chol–GM1 fBLM is significantly red shifted than in the DMPC–chol–GM1 sBLM, indicating that the C=O group is more hydrated in the floating bilayer when GM1 molecules are in the proximal rather than in the distal leaflet. The frequencies of the global maximum in the fBLM with PEG are higher than for the fBLM with GM1 as the spacer, indicating that the hydration of the C=O group is somewhat poorer. However, all these frequencies are much lower than the value of 1744 $cm^{-1}$, which is characteristic for the non-hydrogen-bonded state. Therefore, these results indicate that the PEG and $GM_1$ spacers provide a desirable water-rich region between the bilayer and the metal surface, making the fBLMs good candidates for future transmembrane protein studies.

To better understand the improved hydration in the fBLM containing DMPE-PEG350, the PM–IRRAS technique was used to study the conformation of the PEG spacer. The IR bands of the PEG group are located in the 1350–1100 $cm^{-1}$ region and provide information concerning the conformation of the PEG spacer. These bands overlap with the IR adsorption of the phosphate groups of DMPC, hence Fourier self-deconvolution (FSD) and two-dimensional correlation spectroscopy (2D-COS) were used to determine the total number of peaks and their corresponding frequencies. The PM–IRRAS results showed that the frequencies of the PEG bands in the fBLM are higher than the frequencies of bands corresponding to the crystalline PEG and are close to frequencies observed for amorphous PEG, indicating that the PEG groups of the DMPE-PEG350 molecules are disordered and not coiled. In a helical coil conformation, the oxygen atoms of the $O–CH_2–CH_2–O$ repeat units are turned toward the interior of the coil, and the $CH_2$ groups are directed toward the solution resulting in poor hydration of the PEG group. In the amorphous state, the oxygen atoms of the ethylene oxide repeat units are always exposed to the solvent [143–147]. This conformation aids in the hydration of the PEG group via hydrogen bonding to oxygen atoms. These results confirm that the PEG subunit is assisting with the formation of the water layer between the membrane and electrode surface.

In conclusion, the electrochemical and PM–IRRAS data confirm that the fBLMs are stable at the Au(111) surface and provide a water-rich region between the lipid bilayer and the substrate. This water-rich layer is thick enough to accommodate the exposed hydrophilic protein to prevent it from denaturing. The potential region in which the fBLM is stable can be used to control the open and closed state of voltage-gated ion channel properties of proteins incorporated in the fBLMs. Such fBLMs should be an ideal platform for future voltage-gated ion channel protein research.

## 11.12 SUMMARY AND CONCLUSIONS

We have described the application of in situ PM–IRRAS to study biomimetic membranes supported at a Au(111) electrode surface. These membranes, composed of phospholipids, cholesterol, and glycolipids can be assembled at the atomically flat gold electrode surface either by vesicle fusion or by the LB–LS method. Various models of the membranes have been discussed, including supported bilayer lipid membranes (sBLMs) and floating bilayer lipid membranes (fBLMs). The electrochemical data showed that the bilayers can be adsorbed at the gold surface when the transmembrane potential drop is smaller than 0.4 V and the charge densities lie between $-10\ \mu C\ cm^{-2} < \sigma_M < 10\ \mu C\ cm^{-2}$. In addition, the electric field generated during such electrochemical studies is equivalent to electric fields generated across real cell membranes. PM–IRRAS provided significant molecular-level information about the conformation and tilt angle of the acyl chains of the phospholipid molecules, which is important for interpreting the organization of lipids in such systems. Further, the PM–IRRAS spectra of the polar head region provided information about the conformation, orientation, and hydration of these fragments of lipid molecules.

We have demonstrated that gold-supported biomimetic membranes allow one to study the relation between structure and function of ion channels incorporated into the membrane. PM–IRRAS spectra in the amide I region provide useful information on the secondary structure of proteins and peptides. For the proteins reconstituted into the biomimetic membranes supported at the gold electrode surface, PM–IRRAS offers a unique opportunity to obtain molecular-level information about changes in the orientation and conformation involved in the voltage-gated functions of such proteins. The first PM–IRRAS studies of proteins reconstituted in gold-supported biomimetic membranes by Brosseau [20], Leitch et al. [148], and Brand and co-workers [149, 150] illustrate the tremendous power of PM–IRRAS in providing such structural information. For model membranes supported at a metal electrode surface the options for fundamental research are largely unexplored. These systems constitute ideal platforms for the development of biosensors and for a broad range of biomedical research such as studies of implant biocompatibility, cell adhesion and fusion, drug screening, and amyloid plaque formation [151]. PM–IRRAS is ideally suited to provide significant molecular-level information about the structure of these systems.

## ACKNOWLEDGMENTS

These studies were supported by a Discovery grant from Natural Sciences and Engineering Council of Canada (NSERC). J.L. acknowledges the Canada Research Chair Award.

## REFERENCES

1. Buffeteau, T.; Desbat, B.; Turlet, J. M. *Appl. Spectrosc.* **1991**, *45*, 380–389.
2. Blaudez, D.; Buffeteau, T.; Desbat, B.; Orrit, M.; Turlet, J. M. *Thin Solid Films* **1992**, *210–211*, 648–651.
3. Colorado, R.; Villazana, R. J.; Lee, T. R. *Langmuir* **1998**, *14*, 6337–6340.
4. Saccani, J.; Buffeteau, T.; Desbat, B.; Blaudez, D. *Appl. Spectrosc.* **2003**, *57*, 1260–1265.
5. Moskovits, M. J. *J. Chem. Phys.* **1982**, *77*, 4408–4416.
6. Zamlynny, V.; Lipkowski, J. Quantitative SNIFTIRS and PM IRRAS of organic molecules at electrode surfaces, in R. Alkire, D. M. Kolb, J. Lipkowski, and P. N. Ross, Eds, *Diffraction and Spectroscopic Methods in Electrochemistry*, Wiley-VCH: Weinheim, **2006**, Vol. 9, pp. 315–376.
7. Zamlynny, V.; Zawisza, I.; Lipkowski, J. *Langmuir* **2003**, *19*, 132–145.
8. Zawisza, I; Burgess, I.; Szymanski, G.; Lipkowski, J.; Majweski, J.; Satija, S. *Electrochim. Acta* **2004**, *49*, 3651–3664.

9. Zawisza, I.; Nullmeier, M.; Pust, S. E.; Boukherroub, R.; Szunertis, S.; Wittstock G. *Langmuir* **2008**, *24*, 7378–7387.
10. Roefzaad, M.; Kluner, T.; Brand, I. *Phys. Chem. Chem. Phys.* **2009**, *11*, 10140–10151.
11. Tagliazucchi, M; De Leo, L. P. M.; Cadranel, A.; Baraldo, L. M.; Volker, E.; Bonazzola, C.; Calvo, E. J.; Zamlynny, V. *J. Electroanal. Chem.* **2010**, *649*, 110–118.
12. Zawisza, I.; Wittstock, G.; Boukherroub, R.; Szunertis, S. *Langmuir* **2007**, *23*, 9303–9309.
13. Zawisza, I.; Wittstock, G.; Boukherroub, R.; Szunertis, S. *Langmuir* **2008**, *24*, 3922–3929.
14. Leitch, J.; Kunze, J.; Goddard, J. D.; Schwan, A. L.; Faragher, R. J.; Naumann, R.; Knoll, W.; Dutcher, J. R.; Lipkowski, J. *Langmuir* **2009**, *25*, 10354–10363.
15. Brand, I.; Nullmeier, M.; Klüner, T.; Jogireddy, R.; Christoffers, J.; Wittstock, G. *Langmuir* **2010**, *26*, 362–370.
16. Ramin, M. A.; Bourdon, G. L.; Daugey, N.; Bennetau, B.; Vellutini, L.; Buffeteau, T. *Langmuir*, **2011**, *27*, 6076–6084.
17. Blatt, S.; Roscoe, S. G.; Zamlynny, V. *J. Electroanal. Chem.* **2010**, *649*, 102–109.
18. Barner, B. J.; Green, M. J.; Saez, E. I.; Corn, R. M. *Anal. Chem.* **1991**, *63*, 55–60.
19. Green, M. J.; Barner, B. J.; Corn, R. M. *Rev. Sci Instrum.* **1991**, *62*, 1426–1430.
20. Brosseau, C. L. Ph.D. Dissertation, University of Guelph, **2007**.
21. Abeles, F. in *Advanced Optical Techniques*, A. C. S. Van Heel, Ed., Wiley: New York, **1967**, p. 145.
22. Hansen, W. N. *J. Opt. Soc. Am.* **1968**, *58*, 380–388.
23. Zamlynny, V. Ph.D. Dissertation, University of Guelph, **2002**. Inquiries concerning Fresnel 1 software should be addressed to Vlad.Zamlynny@acadiau.ca.
24. Li, N; Zamlynny, V.; Lipkowski, J.; Henglein, F.; Pettinger, B. *J. Electroanal. Chem.* **2002**, *524–525*, 43–53.
25. Zamlynny, V.; Zawisza, I.; Lipkowski, J. *Langmuir* **2003**, *19*, 132–145.
26. Kauppinen, J. K.; Moffatt, D. J.; Mantsch, H. H; Cameron, D. G. *Appl. Spectrosc.* **1981**, *35*, 271–276.
27. Jones, R. N.; Shimokoshi, K. *Appl. Spectrosc.* **1983**, *37*, 59–67.
28. Noda, I.; Ozaki, Y., in *Two-Dimension Correlation Spectroscopy; Applications in Vibrational and Optical Spectroscopy*, Wiley: Hoboken, NJ, **2004**.
29. Surewicz, W. K.; Mantsch, H. H. *Biochim. Biophys. Acta* **1988**, *952*, 115–130.
30. Griffiths, P. R.; Pariente G. L. *Trac-Trends Anal. Chem.* **1986**, *5*, 209–215.
31. Ozaki, Y.; Murayama, K.; Wu, Y.; Czarnick-Matusewicz, B. *Spectroscopy*, **2003**, *17*, 79–100.
32. Czarnecki, M. A. *Appl. Spectrsosc.* **1998**, *52*, 1583–1590.
33. Laredo, T.; Dutcher, J. R; Lipkowski, J. *Langmuir*, **2011**, *27*, 10072–10087.
34. Mantsch, H. H.; Casal, H. L.; Jones, R. N. Resolution enhance of infrared spectra of biological systems, in R. J. Clark and R. E. Hester, Eds., *Spectroscopy of biological systems*, Wiley: Now York, **1986**, Vol. 13, pp. 1–46.
35. Lafrance, C. P.; Nabet, A.; Prudhomme, R. E.; Pezolet, M. *Can. J. Chem.* **1995**, *73*, 1497–1505.
36. Domingo, J. C.; Madariaga, M. A. D. *Chem. Phys. Lipids* **1996**, *84*, 147–153.
37. Allara, D. L.; Swalen, J. D. *J. Phys. Chem.* **1982**, *86*, 2700–2704.
38. Allara, D. L.; Nuzzo, R. G. *Langmuir* **1985**, *1*, 45–52.
39. Allara, D. L.; Bacca, A.; Pryde, C. A. *Macromolecules* **1978**, *11*, 1215–1220.
40. Palik, E. D. In *Handbook of Optical Constants of Solids*, Academic Press: London, **1998**.
41. Ordal, M. A.; Long, L. L.; Bell, R. J.; Bell, S. E.; Bell, R. R.; Alexander, Jr., R. W.; Ward, C. A. *Appl. Opt.* **1983**, *22*, 1099–1120.
42. Rusk, A. N.; Williams, D.; Querry, M. *J. Opt. Soc. Am.* **1971**, *61*, 895–903.
43. Bertie, J. E.; Ahmed M. K.; Eysel, H. H. *J. Phys. Chem.* **1989**, *93*, 2210–2218.
44. Fringeli, U. P. *Z. Naturforsch.* **1977**, *32C*, 20–45.
45. Umemura, J.; Kamata, T.; Kawai, T.; Takenaka, T. *J. Phys. Chem.* **1990**, *94*, 62–67.
46. Binder, H.; Kohlstrunk, B. *Vibr. Spectrosc.* **1999**, *21*, 75–95.
47. Seelig, J.; Niederberger, W. *J. Am. Chem. Soc.* **1974**, *96*, 2069–2072.

48. Clarke, R. J., *Adv. Colloid Interface Sci.* **2001**, *89–90*, 263–281.
49. Walz, D.; Teissie, J.; Milazzo, G. In *Bioelectrochemistry of Membranes*, Springer: Basel, Switzerland, **2004**, Vol. *6*.
50. Tsong, T. Y.; Astumian, R. D. *Ann. Rev. Physiol.* **1988**, *50*, 273–290.
51. Gennis, R. B. in *Biomembranes, Molecular Structure and Function*, Springer: New York, **1989**.
52. Dickertmann, D.; Schultze, J. W.; Koppitz, F. D. *Electrochim. Acta* **1976**, *21*, 967–971.
53. Burgess, I.; Szymanski, G.; Li, M.; Horswell, S.; Lipkowski, J.; Majewski, J.; Satija, S. *Biophys. J.* **2004**, *86*, 1763–1776.
54. Burgess, I.; Li, M.; Horswell, S. L.; Szymanski, G.; Lipkowski, J.; Satija, S.; Majewski, J. *Colloids Surf. B* **2005**, *40*, 117–122.
55. Lipkowski, J.; Stolberg, L. In *Adsorption of Molecules at Metal Electrodes*, VCH: New York, **1992**, p. 171.
56. Mohilner, D. in *Electroanalytical Chemistry*, A. Bard, Ed., Marcel Dekker: New York, **1966**, Vol. *1*, pp. 241–409.
57. Becucci, L.; Moncelli, M. R.; Herrero, R.; Guidelli, R. *Langmuir* **2000**, *16*, 7694–7700.
58. Zawisza, I.; Bin X.; Lipkowski, J. *Langmuir* **2007**, *23*, 5180–5194.
59. Bin X. M.; Zawisza I; Goddard J. D.; Lipkowski, J. *Langmuir* **2005**, *21*, 330–347.
60. Casal, H. L.; Mantsch. H. H. *Biochim. Biophys. Acta* **1984**, *779*, 381–401.
61. Socrates, G. In *Infrared and Raman Characteristic Group Frequencies*, 3rd ed., Wiley: New York, **2001**.
62. Lewis, R. N. A. H.; McElhaney, R. N. Fourier transform infrared spectroscopy in the study of hydrated lipids and lipid bilayer membranes, in H. H. Mantsch and D. C. Chapman, Eds., *Infrared Spectroscopy of Biomolecules*, Wiley: New York, **1996**.
63. Snyder, R. G. *J. Chem. Phys.* **1967**, *47*, 1316–1360.
64. Mantsch, H. H.; Martin, A.; Cameron, D. G. *Biochemistry* **1981**, *20*, 3138–3145.
65. Mantsch, H. H.; Madec, C.; Lewis, R. N. A. H.; McElhaney, R. N. *Biochemistry* **1985**, *24*, 2440–2446.
66. Mantsch, H. H.; Madec, C.; Lewis, R. N. A. H.; McElhaney, R. N. *Biochemistry* **1987**, *26*, 4045–4049.
67. Snyder, R. G. *J. Mol. Spectrosc.* **1961**, *7*, 116–144.
68. Zawisza, I.; Lipkowski, J. *Langmuir*, **2004**, *20*, 4579–4589.
69. Blume, A.; Hubner, W.; Messner, G. *Biochemistry* **1988**, *27*, 8239–8249.
70. Lewis, R. N. A. H.; McElhaney, R. N.; Pohle, W.; Mantsch, H. H. *Biophys. J.* **1994**, *67*, 2367–2375.
71. Lewis, R. N. A. H.; McElhaney, R. N. *Biophys. J.* **1992**, *61*, 63–77.
72. Mantsch, H. H.; Casal, H. L. *Fresenius Z. Anal. Chem.* **1986**, *324*, 655–661.
73. Fringeli, U. P.; Günthard, H. H. Infrared membrane spectroscopy, in E. Grell, Ed., *Membrane Spectroscopy*, Springer: New York, **1981**.
74. Fringeli, U. P. *Biophys. J.* **1981**, *34*, 173–187.
75. Arrondo, J. L. R.; Goni, F. M.; Macarulla, J. M. *Biochim. Biophys. Acta* **1984**, *794*, 165–168.
76. Goni, F. M.; Arrondo, J. L. R. *Faraday Discuss. Chem. Soc.* **1986**, *81*, 117–126.
77. Wong, P. T. T.; Mantsch, H. H. *Chem. Phys. Lipids* **1988**, *46*, 213–224.
78. Abeles, R. H.; Frey, P. A.; Jencks, W. P. In *Biochemistry*, Jones and Bartlett: Boston, **1992**.
79. Barth, A. *Biochim. Biophys. Acta* **2007**, *1767*, 1073–1101.
80. Kong. J,; Yu, S. *Acta Biochim. Biophys. Sinica* **2007**, *39*, 549–559.
81. Dong, A.; Huang, P.; Caughey, W. S. *Biochemistry*, **1990**, *29*, 3303–3308.
82. Byler, D. M.; SuSi, H. *Biopolymers*, **1986**, *25*, 469–487.
83. Reviakine, I.; Brisson, A. *Langmuir* **2000**, *16*, 1806–1815.
84. Lipowsky, R.; Seifert, U. *Mol. Cryst. Liq. Cryst.* **1991**, *202*, 17–25.
85. Lipowsky, R. *Curr. Op. Struct. Bio.* **1995**, *5*, 531–540.
86. Barenholz, Y.; Gibbes, D.; Litman, B. J.; Goll, J.; Thompson, T. E.; Carlson, R. D. *Biochemistry* **1977**, *16*, 2806–2810.
87. Hallett, F. R.; Craig, T.; Marsh, T.; Nickel, B. *Can. J. Spectrosc.* **1989**, *34*, 63–70.
88. Petty, M. C. In *Langmuir-Blodgett Films: An Introduction*, Cambridge University Press: Cambridge, UK, **1996**.

89. Kycia, A. H.; Wang, J. P.; Merrill, A. R.; Lipkowski, J. *Langmuir* **2011**, *27*, 10867–10877.
90. Horswell, S. L.; Zamlynny, V.; Li, H. Q.; Merrill, A. R.; Lipkowski, J. *Faraday Discussion* **2002**, *121*, 405–422.
91. Garcia-Araez, N.; Brosseau, C. L.; Rodriguez P.; Lipkowski, J. *Langmuir* **2006**, *22*, 10365–10371.
92. Mantsch, H. H.; McEchaney, R. N. *Chem. Phys. Lipids* **1991**, *57*, 213–226.
93. Mitchell, M. L.; Dluhy, R. A. *J. Am. Chem. Soc.* **1988**, *110*, 712–718.
94. Hauser, H.; Pascher, I.; Pearson, R. H.; Sundell, S. *Biochim. Biophys. Acta* **1981**, *650*, 21–51.
95. Hubner, W.; Blume, A. *Chem. Phys. Lipids* **1998**, *96*, 99–123.
96. Li, M.; Chen, M. H.; Sheepwash, E.; Brosseau, C. L.; Li, H. Q.; Pettinger, B.; Gruler, H.; Lipkowski, J. *Langmuir* **2008**, *24*, 10313–10323.
97. Zawisza, I.; Lachenwitzer, A.; Zamlynny, V.; Horswell, S. L.; Goddard, J. D.; Lipkowski, J. *Biophys. J.* **2003**, *85*, 4055–4075.
98. Zawisza, I.; Bin, X. M.; Lipkowski J. *Bioelectrochemistry* **2004**, *63*, 137–147.
99. Zawisza, I.; Nullmeier, M.; Pust, S. E.; Boukherroub, R.; Szunertis, S.; Wittstock, G. *Langmuir* **2008**, *24*, 7378–7387.
100. Hillman, A. R.; Ryder, K. S.; Madrid, E.; Burley, A. W.; Wiltshire, R. J.; Merotra, J.; Grau, M.; Horswell, S. L.; Glidle, A.; Dalgliesh, R. M.; Hughes, A.; Cubitt, R.; Wildes, A. *Faraday Discuss*, **2010**, *145*, 357–379.
101. Madrid E.; Horswell, S. L. Paper presented at the 63rd Annual Meeting of ISE, Prague, **2012**.
102. McMullen, T. P. W.; McElhaney, R. N. *Biophys. Acta* **1995**, *1234*, 90–98.
103. Saez-Cirion, A.; Alonso, A.; Goni, F. M.; McMullen, T. P. W.; McElhaney, R. N.; Rivas, E. A. *Langmuir* **2000**, *16*, 1960–1968.
104. Presti, F. T.; Pace, R. J.; Chan, S. I. *Biochemistry* **1982**, *21*, 3831–3835.
105. Wong, P. T. T.; Capes, S. E.; Mantsch, H. H. *Biochim. Biophys. Acta* **1989**, *980*, 37–41.
106. Robinson, A. J.; Richards, W. G.; Thomas, P. J.; Hann, M. M. *Biophys. J.* **1995**, *68*, 164–170.
107. Brosseau, C. L.; Bin, X.; Roscoe, S. G.; Lipkowski, J. *J.Electroanal.Chem.* **2008**, *621*, 222–228.
108. Bin, X. M.; Horswell S. L.; Lipkowski, J. *Biophys. J.* **2005**, *89*, 592–604.
109. Bin, X.; Lipkowski, J. *J. Phys. Chem. B* **2006**, *110*, 26430–26441.
110. Israelachvili, J. N. In *Intramolecular and Surface Forces*, Academic Press: London, **1991**.
111. Kumar, V. V. *Proc. Natl. Acad. Sci. USA* **1991**, *88*, 444–448.
112. Matyszewska, D.; Leitch, J.; Bilewicz, R.; Lipkowski, J. *Langmuir* **2008**, *24*, 7408–7412.
113. Curatolo, W. *Biochim. Biophys. Acta* **1987**, *906*, 137–160.
114. Cheresh, D. A.; Harper, J. R.; Schulz, G.; Reisfeld, A. *Proc. Natl. Acad. Sci. USA* **1984**, *81*, 5767–5771.
115. Mou, J.; Yang, J.; Shao, Z. *J. Mol. Biol.* **1995**, *248*, 507–512.
116. Vie, V.; Mau, N. V.; Lesniewska, E.; Goudonnet, J. P.; Heitz, F.; Grimellec, C. L. *Langmuir* **1998**, *14*, 4574–4583.
117. Yuan, C.; Johnston, L. J. *Biophys. J.* **2000**, *79*, 2768–2781.
118. Johnston, L. *Langmuir* **2007**, *23*, 5886–5895.
119. Miller, C. E.; Majewski, J.; Kjær, K.; Weygand, M.; Faller, R.; Satija, S.; Kuhl, T. L. *Colloids Surf. B* **2005**, *40*, 159–163.
120. Majewski, J.; Kuhl, T. L.; Kjær, K.; Smith, G. S. *Biophys. J.* **2001**, *81*, 2707–2715.
121. Frey, S. L.; Chi, E. Y.; Arratia, C.; Majewski, J.; Kjaer, K; Lee, K. C. *Biophys. J.* **2008**, *94*, 3047–3064.
122. Brosseau, C. L.; Leitch, J.; Bin, X.; Chen, M.; Roscoe, S. G.; Lipkowski, J. *Langmuir* **2008**, *24*, 13058–13067.
123. Due to an error in data processing a lower value of minimum capacity 0.8 $\mu F cm^{-2}$ was reported in Ref. [122], the value 1.7 $\mu F\ cm^{-2}$ is a correction of that mistake.
124. Liang, C. Y.; Lytton, M. R. *J. Polym. Sci.* **1962**, *61*, S45–S48.
125. Lee, D. C.; Durrani, A. A.; Chapman, D. *Biochim. Biophys. Acta* **1984**, *769*, 49–56.
126. Majewski, J.; Kuhl, T. L.; Kjær, K.; Smith, G. S. *Biophys. J.* **2001**, *81*, 2707–2715.

127. Hille, B. In *Ion Channels of Excitable Membranes*, Sinauer: Sunderland, MA, **2001**, pp. 814.
128. Moran, M. M.; Xu, H.; Clapham, D. E. *Curr. Opin. Neurobiol.* **2004**, *14*, 362–369.
129. Volinsky, R.; Kolusheva, S.; Berman, A.; Jelinek, R. *Biochim. Biophys. Acta* **2006**, *1758*, 1393–1407.
130. Kelkar, D. A.; Chattopadhyay, A. *Biochim. Biophys. Acta* **2007**, *1768*, 2011–2025.
131. Naik, V. M.; Krimm, S. *Biophys. J.* **1986**, *49*, 1131–1145.
132. Kota, Z.; Pali, T.; Marsh, D. *Biophys. J.* **2004**, *86*, 1521–1531.
133. Harroun, T. A.; Heller, W. T.; Weiss, T. M.; Yang, L.; Huang, H. W. *Biophys. J.* **1999**, *76*, 937–945.
134. Killian, J. A. *Biochim. Biophys. Acta* **1992**, *1113*, 391–425.
135. Barth, A.; Zscherp, C. Q. *Rev. Biophys.* **2002**, *35*, 369–430.
136. Koeppe, R. E.; Berg, J. M.; Hodgson, K. O.; Stryer, L. *Nature* **1979**, *279*, 723–725.
137. Guidelli, R.; Aloisi, G.; Becucci, L.; Dolfi, A.; Moncelli, M. R.; Buoninsegni, F. T. *J. Electroanal. Chem.*, **2001**, *504*, 1–28.
138. Lang, H.; Duschl, C.; Vogel, H. *Langmuir* **1994**, *10*, 197–210.
139. Rossi, C.; Chopineau J. *Eur Biophys. J.* **2007**, *36*, 955–965.
140. Kycia, A. H.; Wang, J. P.; Merrill, A. R.; Lipkowski, J. *Langmuir*, **2011**, *27*, 10867–10877.
141. Su, Z. F.; Kycia, A. H.; Leitch, J. J.; Lipkowski, J. *J. Electrochem.* **2012**, *18*, 387–404.
142. Su, Z. F.; Jiang, Y. X.; Velázquez-Manzanares, M.; Leitch, J. J.; Kycia, A. H.; Lipkowski, J. *J. Electroanal. Chem.*, **2013**, *688*, 76–85.
143. Tadokoro, H.; Chatani, Y.; Yoshihara, T.; Tahara, S.; Murahashi, S. *Makromol. Chem.* **1964**, *73*, 109–127.
144. Yoshihara, T.; Tadokoro, H.; Murahashi, S. *J. Chem. Phys.* **1964**, *41*, 2902–2911.
145. Takahashi,Y.; Tadokoro, H. *Macromolecules* **1973**, *6*, 672–675.
146. Matsuura, H.; Miyazawa, T. *J. Polym. Sci., Polym. Phys. Ed.* **1969**, *7*, 1735–1744.
147. Leitch, J.; Kunze, J.; Goddard, J. D.; Schwan, A. L.; Faragher, R. J.; Naumann, R.; Knoll, W.; Dutcher, J. R.; Lipkowski, J. *Langmuir* **2009**, *25*, 10354–10363.
148. Leitch, J.J.; Brosseau, C.L.; Roscoe, S.G.; Bessonov, K.; Dutcher, J.R.; Lipkowski, J. *Langmuir* **2013**, *29*, 965–976.
149. Nullmeier, M.; Koliwer-Brandl, H.; Kelm, S.; Brand I. *J. Electroanal. Chem.* **2010**, *649*, 177–188.
150. Nullmeier, M.; Koliwer-Brandl, H.; Kelm, S.; Zaegl, P.; Koch, K. W.; Brand I. *ChemPhysChem.* **2011**, *12*, 1066–1079.
151. Sinner, E. K.; Ritz, S.; Wang, Y.; Dostalek, J.; Jonas, U.; Knoll, W. *Mater. Today* **2010**, *13*, 46–55.

# Index




*Vibrational Spectroscopy at Electrified Interfaces*, First Edition.
Edited by Andrzej Wieckowski, Carol Korzeniewski, and Björn Braunschweig.

www.ingramcontent.com/pod-product-compliance
Lightning Source LLC
LaVergne TN
LVHW011903200926
848484LV00002B/518